Environmental Physiology III

Publisher's Note

The *International Review of Physiology* remains a major force in the education of established scientists and advanced students of physiology throughout the world. It continues to present accurate, timely, and thorough reviews of key topics by distinguished authors charged with the responsibility of selecting and critically analyzing new facts and concepts important to the progress of physiology from the mass of information in their respective fields.

Following the successful format established by the earlier volumes in this series, new volumes of the *International Review of Physiology* will concentrate on current developments in neurophysiology and cardiovascular, respiratory, gastrointestinal, liver, endocrine, kidney and urinary tract, environmental, and reproductive physiology. New volumes on a given subject generally appear at two-year intervals, or according to the demand created by new developments in the field. The scope of the series is flexible, however, so that future volumes may cover areas not included earlier.

University Park Press is honored to continue publication of the *International Review of Physiology* under its sole sponsorship beginning with Volume 9. The following is a list of volumes published and currently in preparation for the series:

Volume 1: **CARDIOVASCULAR PHYSIOLOGY** (A. C. Guyton and C. E. Jones)
Volume 2: **RESPIRATORY PHYSIOLOGY** (J. G. Widdicombe)
Volume 3: **NEUROPHYSIOLOGY** (C. C. Hunt)
Volume 4: **GASTROINTESTINAL PHYSIOLOGY** (E. D. Jacobson and L. L. Shanbour)
Volume 5: **ENDOCRINE PHYSIOLOGY** (S. M. McCann)
Volume 6: **KIDNEY AND URINARY TRACT PHYSIOLOGY** (K. Thurau)
Volume 7: **ENVIRONMENTAL PHYSIOLOGY** (D. Robertshaw)
Volume 8: **REPRODUCTIVE PHYSIOLOGY** (R. O. Greep)
Volume 9: **CARDIOVASCULAR PHYSIOLOGY II** (A. C. Guyton and A. W. Cowley, Jr.)
Volume 10: **NEUROPHYSIOLOGY II** (R. Porter)
Volume 11: **KIDNEY AND URINARY TRACT PHYSIOLOGY II** (K. Thurau)
Volume 12: **GASTROINTESTINAL PHYSIOLOGY II** (R. K. Crane)
Volume 13: **REPRODUCTIVE PHYSIOLOGY II** (R. O. Greep)
Volume 14: **RESPIRATORY PHYSIOLOGY II** (J. G. Widdicombe)
Volume 15: **ENVIRONMENTAL PHYSIOLOGY II** (D. Robertshaw)
Volume 16: **ENDOCRINE PHYSIOLOGY II** (S. M. McCann)
Volume 17: **NEUROPHYSIOLOGY III** (R. Porter)
Volume 18: **CARDIOVASCULAR PHYSIOLOGY III** (A. C. Guyton and D. B. Young)
Volume 19: **GASTROINTESTINAL PHYSIOLOGY III** (R. K. Crane)
Volume 20: **ENVIRONMENTAL PHYSIOLOGY III** (D. Robertshaw)
Volume 21: **LIVER PHYSIOLOGY** (N. B. Javitt)

Consultant Editor: Arthur C. Guyton, M.D., Department of Physiology and Biophysics, University of Mississippi Medical Center

INTERNATIONAL
REVIEW OF PHYSIOLOGY
Volume 20

Environmental Physiology III

Edited by
David Robertshaw, M.R.C.V.S., Ph.D.
Professor and Chairman
Department of Physiology and Biophysics
Colorado State University
Fort Collins, Colorado

UNIVERSITY PARK PRESS
Baltimore

UNIVERSITY PARK PRESS
International Publishers in Science, Medicine, and Education
233 East Redwood Street
Baltimore, Maryland 21202

Typeset by Action Comp. Co., Inc.

Manufactured in the United States of America by
Universal Lithographers, Inc.,
and The Optic Bindery Incorporated.

Library of Congress Cataloging in Publication Data

Main entry under title:

Environmental physiology III.

(International review of physiology ; v. 20)
Bibliography: p.
Includes index.
1. Body temperature. 2. Animal heat. 3. Adaptation (Physiology) I. Robertshaw, D. II. Series.
QP1.P62 vol. 20 [QP135] 599'.01'08s [599'.01]
ISBN 0-8391-1449-4 79-14496

Consultant Editor's Note

The first volume of the *International Review of Physiology* appeared in 1974, and since that time this new review series has become an important part of physiological literature. One of its most important purposes is to provide a comprehensive learning source for teachers and students of physiology throughout the world.

To explain the reasons for beginning this new publishing venture, we need to repeat once again the philosophy, the goals, and the concept of the *International Review of Physiology.* This Review has the same goals as all other reviews for accuracy, timeliness, and completeness, but it also has policies that we hope and believe will add important qualities often missing in reviews, especially integration of physiological mechanisms and instructiveness. To achieve these goals, the publishing format provides for 1200 to 1500 pages per year, divided into physiological subspecialty volumes organized by experts in their respective fields. This extensive coverage allows consideration of each subject in depth. And to make the review as timely as possible, a new volume in each area of physiology is normally published every two years. In addition, occasional volumes will be published at appropriate times on such topics as the liver, the eye, and other physiological subspecialty areas that might not warrant a new volume every two years.

To help in achieving the goals of the *International Review of Physiology,* special editorial policies have been established. A simple but firm request is made to each author that he utilize his expertise and his judgment to sift from the mass of publications those new facts and concepts that are important to the progress of physiology; that he make a conscious effort not to write a review consisting of an annotated list of references; and that the important material that he does choose be presented in thoughtful and logical exposition, complete enough to convey full understanding as well as being woven into context with previously established physiological principles. Hopefully, these processes will bring to the reader a series of treatises that he will use not merely as a reference but also as an exercise in refreshing and modernizing his whole store of physiological knowledge.

Arthur C. Guyton

Contents

Preface ix

1
Transfer of Heat Through Animal Coats and Clothing 1
K. Cena and J. A. Clark

2
Cold Thermogenesis 43
G. Alexander

3
Hyperthermia and Exercise 157
J. E. Greenleaf

4
Temperature Regulation, Fever, and Disease 209
M. J. Kluger

5
Metabolic Status During Diving and Recovery in Marine Mammals 253
P. W. Hochachka and B. Murphy

6
Physiological Effects of High Altitude on the Pulmonary Circulation 289
J. T. Reeves, W. W. Wagner, Jr., I. F. McMurtry, and R. F. Grover

7
Physiological Effects of Carbon Monoxide 311
G. R. Wright and R. J. Shephard

Index 369

Preface

If scientists are to retain the sympathetic and fiscal support of the lay public then it is important that they be able to demonstrate that the direction of their research can be adjusted to meet current needs and concerns of society. In so far as a physiologist is a scientist so must he also be seen to accept that he has social responsibilities in his research. The chapters in this review are meant to reflect the role that environmental physiology is playing in attempting a) to understand the physiological responses of man and animals to ever-changing environmental conditions, b) to provide a sound scientific base for medicine, and c) to appreciate mechanisms of survival in different surroundings. Environmental physiology therefore has a very broad base that extends into many other branches of physiology. No longer is environmental physiology synonymous with temperature regulation; rather, it relates to the environment in its broadest context. The avid use of fossil fuels has created pollution problems that affect the physiology and health of all living creatures, including man. A chapter is included therefore on the physiological effects of elevated atmospheric levels of carbon monoxide, one of the more important atmospheric pollutants. The recognition that the supply of fossil fuels is finite has led to an awareness of the need for energy conservation. In the design and heating of buildings for man and animals it is necessary to know not only the transfer of heat through the walls of the structure but also the physics of heat transfer through clothing and animal coats. Authors from Poland and England have reviewed present knowledge in this field.

The role of fever in combating infectious disease has not been fully understood, and some new evidence that has accumulated in recent years is reviewed here. This evidence indicates that fever may be an important and possibly essential response that allows the host to overcome the infection. In that fever represents an elevated but controlled body temperature, and so also during exercise is body temperature maintained at a higher level, it has been speculated that both situations may have some common underlying basic mechanism.

The ability to withstand cold by the generation of internal heat is reviewed by an Australian author. A great deal of research in Australia has been directed to understanding cold thermogenesis because of the high incidence of sheep mortality brought about by cold exposure, particularly at shearing time. This research shows how physiologists may be able to contribute to the solution of a farming problem.

Although man is more or less ubiquitous, permanent residence at high altitudes or in marine environments has presented a physiological challenge: lack of oxygen. One chapter reviews the physiological effects of altitude and another chapter is directed toward the adaptations of marine mammals for prolonged survival under hypoxic conditions, i.e., during diving.

Thus, although some seemingly unrelated areas are reviewed, reflecting the diversity of the discipline, the general theme of this volume is the contributions that the study of physiology can make to solve some of the current concerns of man's society.

Environmental Physiology III

International Review of Physiology
Environmental Physiology III, Volume 20
Edited by D. Robertshaw

1
Transfer of Heat Through Animal Coats and Clothing

K. CENA[1] and J. A. CLARK[2]

Institute of Building Science, Technical University, Wroclaw, Poland, and University of Nottingham School of Agriculture, Sutton Bonington, Loughborough, Leics., Great Britain

---

ENERGY BALANCE AND INSULATION 4

METABOLIC HEAT PRODUCTION 7

THE STRUCTURE OF ANIMAL COATS 10

CLOTHING STRUCTURE 11

SENSIBLE HEAT TRANSFER THROUGH COATS AND CLOTHING 13
- **Conduction** 13
- **Fiber Conduction** 15
- **Thermal Radiation** 15
- **Convection** 17
- **Clothing Ventilation** 19
- **Short-wave Radiation** 21

WATER VAPOR TRANSFER 21

---

[1]Recipient of a visiting fellowship from the Science Research Council of Great Britain and a travel grant from the Technical University of Wroclaw, Poland.

[2]Recipient of a visiting fellowship from the Technical University of Wroclaw, Poland, and a travel grant from the British Council.

When invited to contribute to this volume, the authors had recently completed a related review for a British Institute of Physics journal. The different readerships dictate different emphasis, and since completion of the earlier review a good deal of fresh material has become available. The units employed here are also different, in part, from those used previously. However, since a majority of factual material is common, a substantial proportion of the present review is closely based on that which appeared in *Physics in Medicine and Biology* (1).

THE EFFECTS OF WIND 25

SIZE AND INSULATION 28

CLOTHING INSULATION IN HYPERBARIC AND HYPOBARIC ENVIRONMENTS AND IN WATER 30

HUMAN COMFORT AND INSULATION 33

CONCLUSIONS 35

LIST OF SYMBOLS AND UNITS

The symbols used in this review and their units are as follows:
C, convective heat flux (W m^{-2})
E, evaporation rate (G m^{-2} s^{-1})
H_d, sensible heat flux (W m^{-2})
I, thermal resistance, when in m^{2}K W^{-1}
K, skin surface heat load (W m^{-2})
M, net metabolic heat flux (W m^{-2})
R_n, net radiation (W m^{-2})
T, temperature (°C or K)
X, thickness (m)
c_p, constant pressure specific heat (J g^{-1} K^{-1})
d, diameter (m or cm)
k, thermal conductivity (W m^{-1} K^{-1})
l, coat thickness (m or cm)
p, effective hair area per unit of coat depth (cm^{-1})
r, with subscript, transfer resistances in s cm^{-1}; e.g., for water vapor, r_v, and for heat, r_h
u, wind speed (m s^{-1})
λ, latent heat vaporization of water (J g^{-1})
ρ, density (g m^{-3})
σ, the Stefan-Boltzmann constant $= 56.7 \times 10^{-9}$ W m^{-2} K^{-4}
χ, concentration of water vapor in air (g m^{-3})
i_m, water vapor permeability index
Gr, Grashof number
Le, Lewis number
Nu, Nusselt number
Sh, Sherwood number
Other symbols and subscripts are as in text.

Environmental physiologists have devoted many years of study to the mechanisms of homeothermy, which enable mammals and birds to maintain an essentially constant body core temperature. The majority of research work in this field has been concentrated on the physiology of the control of body temperature, including a great deal of detailed study of biochemical processes. However, the mandatory requirement for effective homeothermy is not possession of a thermostat, but possession of adequate insulation. Despite the fact that the physics of heat transfer through animal insulators, including clothing, is not mathematically complex, it has until recently received less attention than a number of the secondary subcutaneous mechanisms concerned in homeostasis. In a previous review in this series Mitchell (2), who has himself contributed to key papers in this field, including one of the few in vivo measurements of clothing insulation (3), remarked that our understanding of the physical processes of heat transfer through mammalian coats, although still imperfect, has improved greatly in recent years, but that clothing is still poorly understood. Mitchell's conclusions are still valid. In addition, reasonably quantitative studies of avian coats can still be counted on the fingers of one hand.

The aim of this chapter is to review current knowledge of the physics of animal insulators, and to describe how their structure controls the processes of energy exchange between the skin surface and the ambient environment (1). Improved knowledge of the physical barrier to heat loss is surely essential for better understanding of homeothermy.

A layer of air, trapped by a fibrous coat, provides the major part of body insulation for all except a few of the larger terrestrial mammals, the cetaceans, and other aquatic mammals. In this chapter we do not consider subcutaneous (tissue) insulation in any detail, since, not only is it possessed by all homeotherms, whether or not they have a coat above the skin, but it is also demonstrable in poikilotherms. For example, cold-blooded animals over the size range from bumble bees (4) to fully grown crocodiles (Bell, personal communication) are able to thermoregulate quite effectively by the control of tissue insulation. The response of the body to environmental demands is obviously both very flexible and precise (5); for example, the body core temperature of most homeotherms can be maintained constant to a few tenths of a degree Celsius when the metabolic rate is constant. However, unless the subject is contained within a calorimeter, it is impossible to estimate total heat loss with comparable precision. Currently it is only possible to measure the insulation of clothing or animal coat specimens to about $\pm 10\%$, even on a calibrated flux plate. In vivo the values may be reduced to less than a half of those on a flux plate, and the errors of insulation estimation are greatly increased even on a realistic manikin. Because of these uncertainties the physical theories of heat transfer that are presented in this chapter have been selected according to the logical Principle of Ockham's Razor; i.e., where alternatives exist, the simplest rational theorem that fits the facts has been selected. Overcomplexity in the formulation of theories of heat transfer through animal coats is not only unnecessary, but can inhibit communication between the different specialists involved in this multidisciplinary subject. It is a disease particularly prevalent in some of the engineering literature, where the use of sophisticated mathematics has sometimes borne little relation to physiological reality.

The problems of clothing military personnel for a wide variety of severe climates during the 1939–1945 war, from the tropics to the arctic and from submarines to the new environments of high-flying airplanes, prompted scientific studies of clothing (6–8). Subsequently, studies of the clothing requirements for men in severe environments have been continued, relevant to, for example, the arctic explorer, the military aircraft pilot, and the diver. The air-conditioning of buildings has also increased interest in human comfort in everyday environments and its relationship to clothing and work (9–12). The development of synthetic fibers has also resulted in much research on the structure and properties of textiles, but this has its own special literature, which is outside the scope of the present review.

Studies of the properties of animal coats have had two stimuli. First, the increasing intensification of animal production in the postwar period and emphasis on its efficiency have led to research on the relationships between the metabolism of farm animals and their environments, including the effects of coat insulation (13, 14). Second, the stimulus for the parallel approach to wild animals, which originated from the same period, has been the desire to understand the thermal and physiological limits that determine the ecological ranges and population densities of wild species of homeotherms (15–17).

ENERGY BALANCE AND INSULATION

The principal function of the coats of homeothermic animals and the clothing of man is the regulation of thermal energy exchange between the body and its environment. Homeothermy is maintained only at the cost of unceasing metabolism, the heat component of which must be dissipated. Thus, the problem of the homeotherm is not so much the conservation of heat as the regulation of dissipation to equal production. Since the ratio between the rate of heat production at rest and that during maximum exercise is usually of the order of 1:10, a variety of solutions may be necessary even for a single species in one environment. Insulating coats play a major role in the regulation of heat dissipation.

Application of the First Law of Thermodynamics to steady heat transfer at the surface of the body gives

$$M = R_n + C + \lambda E \tag{1}$$

where all terms are expressed as watts per square meter of body surface. Because we are interested in external insulation, we have eliminated the respiratory heat fluxes by defining M as the net metabolic heat flux that must be dissipated from the body surface; R_n, C, and λE represent losses of heat by radiation, convection, and evaporation, respectively. λ is the latent heat of vaporization of water and E is the evaporation rate per unit of area. Equation 1 is an approximation, since terms that are usually minor, such as conduction to the ground and the energy required to warm food and water to body temperature, have been neglected. The application of the energy balance equation to homeotherms is discussed in detail by Gates (16), Monteith (18), Mitchell (5), Monteith and Mount (14), Gates and Schmerl (19), and for man in particular by Sibbons (20) and Monteith et al. (21), and in a number of other texts.

The sensible heat flux, $R_n + C$, must pass through the coat thermal resistance, I, and depends on the temperature difference across the insulating layer. Evaporative heat transfer may, to a first approximation, be considered to be driven separately through the coat by the gradient of water vapor concentration, against a diffusive resistance, r_v. Hence,

$$M = \frac{(T_s - T_c)}{I} + \frac{\lambda(\chi_s - \chi_a)}{r_v} \tag{2}$$

where T_s and T_c are the temperatures of the skin below the coat and of the coat surface, respectively, and I has units of $m^2\ K\ W^{-1}$. In the evaporative term, χ_s and χ_a are the concentrations of water vapor at the skin surface and in the air outside the coat, respectively, in units of $g\ m^{-3}$. If λ is in units of $J\ g^{-1}$, then r_v, the resistance to vapor transfer, has units of the reciprocal of velocity, $s\ m^{-1}$ (18, 22, 23). Where the thickness, l, of a coat can be accurately defined, something which is rarely easy, the resistance, r_v, may ideally be derived from first principles, since it is related to the diffusivity for water vapor, D, by

$$r_v = l/D \tag{3}$$

The insulation, I, may be related to the effective thermal conductivity of the coat, k, by the similar relation

$$I = l/k \tag{4}$$

We may use the similarity between heat and mass transfer to show the equivalence between r_v and I. When the Sherwood number and Nusselt number, the dimensionless mass transfer and heat transfer coefficients, are equal, then the Lewis Relation (24) states that the coefficient of mass transfer ($h_m = r_v^{-1}$) is equal to the convective heat transfer coefficient ($H_c = I^{-1}$) divided by the volumetric specific heat of the medium. Hence,

$$h_m = \frac{h_c}{\rho c_p} \quad \text{or} \quad I = \frac{r_v}{\rho c_p} \tag{5}$$

Where sensible heat transfer is not by convection alone, the same form may be used, but r_v must be replaced by a resistance to heat transfer, r_h, in the same units. The sensible heat flux, H_d, may, therefore, be expressed as

$$H_d = \frac{(T_s - T_c)}{I} = \frac{\rho c_p (T_s - T_c)}{r_h} \quad (6)$$

where the product, ρc_p, of the density (ρ) and the specific heat (c_p) is the volumetric specific heat of the medium. The choice of the value for this constant is somewhat arbitrary in this system. Cena and Clark (25, 26) have recently pointed out that, for heat transfer in air, selection of the value for STP (1,298 J m^{-3} K^{-1}) has the advantage of giving a whole number conversion, of 2 s cm^{-1} = 1 *clo*, between resistances in seconds per centimeter and the empirical *clo* unit of clothing insulation, still employed in much of the literature. Because of the advantages of using the same system of units for both heat and mass transfer resistances, values of insulation are presented in both m^2K W^{-1} and s cm^{-1} (the preferred submultiple) in this chapter. Those who still think in *clo* need simply to divide the figures in seconds per centimeter by two. Table 1 presents representative values of thermal resistance in all three units.

In order to estimate the heat loss from an animal, we must consider the internal insulation and that provided by the air layer outside the coat in addition to that of the coat itself. If we define r_s as the skin and tissue resistance and r_a as the air resistance, then r_s can typically take values between about 0.3 and 1 s cm^{-1} for large animals, whereas r_a has a maximum value of around 1 s cm^{-1} (27). Depending on the circumstances we must, therefore, add to the resistances in Table 1 a value between 0.3 and 2 s cm^{-1} if we wish to estimate the conductance of the three resistances in series, $h = \rho c_p (r_s + r_h + r_a)^{-1}$. For example, a man wearing the standard clothing used in the

Table 1. Typical thermal resistances or insulations in m^2 K W^{-1}, the empirical *clo* unit, and the alternative based on the Lewis Relation

Coat	Thermal resistance		
	m^2 K W^{-1}	*clo*	s cm^{-1}
Best bed, sleeping bag, or nest	1.55	10	20
Warmest practical arctic clothing; coats of large arctic mammals	0.77	5	10
Heavy clothing for outdoors in winter; coats of temperate climate animals	0.31	2	4
Normal clothing; coats of small animals	0.155	1	2
Light summer and hot climate clothing	0.077	0.5	1
Shorts and singlet, for exercise	0.037	0.3	0.6
Still dry air (per cm)	0.4	2.6	5.2

definition by Gagge et al. (28) of the *clo*, $r_h = 2$ s cm^{-1}, may have a low tissue resistance (about 0.3 s cm^{-1}) and be in an environment with low air movement ($r_a \simeq 1$ s cm^{-1}). The total resistance to heat loss is, therefore, 3.3 s cm^{-1} and the conductance is $h = 1298 \div 330 = 3.94$ W m^{-2} K^{-1}. At the standard metabolic rate of 58 W m^{-2} the man should, therefore, be comfortable in an environment 15°K below body temperature, i.e., 22°C, which is near enough to Gagge et al.'s figure of 21 °C for the present purposes.

Water vapor is unfortunately usually measured in units of partial pressure (e) rather than concentration. The appropriate S.I. form is the kiloPascal (kPa), although meteorologists insist on persisting with the millibar, disguised as the hectoPascal (hPa). The conversion to the units of concentration employed in Equation 2, which may be derived from first principle, via the Gas Laws, is

$$\chi = \frac{2170}{T} e \tag{7}$$

where T is the absolute temperature in K. Expressed in vapor pressure units of kPa, the conductance (h_e) for evaporative heat transfer is, therefore,

$$h_e = \frac{2170\lambda}{T\, r_v} \tag{8}$$

which gives for 3 s cm^{-1} resistance (of the order of that for normal clothing and the external air layer in series) $h_e \simeq 60$ W m^{-2} kPa at normal temperatures.

METABOLIC HEAT PRODUCTION

Environmental physiologists will be familiar with the classical picture of the responses of the metabolic heat production of homeotherms to their thermal environment. However, we need to present the "metabolic diagram" (Figure 1) in order to show how insulation determines both the energy costs of homeothermy and the lethal limits of the range of environments that an animal may tolerate. If, for simplicity, the environment of an animal is represented by a single environmental temperature, T, a typical graph of metabolic heat production and loss against T will resemble Figure 1.

Most homeotherms show a limited range of temperatures, between the lines C and D on Figure 1, in which their metabolism is a minimum and no effort is required for thermoregulation (29). The width of this zone may vary from tens of degrees in large animals to a few degrees in small neonate animals. It depends on both the size of the animal and its insulation, as well as on other variables, such as food intake. The insulation or thermal resistance of the coat acts in series with the external insulation of the air. For the purposes of measurement and modeling, the flow of heat through the in-

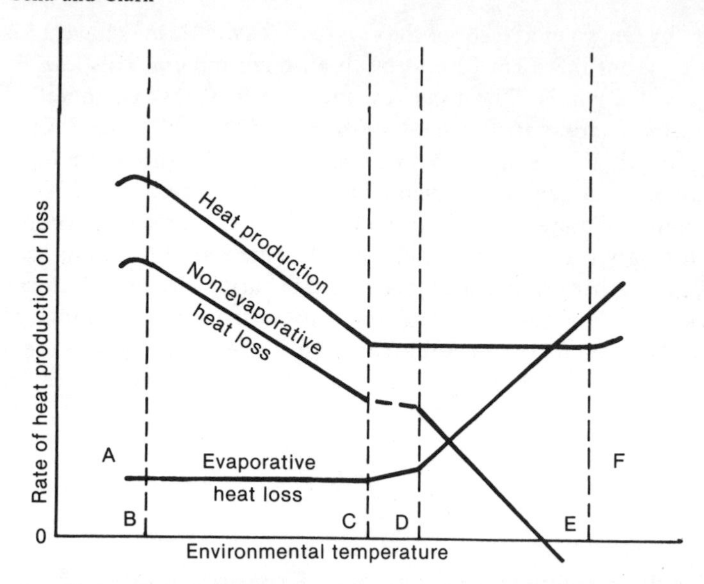

Figure 1. Diagram of relationships between heat production, evaporative and nonevaporative heat loss, and deep-body temperature in a homeothermic animal. A, zone of hypothermia; B, temperature of summit metabolism and incipient hypothermia; C, critical temperature; D, temperature of marked increase in evaporative loss; E, temperature of incipient hypothermal rise; F, zone of hyperthermia; CD, zone of least thermoregulatory effort; CE, zone of minimum metabolism; BE, thermoregulatory range. The zones are as defined by Mount (29). Reproduced by permission.

sulation can be described by using a simple resistance analogue (Figure 2). In the analogue (27), I_c is the coat resistance, I_a the external resistance, and I_r a "resistance" for radiative energy transfer. In the simplest case, when the air and radiant environmental temperatures are equal, we may define a single external resistance $I_a = I_a I_r/(I_a + I_r)$.

Newton's Law of Cooling is often assumed to apply to homeothermic animals. However, Fourier's Law of Heat Loss, which states that the rate of sensible heat exchange between a body and its surroundings is proportional to their temperature difference, is more appropriate. Indeed, neither applies within the zone of "least thermoregulatory effort" (29) where animals control their heat loss by changing body insulation. However, at a lower critical temperature, T' (C in Figure 1), the sensible heat transfer through the coat becomes equal to the net metabolism at the minimum rate, M. This temperature is determined largely by coat insulation. At the lower critical temperature, T', neglecting evaporation,

$$M = (T' - T)/(I_c + I_e) \tag{9}$$

Because in most coated animals and man I_c is usually substantially greater than I_e, the lower critical temperature is determined principally by

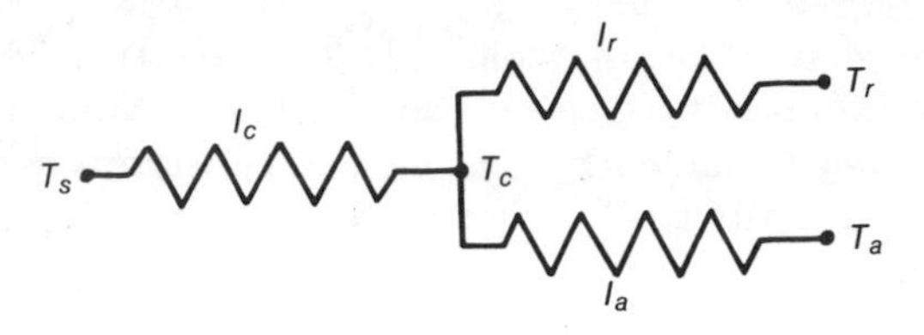

Figure 2. Simplified analogue network for sensible heat transfer in animal coats and clothing. Insulations I_c, I_a, and I_r refer to the coat, air, and radiative insulations, respectively. The thermal potentials are skin temperature, T_s; coat temperature, T_c; air temperature, T_a; and radiative temperature of the environment, T_r.

the coat insulation. Below the lower critical temperature in the zone BC of Figure 1, the rate of increase of metabolism also largely depends on insulation, since for balance M must increase to compensate heat loss as environmental temperature decreases.

The lowest temperature at which a homeotherm may survive indefinitely is again determined by its insulation, since the increase of metabolic heat production in response to cold stress cannot exceed a factor of about three. Lethal temperatures, corresponding to B in Figure 1, range from about 30°C for small naked neonate mammals and young birds to temperatures theoretically below those at which air liquefies, calculated on the basis of insulation for well insulated arctic animals. In general, the larger the animal the greater its available metabolic energy per unit of surface area, and the greater the thickness of insulation it can carry. Lower critical temperatures and wider temperature ranges are, therefore, associated with both greater insulation and size. Insulation is, however, the principal factor. For example, for a 70-kg naked man T' is about 27°C, whereas for a 5-kg arctic fox it is below −40°C. Insulation in smaller animals is, however, closely correlated with size, since geometric considerations limit coat thickness in two ways: first, movement is impeded by coats that are excessively thick (31) and, second, successive layers become increasingly ineffective, for reasons that will be discussed later. Impediment to movement can also be important in thick human clothing (32). Teitlebaum and Goldman (33) found that men walking at 5.6 km hr^{-1} in an arctic uniform weighing 6 kg had a metabolic rate 80 W higher than when they carried the same weight in a belt. This increment alone is equal to the basal rate of metabolism.

The response of homeotherms to high temperatures is more complex. At low temperatures cutaneous evaporation is usually small and constant. For example, in man the rate of evaporative heat loss by diffusion through the skin has a minimum value that is about 6%, only, of the rate from a fully wet surface at the same temperature (34, 35). However, above the upper limit of the zone of least thermoregulatory effort, increased evaporative heat loss is necessary for heat balance. Such an increase occurs even in animals such as the sheep, which are generally considered nonsweating (36). In sweating species, such as man and the Equidae, liquid water is actually secreted onto the skin surface, so that rates of evaporative loss are very much higher. In the limit the whole skin may become wet, and in this condition the rate of heat loss by evaporation is limited by the physical resistance to the transfer of water vapor through the coat, rather than by the physiological control of sweat secretion.

THE STRUCTURE OF ANIMAL COATS

Air trapped by the coat provides the main resistance to heat transfer between the body and its environment; the principal function of animal coats and human clothing is, therefore, to restrain the movement of air close to the skin. Both hair and feather coats are limited in their structure by physiology: they must grow from the skin and are, therefore, anchored to it. Insulation is usually provided by a layer of fine hair or down, which may be supplemented by coarser elements that either project above it or lie over it providing protection from wind and rain—profile feathers in birds and guard hairs in mammals. The structures of the hair coats of mammals, which are the easiest to describe and have received most attention, have been reviewed by Tregear (37) and Ryder (38). Hair densities range from the sparse residual covering on man and the pig, with 10–100 hairs cm^{-2}, to the dense coats of species such as the fox and rabbit with circa 4,000 cm^{-2}. The dominant hair diameter ranges from 20 to 150 μm according to species. Typical values for hair density, diameter, and coat depth are presented in Table 2. This information alone is insufficient to specify the thermal properties of a coat; hair orientation has a considerable influence, in particular on the way in which wind affects insulation (39, 40). More recently Cena and Clark (41) demonstrated that radiative transfer through hair coats also depends on hair orientation. Mammalian coats can be divided into three classes. First are hair coats of coarse straight fibers, with densities of only a few hundred hairs cm^{-2}, such as those of the goat or badger. These are not particularly good insulators, but can be piloerected to increase their insulation. Second is fur, the dense (about 4,000 cm^{-2}) fine coats of the small mammals and arctic species. And, lastly, is wool, the dense and matted coats of crimpled hairs, such as those of the familiar domestic sheep (about 1,000 cm^{-2}). Most mammalian coats are either a mixture of the first two types or intermediate between them.

In contrast to the literature on mammalian coats and their structure there have been few published measurements on feather coats other than the thickness and weight of the feathers (17, 42). The basic structure of most feather coats obviously differs from that of hair coats in that a division into distinct layers is more evident. Adjacent to the skin there is a layer of downy fibers, carried on the bases of large feathers or on separate small quills, and this layer is covered and partially sealed by the overlapping profile feathers.

Work in progress (MacLeod and J. A. Clark, personal communication) suggests that although the geometry of avian coats is more complex than

Table 2. Typical values of the physical coat parameters for various animals

Animal	Parameter[a]					
	n (cm^{-2})	d (μm)	l (cm)	l_s (cm)	$\tan \phi$	p (cm^{-1})
Dorset Down sheep	1,430	48	8.0	10.0	0.75	5.2
Clum Forest sheep	1,460	42	5.0	6.0	0.75	4.6
Welsh mountain sheep	850	65	3.5	4.5	0.75	4.1
Rabbit	4,200	31	2.0	2.5	0.75	9.7
Badger	240	71	1.8	4.5	2.3	3.9
Cow	1,260	44	0.6	2.0	3.2	17.8
Goat	110	83	2.5	3.0	0.75	0.7
Fox	3,600	20	1.4	2.0	1.0	7.2
Deer	520	150	1.5	3.7	2.3	17.9

From Cena and Clark (41).

[a] n, number of hairs per unit of surface area of skin; d, hair diameter; l, coat depth; l_s, stretched length of hairs; $\tan \phi$, tan arc cos (l/l_s); and p, effective hair area.

those of mammals, an equivalent description is possible in terms of fiber area per unit of skin (43).

CLOTHING STRUCTURE

Studies of the environmental physiology of man and of human comfort are bedeviled by the fact that everyday clothing is determined by fashion as well as function. However, we may distinguish three major structural characteristics of clothing which differentiate it markedly from natural coats: first, it is not anchored to the skin; second, it is almost always composed of a number of discrete layers; and third, clothing is assembled from separate garments that individually do not cover the whole body. For the present purposes, other properties of the clothing, such as the thickness, weight per unit of area, and the woven structure of most materials may be regarded as secondary. The air between the layers of fabric provides much of the insulation; it is, therefore, important to distinguish between the intrinsic insulation of fabrics and the total insulation of a clothing assembly (sometimes "ensemble"), which will usually be substantially greater. Of course, it is the total insulation that is relevant to comfort and thermal regulation. For example, Rogers and Sutherland (44) showed that the insulation of multilayer arctic clothing, measured on a copper manikin, may be predicted safisfactorily simply from the number of layers present. For the estimation of total insulation, r, in seconds per centimeter, from the number of layers, n, they give the equation

$$r = 0.16n + 1.02 \quad (10)$$

Windproof outer clothing is given a double weighting. Belding et al. (45, 46) came earlier to a similar conclusion; for example, they measured 19 mm of air and only 9 mm of fabric in one clothing assembly. Fabrics relying on thickness for their insulation, such as "fur" fabrics and the fibrous batting insulation of sleeping bags and anoraks, have been successfully treated as analogies of mammalian coats (47). These show minimum thermal conductivities similar to those of fur and hair coats in Table 2.

The thermal insulation afforded by stacks of fabrics without distinct air layers may also be estimated from the fabric thickness alone, irrespective of other physical parameters. For dry single-layer clothing materials, O'Callaghan and Probert (48) found that the maximum intrinsic insulation is achieved for a bulk density of about 90 kg m^{-3}, the thermal resistance (in m^2 K W^{-1}) being 23 times the clothing thickness in meters. The minimum effective thermal conductivity of a single-layer woven fabric was 40 mW m^{-1} K^{-1}. Under load, thermal behavior of multiple-layer assemblies is identical with that of a single layer of the same material and overall thickness. O'Callaghan and Probert (49) measured the resistance of a stack of eight polyester cloth layers. It fell from 0.46 m^2 K W^{-1} under zero loading to 0.032 m^2 K W^{-1} under 31 N m^{-2}. Azer (50) was able to predict the insulation of single-layer garments from their fabric properties, fit, and the proportion of the body covered. His model also allowed for the increase in surface area that is caused by the finite thickness of the garment, and which reduces the effective insulation of any garment to a value less than that of the same fabric measured on a flat plate. Typical values for fabric insulation are given in Table 3.

While specialized clothing for military purposes and severe environments may be designed to provide a known insulation, everyday clothing is assembled by the individual to satisfy both his thermal and his social needs. Layers come as integer multiples but garments cover fractions of the body. However, work by Seppanen et al. (53) and by Sprague and Munson (54) has shown that separate measurements of the insulation of

Table 3. Thermal insulation of selected fabrics

Fabric	Thickness (mm)	Insulation	
		m^2 K mW^{-1}	s cm^{-1}
Cotton poplin shirting	0.5	8	0.1
Wool serge	1.0	17	0.22
Light woolen coating	2.2	46	0.6
Heavy woolen coating	4.3	93	1.2
Cellular woolen blanket	6.6	147	1.9
Anorak lining (raised wool)	8.3	201	2.6
Mohair pile	1.27	310	4.0

From Auliciems and Hare (51) and Moote (52).

items of clothing, made on a copper manikin model, can be used satisfactorily to predict the insulation of clothing assemblies on the model, although the effects of compression and fit reduce the total to less than the sum of the individual insulations. The equations they present give the total insulation for men (in seconds per centimeter) as

$$r = 1.454\, \Sigma r' + 0.226 \tag{11}$$

and for women

$$r = 1.54\, \Sigma r' + 0.1 \tag{12}$$

where r' is the insulation of an individual garment. The standard error is about 0.1 s cm^{-1}. Although more soundly based, these equations do not produce results very different from those of the simpler approach of Rogers and Sutherland (44) (Equation 10) even in the insulation range of normal clothing. Moreover, Equations 11 and 12 would lead to overestimates of total insulation if applied to multilayer cold-climate clothing. More recently, Nishi et al. (55) tested the predictions of Sprague and Munson against in vivo measurements of comfort indicators such as skin temperatures and found them satisfactory for the practical purposes of comfort prediction in normal environments. However, values obtained from Equations 11 and 12 should be reduced by about 30% according to later information from Nishi et al. (139) and Gagge (personal communication). Estimations of clothing insulation in vivo, from the metabolic rates of men, remain rare, although Mitchell and van Rensburg (3) conclude that the technique is not appreciably more difficult than the use of manikins. Measured values of insulation for representative garments are presented in Table 4. These are taken largely from Seppanen et al. (53). If the reader uses this table to estimate his or her own insulation using Equations 11 or 12 they will almost certainly find that it comes to substantially less than 2 s cm^{-1} (1 *clo*). This is the consequence of changes in fashion since the *clo* unit was proposed by Gagge et al. (28).

SENSIBLE HEAT TRANSFER THROUGH COATS AND CLOTHING

Conduction

For large animals, layers of insulation can be approximately considered as plane sheets, but effects of geometry become important when the ratio of coat thickness to body radius exceeds about 1:10. Because conduction in the air of the coat is not the only process contributing to heat transfer in clothing and animal coats, their measured thermal conductivities are always greater than that of still air. The extensive literature on the thermal conductivities of animal coats is summarized in Table 5, together with some values for fabrics. Many publications unfortunately present only the conductance of the insula-

Table 4. Thermal insulation of garments

Item/Assembly	Insulation[a]	
	m^2 K mW^{-1}	s cm^{-1}
Socks	1.5–5	0.02–0.06
Light underwear	8	0.1
Winter underwear (cotton briefs and short sleeve vest)	30	0.4
Shirt or blouse, short sleeve	30–40	0.38–0.44
Shirt or blouse, long sleeve	39–45	0.5–0.58
Light trousers	40	0.5
Heavy trousers	50	0.64
Sweater, long sleeve	26–57	0.34–0.74
Jacket	26–57	0.34–0.74
Heavy jacket or quilted anorak	76	1.0
Boiler suit or long winter underwear	77–116	1.0–1.5
Tights	1.5	0.02
Skirt	15–34	0.2–0.44
Light dress	26	0.34
Winter dress	100	1.3
Track suit[b]	77	1.0
Arctic combat assembly[c]	666	8.6

[a]Values measured on manikin from Seppanen et al. (53) and Sprague and Munson (54), except as indicated. This table may be used for estimation of the insulation of assemblies in conjunction with Equations 11 and 12 (see text).

[b]Measured in vivo while working (3).

[c]About the warmest practical assembly, total weight approximately 16 kg (56).

Table 5. Thermal conductivities of animal coats

Species and source	k ($mW\ m^{-1}\ K^{-1}$)
Still dry air	25
Arctic mammals (15, 39)	36–106
Various wild mammals (57, 52)	38–51
Merino sheep (40)	37–48
Newborn merino, Down, cheviot, and Scottish blackface sheep (58–61)	65–107
Cattle (62, 63)	76–147
Rabbit (52, 64, 65)	38–100
Kangaroo (66)	43–64
Harp seal pups (67)	47–65
Grouse feathers (68)	29–58
Penguin (42)	31–46
Gosling (69)	36–46
Artificial fur (47, 52)	40–67
Woven fabric (48)	40

tion, without stating the thickness, so that conductivities cannot be derived from all published measurements. However, it is apparent that none of these natural insulators achieve the thermal conductivity of still air (25 mW m^{-1} K^{-1}). Minimum values of conductivity measured for a wide range of insulators lie in the range of 30–45 mW m^{-1} K^{-1}. Molecular diffusion, therefore, cannot be the only process of heat transfer, as there is some irreducible minimum conductivity in animal coats and clothing that is greater than that of still air. Since some measured conductivities exceed the minimum for the particular coat by a factor of 2–4, additional factors must contribute to the enhancement of heat transfer from the minimum, even in the absence of wind.

Three additional mechanisms could contribute to sensible heat transfer in insulating layers: conduction along the fibers, radiative energy transfer, and free convection within the insulation.

Fiber Conduction

Fiber conduction may be quickly dismissed. Tregear (37) calculated that rates of heat conduction along the fibers of animal coats were trivial, since the cross-section of the hairs is at most 10^{-2} of the skin surface area, and usually closer to 10^{-3}, whereas the conductivity of keratin, the main constituent of the fibers, is unlikely to be greater than 5 times that of still air. Fiber densities may be higher within the fabric plane of a clothing layer, but since the principal orientation of the fibers lies parallel to the skin and air separates the fabric layers, fiber conduction is again trivial except in cotton and artificial fiber fabrics under high compression loads (49, 70).

Thermal Radiation

The exchange of thermal energy by radiative transfer within fur and hair coats has until recently been neglected. However, Cena and Clark (41) and Cena and Monteith (71) showed that thermal radiation plays a significant role in sensible heat transfer. The structure of the hair coat affects radiation transfer. For this purpose we need to consider the orientation of the hairs as well as their number and size. The angle of repose of the hairs for coats with straight hair (Figure 3), or the arc cos of the ratio between the coat depth, l, and the hair length, l_s, for crimpled (wavy) coats, gives a coat parameter for radiation transmission

$$p = n\,\mathrm{d}\tan(\arccos l/l_s) \tag{13}$$

Physically, p is the projected area of hair on the plane of the skin per unit of skin area, for unit depth of coat, and has units of m^{-1} (or cm^{-1}). The fraction (τ) of radiation transmitted through a layer of coat of thickness z, normal to the surface, then follows a Beer's Law form with

$$\tau = \exp(-pz) \tag{14}$$

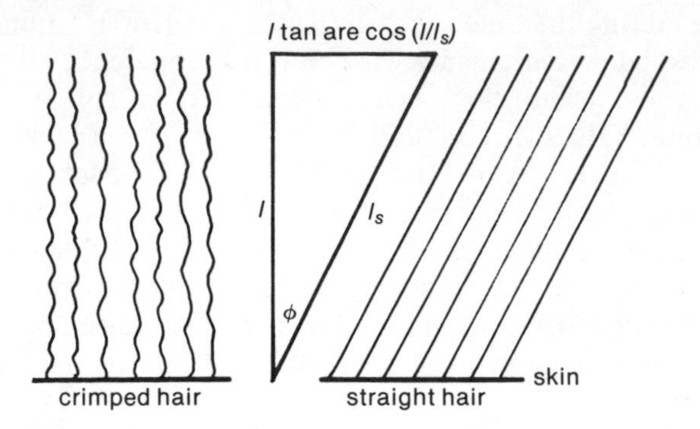

Figure 3. Simplified geometry for animal coats with crimpled and straight hair. l, coat depth; l_s, stretched hair length; ϕ = arc cos (l/l_s). Reproduced from Cena and Clark (41) with permission of the author.

where p is the coat parameter from Equation 13. Linearization of the Stefan-Boltzmann relation for small temperature intervals leads to the "radiant conductivity" of a fibrous insulating material, in the form

$$k_r = 16\sigma \overline{T^3}/3p \tag{15}$$

where σ is the Stefan-Boltzmann constant and $\overline{T}$ is the mean temperature of the coat layer. At 303 K, k_r is, therefore, $8.4/p$ in units of W m^{-1} K^{-1}. The form of the radiation transfer equation is similar to that established in engineering studies of heat transfer through insulation composed of randomly oriented inorganic fibers by Hager and Steere (72). The corresponding radiant conductivities range from about 20 mW m^{-1} K^{-1} for sheep fleece (p = 4 cm^{-1}) to a minimum of about 5 mW m^{-1} K^{-1} for cattle coats (p = 18 cm^{-1}). The addition of these conductivities to that of still air leads to minimum coat conductivities of between 30 and 45 mW m^{-1} K^{-1}. Similar conclusions are reached by Davis and Birkebak (73) on the basis of a much more complex model. A recent publication of Özil and Birkebak (74) in the engineering literature, however, shows that, in fact, their results can be simplified to the same form as that derived by Cena and Monteith (71). The addition of radiant transfer alone to conduction through still air, therefore, accounts for the measured base levels of conductivity in the hair coats of mammals. The relative contributions are probably similar in clothing materials that depend on thickness for their insulation and possess a fibrous structure. Studies of heat transfer through simulated fur fabrics by Moote (52) and Skuldt et al. (47) also show that natural and artificial furs may be treated in the same way. Özil and Birkebak (74) show curved temperature profiles, measured in artificial fur, that are consistent with those measured in natural furs by Cena and Monteith (71). When the artificial fur sample was placed in vacuum, the dependence of the dimensionless radiative heat flux through the insulation on their dimensionless temperature index appears to be more linear than in air. This is more likely to be due to thermal coupling of the fibers to the air than to the influence of the external radiation field.

Both animal coats and conventional clothing are black body emitters in the thermal radiation spectrum according to Hammel (57) and Dunkle et al. (75). However, the transmission of thermal radiation through layers of clothing will follow a different form from that in hair coats. For the simplest case of n equally spaced black layers with temperatures T_1, $T_2 \cdots T_n$, situated in a linear temperature gradient, the thermal radiation flux density, R_L, in W m^{-2} is given by

$$R_L = \sigma(T_s^4 - T_1^4) = \sigma(T_1^4 - T_2^4) = \cdots \sigma(T_{n-1}^4 - T_n^4) = \sigma(T_s^4 - T_n^4)/n \quad (16)$$

which for small temperature intervals may be linearized as

$$R_L = 4\,\sigma\,\overline{T^3}\,(T_s - T_n)/n \quad (17)$$

Therefore, if we consider radiation transfer outside the coat as a separate process, then the radiant conductance of a clothing assembly will be proportional to $1/n$. For a typical three-layer assembly at 300 K this gives a radiant conductance equal to 200 mW m^{-2} K^{-1}, very similar to values that may be derived for the coats of mammals. For example, a 5-cm coat with a radiant conductivity of 10 mW m^{-1} K^{-1} will have the same radiant conductance. Thermal radiation has also been recognized as contributing to heat transfer in nonconventional clothing materials (32); e.g., Crockford and Goudge (76) found that it was an important mode of heat transfer in a cellular plastic insulation.

In theory the emissivity of natural coats could help in the control of radiant heat exchange. But Hammel (57) concluded that no animal has developed a low emissivity coat to minimize radiant exchange. Specialized clothing fabrics with low emissivities have, however, been developed (77) but are used more for the concomitant high reflection, to minimize radiant loads on the wearer in firefighting in particular, than to control radiation transfer of normal clothing. The much advertized "space blanket" is a notable exception, and perhaps the low emissivity of 0.6 reported for Lamé fabrics may contribute to the comfort of the otherwise inadequately clothed wearer. Goldman (78) concludes that the application of low emissivity fabrics is limited principally by the restriction of water vapor transfer that accompanies an effective barrier to radiation. Coating the fibers of a fabric has little effect on its emissivity because the spaces between the fibers act as black body cavities.

Convection

There still remains unaccounted for a wide range of enhanced conductivities measured in still air. In vivo the disturbance of animal coats by movement and the internal ventilation that accompanies breathing and bypasses human clothing cause much of the degradation of insulation. However, thermal conductivities of over 100 mW m^{-1} K^{-1}, over 4 times that of air, have been measured on flux plates. Undoubtedly some of the reported measurements were made in poorly controlled conditions, but one of the major variables is simply the orientation of the sample. Cena and Monteith (65) pursued this clue and compared rates of heat transfer through vertical and horizontal samples of sheep fleece, measured on a flux plate, with rates of free convection predicted for uninsulated flat plates with the same substrate to air temperature difference. The dependence of the dimensionless heat transfer number, the Nusselt number, Nu, on the Grashof number, Gr, which describes free convection, may be considered the index of the heat transfer processes. The Nusselt number is $Hd/k\ \Delta T$, where H is the heat flux, d the diameter of the flux plate, k the conductivity of the air, and ΔT is the temperature differential. The Grashof number is given by $agd^3\ \Delta T/\gamma^2$, where γ is the coefficient of kinematic viscosity of air, g is the acceleration due to gravity, and $a = 1/T$ is the coefficient of thermal expansion of air at the temperature T (K). For the conditions of measurement, free convection from a vertical plate should follow the relation:

$$\mathrm{Nu} = 0.58\ \mathrm{Gr}^{0.25} \qquad (18)$$

Similar expressions predict the heat transfer from other shapes and orientations. Cena and Monteith found that for fleece samples on a flux plate the heat flux was only 30–40% lower than that predicted by Equation 17. For a 2-cm-thick sample on a vertical flux plate they obtained the empirical relation

$$\mathrm{Nu} = 1.66\ \Delta T^{0.7} \qquad (19)$$

Figure 4 shows the heat loss calculated from Equation 18, i.e., that expected in the absence of a coat. The measurements of heat loss from a 4-cm fleece occupy an intermediate position on the graph between this and the corresponding relation for still air. Similarity between the predicted and measured heat loss is even closer for a 2-cm fleece. The equivalent thermal conductivity varies from about 2 to 3 times that of still air, depending on the temperature differential, for a sample in a vertical plane. When the sample was horizontal, the heat flux was much closer to that predicted for conduction through still air.

Free convection, therefore, plays a major role in heat transfer through animal coats, although it has often been thought unimportant (79). The required air velocities are, in fact, very small; Cena and Monteith calculated that the rates of additional heat transfer they measured in fleece samples less than 2 cm deep could have been produced by a uniform vertical air velocity of 1 cm s^{-1} within the insulation. Analogous transfer mechanisms may be expected to occur within thick layers of clothing. Nishi (80) demonstrated a

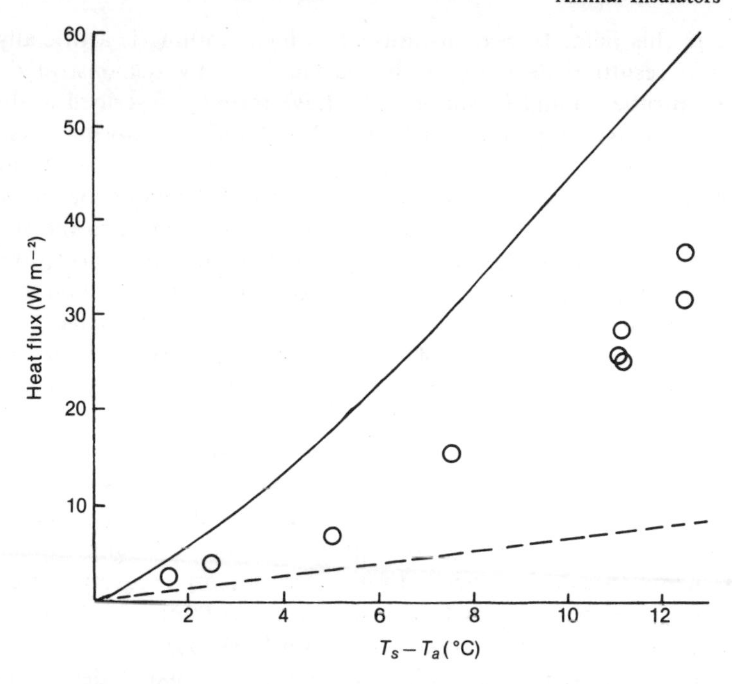

Figure 4. Heat flux through 4 cm of fleece mounted vertically. The solid line is calculated from Equation 14. The *broken line* is the calculated heat flux for pure conduction in still air. $T_s - T_a$ is the temperature difference between the skin and air. Wind speed is 1 m s^{-1}. Reproduced from Cena and Monteith (71) with permission of the author.

close correspondence between naphthalene sublimation through fabrics and heat and water vapor transfer, confirming that similar processes drive all three fluxes. This would not be expected if molecular diffusion were the principal mechanism of transport.

Clothing Ventilation

For human clothing there is further enhancement of heat transfer by the "short circuit" of direct ventilation (81), which has been mentioned briefly already. This is usually of advantage to man because, except during shivering, high rates of ventilation accompany activity and high rates of metabolism. Then enhanced heat dissipation is needed to maintain comfort. For example, Belding et al. (46) found that the insulation of a military clothing assembly for cold climates was halved when the wearer ran at 3 m s^{-1} compared with standing. However, although the importance of ventilation is recognized, there have been very few other measurements of the effects of activity on clothing insulation. As noted by Mitchell (2), in the preceding volume, the thermophysical properties of clothing are still poorly understood. Improved knowledge of the relative importance of the heat transfers through the fabric and by ventilation is obviously the key to further advance in this field. In circumstances in which clothing is artificially ventilated the resulting decrease in insulation is easily demonstrated (82). However, Birnbaum and Crockford (83) have recently described in detail a method, previously employed by Crockford (84) and Crockford and Goudge (76), which allows direct measurement of the ventilation rate of clothing assemblies. They have successfully adapted a tracer technique for the estimation of ventilation rates, originating from building science, to the study of clothing. In their version they use atmospheric oxygen as the tracer. The concentration of oxygen on the clothing is reduced by the introduction of an excess of nitrogen gas. The ventilation rate can then be estimated from the subsequent exponential return of the oxygen concentration to its natural value. The main source of error is that a measurement of the volume of the clothing assembly is also required. Figure 5 shows the curve of oxygen concentration against time obtained by Birnbaum and Crockford (83) in a typical measurement. The fractional ventilation rate, in units of $[\text{time}]^{-1}$, is obtained as the slope of a graph of the natural logarithm of the difference from the atmospheric oxygen concentration, plotted against time. The measurements on clothing appear to give very smooth recovery curves, and hence high accuracy, compared with the present authors' own experience of applying the technique to buildings. Crockford (84, 85) first used this technique to show differences in ventilation between alternative designs of foul

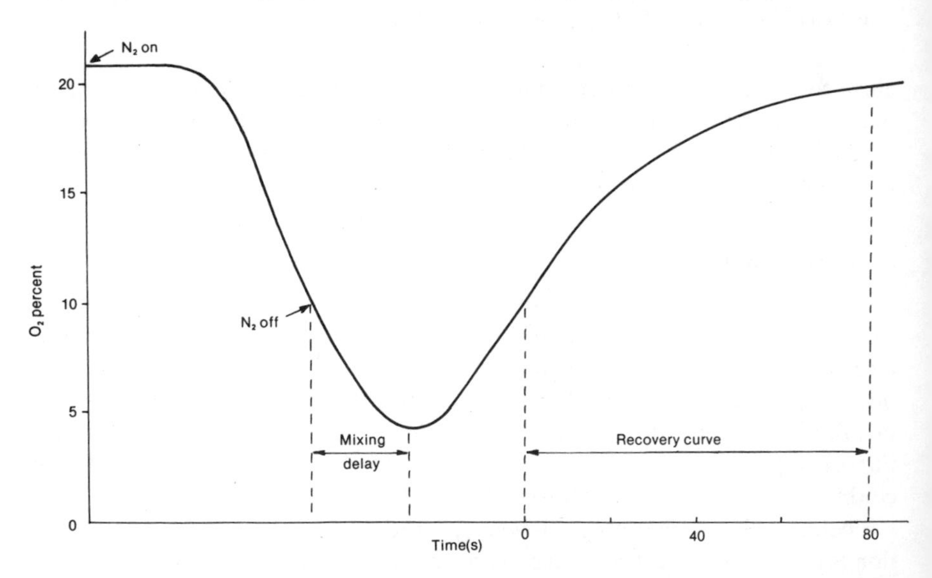

Figure 5. Typical curve of oxygen concentration against time obtained during measurements of the ventilation rate of clothing. The fractional ventilation rate may be estimated from the recovery phase (*right* of the curve, with the time scale indicated below. This example gives 1.8 air changes per minute (0.03 s^{-1}). Redrawn from data presented by Birnbaum and Crockford (83).

weather clothing for fishermen. Wider application of this technique is obviously warranted.

Conventional clothing depends on the movement of the wearer to "pump" the ventilation during work. In more severe environments artificially ventilated suits are necessary. Colin and Houdas (86) reviewed the methods for protecting man against heat and concluded that only artificial means of cooling were capable of removing the metabolic heat produced while working for long periods in protective clothing, such as that designed for firefighting.

Short-wave Radiation

It is necessary to mention briefly the absorption of short-wave radiation ($\lambda <$ 1 μm) by animal coats and clothing, which often affects the energy balance. Like the transmission of thermal radiation it depends on structure and in animal coats follows a form similar to Equation 14 (71, 86). Radiation is, therefore, absorbed throughout a finite depth of coat, and this may produce temperature gradients that modify heat transfer. The extent of absorption by both clothing and natural coats depends in a complex way on color; the minimum heat load corresponds to neither black nor white but some intermediate shade. There have consequently been a number of studies of their spectral properties in the 0.1–1.0 μm band (e.g., refs. 75, 88). The effects of radiation absorption in this band on the energy balance of man and animals have been reviewed by Cena (89).

Short-wave radiation can be considered simply as a contribution to the net radiant load, although it has rarely been measured in this way. Clark and Cena (90) recently considered the influence of changes in net radiation produced by quartz halogen lamps on the heat fluxes through clothing. Cotton shirts of various colors were worn, and the heat flux at the skin measured with flux plates. For the same incident short-wave irradiance, the heat loads under black and white shirts differed by 190 W m^{-2}, when the net radiation to the black shirt was 780 W m^{-2}. The effects of color on the surface temperature of clothing are illustrated in Figure 6, which shows the temperature difference between a black number and the surrounding surface of a white cotton shirt due to differential absorption of short-wave radiation. Temperature differences of up to 7°C were observed between adjacent black and white areas in an irradiance of 800 W m^{-2}, equivalent to full sunshine.

WATER VAPOR TRANSFER

Latent heat transfer by evaporation from the respiratory tract and at the surface of the skin is an essential mechanism for regulation of the energy balance by homeotherms. However, there are large interspecies differences both in the importance of evaporative heat transfer and in the relative importance of the two routes (91). Man is the most dependent on cutaneous evaporation for heat dissipation, and man's hairlessness is probably an adaptation to

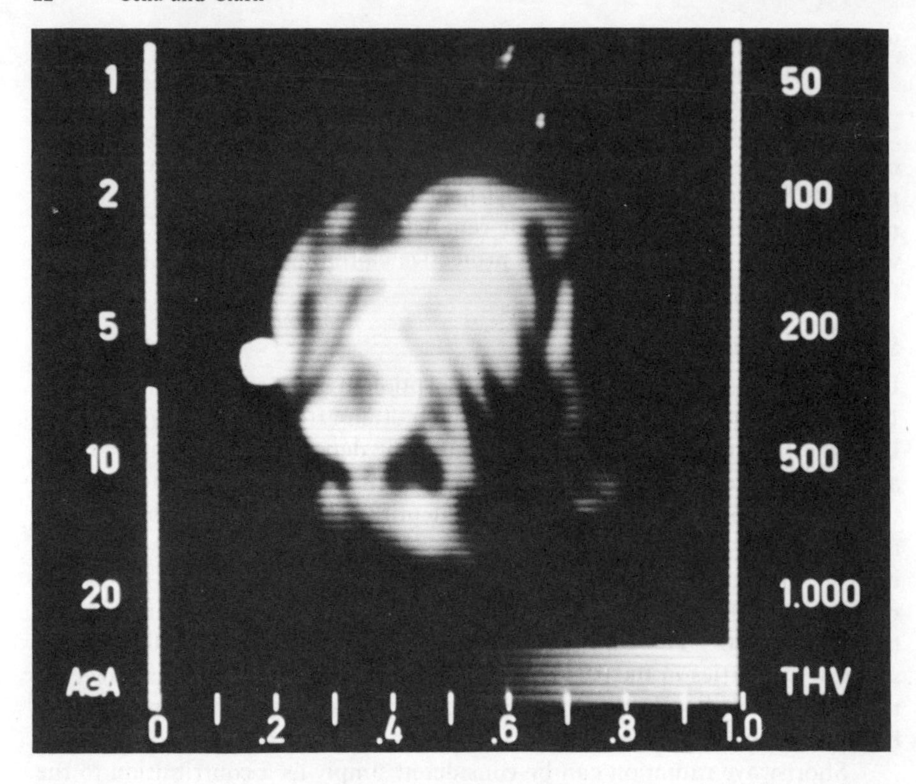

Figure 6. Thermogram of a subject wearing a white sports shirt with a black "3" on the back. The black "3" appears hotter (lighter on the thermogram) than the shirt surface because of radiant heating.

facilitate sweating. Evaporation must normally take place from the skin because the conduction of water along the fibers of coats is likely to be low in natural conditions. For example, Allen et al. (92) found that the hair close to the skin of cattle rarely contained free water. Cena and Monteith (93) compared the rates of diffusion of water vapor through samples of fiberglass wool with those through both natural and cured samples of sheep fleece. They found that for either material the resistance to diffusion was close to that for still air (i.e., $r_v = 4.2$ s cm^{-1} per centimeter of path), as long as the system was isothermal.

Fur-like fabrics and fibrous batting should have diffusion resistances for water vapor similar to those for the natural coats. For normal fabrics Behmann (94) showed that rates of evaporation from man decreased with clothing thickness. Nishi and Gagge (95) and Nishi (96) have used the naphthalene sublimation method to investigate diffusion through clothing layers and in still air obtained results consistent with transfer by molecular diffusion. According to Levell et al. (97) the normal laundering processes of washing and starching affect neither the insulation nor vapor permeability of conventional fabrics. In contrast it has long been known that tightly woven fabrics, especially those composed of artificial or coated fibers, may present a resistance to vapor transfer considerably greater than that of an equivalent layer of air, so that such fabric layers inhibit latent heat transfer (77, 98, 99). When totally impermeable clothing is worn the effects on thermoregulation may be severe. Joy and Goldman (100) found that, because they were unable to dissipate metabolic heat sufficiently quickly, over one-half of a group of soldiers suffered from heat exhaustion when asked to perform routine tactical tasks while wearing protective clothing.

The standard presentation for the heat balance of a sweating man is the psychrometric chart. However, here we are concerned with the second term of Equation 2 only, and the effects of clothing resistance and environmental humidity on evaporative heat loss can be displayed in a much clearer form (23). For example, Figure 7 shows the rates calculated for various total

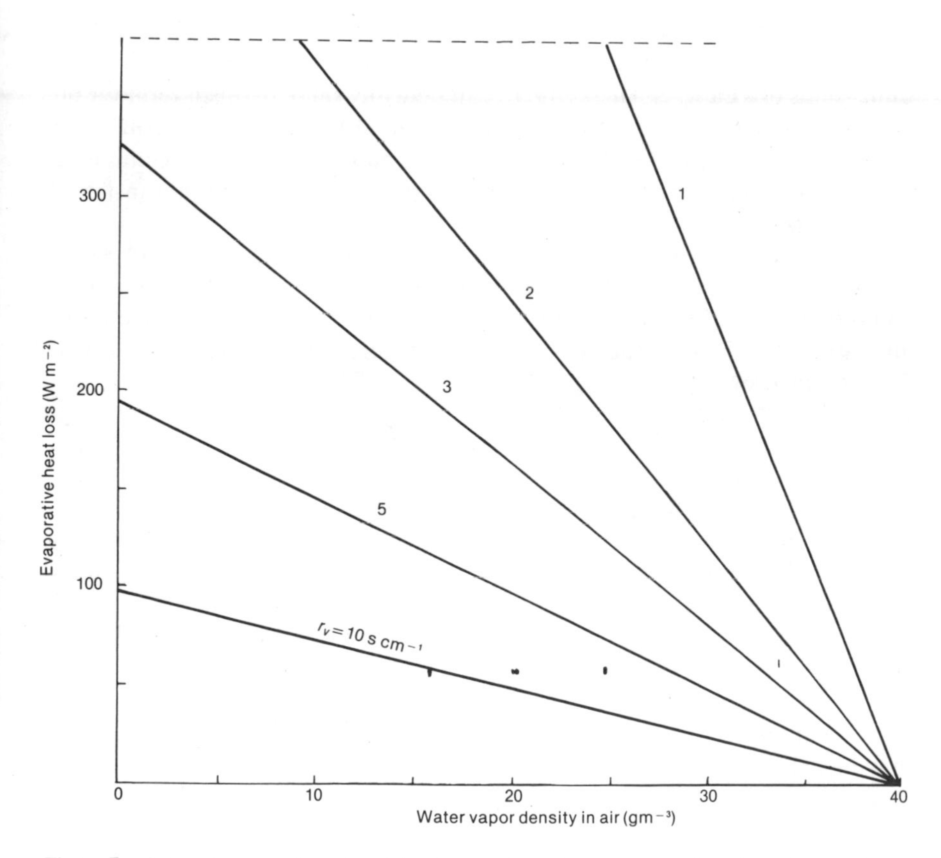

Figure 7. Latent heat loss from the skin of heat-stressed humans, as a function of atmospheric vapor density and boundary layer plus clothing vapor diffusion resistance. Conditions assumed are a skin temperature of 35°C and a maximum sweat rate of 1 kg h^{-1}, corresponding to the *dashed line* parallel to the abscissa at 380 W m^{-2}. The lines were calculated using relationships derived by Campbell (23).

resistances to vapor transfer. The lines represent maximum rates for each condition, which are physically limited. Any rate below the lines is possible and within this range will be physiologically controlled. The alternative is to adjust the vapor conductance of fabrics for their permeability. Breckenridge and Goldman (101) compared vapor-permeable and impermeable raincoats and found that the former, although tightly woven and treated for water repellence, did not significantly inhibit the passage of water vapor evaporated from the skin. Goldman's group (32, 102) use a dimensionless index (i_m) to specify the clothing permeability, defined as the ratio of the rate of evaporation from a wet-skinned flat plate covered by the cloth sample to that without the sample. They then write the evaporative heat conductance (h_e) for the sample (in W m^{-2} kPa^{-1}) as

$$h_e = 16.5\, i_m\, h_I \tag{20}$$

where h_I is the fabric conductance in W m^{-2} K^{-1} ($h_I = 1/I$) and the constant 16.5 is the reciprocal of the psychrometric constant, and therefore has the dimensions of kPa K^{-1}. Therefore, i_m can take any value between zero, for no evaporation, and unity, for a perfect wet bulb. However, at intermediate values of i_m this equation is likely to overestimate h_e, because h_I contains a contribution from radiative transfer, which has no analogy in the transfer of water vapor.

Cena and Monteith (93) have pointed out that, when heat and water vapor transfer occur together in an insulating coat, free convection will be controlled by the gradient of "virtual temperature" between the skin and the air, rather than by the gradient of air temperature. The virtual temperature, T_v, is defined by

$$T_v = T(1 + 0.38\, e/P) \tag{21}$$

where T is the air temperature in degrees Kelvin, P is the atmospheric pressure, and e the partial pressure of water vapor expressed in the same units. Replacing ΔT in Equation 19 by ($T_{vs} - T_{va}$), where T_{vs} and T_{va} are the virtual temperatures of the skin and the air, respectively, gives

$$\mathrm{Nu} = 1.66\,(T_{vs} - T_{va})^{0.7} \tag{22}$$

The dimensionless mass transfer coefficient, the Sherwood number, Sh, may be expressed in a similar form

$$\mathrm{Sh} = 1.66\,(T_{vs} - T_{va})^{0.7}\,(\mathrm{Le})^{0.7} \tag{23}$$

where Le is the Lewis number, the ratio of the molecular diffusivities for heat and water vapor in air, i.e., Le = 0.89. Water vapor transfer and latent heat transfer through fibrous insulation should, therefore, be enhanced by free convection processes in the same ratio as the convective component of sensible heat transfer.

Stewart and Goldman (103) have recently measured the heat flux through clothing on a manikin, with and without a wet skin. They found that the conductance for sensible heat was increased by over 30% in the presence of evaporation. This proportion is consistent with the theory of Cena and Monteith (93). It appears, therefore, that in still air conditions, when free convection is important in a coat, it is invalid to use the usual approximation that sensible and insensible heat transfer can be considered as separate processes. Instead, the Grashof number for free convection should be estimated by using the virtual temperature of the atmosphere in place of the dry bulb temperature.

Latent heat transfer through animal coats or clothing cannot be estimated without reliable measurements of vapor concentrations near the skin, which are almost entirely lacking. Cena and Monteith (93) used values measured by Eyal (104) to calculate rates of latent heat transfer through sheep fleeces in hot desert conditions. They showed that rates of up to 200 W m^{-2} were possible. Hofmeyer et al. (105) have measured 100 W m^{-2}. Latent heat loss will, however, more often be controlled by the rate of sweat evolution, which determines skin wetness (11), rather than by the diffusion resistance of the coat, except when the whole skin is wet.

Sweating may be made less effective as a mode of heat dissipation when the clothes or coat are permeated by liquid sweat, so that evaporation takes place away from the skin (8). In cold conditions dissipation of heat during exercise often results in condensation within the outer layers of the clothing (78); in this case much of the heat dissipated from the skin is released within the clothing on condensation, and the wearer is susceptible to exposure when exercise ceases. The presence of water in the clothing lowers its insulation and increases the temperature gradient by evaporative cooling, as occurs also when clothing is deliberately wetted (106).

THE EFFECTS OF WIND

There have been almost as many studies of the effect of wind on coat and clothing insulation as on insulation in still air. The effects differ between furs and open weave clothing on the one hand and clothing composed of impermeable layers on the other. Insulation is always decreased by the effects of wind, but the extent of the decrease depends on both the type of coat and the wind direction. Lentz and Hart (39) found that heat transfer through the pelts of newborn caribou increased by a factor of 2–3, depending on the wind direction, when a wind of 24 m s^{-1} was applied. In flow parallel to the hair axis the conductivity increased from 38 mW m^{-1} K^{-1} to 106 mW m^{-1} K^{-1}. Typical measurements on samples of fur and an artificial fur, made by Moote (52), are presented in Figure 8. For adult caribou fur she found conductivities of 30–49 mW m^{-1} K^{-1} in still air, almost doubling at 14 m s^{-1};

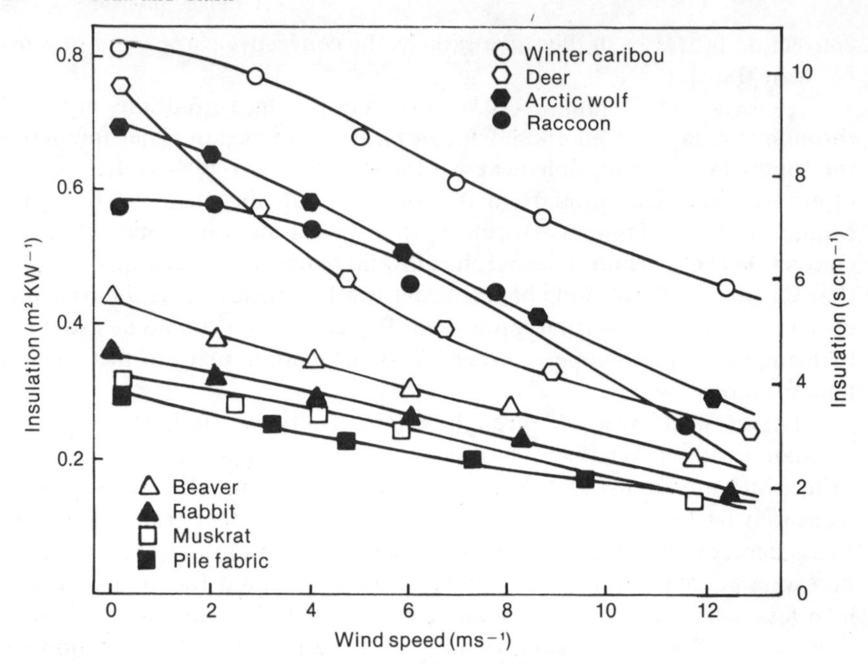

Figure 8. Variation of the insulation of representative fur samples and a pile fabric with wind speed. Measurements were made in a wind tunnel by Moote (52).

similar changes occurred in other samples. Tregear (64) found that the conductivity was increased proportionately more for sparse coats, like that of the pig, than for those with dense fur. According to Tregear, published measurements are consistent with a dependence of insulation on wind speed, u, in the form

$$I = \mathrm{A} - \mathrm{B}\, u^{1/2} \tag{24}$$

where A and B are constants for a particular coat, u is in meters per second, and I is in $m^2\,K\,W^{-1}$. Tregear's results for rabbit fur give A = 0.22 and B = 0.035, whereas those of Blaxter (13) for sheep give A = 0.11 and B = 0.023.

Most authors have started from the reasonable premise that the reduction of insulation depends on the square root of wind speed, since this is the form in which wind speed is expressed in the Nusselt number for forced convection. However, within the coat other processes are also operating. For example, pressure gradients, which depend on the square of the wind speed, may drive flow within the coat. McArthur (107) found that his recent measurements of the insulation of sheep fleece gave a better fit to a linear decrease of conductance ($h = I^{-1}$) with wind speed than to Equation 24. The insulation change with wind speed in this form is described by the equation

$$I = h^{-1} = \frac{1}{A + Bu} \tag{25}$$

Measurements of the insulation changes of sparrow coats with wind speed by Robinson et al. (108) are also now considered by Campbell (personal communication) to be a good fit to a linear relation. G. S. Campbell, A. J. McArthur, and J. L. Monteith (personal communication) have also examined many of the published measurements for animal coats and find that most give a better fit to a simple linear increase of coat conductance with wind speed than the more complex relationships originally assumed. The same is also true of the measurements on clothing by Goldman, referred to below and presented in Figure 9.

Some coats are little affected by wind. For example, Bennett (62) found little change in the insulation of cattle coats at wind speeds up to 5 m s^{-1}, and Dawson and Brown (66) also found the coats of some kangaroos resistant to disturbance by wind. Certain types of fabrics are also highly resistant to wind penetration, and men likely to be exposed to wind chill conditions elect to wear such fabrics as outer layers to their clothing assembly. The principal problem for man may then be restated as one already considered, the dissipation of water vapor through the fabrics during work, and control of the openings of the clothing to regulate ventilation and heat dissipation.

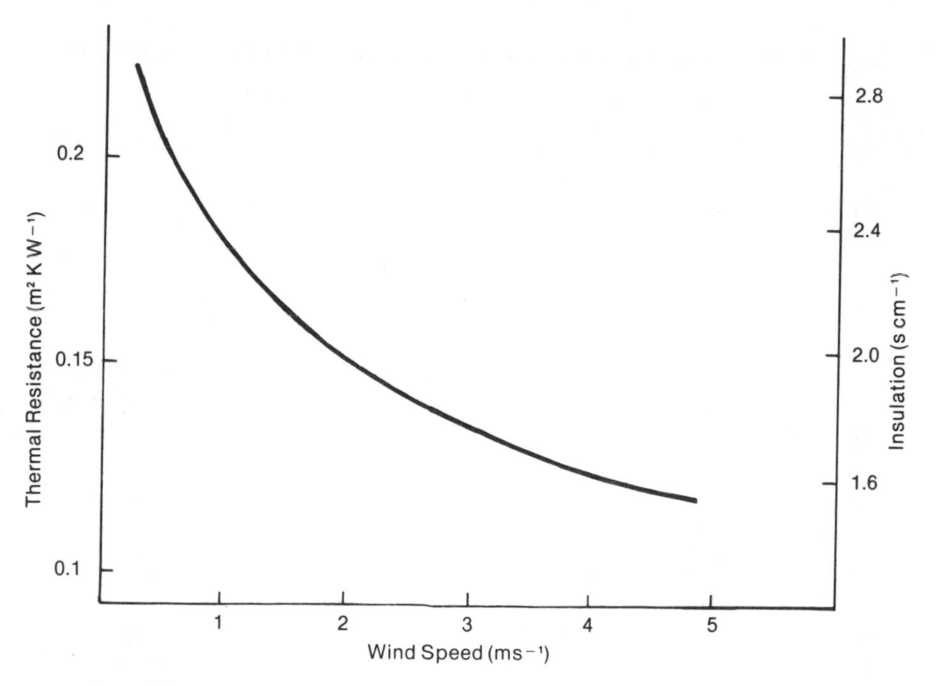

Figure 9. Variation of thermal insulation of United States Army tropical fatigues with wind speed. Adapted from Goldman (78).

We might expect the insulation of other fabric types to vary in an analogous way to that of natural coats, but according to Goldman (78) there is surprisingly little data available on the effects of wind on everyday clothing for temperate climates. We may, however, infer the influence of wind speed on the insulation of such clothing from Goldman's own measurements on a light tropical uniform, redrawn here as Figure 9. The figure shows that insulation is reduced considerably by wind, being halved at only 2 m s^{-1}, while the moisture permeability was proportionately increased. Measurements of naphthalene vapor sublimation through flat samples of fabric by Nishi and Gagge (95) show similar increases in rates of transfer with wind speed.

SIZE AND INSULATION

Geometric control of insulation has a major influence on homeotherms; it is a principal determinant both of the minimum size for viable homeothermic animals and of the metabolic energy expenditure of animals in the cold. Only simple mathematics is required to show that the insulation available to an animal is limited by its body size (31). The addition of a thickness x of insulation of conductivity k to a plane surface always increases the insulation by an amount

$$I = x/k \tag{26}$$

However, when a thickness x is added to a cylinder of radius r the increase is

$$I = \frac{r}{k} \ln\left(1 + \frac{x}{r}\right) \tag{27}$$

and for a sphere

$$I = \frac{r}{k}\left(\frac{x}{x + r}\right) \tag{28}$$

Therefore, when $x = r$ a thickness increment is only half as effective on a sphere, and 70% as effective on a cylinder, as the same increment of insulation on a plane surface. Indeed, it is obvious that for the sphere there is a limiting value of $I = r/k$ as an $x \to \infty$. The effectiveness of insulation also falls rapidly on a cylinder at increasing x/r, but when $x \gg r$ we may consider any animal as approximating to a sphere. The consequences of this are evident from Figure 10. The smaller the object the more difficult it is to provide insulation; indeed, small animals may only survive in cold climates by behavioral responses that avoid severe stress and by having very high rates of metabolism. A particular example of medical interest is the human baby (109). Neonate infants, especially those premature babies at weights close to 1 kg, have a very narrow range of thermoneutrality (110) although their

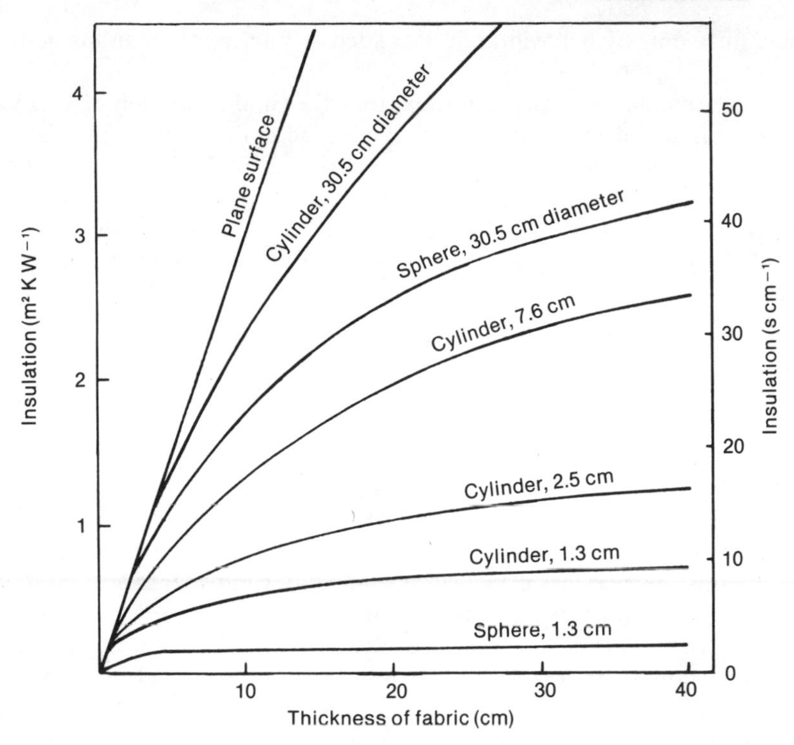

Figure 10. Substrate geometry and the effectiveness of insulation: dependence of insulation on depth for a flat plate and different sized cylinders and spheres. The insulation is assumed to have the conductivity of still air. The curves are those predicted by the theory given by Van Dilla, Day, and Siple (112), but are calculated using current units.

metabolic rates are not dissimilar to those of animals of the same size which can survive in arctic climates (111). The difference is partly due to the difficulty in providing good insulation in the layers close to the surface of the body, where it is most effective, and partly because the geometry of the human, which as an adult is a comparatively large animal who does not need to be adapted for heat conservation, limits postural control of heat loss in the neonate. Man's respiratory passages are also ill-adapted for heat conservation.

The insulation of parts of animals may be considered in the same way. For example, van Dilla et al. (112) showed that it is impossible to provide adequate insulation for a man's fingers in the cold except when he is working at a high rate. The fingertip has a radius of approximately 5 mm. Even an infinite thickness of insulation with the properties of air provides only 0.124 m^2 $K\ W^{-1}$ (1.6 $s\ cm^{-1}$) for a sphere of this radius, so that no practical glove can prevent the fingers from falling to temperatures that impair their function and cause pain in prolonged exposures to temperatures much below 0 °C. In such conditions a resting man can maintain his hands at a tolerable temperature only by behavioral devices such as burying them in his body insulation.

There remains a further paradox, that the total insulation of small objects may initially decrease as insulation is added to the surface. The external insulation of the boundary layer is proportional to $Re^{-1/2}$, as the Nusselt number for cylinders in the range of interest is approximately $Nu = 0.68\ Re^{0.5}$ (18). Monteith showed that differentiation of the formulas presented by van Dilla et al. (112) leads to a critical value of the Nusselt number

$$Nu = 2n\,k'/k \tag{29}$$

where k' is the conductivity of the insulation, k is the conductivity of air, and n is 0.5 for forced convection or 0.75 for free convection. Since the characteristic dimension of the body appears in the Nusselt number, the critical value changes with body size. When Nu is greater than the above value, heat loss decreases as the insulation thickness is increased; for values less than this the heat loss is increased. For small wind speeds and radii, this results in an initial decrease of total insulation as the reward for adding insulation to the surface. The effect is likely to be significant for objects of less than 5-mm radius, say the fingers of children and small adults (8). We may also conclude that this is one of the reasons why the thin legs of small birds are devoid of insulation, and perhaps why only the larger insects have developed hair insulation to aid the regulation of thorax temperature often required for flight (113, 114).

CLOTHING INSULATION IN HYPERBARIC AND HYPOBARIC ENVIRONMENTS AND IN WATER

The operational requirements for clothing men in the special environments of aircraft, deep mines, high pressure diving equipment, and pressurized tunneling machines prompt consideration of these conditions. The effects of exotic environments may be severe; Timbal et al. (115) measured large increases in heat loss and raised metabolism in men exposed to hyperbaric helium environments.

A large part of the increased heat load is due to the much greater respiration heat loss at high pressures. At one atmosphere pressure, respiratory convection accounts for only about 3% of metabolic heat production at normal temperature differentials (116). However, it follows that at 30 atm (=300 m of water) respiratory convection alone demands the whole of the normal metabolic heat production. Moreover, since respiratory displacement is almost proportional to metabolic rate, the situation cannot be alleviated by exercise. At high pressures the temperature of the breathing gases must, therefore, be controlled. The coat insulation is also altered, so that thermoregulation is made more difficult. Changes may take place in both the intrinsic insulation of the clothing layer and in the external insulation, and may be equally important. Convection in the air layer external to the clothing is governed by the Nusselt number for the flow regime, which for $Gr < 10^5$ laminar free convection is related to the Grashof number (Gr) and Prandtl number (Pr) by

$$Nu \propto (Gr\ Pr)^{0.25} \tag{30}$$

Equation 30 is an expanded form of Equation 11. Hayes and Toy (117) have recently measured free convection from cylinders in hyperbaric atmospheres of air, argon, helium, and SF_6 at pressures of up to 30 atmospheres, equivalent to a 300-m depth of water, to simulate diving conditions. Their measurements for all the gases, shown in Figure 11, are a good fit to Equation 30, which in dimensional form may be expressed as

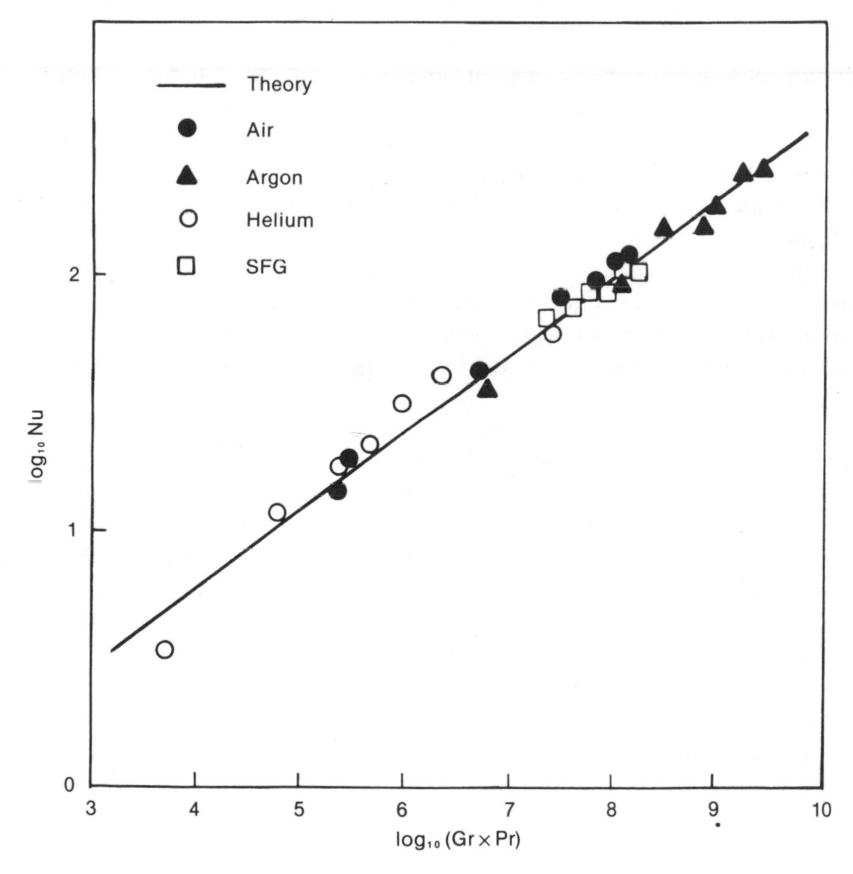

Figure 11. Logarithmic plot of Nusselt number against Grashof number for free convection heat transfer in different gases, as indicated, and pressures of 1–30 atm. Reprinted by permission of the Council of the Institution of Mechanical Engineers from *Engineering in Medicine* (Ref. 117).

$$I = (\mu/k^3c_p)^{0.25}p^{\ \ 0.5} \tag{31}$$

where k is the thermal conductivity of the gas, c_p its specific heat at constant pressure, μ its dynamic viscosity, and p the applied pressure. The external insulation I, therefore, strongly depends on the thermal conductivity of the gas, which is a function primarily of its molecular weight, and also on the pressure. The insulation of a helium atmosphere is little over half that of air at the same pressure, and at 30 atm of helium the heat transfer rate per unit of temperature difference will be over 10 times that in air at sea level. Differences of similar magnitude may be expected in forced convection. In comparison, the variation in external insulation accompanying the changes of atmospheric pressure experienced by climbers, aircraft pilots, and men working in deep mines are trivial. For example, the forced convection heat transfer coefficient changes by about 12% only between 100 and 130 kPa (118). The intrinsic insulation of clothing responds rather differently to pressure and the nature of the gas. Since the thermal conductivity of gases is almost independent of pressure over a realistic range, changes of clothing insulation with pressure for a particular gas are small, provided the insulation is not compressed. However, because the thermal conductivity of a gas is a function of its molecular weight, the insulation of clothing is influenced by the composition of the gases present. In particular, replacing the nitrogen in air by helium results in a decrease of insulation. If the partial pressure of oxygen in a helium atmosphere is maintained at one-fifth atm then at 30 atm total pressure the insulation of clothing is around 60% of its value in air. The opposite effect is also valid; Hammel (57) replaced air in the hair coats of mammals by Freon, which has a density 4.2 times that of air, and found the increase in insulation was consistent with the difference between the molecular weights.

The consequences of such changes in insulation are likely to be most serious in helium atmospheres at high pressures. A diver clothed comfortably in air would find his total insulation approximately halved on exposure to such an atmosphere. Most critically, transfer rates from the 15% of the body that is usually exposed, the face and hands, could be an order of magnitude greater than in air. Compensation by increasing the thickness of insulation obviously impedes movement in the restricted space of diving apparatus, and inhibits evaporative heat dissipation during work. The diffusion of water vapor is also greatly reduced at high pressures (119), so that evaporative cooling becomes ineffective as a means of thermoregulation. This is because the Lewis number is inversely proportional to pressure. The best solutions to the problem, therefore, are almost certainly to seek suitable alternative gases of higher molecular weight than helium, or to increase the ambient temperature in the environment in order to avoid cold stress. During immersion in water, rates of heat loss from exposed tissue are greatly in excess of those in air, because of the higher thermal capacity of water. Convective heat transfer coefficients in water are of the order of 100 W m^{-2} K^{-1} at rest, and 500 W

$m^{-2} K^{-1}$ when swimming. Water, therefore, offers virtually no external insulation, and during immersion man must depend on his tissues and clothing for thermal protection (116). Thus, suits for divers, or for use by aircrew during involuntary immersion, must minimize this exposed area. For the covered areas the wet suit is a much better insulator than alternative dry suits. Water is incompressible, but air-filled spaces quickly collapse unless completely sealed (120). Even the insulation of the wet suit is approximately halved on immersion. The effect of enhanced heat transfer in both aquatic and hyperbaric environments is consequently to narrow the range of conditions for comfort or survival. The best way of preserving comfort with mobility is almost certainly that of using active protection, by heating (or cooling) the layer next to the skin to maintain the skin temperature within the comfort range (121). There are analogues of both the wet and dry suits in the animal kingdom. According to Frisch et al. (122) water penetrates the fur of seals and polar bears, but a 2–10-mm-thick water layer adjacent to the skin acts as the insulator, just as in the neoprane wet suit. In contrast, the feather pelt of penguins retains its air layer during immersion, although its insulation is reduced because of compression during dives (42). The penguin seems to do better than clothing technologists.

HUMAN COMFORT AND INSULATION

The environments that man finds comfortable depend on both his level of activity and insulation (9, 12, 123, 124). Not only must his overall insulation be adequate, but that of his peripheries must be sufficient to maintain them at a comfortable temperature (125). McIntyre and Griffiths (126) found that people with cold feet remain uncomfortable even when given an extra sweater. Obviously, increased metabolism and insulation result in a preference for cooler temperatures, decreased metabolism, and insulation for higher temperatures. Physiological reactions to stress also depend on both work rate and insulation (127, 128). Measurements of human comfort are principally of value to the air-conditioning engineer, architect, and designer of transport vehicles, since the normal adult may change his dress, or, over a longer term, his habits, to alleviate discomfort, although Humphreys (129) finds people surprisingly conservative in their seasonal clothing adjustments to microclimate; longer term habituation is evidently effective (130). However, Gagge et al. (125) point out that educating the public to wear clothing more suited to the season could make a major contribution to economies in the use of energy for the heating and air-conditioning of buildings by allowing the lowering of temperature differentials from the outdoor world. The link between insulation and comfort preferences has been quantified by Fanger (9) and by Rohles et al. (131). According to these authors the preferred temperature for an average subject decreases linearly as his insulation is increased, from approximately 30 °C when naked to 23 °C when wearing clothes with an insulation of 0.155 $m^2 K W^{-1}$ (1 *clo*). This temperature is higher by

2°C than that used in the original definition of the *clo* unit (28). Social criteria determine the acceptable minimum clothing, and the impediment of movement determines the indoor maximum (132). In summer working naked would allow comfort in sedentary occupations up to 30°C, but the social minimum is a very light dress or shorts and shirt, which corresponds to an insulation of about 0.6 s cm^{-1} and a comfort temperature of 27°C. In winter the indoor temperature could be lowered to about 15°C by wearing much heavier clothing than is at present usual in Northern America or Europe, with $r = 3$ s cm^{-1}. The principal limit then becomes local cooling of the hands, which must be left uninsulated to allow tasks involving manipulation. Even if we exclude space, men must work in many severe environments where the requirements of comfort or even survival cannot be met by conventional clothing. In some circumstances high rates of work require rapid heat dissipation through clothing that must also protect against external insults; in others the provision of adequate insulation against cold is precluded by the requirement of movement or the limitations of space. The cockpit of high-flying aircraft, for example, may present both problems within half an hour. In such cases man's physiological needs may only be met by bypassing the thermal barrier of his clothing and providing a heat source or sink close to his skin. High technology solutions include electrically heated clothing (133) and liquid-conditioned suits (134–136). Both require a reliable power source. Attempts to design such clothing to be independent of power supplies have rarely proved successful, although Strydom et al. (137) constructed a simple ice jacket for use in the deep gold mines of South Africa, which allows several hours of useful work in air temperatures close to body temperature and almost saturated atmospheres. In this case the metabolic heat production during work is removed as latent heat absorbed by the ice jacket.

There are also some areas of clinical interest that are concerned with comfort. The elderly and infirm, for example, show comfort preferences significantly different from those of normal adults (138). The sensitivity of the old to both cold and heat is reduced, but this alone is insufficient reason for the increased risk of hypothermia in the aged, since autonomous metabolic compensation for cold stress could operate without sensation. The old and infirm are at risk from hypothermia for reasons that are part physical and part physiological (109). First, their rates of metabolism and the ability to raise the metabolic rate in response to cold stress are depressed relative to those of a healthy adult in midlife. In illness the depression may be 50% or more. Second, their clothing habits are determined by those 50 years of adult life. Competence in the sweating response to heat stress is also reduced. A sitting healthy adult is comfortable at 21°C in 2 s cm^{-1} (1 *clo*, the basis of the definition of Gagge et al. (28)) only because his metabolic rate is significantly above the basal level. This insulation would not be adequate below 25°C for a sleeping adult at basal metabolism (i.e., with an empty stomach). For the sick elderly person with a low food intake the lower critical temperature clothed is little different from that given by Hey (110) for naked full-term babies, 33 °C. Even the suit of an arctic explorer or a well-covered bed keeps such an elderly person thermoneutral down to only about 17 °C. It is obviously impractical to expect a frail old person either to remain permanently in bed or to adopt the unfamiliar encumbrance of bulky clothing. A poorly heated house *is* the arctic for the elderly infirm human, and only by raising the ambient temperature can the thermal demands on their metabolism be brought within their physiological range.

CONCLUSIONS

It should be evident from the range of literature covered in this review that the topic has wider implications than are evident in the title alone. In purely physical terms, it is relatively easy to write equations that contain all the components necessary to describe the heat balance of a man or other homeothermic animals. What is difficult is to obtain the correct values of the physical and physiological parameters to insert in the equations. The test of reality is whether the results accord with the physiological responses of the organism, which is the integrator of its thermal environment.

Physiologists and environmental scientists with a wide variety of specialized interests have studied the properties of animal insulators. However, they often appear to be unaware of related work outside their own field of endeavor. Human and animal physiologists can learn much from studies of the coat properties of the other subjects. We hope we have brought together in this chapter enough variety of material to provide some useful cross-fertilization of ideas.

To summarize, we now have quite a good description of the structures of natural mammalian coats and how they control heat exchange. However, some points require further study. In particular, the simultaneous transfer of heat and mass by free convection has a good theoretical basis, but there is a need for measurements made in a form which can test its validity. Our knowledge of heat loss through mammalian coats is probably now adequate for cold, dry conditions, but further work is needed on the effects of wetting on coats in the cold and on evaporation through them in the heat. Avian coats are almost unexplored territory. Despite the long history of measurements of the thermal properties of the clothing of man, our present knowledge in this area is almost purely empirical. There is an obvious need for a coherent theory to describe heat loss both through clothing and via the short circuit of ventilation. Perhaps one has not been constructed because measurement of the ventilation rate of clothing has seemed an intractable problem. The development of a technique for measurement of the natural ventilation of clothing (83) appears to be a major advance in this field, and enables the possibility of the experimental validation of a theory on the lines we propose.

ACKNOWLEDGMENT

The authors thank Miss Edna Lord for preparation of the manuscript.

REFERENCES

1. Cena, K., and Clark, J. A. (1978). Thermal insulation of animal coats and human clothing. Phys. Med. Biol. 23:565.
2. Mitchell, D. (1977). Physical basis of thermoregulation. *In* D. Robertshaw (ed.), MTP International Review of Physiology, Vol. 15, Environmental Physiology II, pp. 1-27. University Park Press, Baltimore.
3. Mitchell, D., and van Rensburg, A. J. (1973). Assessment of clothing insulation: the problem and its solution using direct calorimetry on exercising men. Arch. Sci. Physiol. 27:A149.
4. Heinrich, B. (1976). Heat exchange in relation to blood flow between thorax and abdomen in bumblebees. J. Exp. Biol. 64:561.
5. Mitchell, D. (1974). Physical basis of thermoregulation. *In* D. Robertshaw (ed.), MTP International Review of Physiology, Vol. 7, Environmental Physiology I, pp. 1-32. Butterworths, London.
6. Nielsen, M., and Pedersen, L. (1952). Studies on the heat loss by radiation and convection from the human body. Acta Physiol. Scand. 76:137.
7. Newburgh, L. H. (1968). Physiology of Heat Regulation and the Science of Clothing. Fascimile of the 1949 edition. Hafner, New York.
8. Burton, A. C., and Edholm, O. G. (1969). Man in a Cold Environment. Fascimile of the 1955 edition. Hafner, New York.
9. Fanger, P. O. (1970). Thermal Comfort. Danish Technical Press, Copenhagen.
10. McIntyre, D. A. (1973). A guide to thermal comfort. Appl. Ergonomics 4:66.
11. Gagge, A. P., and Nishi, Y. (1977). Heat exchange between human skin surface and thermal environment. *In* D. H. K. Lee (ed.), Handbook of Physiology, Section 9, Reaction to Environmental Agents, pp. 69-92. American Physiological Society, Bethesda.
12. Cena, K., and Clark, J. A. (eds.). (1979). Bioengineering, Thermal Physiology and Comfort. Wroclaw Technical University Press, Wroclaw.
13. Blaxter, K. L. (1969). The Energy Metabolism of Ruminants. Hutchinson, London.
14. Monteith, J. L., and Mount, L. E. (eds.). (1974). Heat Loss from Animals and Man. Butterworths, London.
15. Scholander, P. F., Walters, V., Hock, R., and Irving, L. (1950). Body insulation of some arctic and tropical mammals and birds. Biol. Bull. 99:225.
16. Gates, D. M. (1973). Energy Exchange in the Biosphere. Harper and Row, New York.
17. Moen, A. N. (1973). Wildlife Ecology. W. H. Freeman and Company, San Francisco.
18. Monteith, J. L. (1973). Principles of Environmental Physics. Arnold, London.
19. Gates, D. M., and Schmerl, R. B. (1975). Perspectives of Biophysical Ecology. Springer-Verlag, New York.
20. Sibbons, J. L. H. (1966). Assessment of thermal stress from energy balance considerations. J. Appl. Physiol. 21:1207.
21. Monteith, J. L., Clark, J. A., McArthur, A. J., and Wheldon, A. E. (1979). The physics of the microclimate. *In* K. Cena and J. A. Clark (eds.), Bioengineering, Thermal Physiology and Comfort. Wroclaw Technical University Press, Wroclaw.

22. Clark, J. A., Cena, K., and Monteith, J. L. (1973). Measurements of the local heat balance of animal coats and human clothing. J. Appl. Physiol. 35:751.
23. Campbell, G. S. (1977). An Introduction to Environmental Biophysics. Springer-Verlag, New York.
24. Ede, A. J. (1967). An Introduction to Heat Transfer. Pergamon Press, Inc., New York.
25. Cena, K., and Clark, J. A. (1978). Thermal resistance units. J. Thermal Biol. 3:173.
26. Cena, K., and Clark, J. A. (1979). Physics, physiology and psychology. *In* K. Cena and J. A. Clark (eds.), Bioengineering, Thermal Physiology and Comfort. Wroclaw Technical University Press, Wroclaw.
27. Cena, K., and Clark, J. A. (1974). Heat balance and thermal resistances of sheep's fleece. Phys. Med. Biol. 19:51.
28. Gagge, A. P., Burton, A. C., and Bazett, H. C. (1941). A practical system of units for the description of the heat exchange of man with his environment. Science 94:428.
29. Mount, L. E. (1974). The concept of thermal neutrality. *In* J. L. Monteith and L. E. Mount (eds.), Heat Loss from Animals and Man. Butterworths, London.
30. Strunk, T. H. (1971). Heat loss from a Newtonian animal. J. Theor. Biol. 33:35.
31. McNeil-Alexander, R. (1971). Size and Shape. Arnold, London.
32. Goldman, R. F. (1979). Evaluating the effects of clothing on the wearer. *In* K. Cena and J. A. Clark (eds.), Bioengineering, Thermal Physiology and Comfort. Wroclaw Technical University Press, Wroclaw.
33. Teitlebaum, A., and Goldman, R. F. (1972). Increased energy cost with multiple clothing layers. J. Appl. Physiol. 32:743.
34. Gagge, A. P. (1979). Rational indices of thermal comfort. *In* K. Cena and J. A. Clark (eds.), Bioengineering, Thermal Physiology and Comfort. Wroclaw Technical University Press, Wroclaw.
35. Nishi, Y. (1979). Measurement of the thermal balance of man. *In* K. Cena and J. A. Clark (eds.), Bioengineering, Thermal Physiology and Comfort. Wroclaw Technical University Press, Wroclaw.
36. Johnson, K. G. (1971). The discharge of sweat in Welsh mountain sheep. J. Physiol. (Lond.) 215:743.
37. Tregear, R. T. (1965). Hair density, windspeed and heat loss in mammals. J. Appl. Physiol. 20:796.
38. Ryder, M. (1973). Hair. Arnold, London.
39. Lentz, C. P., and Hart, J. S. (1960). The effect of wind and moisture on heat loss through the fur of newborn caribou. Can. J. Zool. 38:679.
40. Bennett, J. W., and Hutchinson, J. C. D. (1964). Thermal insulation of short lengths of Merino fleece. Aust. J. Agric. Res. 15:427.
41. Cena, K., and Clark, J. A. (1973). Thermal radiation from animal coats: coat structure and measurements of radiative temperature. Phys. Med. Biol. 18:432.
42. Kooyman, G. L., Gentry, R. L., Bergman, W. P., and Hammel, H. T. (1976). Heat loss in penguins during immersion and compression. Comp. Biochem. Physiol. 54A:75.
43. Mahoney, S. A., and King, J. R. (1977). The use of the equivalent black-body temperature in the thermal energetics of small birds. J. Thermal Biol. 2:115.
44. Rogers, A. F., and Sutherland, R. J. (1974). Clo values of polar clothing and their relation to total number of layers counts. J. Physiol. (Lond.) 238:22P.
45. Belding, H. S., Russell, H. D., Darling, R. C., and Folk, G. E. (1947). Thermal responses and efficiency of sweating when men are dressed in Arctic clothing and exposed to extreme cold. Am. J. Physiol. 149:204.

46. Belding, H. S., Russell, H. D., Darling, R. C., and Folk, G. E. (1947). Analysis of factors concerned in maintaining energy balance for dressed men in extreme cold; effects of activity on the protective value and comfort of an Arctic clothing. Am. J. Physiol. 149:223.
47. Skuldt, D. J., Beckman, W. A., Mitchell, J. W., and Porter, W. P. (1975). Conduction and radiation in artificial fur. *In* D. M. Gates and R. B. Schmerl (eds.), Perspectives of Biophysical Ecology. Springer-Verlag, New York.
48. O'Callaghan, P. W., and Probert, S. D. (1976). Thermal insulation provided by dry, single-layer clothing materials. Appl. Energy 2:269.
49. O'Callaghan, P. W., and Probert, S. D. (1977). Thermal resistance behaviour of single and multiple layers of clothing fabrics under mechanical load. Appl. Energy 3:3.
50. Azer, N. Z. (1976). The prediction of thermal insulation values of garments from the physical data of their fabrics. ASHRAE Trans. 82:87.
51. Auliciems, A., and Hare, F. K. (1973). Weather forecasting for personal comfort. Weather 28:118.
52. Moote, I. (1955). The thermal insulation of caribou pelts. Textile Res. J. 25:832.
53. Seppanen, O., McNall, P. E., Munson, D. M., and Sprague, C. H. (1972). Thermal insulating values for typical indoor clothing ensembles. ASHRAE Trans. 78:120.
54. Sprague, C. H., and Munson, D. M. (1974). A composite ensemble method for estimating thermal insulating values of clothing. ASHRAE Trans. 80:120.
55. Nishi, Y., Gonzalez, R. R., and Gagge, A. P. (1975). Direct measurement of clothing heat transfer properties during sensible and insensible heat exchange with thermal environment. ASHRAE Trans. 81:183.
56. Goldman, R. F. (1965). The arctic soldier: possible research solutions for his protection. *In* G. Dahlgre (ed.), Science in Alaska. American Association for the Advancement of Science, College, Alaska.
57. Hammel, H. T. (1955). Thermal properties of fur. Am. J. Physiol. 182:369.
58. Alexander, G. (1961). Temperature regulation in the newborn lamb. III. Effect of environmental temperature on metabolic rate, body temperatures, and respiratory quotients. Aust. J. Agric. Res. 12:1152.
59. Blaxter, K. L., Graham, N. McC., and Wainman, F. W. (1959). Environmental temperature, energy metabolism and heat regulation in sheep. III. The metabolism and thermal exchanges of sheep with fleeces. J. Agric. Sci. 52:41.
60. Doney, J. M. (1963). The effects of exposure in Blackface sheep with particular reference to the role of the fleece. J. Agric. Sci. 60:267.
61. Joyce, J. P., Blaxter, K. L., and Park, C. (1966). The effect of natural outdoor environments on the energy requirements of sheep. Res. Vet. Sci. 7:342.
62. Bennett, J. W. (1964). Thermal insulation of cattle coats. Proc. Aust. Soc. Anim. Prod. 5:160.
63. Gonzalez-Jimenez, E., and Blaxter, K. L. (1962). The metabolism and thermal regulation of lambs in the first month of life. Br. J. Nutr. 16:199.
64. Tregear, R. T. (1965). Hair density, windspeed and heat loss in mammals. J. Appl. Physiol. 20:796.
65. Cena, K., and Monteith, J. L. (1975). Transfer processes in animal coats. II. Conduction and convection. Proc. R. Soc. Lond. (Biol.) 188:395.
66. Dawson, T. J., and Brown, G. D. (1970). A comparison of the insulative and reflective properties of the fur of desert kangaroos. Comp. Biochem. Physiol. 37:23.
67. Øritsland, N. A., and Ronald, K. (1978). Aspects of temperature regulation in harp seal pups evaluated by *in vivo* experiments and computer simulation. Acta Physiol. Scand. 103:263.

68. Evans, K. E., and Moen, A. N. (1975). Thermal exchange between sharp-tailed grouse (*Pedioecetes phasianellus*) and their winter environment. Condor 77:160.
69. Poczopko, P. (1972). Thermal insulation in goslings. Acta Physiol. Pol. 23:843.
70. Bogaty, H., Hollis, N. R. S., and Harris, M. (1957). Some thermal properties of fabrics. Textile Res. J. 27:445.
71. Cena, K., and Monteith, J. L. (1975). Transfer processes in animal coats. I. Radiative transfer. Proc. R. Soc. Lond. (Biol.) 188:377.
72. Hager, N. E., and Steere, R. C. (1967). Radiant heat transfer in fibrous thermal insulation. J. Appl. Phys. 38:4663.
73. Davis, L. B., and Birkebak, R. C. (1974). On the transfer of energy in layers of fur. Biophys. J. 14:249.
74. Özil, E., and Birkebak, R. (1977). Effects of environmental thermal radiation on the insulative property of a fibrous material. *In* A. Cezairliyan (ed.), Proceedings of the Seventh Symposium on Thermophysical Properties. American Society of Mechanical Engineers, New York.
75. Dunkel, R. V., Ehrenburg, F., and Gier, J. T. (1960). Spectral characteristics of fabrics from 1 to 23 microns. J. Heat Transfer 82:64.
76. Crockford, G. W., and Goudge, J. (1970). Dynamic insulation: an equation relating airflow, assembly thickness and conductance. Proc. R. Soc Med. 63:1012.
77. Fourt, L., and Harris, M. (1968). Physical properties of clothing fabrics. *In* L. H. Newburgh (ed.), Physiology of Heat Regulation and the Science of Clothing. Fascimile of the 1949 edition. Hafner, New York.
78. Goldman, R. F. (1973). Clothing, the interface between man and his environment: resistance against meteorological stimulus. *In* S. W. Tromp (ed.), Progress in Biometeorology, Vol. 1, Part 1. Swets and Zeitlinger, Lisse.
79. Davis, L. B., and Birkebak, R. C. (1975). Convective energy transfer in fur. *In* D. M. Gates and R. B. Schmerl (eds.), Perspectives of Biophysical Ecology. Springer-Verlag, New York.
80. Nishi, Y. (1973). Vapour permeation efficiency of clothing by naphthalene sublimation. Arch. Sci. Physiol. 27:A163.
81. Kerslake, D. McK. (1972). The Stress of Hot Environments. University Press, Cambridge.
82. Fonseca, G. F. (1978). A biophysical model for evaluating auxiliary heating and cooling systems. Proceedings of the VIIIth Intersociety Conference on Environmental Systems, San Diego.
83. Birnbaum, R. R., and Crockford, G. W. Measurement of the clothing ventilation index. Appl. Ergonomics. In press.
84. Crockford, G. W. (1970). Protective clothing for fishermen: assessment of design. Proc. R. Soc. Med. 63:1007.
85. Crockford, G. W. (1970). Protective clothing for fishing crews. Text. Inst. Industry 8:121–124.
86. Colin, J., and Houdas, Y. (1967). La protection individuelle contre la chaleur. Le Travail Humain 30:233.
87. Kovarik, M. (1973). Radiation penetrance of protective covers. J. Appl. Physiol. 35:562.
88. Stewart, R. E. (1963). Absorption of solar radiation by the hair of cattle. Agric. Eng. 34:235.
89. Cena, K. (1974). Radiative heat loss from animals and man. *In* J. L. Monteith and L. E. Mount (eds.), Heat Loss from Animals and Man. Butterworths, London.
90. Clark, J. A., and Cena, K. (1978). Net radiation and heat transfer through clothing: the effects of insulation and colour. Ergonomics 21:691.

91. McLean, J. A. (1974). Loss of heat by evaporation. *In* J. L. Monteith and L. E. Mount (eds.), Heat Loss from Animals and Man. Butterworths, London.
92. Allen, T. E., Bennett, J. W., Donegan, S. M., and Hutchinson, J. C. D. (1970). Moisture, its accumulation and site of evaporation in the coats of sweating cattle. J. Agric. Sci. Camb. 74:247.
93. Cena, K., and Monteith, J. L. (1975). Transfer processes in animal coats. III. Water vapour diffusion. Proc. R. Soc. Lond. (Biol.) 188:413.
94. Behmann, F. W. (1976). Maximum of evaporative heat loss in relation to clothing thickness. J. Physiol. (Paris) 63:201.
95. Nishi, Y., and Gagge, A. P. (1970). Moisture permeation of clothing—a factor governing thermal equilibrium and comfort. ASHRAE Trans. 76:137.
96. Nishi, Y. (1973). Vapour permeation of clothing by naphthalene sublimation. Arch. Sci. Physiol. 27:A163.
97. Levell, C. A., Breckenridge, J. R., and Goldman, R. F. (1970). Effect of laundering and starching on insulation and vapour permeability of standard fatigues. Textile Res. J. 40:281.
98. Goldman, R. F. (1963). Physiological effects of wearing a rubberized net overgarment in the heat. Textile Res. J. 33:764.
99. Haisman, M. F., and Goldman, R. F. (1974). Physiological evaluations of armoured vests in hot-wet and hot-dry climates. Ergonomics 17:1.
100. Joy, J. T., and Goldman, R. F. (1968). A method of relating physiology and military performance: a study of some effects of vapour barrier clothing in a hot climate. Milit. Med. 133:458.
101. Breckenridge, J. R., and Goldman, R. F. (1977). Effect of clothing on bodily resistance against meteorological stimuli. *In* S. W. Tromp (ed.), Progress in Biometeorology, Vol. 1, Part 2. Swets and Zeitlinger, Amsterdam.
102. Woodcock, A. H. (1962). Moisture transfer in textile system. Textile Res. J. 32:628.
103. Stewart, J. M., and Goldman, R. F. (1977). Development and evaluation of heat transfer equations for a model of clothed man. Proceedings of the International Conference on Bioengineering, Cape Town.
104. Eyal, E. (1963). Shorn and unshorn Awassi sheep. J. Agric. Sci. 60:159.
105. Hofmeyer, H. S., Guidry, A. J., and Waltz, F. A. (1969). Effects of temperature and wool length on surface and respiratory evaporative losses of sheep. J. Appl. Physiol. 26:517.
106. Craig, F. N., and Moffit, J. T. (1974). Efficiency of evaporative cooling from wet clothing. J. Appl. Physiol. 36:313.
107. McArthur, A. J. (1978). Heat loss from sheep. Ph.D. dissertation, University of Nottingham.
108. Robinson, D. E., Campbell, G. S., and King, J. E. (1976). An evaluation of heat exchange in small birds. J. Comp. Physiol. 105:153.
109. Robertshaw, D. (1979). Man in extreme environments, problems of the newborn and elderly. *In* K. Cena and J. A. Clark (eds.), Bioengineering, Thermal Physiology and Comfort. Wroclaw Technical University Press, Wroclaw.
110. Hey, E. N. (1974). Physiological control over body temperature. *In* J. L. Monteith and L. E. Mount (eds.), Heat Loss from Animals and Man. Butterworths, London.
111. Ingram, D. L., and Mount, L. E. (1975). Man and Animals in Hot Environments. Springer-Verlag, New York.
112. Van Dilla, M., Day, R., and Siple, P. A. (1968). Special problems of hands. *In* L. H. Newburgh (ed.), Physiology of Heat Regulation and the Science of Clothing. Fascimile of the 1949 edition. Hafner, New York.
113. Church, N. S. (1960). Heat loss and the body temperatures of flying insects. II.

Heat conduction within the body and its loss by radiation and convection. J. Exp. Biol. 37:186.

114. Cena, K., and Clark, J. A. (1972). Effect of solar radiation on temperatures of working honeybees. Nature (New Biol.) 236:222.

115. Timbal, J., Wieillefond, H., Guenard, H., and Varene, P. (1974). Metabolism and heat losses of resting man in a hyperbaric helium atmosphere. J. Appl. Physiol. 36:444.

116. Holmer, I., and Bergh, U. (1979). Thermal physiology of man in the aquatic environment. *In* K. Cena and J. A. Clark (eds.), Bioengineering, Thermal Physiology and Comfort. Wroclaw Technical University Press, Wroclaw.

117. Hayes, P. A., and Toy, N. (1976). Convective heat losses in the hyperbaric environment. Eng. Med. 5:69.

118. Mitchell, D. (1973). Prediction of heat stress from heat transfer. Arch. Sci. Physiol. 27:A285.

119. Nishi, Y., and Gagge, A. P. (1977). Effective temperature scale useful for hypo- and hyperbaric environments. Aviat. Space Environ. Med. 48:97.

120. Goldman, R. F., Breckenridge, J. R., Reeves, E., and Beckman, E. L. (1966). "Wet" versus "dry" suit approaches to water immersion protective clothing. Aerospace Med. 37:485.

121. Boutelier, C., Timbal, J., and Colin, J. (1973). Conception des vĕtements anti-immersion et évaluation de leur efficacité. Le Travail Humain 36:313.

122. Frisch, J., Øritsland, N. A., and Krog, J. (1974). Insulation of furs in water. Comp. Biochem. Physiol. 47A:403.

123. Gonzalez, R. R. (1979). Exercise physiology and sensory responses. *In* K. Cena and J. A. Clark (eds.), Bioengineering, Thermal Physiology and Comfort. Wroclaw Technical University Press, Wroclaw.

124. McIntyre, D. A. (1979). Design requirements for a comfortable environment. *In* K. Cena and J. A. Clark (eds.), Bioengineering, Thermal Physiology and Comfort. Wroclaw Technical University Press, Wroclaw.

125. Gagge, A. P., Nishi, Y., and Nevins, R. G. (1976). The role of clothing in meeting FEA energy conservation guidelines. ASHRAE Trans. 82:234.

126. McIntyre, D. A., and Griffiths, I. D. (1975). The effects of added clothing on warmth and comfort in cool conditions. Ergonomics 18:205.

127. Givoni, B., and Goldman, R. F. (1972). Predicting rectal temperature response to work, environment and clothing. J. Appl. Physiol. 32:812.

128. Givoni, B., and Goldman, R. F. (1973). Predicting heart rate response to work, environment and clothing. J. Appl. Physiol. 34:201.

129. Humphreys, M. A. (1977). Clothing and the outdoor microclimate in summer. Building Environ. 12:137.

130. Humphreys, M. A. (1979). The dependence of comfortable temperature upon indoor and outdoor climates. *In* K. Cena and J. A. Clark (eds.), Bioengineering, Thermal Physiology and Comfort. Wroclaw Technical University Press, Wroclaw.

131. Rohles, F. H., Woods, J. E., and Nevins, R. G. (1973). The influence of clothing and temperature on sedentary comfort. ASHRAE Trans. 79:71.

132. Goldman, R. F. (1977). The role of clothing in modifying the human thermal comfort range. Proceedings of the Symposium on Thermal Comfort, University Paris South.

133. Beckman, E. L., Reeves, E., and Goldman, R. F. (1966). Current concepts and practices applicable to the control of body heat loss in aircrew subjected to water immersion. Aerospace Med. 37:348.

134. Bewley, A. D., and Short, B. C. (1976). Liquid conditioned system for aircraft. Eng. Med. 5:64.

135. Fonseca, F. (1976). Effectiveness of four water-cooled undergarments and a water-cooled cap in reducing heat stress. Aviat. Space Environ. Med. 47:1159.
136. Brengelman, G. L., McKeag, M., and Rowell, L. B. (1977). Temperature control system for water-perfused suits. J. Appl. Physiol. 42:656.
137. Strydom, N. B., Mitchell, D., Van Rensburg, A. J., and Van Graan, C. H. (1974). The design, construction and use of a practical ice-jacket for miners. J. S. Afr. Inst. Min. Metall. 74:22.
138. Rohles, F. H., and Johnson, M. A. (1972). Thermal comfort in the elderly. ASHRAE Trans. 78:131.
139. Nishi, Y., Ganzalez, R., Nevins, R., and Gagge, A. P. (1976). Field measurements of clothing thermal insulation. ASHRAE Trans. 82:248.

International Review of Physiology
Environmental Physiology III, Volume 20
Edited by D. Robertshaw

2 Cold Thermogenesis

G. ALEXANDER
C.S.I.R.O., Ian Clunies Ross Animal Research Laboratory, Prospect, Australia

---

TEMPERATURE-METABOLISM CURVES 46

SUMMIT METABOLISM 49
Definitions and Measurement 49
Effect of Body Temperature 51
Effect of Prolonged Cold Exposure and Fasting 52
Effect of Posture 53
Exercise in the Cold: Additive or Substitutive? 53
Effect of Anesthesia and Restraint 54
Relationship with Body Weight 54

CLIMATIC LIMITS TO HOMEOTHERMY 56

COMPONENTS OF COLD THERMOGENESIS 57
Historical 57
Quantitation of Shivering and Nonshivering Thermogenesis 58

OCCURRENCE OF NONSHIVERING THERMOGENESIS 60
Newborn Animals 60
Adult Nonhibernating Animals 64
Hibernators 66
Relationship to Body Weight 67

BROWN FAT 68
Historical 68
Identification 68
Distribution 71
Ontogenic Changes: Conversion to White Fat 72
Factors Affecting the Amount of Brown Fat in the Body 75
Contribution to Nonshivering Thermogenesis 75
Mechanisms of Thermogenesis in Brown Fat 78

OTHER SITES OF NONSHIVERING THERMOGENESIS 81

SYMPATHETIC CONTROL OF NONSHIVERING THERMOGENESIS 85
Brown Fat 85
Skeletal Muscle 92

SHIVERING 93
Role in Homeothermy 93
Magnitude of Shivering Thermogenesis 94
Nature of Shivering 95
Neural Control 97
Heat Production and Substrate Use 98
Energy Source for Shivering 99

CIRCULATORY ADJUSTMENTS DURING COLD THERMOGENESIS 100

NEURAL CONTROL OF COLD THERMOGENESIS 106

SUBSTRATES FOR COLD THERMOGENESIS 107
Tissue Substrate Stores 107
Substrates in Blood 109
Free Fatty Acids 109
Glucose 110
Lactate 111
Glycerol 111
Ketone Bodies 112
Dietary Substrate 112

POSSIBLE CONSTRAINTS ON COLD THERMOGENESIS 112
Effect of Substrate Limitation 112
Drugs, Hormones, and Other Substances 113
Catecholamines 114
Serotonin 114
Hypophyseal Peptide Hormones 114
Adenyl Cyclase–Cyclic AMP System 115
Carnitine 115
Glucagon 115
Insulin 116
Corticosteroids 117
Thyroid Hormones 117
Respiratory Gases 118
Acidosis and Pulmonary Ventilation 120
Blood Pressure 120
Trauma 121

Endotoxins 121

CAN RESTRAINTS BE LIFTED? 121

CONCLUSION 123

The rates of biochemical and physiological processes of animals depend markedly upon the temperature of their tissues, and there are upper and lower limits that tissue temperature cannot pass without resulting in permanent tissue damage or in death. Hence, the activity, competitive success, and even survival of a species depends upon the degree to which its tissue temperature and metabolism are independent of the thermal environment.

Evolution has provided the animal kingdom with a gamut of defense mechanisms that permit various degrees of thermal independence both in hot and in cold environments (1). Among these mechanisms are metabolic processes that permit animals to sustain activity at the cooler end of the range of thermal conditions of their habitat (2).

At the lower end of the evolutionary scale, some aquatic invertebrates show metabolic compensation (shifts in metabolic pathways) as water temperature declines; their rate of metabolism falls more slowly than would be expected from the van't Hoff Arrhenius effect, i.e., the Q_{10} is less than 2 (3, 4). Fish and amphibia also show metabolic compensation, and some antarctic fish have as high a metabolic rate as their tropical counterparts, although their tissue temperature is very different, approximating that of seawater (5). Thermogenesis in frogs can be stimulated by catecholamines (6) but to a much smaller extent than in many mammals (see under "Components of Cold Thermogenesis"). Metabolic heat production, from the red muscle groups that are used in steady swimming of tunas and some sharks, is conserved through efficient countercurrent heat exchange in the circulatory system. The mechanism is particularly effective in bluefin tuna, so that the body temperature is 25–30 °C and can be as much as 15 °C higher than water temperature (5).

In solitary insects metabolic heating is associated with movement of flight muscles, and a shivering type of activity apparently serves as a preflight warm-up (7). Optimal thoracic temperature for flight appears to be 30–40 °C, and during flight of the larger insects the maintenance of substantial temperature gradients between thorax and air is facilitated by a hairy pelage that protects against convective heat loss. There is no fine temperature control; after flight, tissue temperature stabilizes close to ambient temperature. Social insects, such as bees and wasps, control the temperature of their developing brood both by metabolic heating and by behavioral means in the form of clustering (7); the heat appears to be produced by the muscular activity of abdominal pumping (8) or thoracic shivering (9).

Some reptiles, too, show metabolic compensation (10, 11), and muscular activity in large reptiles can raise body temperature above ambient temperature (12). The Indian python uses muscular contraction to elevate body temperature when brooding eggs (13). There is no evidence of a noradrenaline- (norepinephrine-) responsive thermogenesis (see under "Components of Cold Thermogenesis") in reptiles (14).

However, in general, body temperature of submammalian or subavian orders broadly depends on the environmental thermal conditions and is little affected by the heat of metabolism; these animals are termed ectotherms or poikilotherms. On the other hand, the deep body temperature of mammals and birds generally shows little or no dependence on environmental temperature over the range that is usual for the habit of the species, and is maintained within a degree or so of a temperature, usually 35–40°C, that is typical of the species. The heat of metabolism sustains the body temperature, and the animals are termed endotherms or homeotherms. There is a fine degree of neural control of mechanisms concerned with conservation and dissipation of heat, but the keystone of this thermostability is thermoregulatory heat production (cold thermogenesis), over and above that produced by exercise, by digestion of food, or by the tissues generally, under resting or basal conditions.

TEMPERATURE-METABOLISM CURVES

The effectiveness of the various nonbehavioral mechanisms by which an animal controls its thermal state is conveniently illustrated by a set of curves or equations relating ambient temperature (T_a) to deep body temperature, usually rectal temperature (T_{re}), metabolic heat production (M) per unit of surface area (A), and nonevaporative and evaporative heat loss (H and E, respectively) (Figure 1); these curves also provide a basis for comparison of the capacity of animals to regulate their body temperatures.

Curves have been produced over the last 100 years for a wide variety of domestic, laboratory, and wild mammals and birds, but the physiological interpretation dates from the classical comparison of tropical with arctic animals by Scholander and colleagues, reported in 1950 (15). The physiology, physics, and mathematics were popularized by Burton and Edholm in 1955 (16); and salient points, together with experimental methods, were summarized in Volume 7 of this series (17, 18). Depending on the range of environmental temperature covered, the curves provide information on the effectiveness of two important physiological mechanisms for the control of body temperature in cool conditions, thermal insulation (I) of the whole animal (16, 18) and thermogenesis (M/A). With animals showing well-

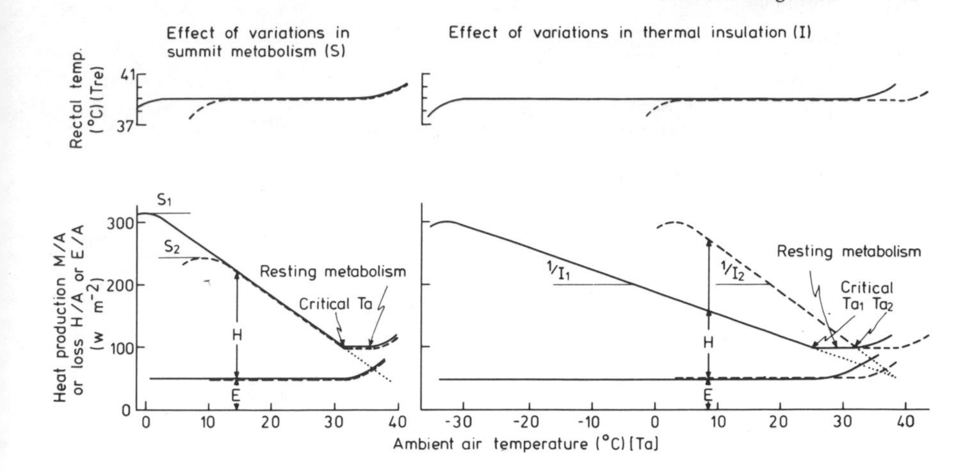

Figure 1. Stylized homeothermic relationships between ambient temperature (T_a), rectal temperature (T_{re}), metabolic heat production (M/A), and nonevaporative and evaporative heat loss (H/A and E/A, respectively), indicating the effect of variations in maximum or summit heat production (S) and in total body thermal insulation (I). The advantages of high S and I values in promoting thermostability at low ambient temperatures are obvious. Relationships at the hot end of the scale are discussed elsewhere (28).

developed homeothermy, the experimentally derived curves conform with the approximate formula $H/A = (M - E)/A = (T_{re} - T_a)/I$, where E is evaporative heat loss, T_{re} and T_a are the temperatures of the rectum and ambient air, respectively, and H is the nonevaporative heat loss ($M - E$). Since I, A, T_{re}, and E tend to remain constant at various values of T_a, the line relating H or ($M - E$) to T_a is straight and on extrapolation passes through $T_a = T_{re}$ (15, 19, 20) (Figure 2). The T_a (critical temperature) below which M must increase depends on the insulation of the animal as well as on its minimum resting metabolic rate (Figure 1). These strictly homeothermic animals are said to obey Newton's Law of cooling (the rate of heat loss is proportional to the temperature gradient from core to ambient air) (16), and any curvature of the line relating H to ($T_{re} - T_a$) indicates variations in total insulation or in exposed surface area due to postural changes. However, in relatively underdeveloped newborn animals, such as rats, rabbits, mice, and premature humans, and in some species with a low mature size that results in rapid heat loss (area to weight ratio is high), the maintenance of strict homeothermy may be energetically very expensive; as T_a declines they allow T_{re} to decline to a new level, so that ($T_{re} - T_a$) and M are lower than Newton's law would predict (21–24). This decline in T_{re} in newborn animals is not necessarily due to immaturity of the neural or thermogenic mechanisms; in newborn rats and rabbits, for example, the level of heat production in moderate cold can be exceeded immediately after a hypoxic episode at the same ambient temperature (25, 26).

In natural environments, and indeed unintentionally in some laboratory

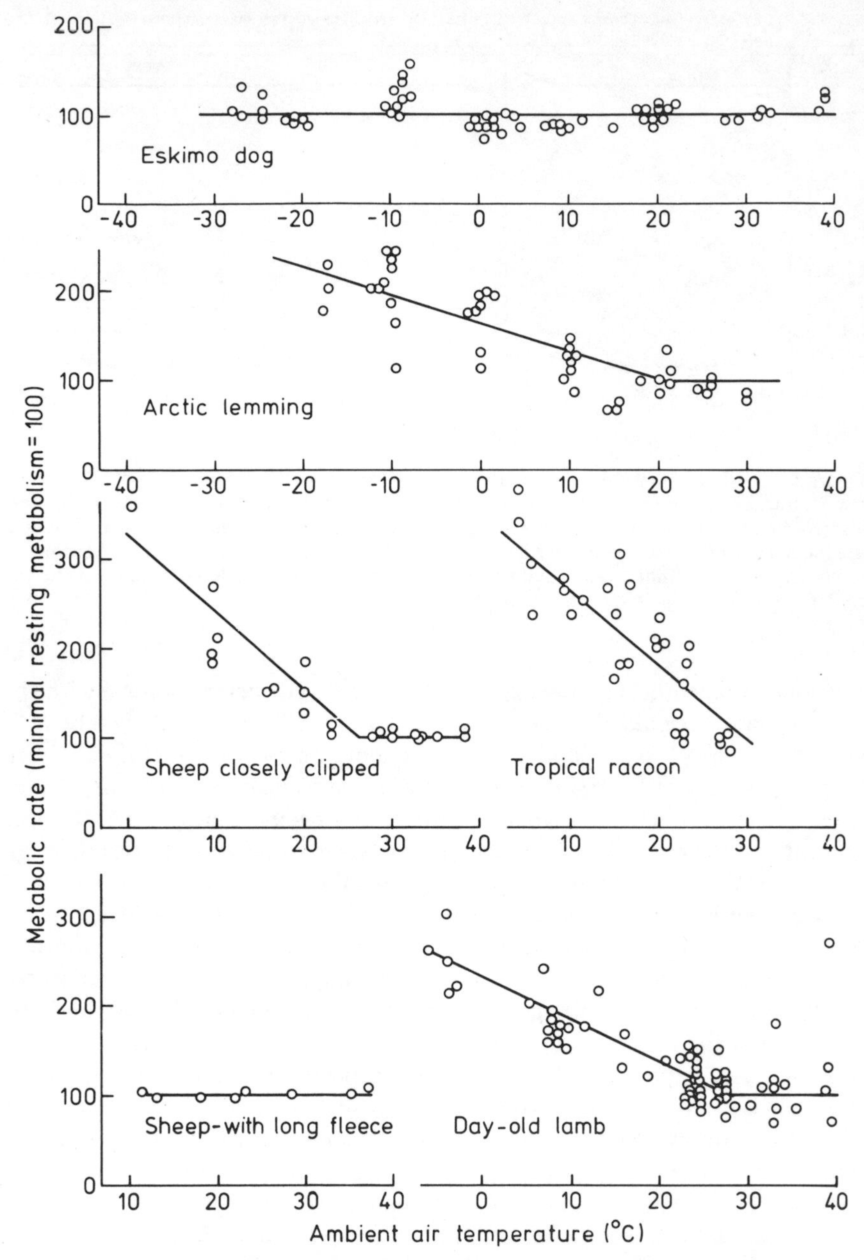

Figure 2. Examples of experimental data relating metabolic rate in number of individuals to ambient air temperature in Eskimo dog, arctic lemming, and tropical racoon (15), sheep with long or closely clipped fleece (ref. 632 and J. W. Bennett, unpublished data), and one-day-old lambs (20). The relative slopes of the lines of best fit do not necessarily affect the relative thermal insulations because the absolute magnitude of the minimum or resting metabolic rate varied between sets of data. Ambient temperatures were insufficiently low to reveal unequivocal cold thermogenesis in the very well insulated Eskimo dog and well fleeced sheep.

experiments, the thermal environment is complex, and the components of H may be higher or lower than predicted by T_a alone; examples include radiative heat losses from human infants to cool nursery surroundings (27) or from cattle to a winter sky (18), absorption of solar radiation by domestic herbivores (18), convective losses from lambs born in windy outdoor conditions (28), and conductive losses from piglets lying on a cold concrete floor (29). Thus, defining the environment solely in terms of T_a ignores heat exchange by radiation, conduction, and forced convection. However, few studies on relationships between ambient temperature and metabolic rate have included these additional important climatic factors, in particular, the effects of a wet coat (30–32) on increasing the rate of heat loss. The relevant physics are complex and not well understood (31, 33, 34).

Most studies are further limited by the range of values of T_a examined; they do not include temperatures where thermoregulatory defenses are stretched to the limit so that heat loss exceeds the maximum thermogenic response and hypothermia ensues. Studies of this part of the curve appear to have been inhibited by the fact that steady state conditions no longer prevail, and by the very low ambient temperatures that may be required. The maximum thermogenic response has become known as summit metabolism (Figure 1) but has also been referred to as "peak metabolic effort" (19). Summit metabolism and its estimation are important if the climatic limits to homeothermy are to be defined. These limits can be of real importance for survival of an animal in inclement weather (see under "Climatic Limits to Homeothermy").

SUMMIT METABOLISM

Definitions and Measurement

Summit metabolism received considerable attention from European physiologists in the 1920's and 1930's. Two measures of this maximum metabolic response to cold were described. The term "summit metabolism" was used by Giaja (35) and Gelineo (36) to describe the highest oxygen consumption obtainable at normal body temperature, without voluntary muscular activity. It was distinguished by Gelineo from "maximum metabolism," which develops briefly in the initial stages of intense body cooling and reaches a higher level than summit metabolism before heat production declines as body temperature continues to fall.

The magnitude of maximum metabolism must depend on the rate of fall in body temperature and on the duration of the measurement and has little practical significance, particularly if body temperature is falling rapidly. On the other hand summit metabolism as defined above can be measured only by the cumbersome method of trial and error, because the only criterion that a maximum response is obtained is that body temperature falls. In the early studies on these intense metabolic responses, animals were exposed to low constant temperatures, sometimes after the removal of the pelage and immersion in water to facilitate heat loss; changes in rectal temperature thus depended on the conditions selected initially, and rectal temperature either did not fall at all or fell at various uncontrolled rates. The period for which such a high metabolic rate can be maintained is also uncertain, and there is doubt about the effects of recent exposure to cold, such as during experiments in which the maximum response is determined by exposing individuals to consecutive periods of successively lower temperatures; this could reduce the maximum response through exhaustion (see under "Effect of Prolonged Cold Exposure and Fasting").

These difficulties can be largely avoided by measuring the metabolic response to cold when rectal temperature is induced to fall at a small, controllable rate and the measurement is continued for a period, say 20 min, that has some practical significance. The controlled fall in rectal temperature is readily accomplished, for example, by varying air movement over the animal, held at a fixed low T_a, and perhaps with the coat clipped and wetted to produce a level of heat loss close to, but below, that necessary to induce summit metabolism (37). Immersion in water, or even in air, cooling at a controlled rate, can also be used (J. Slee, unpublished data and ref. 38). The selection of the appropriate conditions is facilitated by a moderate amount of experience. These methods of measuring summit metabolism have not been widely used except in adult sheep (39) and lambs (J. Slee, unpublished data and ref. 37); in this species the results are reasonably repeatable with a coefficient of variation around 6%.

Summit metabolism of an individual can vary with its physiological state; the term is not meant to apply to only the maximum response under the best physiological conditions. Summit metabolism does not appear instantaneously on exposure to cold; adult sheep require more than one-half hour to reach summit (39), lambs 5–10 min (37), and rats only 3 min (40). Objections have been raised to the use of subfreezing temperatures on the grounds of risk from cold injury, especially in small animals, and clipping of the coat may disqualify the animal from other studies (41, 42). These objections may be overcome by the use of helium in the place of nitrogen in the atmosphere to which the animal is exposed (41, 42). The helium-oxygen mixture has a thermal conductivity 4 times that of air (43), so the insulations of the pelage, which depends on trapped gas, and of the surrounding atmosphere are reduced. The overall effect, at least in small animals, is a halving of total insulation so that summit metabolism is evoked at a relatively mild T_a (Figure 3). The response of animals to a variable wind in a helium-oxygen mixture may be the method of choice for the measurement of summit metabolism, but helium has yet to be used in this way. Summit oxygen consumption is most readily measured with an open circuit system, but allowance can be made for rapidly changing thermal conditions that may be encountered during measurement of summit metabolism in a closed circuit

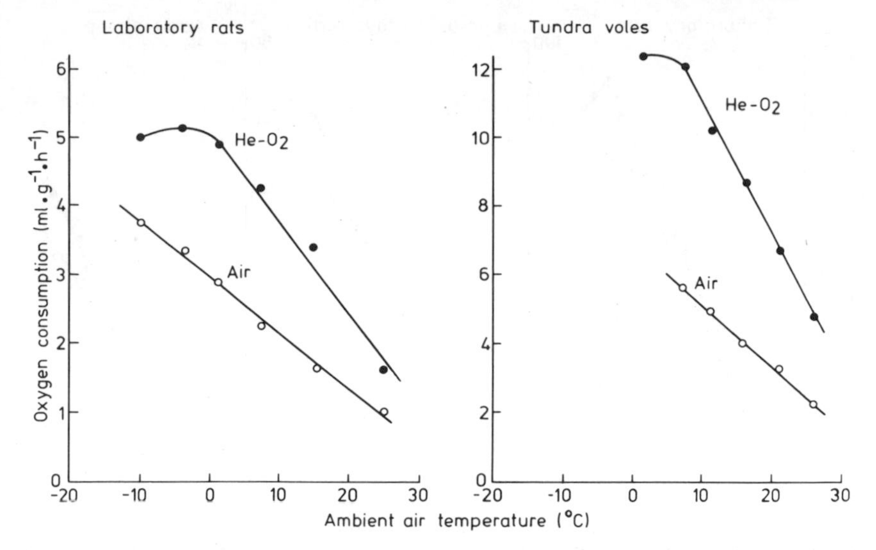

Figure 3. Oxygen consumption of two species of mammals in air and in an atmosphere of 80% helium and 20% oxygen. Each *point* is the mean of several animals. The high thermal conductivity of helium facilitates heat loss and the attainment of summit metabolism without recourse to excessively low temperatures. Adapted from Rosenmann and Morrison (41) by permission of the publishers (American Physiological Society).

system, due to insufficient time for equilibration, or to varying wind speed (44).

The individual measurement of summit metabolism in young animals of some species and adults of others, which do not show rigid homeothermy and allow body temperature to drift downward in response to even moderate cold (see under "Temperature-Metabolism Curves"), presents difficulties. Estimates can be made from temperature-metabolism curves derived from a number of individuals (22) but it is not certain that these estimates would represent thermogenic capacity (25, 26). However, the magnitude of summit metabolism in these types has less relevance to survival than in strict homoetherms (see under "Climatic Limits to Homeothermy").

Effect of Body Temperature

The level of summit metabolism in strict homeotherms appears to be unrelated to rectal temperature when this is close to normal; no relationship has been demonstrated in lambs and adult sheep with rectal temperatures greater than 36°C or 37°C (normal about 39.5°C for lambs and about 39.0°C for adults (37, 39)). Thus, summit metabolism would be little affected by a decline of rectal temperature of 1°C or 2°C during its measurement. However, at lower rectal temperatures, metabolic rate declines in a linear fashion with declining rectal temperature (Figure 4) in both the adult sheep and lamb (37, 39) and also in guinea pigs (45) and rats (46). Clearly,

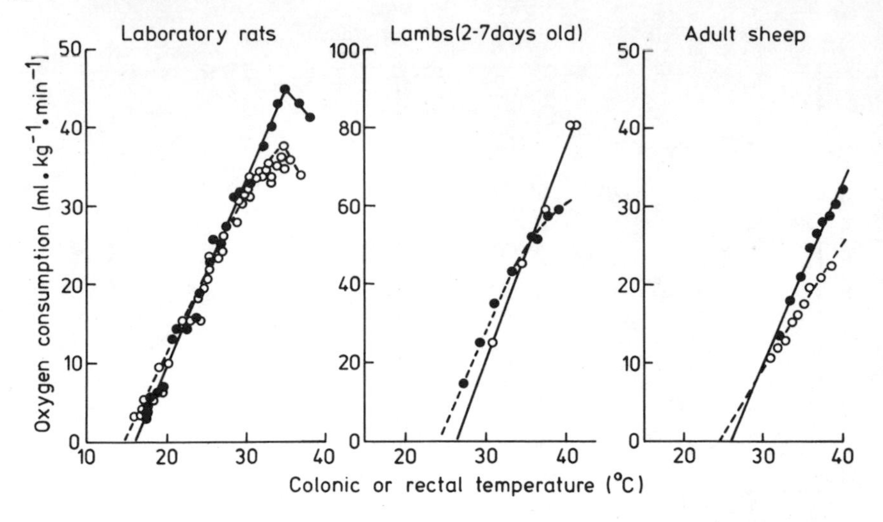

Figure 4. Relation of maximum oxygen consumption to deep body temperature during body cooling in two laboratory rats (46), two lambs (37), and two adult sheep (39). Clearly, reduction of body temperature reduces summit metabolism.

this is not a situation in which the van't Hoff Arrhenius law can be applied. The rate of decline in metabolic rate is about 2 ml of O_2 kg^{-1} min^{-1} $°C^{-1}$ in all but the lamb, in which it is more than twice this value. This linear decline in metabolic rate enables the temperature of zero metabolism to be estimated by extrapolation; it is about 25°C in adult sheep, around 23°C in lambs, 19°C in guinea pig, and 15°C in the rat.

Effect of Prolonged Cold Exposure and Fasting

Although summit metabolism in adult sheep and lambs is usually well maintained throughout several hours of cold exposure (37, 39), a reduction of more than 30% has been occasionally observed in lambs. Similarly, fasting for 2 or 3 days results in a substantial decline in summit metabolism, especially in young lambs (37, 39). These findings indicate that the level of energy reserves could play an important role in the maintenance of summit metabolism (see under "Tissue Substrate Stores"). However, summit metabolism of newborn lambs is not increased by prior ingestion of milk (37).

The inability of animals to maintain a full summit response for prolonged periods raises questions about the magnitude of elevated metabolic rates, as a proportion of summit metabolism, that can be maintained indefinitely; in adult sheep, metabolism of 80% of summit metabolism can be maintained for up to 8 hr (39). This is a neglected area of investigation, complicated by the inhibitory effects that cold exposure may have on the ability of the animal to obtain food (47) or by impairment of absorption because of redistribution of blood away from the gastrointestinal tract (48).

Effect of Posture

Little attention has been paid to the effects of posture on thermoregulation. In sheep, summit metabolism is about 27% lower during lying than during standing (39), and even though the effective surface area for heat loss is reduced by lying, the rectal temperature during lying tends to decrease at a much faster rate than during standing. The mechanisms are obscure, but are probably associated with effects on the ability of the muscle of the legs to produce heat by shivering (see under "Shivering"); leg muscle in sheep represents about half of total body muscle. Sheep prefer to stand rather than to lie during exposure to severe cold, both in the laboratory and in natural environments (39, 49); this may be associated with the much higher summit metabolism during standing than lying.

Exercise in the Cold: Additive or Substitutive?

Animals exposed to severe cold frequently become restless and occasionally struggle violently if restrained; consequently, it is important to know whether cold and exercise thermogenesis are additive or substitutive. It is sometimes assumed that maximum oxygen consumption ($\dot{V}_{O_2\,max}$) induced by cold (i.e., summit metabolism) should have the same magnitude as that induced by exercise (often referred to as maximum aerobic capacity) (50). This could be so if the same tissues were involved in both processes and both were subject to the same limitations, say of oxygen or substrate supply. However, thermoregulatory heat is produced by tissues other than shivering muscle, especially in some species or age groups (see under "Components of Cold Thermogenesis" and "Occurrence of Nonshivering Thermogenesis"), and muscles involved in maximum exercise of various types (51) and in shivering (see under "Shivering") are not necessarily the same. In addition, factors limiting either process are ill defined (see under "Possible Constraints on Cold Thermogenesis").

Difficulties of interpretation in experimental studies with laboratory animals include doubts about whether or not the exercise was indeed maximum and whether or not sufficient time was allowed for development of the maximum metabolic response to cold (52). Experiments on laboratory rats that show significant nonshivering thermogenesis (53, 54) provide conflicting evidence about whether thermogenesis due to exercise and shivering is substitutive or additive. Differences between species have also been reported (52). In man, whose response to cold appears to rely almost solely on shivering (see under sections "Components of Cold Thermogenesis," "Occurrence of Nonshivering Thermogenesis," and "Brown Fat"), $\dot{V}_{O_2max}$ during exercise (55–57) was consistently 2–3 times larger than the infrequently measured summit metabolism (58, 59). Although exercise appears to substitute for shivering in man unless body temperature is depressed (60), shivering can be additive to submaximum exercise; the oxygen cost of swimming submaximally increases as water temperature decreases (61). In adult sheep also, occasional voluntary muscular activity appears to increase oxygen consumption momentarily during the measurement of summit metabolism; but the effect may be only partly additive, since shivering appears to be inhibited by voluntary movement (59, 62). The effect of sustained exercise in sheep is not known. It is, therefore, not possible at present to be certain that voluntary movement does not interfere with the measurement of summit metabolism.

Effect of Anesthesia and Restraint

Anesthetics of various types depress shivering and cold- or catecholamine-induced nonshivering thermogenesis and can produce hypothermia even under quite mild thermal conditions (63–67). This is perhaps partly due to diversion of as much as 25% of cardiac output through arteriovenous anastomoses with concomitant reduction in nutrient blood flow (68). Restraint coupled with anesthesia has also been suggested as depressing thermogenesis, at least in newborn guinea pigs and rabbits (64), but this could result from accelerated induction of hypothermia due to the restraint posture facilitating heat loss (69).

Relationship with Body Weight

The relation between metabolic rate (M) and body weight (W) in a population of animals is usually expressed in the form $M = aW^b$, where a and b are constants. The magnitude of the exponent (b) for basal or minimum resting metabolic rate has been the subject of a continuous debate (29, 70, 71), and the usually accepted value is 0.75.

The relation of summit metabolism to body weight appears to have been examined in sheep and pigs only. The exponent for all practical purposes was unity; groups examined included adult sheep and four lots of lambs examined by the variable wind method (37, 39, 72, 73) (Figure 5), several breeds of lambs examined by the variable water temperature method (J. Slee, personal communication), and piglets examined in falling air temperature (38). A lower exponent applies when different species are considered together and the range of body weights extends over several orders of magnitude (Figure 6). However, in homogenous groups, summit metabolism expressed as a rate per kilogram of body weight tends to be constant over a range of body weights.

An exponent of 1 for summit metabolism has an important practical implication. Since surface area is also related to body weight by the theoretical equation $A = aW^{0.67}$, and empirically derived values of the exponent can be even lower ($A = 0.121\ W^{0.59}$ for sheep) (74), it follows that surface area to weight ratio is higher in small than in large individuals, so summit metabolism per unit of surface is lower in small than in large individuals (Figure 7). This has major implications about body weight and survival in cold climates (see under "Climatic Limits to Homeothermy").

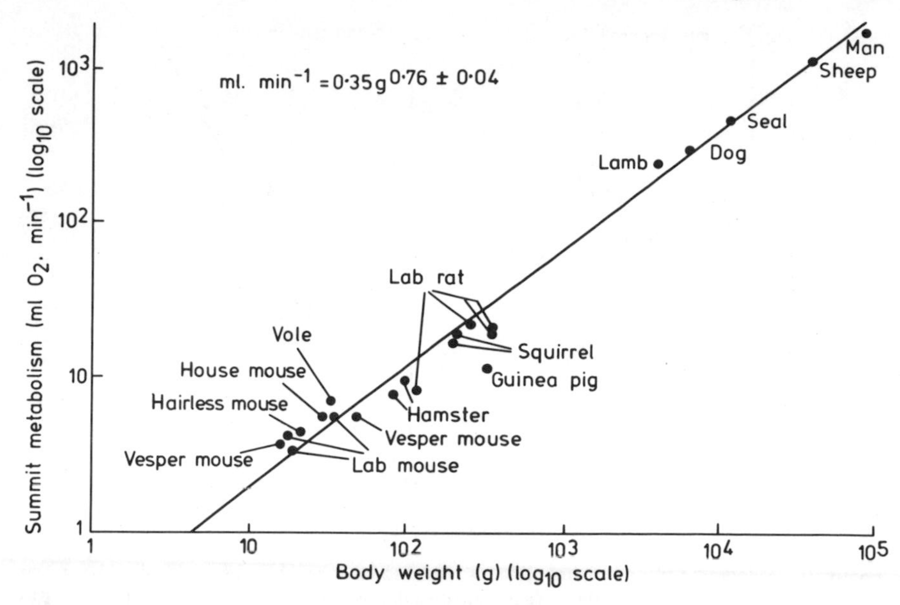

Figure 5. The relation of summit metabolism to body weight in adult sheep (39) and lambs (73).

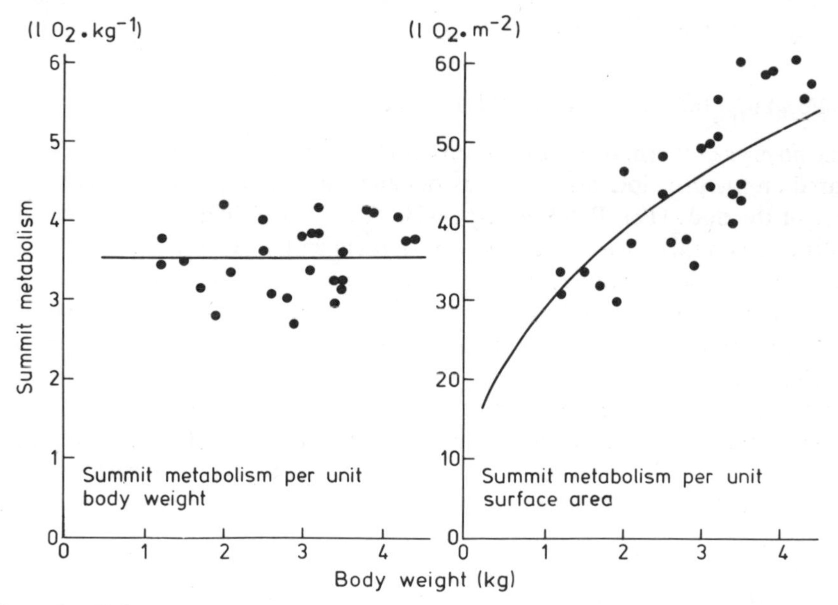

Figure 6. Estimation of the relationship of summit metabolism (S) to body weight (W) across a variety of mammals with body weights ranging from 16 g to 90 kg. The value of the exponent (b) in the relationship $S = KW^b$ is clearly less than unity. The data were derived from a variety of sources (28, 35, 39, 41, 45, 52, 58, 178, 264, 633, 634) in most of which the estimate of summit metabolism was obtained from temperature-metabolism curves. The data were collated by J. W. Bennett.

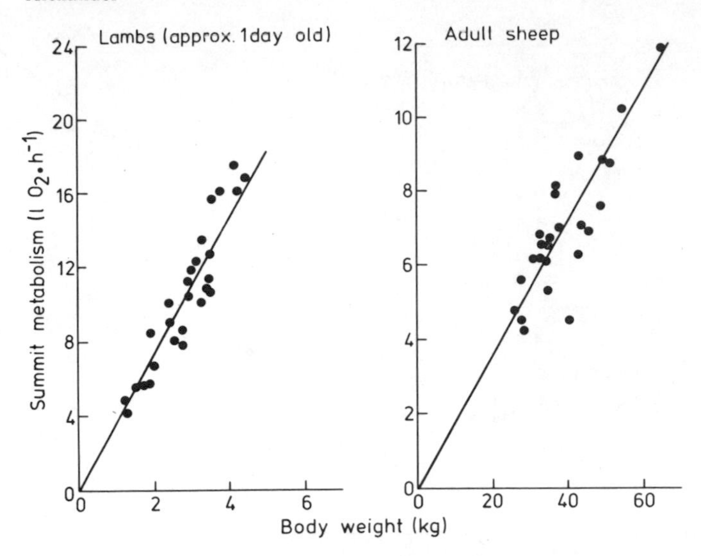

Figure 7. Relation of summit metabolism of lambs to body weight (*right*), showing the strong dependence on body size when the oxygen consumption is expressed as l O_2/m^2, in contrast to the relative independence from body size when oxygen consumption is expressed as l O_2/kg (*left*). The figures are derived from the data for lambs in Figure 5 and the formula of Peirce (74) ($m^2 = 0.121$ $kg^{0.59}$). The *trend lines* were calculated from the mean figure of 3.54 l O_2/kg for summit metabolism obtained from several series of lambs.

CLIMATIC LIMITS TO HOMEOTHERMY

The physics of thermal exchanges between an animal and its environment are based on the principle that heat loss depends on the area of the exposed surface of the body (17). Relationships of heat loss to ambient temperature are calculated on this basis and appear to conform with it. If summit metabolism and rate of heat loss are expressed in the same units (say W m^{-2}), then the ambient temperature at which they are equal provides an estimate of the temperature below which the animal would begin to chill and hypothermia would ensue; a zone of potentially lethal thermal environments can be defined in this way. Estimates of these temperatures have been made for several combinations of the climatic variables, ambient temperature, wind speed, and wetness of the coat, both in adult sheep and lambs of various body weights (Table 1); it is clear that the lighter animals are at a marked disadvantage.

Since summit metabolic rates cannot be maintained indefinitely (see under "Effect of Prolonged Cold Exposure and Fasting"), estimates of these critical cold temperatures must, therefore, be modified upward if the cold exposure were to continue much beyond the 20-min period adopted for the measurement of summit metabolism. These findings emphasize the susceptibility of the newborn to death from hypothermia, especially the newborn of mammalian species such as ungulates, which may be considered to show

Table 1. Environmental temperatures at which heat loss would equal summit metabolism and below which body temperature would fall

			Environmental temperatures (°C)		
			Still air (0.4 km/hr)	Wind (20–25 km/hr)	
Animals	Body weight (kg)	Summit metabolism (W)	fleece dry	fleece dry	fleece wet
Adult sheep, freshly shorn (fleece 7 mm long)	25	260	−45	−5	15
	50	493	−62	−14	10
Lambs	2	40	−32	−4	13[a]
(one day old)	5	100	−69	−25	−7[a]

Adapted from Alexander (37) and Bennett (39) by permission of the authors and publisher (CSIRO).

[a]Add 10 °C for newborn lambs naturally wet with fetal fluids.

precocious development at birth. They are frequently born in exposed windy environments and the coat is saturated with fetal fluids, so that, for example, chilling of a small lamb would commence at an ambient temperature above 20 °C, although a large lamb would not be at risk unless the temperature was around freezing point.

Hypothermia can also be a problem in man; the newborn and the elderly are particularly susceptible, even indoors (75, 76). In other age groups, fatal exposure to cold conditions during outdoor recreation is occasionally reported (77), and most humans would die from hypothermia if accidentally immersed in water at 0 °C for more than 2 hr (59). The importance of efficient mechanisms for cold thermogenesis is obvious.

COMPONENTS OF COLD THERMOGENESIS

Historical

The recognition that shivering was associated with an increase in metabolic heat production must be very old, but the recognition that there are other heat-producing defences against cold that are associated with comparatively modern names, such as Claude Bernard and Rübner, and convincing proof of nonshivering forms of thermogenesis, are quite recent (16, 78, 79). Even up to 1955 the concept of cold thermogenesis from sources other than shivering was regarded skeptically (for review see ref. 80). Increased heat production was observed in the absence of overt shivering, but the stumbling block to acceptance of this as nonshivering thermogenesis was the demonstration of concomitant active electromyograms; it was thought that increased muscle tone could account for the heat produced (16).

The skepticism was dispelled by a series of experiments in the mid-1950s. First, cold thermogenesis was observed in cold-acclimated rats without any increase in the electrical activity of skeletal muscle (81); and cold-acclimated rats were shown to increase their heat production in the cold even when skeletal muscles were paralyzed with curare (82). In addition, injections of noradrenaline, the catecholamine released at sympathetic nerve endings, were shown to stimulate a much higher heat production in cold-acclimated than in warm-acclimated rats (83); and the sympathetic nervous system was further implicated by the demonstration that ganglionic blockade with hexamethonium reduced the thermogenic effect of cold exposure on cold-adapted rats (84). Shortly afterward, a significant thermogenic response to noradrenaline administration was demonstrated in newborn cats, rats, mice, and rabbits (85), and evidence of nonshivering thermogenesis has been obtained subsequently in a wide variety of newborn mammals (see under "Newborn Animals").

Once the existence of nonshivering thermogenesis was accepted, a new controversy arose about which tissues were responsible. The liver and gut were apparently eliminated as major sites by experiments with cold-adapted rats in which the circulation of the liver and gut had been tied off (86); increased thermogenesis still occurred in response to cold exposure or noradrenaline administration. Interest returned briefly to skeletal muscle, but in the early 1960's attention was diverted to brown adipose tissue (brown fat) as a major source of cold thermogenesis in cold-adapted animals (87), in hibernators (88, 89), and in newborn mammals (90). Since that time, the literature on brown fat and the biochemical basis of nonshivering thermogenesis has swamped investigations on other sites of nonshivering thermogenesis. Nevertheless, there is evidence that muscle and other organs, including liver, do play some role in nonshivering thermogenesis, at least in cold-acclimated rats and dogs (91–93) (see under "Other Sites of Nonshivering Thermogenesis").

It also became clear that, in cold exposure, mechanisms for nonshivering thermogenesis were called into play before electrical activity of the muscle increased appreciably or shivering appeared (see under "Neural Control of Cold Thermogenesis"). If nonshivering thermogenesis were prevented by sympathetic blockade, shivering appeared immediately (94). Shivering was, therefore, regarded as a second line of defense and tended to be dismissed as an insignificant source of heat in the most used experimental animal, the cold-adapted rat (95), as well as in the newborn of many species, such as the dog (96), cat (97), rat (98), man (99), rabbit (100), and guinea pig (101).

Quantitation of Shivering and Nonshivering Thermogenesis

The partitioning of cold thermogenesis into its shivering and nonshivering components presents several difficulties, and there is doubt about the general applicability and validity of some of the methods that have been used. The cumbersome method of Hart and Janský (53) unjustifiably assumed that metabolic rate in maximum exercise, at thermoneutral temperatures, was equal to the thermogenic capacity of shivering (see under "Exercise in the Cold: Additive or Substitutive") and that a cold-induced increase in metabolic rate above this level represented nonshivering.

The magnitude of nonshivering thermogenesis has also been taken as the maximum metabolic rate that could be achieved in the cold before the onset of electrical activity indicative of shivering (102). However, this method unjustifiably assumes that shivering does not begin until nonshivering has reached its maximum.

Other less cumbersome methods that have been used for estimating nonshivering thermogenesis have a sounder physiological base, although there are associated practical difficulties. The increase in metabolic rate due to administration of noradrenaline by injection, or preferably by infusion, under thermoneutral conditions, has been equated with the potential for nonshivering thermogenesis, for example, in young guinea pigs (103), lambs (62), and in cold- or warm-acclimated rats (102). The procedure requires some preknowledge of the noradrenaline dose-response curve and precautions to ensure that body temperature does not rise sufficiently to elevate metabolic rate (62); rapid and fatal hyperthermia can readily occur with this method.

The decrease in summit metabolic rate due to sympathetic blockade of the β receptors (see under "Sympathetic Control of Nonshivering Thermogenesis") with the subject under cold conditions can also indicate the magnitude of nonshivering thermogenesis (62). However, if the animal is not at summit metabolism both before and after blockade, the nonshivering capacity will be underestimated, because the initial value will be too low and the final value too high due to an increase in shivering. This is likely to have occurred in studies with cold-acclimated rats (102) and in young guinea pigs during tests at a constant 8°C (94). It has been suggested (102) that the objection can be overcome if shivering is first prevented by paralysis with curariform drugs (84), but there would then be no need for sympathetic blockade since metabolic rate should now reflect nonshivering capacity.

The decrease in summit metabolic rate due to inhibition of shivering by curariform drugs provides an estimate of the potential for shivering thermogenesis (62); but if the animals are not in summit metabolism, the decrease in shivering would be submaximal. This method has not been widely used.

The contribution of shivering to cold thermogenesis has also been estimated by determining the relationship between electrical activity of muscle and oxygen uptake when nonshivering thermogenesis has been selectively blocked with a β-sympatholytic agent (104); the fraction of shivering and nonshivering thermogenesis was then estimated in cold-exposed untreated animals from the electrical activity and metabolic rate. However, there are uncertainties about the repeatability of this method due to variation in the placement of electrodes (79, 105, 106) (see under "Shivering").

Objections can be raised to the use of pharmacological agents on grounds that they may affect metabolic rate other than in the way envisaged. Effects that may be undesirable have been observed in studies on lambs (62); phentolamine (an α-sympatholytic agent) and hexamethonium (a ganglionic blocking agent) produced marked reductions in blood pressure, propranolol (a β-sympatholytic agent) and suxamethonium (a curariform drug) slowed the heart, propranolol and phentolamine sometimes reduced shivering, and the catecholamines increased blood pressure and frequently increased rectal temperature and voluntary activity.

Because of these uncertainties, several simultaneous approaches to the partitioning of cold thermogenesis are desirable, but few systematic studies have been done (62, 102), and in only one study have the animals been examined under conditions of summit metabolism (62). Therefore, in other studies the importance of nonshivering, and especially of shivering, may have been underestimated.

In the study conducted under summit conditions, newborn lambs were treated with propranolol to inhibit nonshivering thermogenesis, or with suxamethonium to inhibit shivering, and their summit metabolism was measured by using the criteria outlined under "Definitions and Measurement." The results were compared with the control summit metabolism and with the increase in metabolic rate due to infusion of noradrenaline at thermoneutrality (Table 2). There were no major inconsistencies in the data, which indicated that, in newborn lambs, shivering is quantitatively at least as important as nonshivering metabolism.

While a similar approach could be made in other animals that show well developed homeothermy it may be difficult to use with animals whose body temperatures are labile (see under "Definitions and Measurement").

OCCURRENCE OF NONSHIVERING THERMOGENESIS

Newborn Animals

Few of the numerous reports on the development of thermogenesis in neonatal mammals and birds indicate whether or not cold thermogenesis occurs in the absence of overt shivering. Stringent tests based on electrical activity of muscle, as a criterion for the absence of shivering, or on determination of metabolic rate during curariform paralysis have been confined to few species, including the newborn rabbit (64), guinea pig (94), and lamb (62). However, the presence of nonshivering thermogenesis among newborn animals appears to be widespread, as judged by the stimulation of metabolic rate by catecholamines (107) and by the presence of brown fat (90) (see under "Brown Fat").

Unfortunately, the only systematic search of the newborn for the occurrence of brown fat (108) is unacceptable (see under "Identification"). Nor has there been any systematic study in newborns for the presence of a type of

Table 2. Estimation by pharmacological means of the contribution of shivering and nonshivering thermogenesis to summit metabolism of newborn lambs

	Estimated components of summit metabolism (l O_2/kg/hr)			
Experimental agent	Shivering	Minimum resting	Nonshivering	Summit metabolism
Propranolol	2.3 (residue)		1.2 (reduction)	3.5 (initial)
Suxamethonium	1.9 (reduction)	1.6 (residue)		3.5 (initial)
Propranolol plus suxamethonium		0.8 (residue)		3.5 (initial)
Catecholamines		1.0 (initial)	1.4 (increase)	
Calculated values, assuming minimum metabolism = 0.8 and summit metabolism = 3.5				
Propranolol	1.5	0.8	1.2	3.5
Suxamethonium	1.9	0.8	0.8	3.5
Catecholamines	1.3	0.8	1.4	3.5
Mean	1.6	0.8	1.1	3.5
Percentage of summit metabolism	46	23	31	100

Adapted from Alexander and Williams (62) by permission of the authors and publisher.

nonshivering thermogenesis other than that associated with brown fat; such a mechanism has been implicated in the newly hatched domestic chicken, which is believed to have no brown fat (109), in common with other birds (110). However, the absence of brown fat from birds also requires confirmation by electron microscopy, and shivering appears to be an important mechanism in hatchlings of at least some birds (106).

The pig is the only newborn mammal in which the absence of brown fat and associated nonshivering thermogenesis has been confirmed (21, 111–114). However, other forms of nonshivering thermogenesis cannot yet be ruled out in this species. The newborn calf (*Bos taurus*) was originally reported to have no brown fat or nonshivering thermogenesis (115). However, virtually all of the fat is now known to be brown, although much of it has the appearance of white fat under the light microscope, and nonshivering thermogenesis in the form of a response to noradrenaline has been demonstrated in calves less than a day old (116, 117). Heat production in brown fat of calves with curarized muscles has also been demonstrated (118).

In some species with small newborns and a short gestation, such as the hamster, nonshivering thermogenesis is scarcely evident at birth and the brown fat is in the rudimentary state; both develop postnatally (119). In the rat, brown adipose tissue contains little or no lipid at birth and does not become functional until lipid has accumulated after the consumption of milk (120); thus maturation can be delayed by nutritive restriction (121).

Other species in which there is evidence of nonshivering thermogenesis in the newborn or suckling young include man (122–126), goat (127), mouse, cat, dog (85), harp seal (128), lemming (129), shrew and bats (130), and coypu (131). There is no doubt that as more and more species are examined by adequate criteria the list will extend into the majority of orders of eutherian mammals and presumably include those species such as rhesus monkey (132), red squirrel (133), and various species of Peromyscus (134) that are known to have brown fat as adults, but in which the neonate does not appear to have been studied. There appears to be no ultrastructural evidence about the existence of nonshivering thermogenesis in monotremes or marsupials.

The magnitude of cold-induced nonshivering thermogenesis, at least as judged by the magnitude of the metabolic response to noradrenaline, appears to be of the order of 1–2 times the minimum resting metabolic rate in the neonates of most species for which there are data (79), including the newborn calf (116), which may weigh 30–40 kg. Thus, the generalization for adult animals that, across species, nonshivering thermogenesis is negatively correlated with body weight, and that mammals with an adult body weight of more than 10 kg have little or no nonshivering thermogenesis (135) does not hold for newborn mammals.

In mammals that are born at a relatively advanced stage of development, such as the sheep (Table 2) and guinea pig, nonshivering thermogenesis is maximum just after birth and normally declines to the point of insignificance by the time the animal is 3–4 weeks old (62, 101), leaving shivering as the main thermogenic defense. This decline is accompanied by the disappearance of brown fat (136, 137). However, the rate of decline in thermogenesis and disappearance of brown fat can be retarded, although not completely prevented, by exposing the animal to cold during the first weeks of life (103, 138). Since the cold resistance of the lamb normally increases rapidly during this period due to rapid body and wool growth, it is necessary to partially shear them and maintain them at temperatures close to 0°C in order to demonstrate this effect.

In other species, such as the rat, mouse, lemming, and hamster, that are born at an earlier stage of development, nonshivering thermogenesis is insignificant at birth, increases to a maximum during the 2nd and 3rd week of life and then subsequently decreases (129, 139–142).

Some of the more mature species show evidence that the potential for nonshivering thermogenesis is developed well before birth. For example, brown fat is present and its oxidative enzymes are active in human fetuses during the 5th month of the 9-month gestation. Also, recognizable brown fat is present in fetal lambs by the 100th day of the 150-day gestation (143). Indeed, the ability to produce heat by nonshivering thermogenesis, as indicated by the response to noradrenaline, is already present in lambs delivered some 20 days prematurely. However, this response is below, and remains below, that of full term lambs (Figure 8), so that the development of the nonshivering thermogenic mechanisms between conceptual age of 130 days and full term appears to depend on continued intrauterine existence (144); presumably the events around birth terminate the development phase.

Two advantages to the newborn of nonshivering as compared with shivering thermogenesis are apparent. Shivering could increase air movement round the animal and so reduce the external insulation, therefore being uneconomical, especially in small animals that have a high surface to mass ratio. Secondly, shivering could interfere with fine movement, especially immediately after birth when the young have to find the teats. In addition, in animals such as the newborn lamb, in which shivering thermogenesis is well developed (62), nonshivering thermogenesis greatly increases the range of climatic conditions in which homeothermy can be maintained. For example,

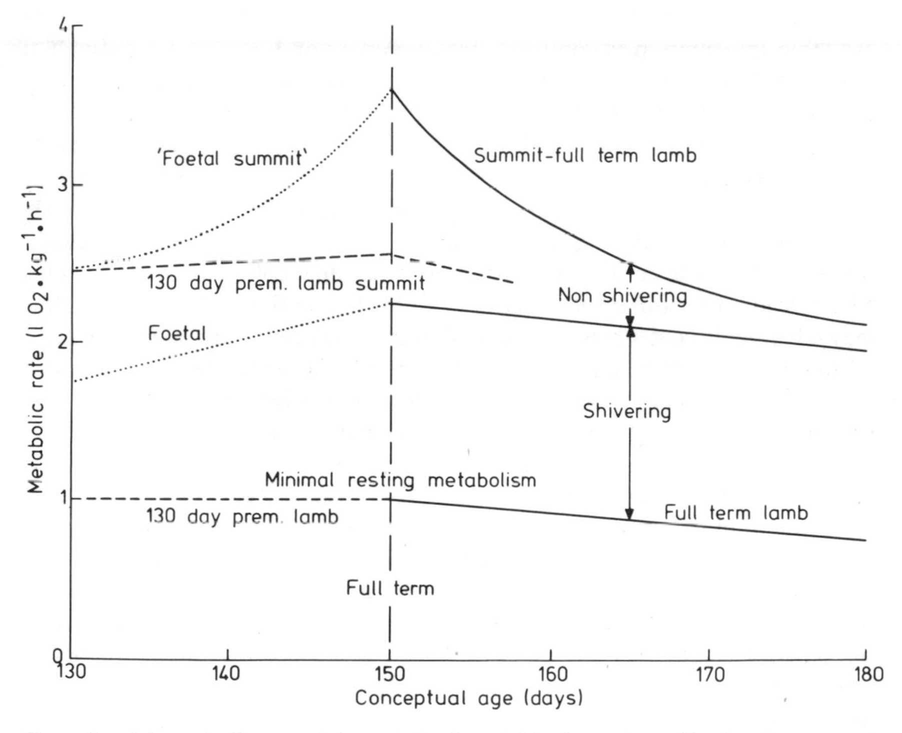

Figure 8. Schematic diagram of the contribution of shivering and nonshivering thermogenesis to summit metabolism in lambs according to conceptual age. The *dotted lines* represent the situation for a number of premature lambs delivered at different ages and each examined soon after delivery. The *broken line* represents the progressive changes in summit metabolism and minimum resting metabolism of premature lambs born at conceptual age 130 days. The *solid lines* represent the progressive changes in a lamb born at full term (150 days). Reproduced from Alexander et al. (144) with permission of the authors and publisher (Cambridge University Press).

it can be shown (62) that no newborn lamb would maintain homeothermy, in the absence of nonshivering thermogenesis at 20°C, in a wind of 15 km hr^{-1} and with water evaporating from a coat saturated with fetal fluids, whereas lambs do in fact survive similar conditions, but at temperatures just above freezing (37). Indeed, the possession of brown fat by precocious newborn animals that do not have the advantage of a nest may be a special adaptation for survival during the first critical hour or two of life when the birth coat is drying.

Adult Nonhibernating Animals

In adult animals, also, there has been no systematic survey, across the various mammalian orders, of the occurrence of nonshivering thermogenesis or of brown fat, and except in very few species there is little information about the importance of nonshivering thermogenesis throughout life. Most estimates of the capacity for nonshivering thermogenesis are derived indirectly, by measuring the calorigenic response to noradrenaline (79), and the available information is confused, fragmentary, and sometimes difficult to interpret. Most is derived from studies on rats that have been adapted, or acclimated, to low temperature in the laboratory; this treatment should be distinguished from acclimatization, which is physiological adjustment to the complex outdoor environment (145). Confusion arises over several issues.

First, there are apparently large differences between species in the magnitude of nonshivering thermogenesis and its response to cold acclimation (79). It has been usual, in attempts to induce cold acclimation, to use 5°C continuously for several weeks (79, 146), but thermal conditions that may adequately cold stress one class of animal may represent no cold stress at all to another, better insulated species, or may be so severe for others that debilitation results and the acclimation response is inhibited (92, 113, 138). For example, it is difficult to demonstrate cold acclimation in the rabbit, unless its coat has been clipped (147). Thus, thermal conditions for cold acclimation experiments would be better standardized in terms of the metabolic rate induced by the acclimation conditions (say two-thirds of summit metabolism) than in terms of ambient temperature.

There is also confusion over whether the calorigenic responses to exogenous catecholamines add to or substitute for cold-induced nonshivering thermogenesis (79). The degree to which the metabolic responses to cold and to catecholamines add or substitute depends on whether or not the animal has been maximally stimulated by one or the other and whether thermogenesis by muscular contraction is prevented; these conditions have been attained in very few studies. In addition, anesthesia has been widely used (67, 102, 148); this reduces the response to catecholamines and could affect the level of resting metabolism with which the response to noradrenaline is often compared (see under "Effect of Anesthesia and Restraint"). In addition, the effectiveness of restraint employed in some studies to prevent thermogenesis from muscular movement (67, 113) is dubious.

As judged by the calorigenic response to noradrenaline, there are major species differences in the magnitude of nonshivering thermogenesis among animals reputed to be living at thermoneutrality. There appears to be no response in man (149–151), rhesus monkey (152), rabbit (147), sheep (153), pigs (113), and birds (154). However, it is possible that at least some of these animals possess a nonshivering mechanism that is not responsive to noradrenaline (153) (see under "Other Sites of Nonshivering Thermogenesis"). Nonhibernating species that show a significant response to noradrenaline while unacclimated, include the laboratory rat (142), dog (155), white mouse (67, 156), and guinea pig (113). The magnitude of the increase appears to be of the order of 50% of resting metabolism or less, although much higher responses have also been reported to occur in the mouse. Whether these wide differences represent strain of animal used, age, or environmental differences associated with housing or nutrition is not clear.

While the magnitude and significance of nonshivering thermogenesis in animals kept at thermoneutrality is debatable, most species do show a significant and often marked increase in nonshivering thermogenesis and in the amount of brown fat in the body during cold acclimation. Among nonhibernating species, this can represent twice the resting metabolic rate in mice (67) and rats (79); about one-half the resting metabolism in rabbits, guinea pigs (147), and dogs (155); about one-quarter in man (150) and rhesus monkey (152); and no response at all in sheep (153) and pig (113). However, as already indicated, the effectiveness of the acclimation procedures is not always certain.

The response to noradrenaline increases as the acclimation conditions become more severe and appears to stabilize, regardless of the acclimation conditions, after about 3 weeks, at least in rats (157–159). When the animals are returned to warm conditions the response declines to reach control levels after 3–4 weeks.

Improved thermogenic response to noradrenaline also results from repeated short exposures to cold in the rat (+5°C for 6–12 hr daily for 3 weeks) (159, 160), but the effect appears to depend on the total number of hours in the cold. However, in mice, the effect of exposures for only 2.5 hr to 0°C daily, for only a week, appears similar to that observed in chronically exposed mice, at least as judged by the development and ultrastructure of brown fat (161, 162). In addition, young pigs, which appear to lack brown fat, show a significant increase in metabolic rate (33–73%) in response to noradrenaline injection after 5 consecutive days of exposure to −7°C for only 2.5 hr. daily (146).

Even shorter term exposure to more severe cold (−20°C) for 10 min every few hours over 2 or 3 days improves cold resistance in mice (156) and rats (163, 164), but this does not appear to be associated with significant improvement in nonshivering thermogenesis. Similarly, exposure of near-naked human subjects to 5°C for 1 hr four to seven times over 2 weeks improves cold tolerance, but produces no obvious development of nonshivering thermogenesis (165).

The stimulus for development of the potential for nonshivering thermogenesis in response to continued cold exposure appears to lie, at least in part, in the central nervous system; prolonged and repetitive cooling of the preoptic-anterior hypothalamic area, or of the cervical region of the spinal cord, stimulates the development of nonshivering thermogenesis in rats (166, 167).

The effects of cold acclimation can also be simulated in rats by noradrenaline infused daily for 3 hr over 21 days (1.1 μg/min) (168) or injected daily for 45 days (300 μg/kg) (169) or by the specific β-agonist isoproterenol, injected daily for 20 days (300 μg/kg) (170). The mechanism of action is not clear, but is likely to be associated with increases in amount of brown fat or in its activity (171) (see under "Brown Fat") rather than with increased sensitivity of the sympathetic receptors to noradrenaline, as is often implied (170).

The effectiveness of interrupted or periodic cold exposure is not surprising since a free-living animal experiences fluctuating heat loss throughout the day; improved nonshivering thermogenesis has been recorded in white rats (95) and Norway rats (172) kept out-of-doors in winter.

Hibernators

Nonshivering thermogenesis is of special significance to the group of mammals, mostly of small size, including bats and some rodents, that are loosely called hibernators, which economize in their energy usage by allowing their body temperature to cool to near ambient temperature, even when this is close to freezing point (173–175). Metabolic rates are correspondingly lowered, and the resultant state of dormancy or torpidity can persist through periods of seasonal food or water shortage, or can occur over shorter periods, more or less on a daily basis, in small animals such as insectivorous bats that have a high energy requirement per unit of body weight while in the active state.

In general, these animals, whether cold or warm adapted, have much greater proportions of brown fat in their bodies than rigidly homeothermic species or age groups (135, 176), and their potential for nonshivering thermogenesis, as indicated by noradrenaline injection, is as high as that of the cold-acclimated rat (2 or 3 times resting metabolic rate). This has been observed in unacclimated hedgehogs (142), hamsters (148, 177), ground squirrels (178), and lemmings (179). Nevertheless, there is a significant nonshivering thermogenic response to cold acclimation, and cold-acclimated hamsters, for example, can maintain a normal body temperature at ambient temperatures as low as −40°C without shivering. However, the metabolic response to noradrenaline in these species is substantially less than in some species of hibernating bats in the normothermic phase (180, 181). Whether the increase in these animals is as high as 10 times the resting metabolism, as claimed by Haywood (180), needs confirmation with unanesthetized active bats.

The potential for nonshivering thermogenesis in hibernating species is particularly important during arousal from dormancy (79, 174, 182, 183) and appears to be the major, if not sole, means of heat production in the initial phases when active circulation and rewarming is confined to the body core (heart and central nervous system). At that stage, sympatholytic drugs that prevent nonshivering thermogenesis also prevent arousal. Shivering becomes more important when hypothalamic temperatures have increased and active circulation is returning to the whole body; at this stage sympatholytic drugs have little effect on rewarming. Estimates of the relative contribution of shivering and nonshivering thermogenesis to the heat of arousal vary according to the species and investigator. For example, in at least one species of bat about 80% of the heat of rewarming appears to be from nonshivering thermogenesis, since it rewarms rapidly even when shivering is blocked; in contrast, the hamster similarly treated arouses only very slowly, and only about 30% of the heat required to rewarm appears to derive from nonshivering thermogenesis (184). These differences have been related to the differing proportions of brown fat in the body.

Incomplete rewarming to flight temperature has been observed in fan-tailed bats (185) that have daily periods of torpidity, and successive incomplete warm-ups appear to reduce the subsequent ability to rewarm completely. This could be associated with the depletion of energy reserves for thermogenesis (see under "Effect of Substrate Limitation").

Relationship to Body Weight

Almost all of the data on nonshivering thermogenesis in hibernating or nonhibernating adult mammals derive from species with an adult weight of less than 5 kg. The few animals with substantially higher weights, sheep and man, are the only species with no, or insignificant, nonshivering thermogenesis. Heldmaier (135) has derived a highly significant inverse relationship between body size and capacity for nonshivering thermogenesis as indicated by the calorigenic response to noradrenaline. The regression relationship predicts that nonshivering thermogenesis would be insignificant in animals of more than 10 kg. This analysis indicates an interesting evolutionary situation. Animals with a surface to mass ratio larger than about 0.04 $m^2\ kg^{-1}$ have, in addition to shivering, a metabolic defense associated with the presence of brown fat for maintaining body temperature and activity at the cooler end of the thermal habitat. Presumably, the relative ability to shiver ($W\ kg^{-1}$ of body or muscle weight) does not increase sufficiently with a decline in body weight to allow thermal independence over an adequate range of environmental conditions, even though there is an inverse relationship between the frequency of shivering and body weight between a variety of species (186). In addition, the larger animals, which show minimum non-shivering thermogenesis and minimum brown fat as adults, do have brown fat as infants and show substantial nonshivering thermogenesis in infancy, so the potential for nonshivering appears to be lost as the need for it declines. However, the critical experiments on the reactivation by cold acclimation of nonshivering mechanisms and reappearance of brown fat in the larger species (sheep and man) have not been done.

BROWN FAT

Historical

Brown fat is a tissue with a marked capacity for producing heat (see under "Contribution to Nonshivering Thermogenesis") and has a wide distribution in mammals. The tissue was first described by anatomists several centuries ago and a variety of functions was proposed (for reviews see refs. 187–189). Speculation about its role was not settled until observations on the hypertrophy and changes in microscopic appearance that occurred in the tissue on acclimation of rats to cold led R. E. Smith and colleagues in the early 1960s to measure the oxygen consumption of brown fat in vitro, before and after rats had been acclimated to cold (87), and to make direct temperature comparisons between brown fat and the rest of the body in arousing marmots (88). From the increased metabolic rate and mass of the tissue it was calculated that heat production from brown fat would increase 3–6-fold in cold-acclimated rats (190). Direct evidence, based on temperature measurements, of the thermogenic capacity of brown fat has also been obtained in the newborn rabbit (191), guinea pig (137), rat (192), and coypu (193). This demonstration of nonshivering thermogenesis in brown fat provided an exciting new area for physiological and biochemical investigations, which have expanded enormously in the last 15 years, and have been the subject of several reviews (188, 194, 195). The purpose of this section is to concentrate on areas of confusion and new developments.

Identification

The anatomical distribution of brown fat throughout the body in various species was actively investigated in the 1920s (196, 197), but it is a confused field (198), because many investigators have been unaware of pitfalls in distinguishing white from brown fat, despite major differences in their physiology and biochemistry (199, 200). Confusion has also arisen over failure to recognize that there are major species differences in the ontogeny of adipose tissue (188) (see under "Ontogenic Changes: Conversion to White Fat").

The distinction between the two types of tissue cannot be made on the basis of color or gross anatomy, but rather on the morphology of their cell types. In contrast to the single large lipid locule that almost fills the normal white fat cell, brown adipocytes tend to be small, their cytoplasmic volume relative to their lipid content is large, the concentration of mitochondria is high, and usually there is a number of small lipid inclusions in the cytoplasm (188). Unfortunately, this last characteristic, as revealed by light microscopy, usually at low power, has been commonly used for distinguishing tissue types; multilocular cells are identified as brown fat, and unilocular cells as white fat (194). Evidence that this characteristic is fallible has been available for some years. For example, it is clear that the degree of repletion of the cell with lipid affects the intracellular disposition of the lipid (198). When the lipid content of white fat is low, such as during development of starvation, the cells become multilocular, whereas brown fat cells can become unilocular if the animal is particularly well fed, if the tissue is denervated, or if the cells are cultured in vitro. Nevertheless, it was widely accepted that in their typical forms brown fat cells are multilocular and white fat cells unilocular (198).

To overcome difficulties in distinguishing white from brown fat, it was suggested (188, 201) that the number of mitochondria in the cell and the mitochondrial morphology, as revealed by electron microscopy (199, 202, 203), would provide an infallible distinction between brown and white fat cells; and the use of this criterion has revealed normal situations in which a high proportion of brown fat cells appear unilocular under the light microscope. This appears to be typical of ruminants. For example, the perirenal adipose tissue of the newborn lamb appears, under the light microscope, to be a mixture of unilocular and multilocular cells, but electron microscopy reveals that they are all brown fat cells (Figure 9) (136). The situation can be even more misleading in the newborn calf, in which most brown fat cells appear unilocular (116) (Figure 9).

Extreme caution must, therefore, be used in identification of adipose tissue in novel situations or in species that have not been examined previously (204). For example, the suggestion that some adipose tissue depots in newborn guinea pigs contain both brown and white fat cells (114) is unacceptable without evidence based on electron microscopy. A great deal of work on the distribution of brown fat in man has not been subject to confirmatory tests with the use of electron microscopy. This may be difficult because of problems in obtaining the necessary fresh material. In addition, most samples have come from subjects who have died and cannot be regarded as normal (126, 205, 206); this is particularly so for elderly persons or babies who have died from hypothermia. The same objection can be made to an extensive survey for the presence of brown fat in 285 individual neonates representing 17 orders of mammals (108); in addition, some of the subjects were more than a few days old, so apparent absence of brown fat could have been due to rapid postnatal replacement of brown by white fat (see under "Ontogenic Changes: Conversion to White Fat"). Detailed anatomical studies of the distribution of brown fat in a small nonhibernator (deer mouse) and in a hibernating bat (207, 208) were also spoiled by the lack of

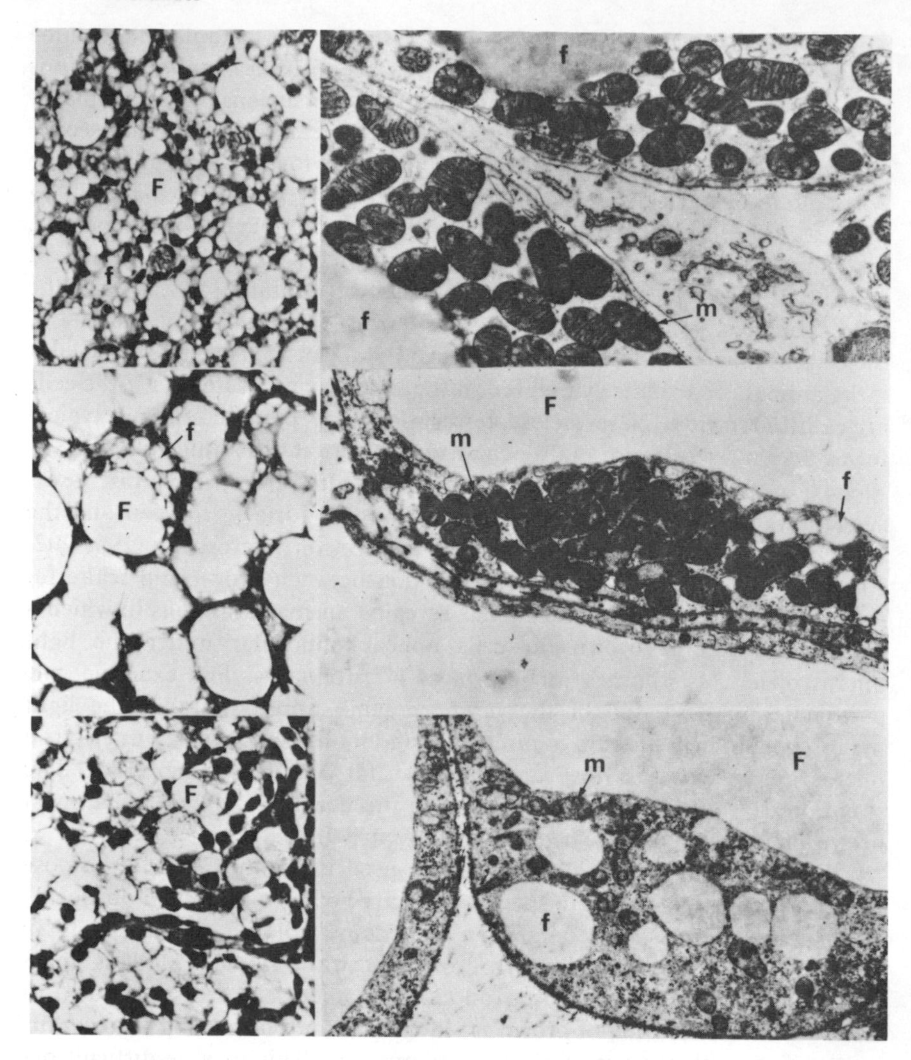

Figure 9. Light and electron micrographs of perirenal adipose tissue from newborn lamb (*top*), calf (*center*), and piglet (*bottom*). Light micrographs × 500, electron micrographs × 10,000 for lamb and calf and × 5,000 for piglets. Despite the presence of large fat locules (F) in many adipocytes in the lamb and calf adipose tissue, all the cells have the large numbers of mitochondria (m), with distinct cristae that are characteristic of brown adipose tissue. There are fewer small fat locules (f) in calf than in lamb adipocytes. In the piglet the mitochondria are sparse and lack cristae; the adipose tissue is clearly white. Photographs supplied by R. Gemmell.

histological confirmation, either from light or electron microscopy, that each of the depots described was indeed brown fat. The same criticism can be leveled at biochemical studies on adipose tissue glibly stated to be brown, for example, in the rhesus monkey, but without supporting evidence (132, 209). Although the conclusions in these various studies may well be correct, the stringent confirmatory evidence from electron microscopy is still required.

Distinction of brown from white fat may also be possible on a biochemical basis (210), but this may be complicated by biochemical differences between brown fat of various species (211).

Distribution

Smith and Roberts (190) placed particular emphasis on the relation of the sites of brown fat depots to the circulation. In the rat, the main sites are deep in the mid-dorsal superior cervical region, the interscapular region, and the dorsal region of the thorax and abdomen; thus, the resulting heat is applied directly to the thoracocervical regions of the spinal cord (which is peculiarly thermosensitive) to the heart, to the sympathetic chain, and to peripheral venous return from the superficial tissues. Perhaps because of these studies and the accessibility of the interscapular depot in rats and hibernators, many investigators have confined their attention to interscapular and cervical depots (188). These sites are widely regarded as the major and most important depots (211). However, 13 apparent depots located for efficient transfer of heat to blood have been described both in a hibernating bat and a small nonhibernator (deer mouse) (207, 208). In addition, in the newborn of some species, such as the sheep and the ox, there is little or no interscapular depot, and the major depot lies in the abdominal cavity, covering the lymph nodes as well as the kidneys and extending caudally from the diaphragm to the pelvis (116, 212). There is also a substantial but smaller depot in the prescapular-cervical region; this depot consists of a discrete mass around the prescapular lymph nodes and extends beneath the cervical trapezius, omotransversarius, and brachiocephalicus muscles; there is also a more diffuse mass extending posterially from the mid-cervical region along the major cervical vessels to the brachial plexus. There are smaller amounts in the orbit of the eye, on the pericardium, along the cardiac groove on the surface of the heart, around other lymph nodes, and along the major blood vessels. Indeed, in the newborn of these two species all of the fat is brown, with the exception of subcutaneous adipose tissue, which is white; this is usually present in small amounts only, and is virtually absent from the newborn lamb (136, 143). Since all of the internal fat in these species is brown, quantitation of their various brown fat depots is not difficult; 1.5–2% of their body weight is brown fat (Table 3). In other species, such as the newborn rabbit, there are adjacent lobes of brown and white fat, but at least 4% of body weight is brown fat (193). About 5% of body weight of newborn guinea pigs appears to be brown fat (114).

In contrast to lambs and calves, the newborn harp seal pup has a convoluted subcutaneous layer of brown fat (128). Heat production in this depot is inhibited by β-sympatholytics in the expected fashion, and a high thermogenic potential is maintained when the pup is paralyzed with a curariform drug. The hedgehog also has a layer of brown fat, apparently subcutaneously, over the back, but this appears to respond thermogenically to corticosteroids rather than noradrenaline (213).

Table 3. Average brown fat content of 10 lambs (mean body weight 3.5 kg) and 3 calves (mean body weight 33 kg)

	Weight of brown fat as % of body weight	
	Lamb	Calf
Perirenal-abdominal	0.54	0.56
Prescapular-cervical	0.36	0.13
Inguinal	0.14	0.18
Orbital	0.12	0.07
Lumbar	0.07	0.03
Pericardial	0.06	0.06
Costal	0.06	
Mesenteric	0.05	0.79
Coxal	0.03	
Sacral	0.03	
Popleteal	0.02	0.03
Total	1.48	1.85

From Alexander and Bell (212) and Alexander et al. (116), reproduced by permission of authors and publishers.

The physiological advantages in having major brown fat depots subcutaneously or intra-abdominally are not clear. However, the widely differing patterns of brown fat distribution warn against generalization from species to species.

Ontogenic Changes: Conversion to White Fat

It is controversial whether brown fat represents an early stage in the development of white fat. Much of the evidence is derived from inadequate morphological information provided by the light microscope (188). Brown fat first appears in human and sheep fetuses early in the second half of gestation (143, 214). In the sheep, at least, brown and white adipose tissue appears to originate from cells of similar ultrastructural appearance, but at separate sites. Lipid begins to accumulate in both at about day 70 of gestation, but thereafter there are major increases in the mitochondrial density of brown fat. During development, cells of both types may be unilocular or multilocular, which further demonstrates that the deposition of lipid within the cells is not a reliable index of the type of adipose tissue (see under "Identification"). Electron microscopy has also indicated that brown and white fat differentiate independently in the young rabbit (199). Thus, in the sheep and rabbit, brown fat does not seem to be identical with immature white fat, a possibility that was suggested from a study of rat brown fat cells (215). Immature brown fat cells in the hamster were also described as developing in a matrix of unilocular cells that had developed several days previously (216). However, in the absence of ultrastructural studies it is not possible to clarify apparent species differences in the development of brown fat.

The postnatal decline in nonshivering thermogenesis in newborn mammals (see under "Newborn Animals") is paralleled by replacement of brown fat with white fat (114, 199, 128, 136). In lambs, the rate of replacement varies from site to site, occurring earliest in cardiac and intermuscular fat; changes are first apparent near the periphery of the adipose tissue lobes (136). Hull (199) argues that this replacement may result from hyperplasia of white adipose tissue. This argument might apply to newborn rabbits, but it does not apply to newborn lambs or calves, in which the postnatal disappearance of brown fat represents a complete conversion of brown to white adipose tissue; during the period of conversion a continuous range of cell types between the typical multilocular adipose cell of the newborn and the typical white fat cell is seen under the electron microscope (Figure 10), and there are corresponding changes in the biochemical characteristics of the adipocytes (217). Degenerating brown fat cells are not found.

The rate of conversion varies between species and also depends on the degree of cold stress to which the neonate is exposed (218). It appears to be slower in the rabbit (219) than in the lamb (136), and is particularly rapid in calves, being well advanced within a week of birth (116). In warm conditions it is complete within several weeks, but in animals reared in the cold, brown fat may still be present after a month or more (114, 136). The mechanism of conversion is not understood, but is presumably related to the absence of stimulation of the tissue by cold.

Little is known about subsequent function or reactivation of areas of brown fat that have converted to white fat. In the rabbit, the adipose tissue lobes that were brown or white at birth are still anatomically distinguishable in adult life and the tissue retains the higher degree of vascularity characteristic of brown fat, as compared with white fat, as well as the marked blood flow response to noradrenaline and the ability to mobilize lipid (219, 220). Although the physiological significance of this is obscure, the finding could account for differences reported in the properties of white fat from different anatomical locations (221). In sheep and cattle it is not known how much of the internal white fat was originally brown fat, so the comparison is not readily made. Some degree of reconversion to brown fat appears to occur in the guinea pig during cold acclimation (114), but it does not seem to occur in the sheep (153). The changes in nature and distribution of the two forms of fat, in species such as laboratory rodents and hibernating animals that retain some brown fat in later life, have received remarkably little attention, especially since the ability of rats and mice to increase their nonshivering thermogenesis in response to cold acclimation declines after the animals are a few months old (222, 223). The decline has been associated with a progressive loss of brown fat cells that can be reactivated by cold acclimation (223). It appears that the loss may be due to a slow progressive conversion to white fat cells, but these changes require elucidation by electron microscopy.

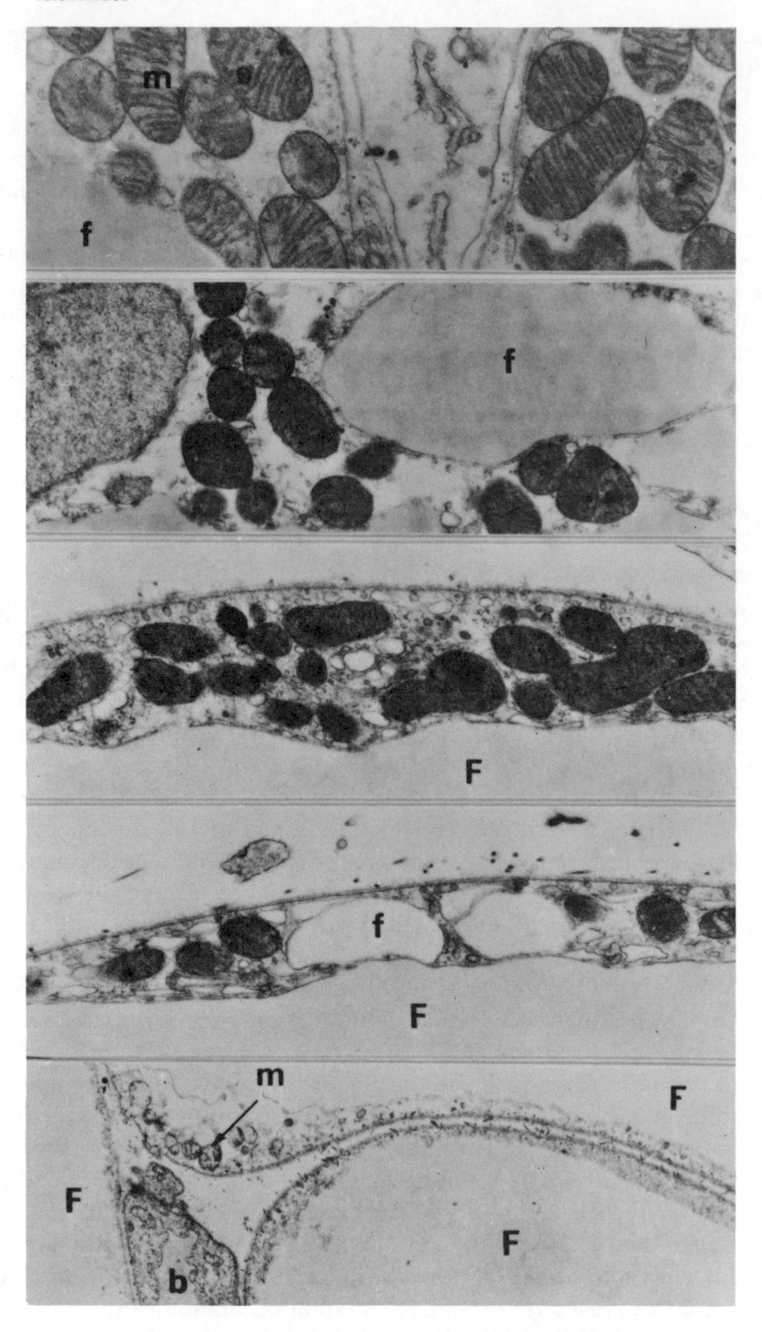

Figure 10. Electron micrographs of perirenal adipose tissue from lambs, showing stages in the conversion of brown to white adipose cells. From top to bottom: newborn lamb × 24,000, 2-day-old lamb × 19,000, 8-day-old lamb × 19,000, 8-day-old lamb × 23,000, and 32-day-old lamb × 5,400. The mitochondria (m) decrease in number and lose their distinct cristae, the cytoplasm thins out, and the fat droplets become very large. F, large fat locule; f, small fat locule; b, blood vessel. Adapted from Gemmell et al. (136) by permission of the authors and publisher (The Wistar Press).

Factors Affecting the Amount of Brown Fat in the Body

The proportion of brown fat in the body of species that retain this tissue beyond infancy shows significant seasonal fluctuations as well as changes associated with acclimation to heat or cold in the laboratory (146, 188). Cold exposure for several weeks induces increases of severalfold in the enzyme content per unit of fat-free weight and thermogenic capacity of the tissue, but the responses vary markedly from species to species (224, 225) and represent only part of an increase in the animal's tolerance to cold. The mechanisms that control the changes in brown fat are incompletely known, and it has not been possible to duplicate the effect in rats by administration of the various hormones, such as thyroid hormones and catecholamines, that are known to increase the body's content of brown fat (170, 226). The sympathetic nervous system appears to be involved, but denervation does not completely abolish the changes in brown fat metabolism occurring in cold acclimation (227). A pineal hormone, melatonin, appears to be involved in promoting seasonal increases in the brown fat content of hamsters and mice (228, 229).

In the rat, during cold acclimation and neonatal development, it appears that new brown fat cells arise from reticuloendothelial progenitor cells, and not by proliferation of existing brown fat cells (230-232). In warm conditions, the adipose cells of the brown fat tend to become unilocular (231) but retain a high density of mitochondria; during cold acclimation the cells all become multilocular, the mitochondria hypertrophy, their cristae restructure, and their oxidative capacity increases (171, 233-235). Two types of brown adipose cells with differing ultrastructural appearance, particularly of the mitochondria, have been observed in rats, but their significance is unknown (236).

Factors affecting the brown fat content of the neonate have, by contrast, been largely neglected. However, changes in the weight of brown fat can be induced by manipulation of the diet of the pregnant mother. Short periods of starvation toward the end of gestation increase the lipid content of brown fat in newborn rabbits with a concomitant increase in thermogenic capacity (237) and exposure of pregnant rats to cold increases the proportion of brown fat and its lipase activity in their newborn (238). The adenyl cyclase activity of brown fat of newborn rats can also be increased by feeding the pregnant mothers with a high linoleic acid diet (239). By contrast, experimentally induced intrauterine growth retardation in rats and prolonged maternal undernutrition, which inhibits fetal growth in sheep, greatly reduce the proportion of brown fat in the newborn (73, 240).

Contribution to Nonshivering Thermogenesis

Estimation of the proportion of nonshivering thermogenesis that arises from heat production by brown fat, as opposed to other organs and tissues, has been approached in a number of ways. In an elegantly executed approach by Heim and Hull (241) with the use of newborn rabbits, the blood flow through a portion of the cervical brown adipose tissue, estimated to represent one-quarter of the total brown fat, was measured directly by observing flow rate from a catheter in the vessel carrying most of the venous drainage, and oxygen content of the venous and arterial blood was determined: During noradrenaline infusion, the flow increased from 90 to 360 ml/100 g/min and the venous oxygen saturation fell from a mean of 43% to 18%; the oxygen consumption of the tissues was calculated to rise from 9 to a maximum of 60 ml/min/100 g of tissue. From a knowledge of the maximum oxygen consumption of the rabbit in the cold, Heim and Hull estimated that brown fat accounted for more than two-thirds of the extra oxygen consumed by the young rabbit in response to cold; it was assumed that the arteriovenous difference in oxygen content of the blood was the same in the cold as in stimulation with noradrenaline. In view of the possible effects of surgery and anesthesia, the possibility that the brown fat content of the rabbit was underestimated, and the fact that venous drainage was incomplètely collected, it is likely that the estimate should be higher than two-thirds. The experiment did not rule out the possibility that there was some nonshivering thermogenesis in other tissues.

Calculations indicating that brown fat is the major source of nonshivering thermogenesis can also be made for newborn lambs (212) on the assumption that the oxygen saturation of the mixed venous blood, often as low as 14% (242) in the vena cava posterior to the renal veins, is representative of that in the venous blood draining from the brown fat. The blood flow through brown fat of lambs in summit metabolism is approximately 800 ml/100 g/min, as estimated by the use of radioactive microspheres (48) (see under "Circulatory Adjustments During Cold Thermogenesis"), and the proportion of brown fat in lambs is approximately 1.5% (212). From these figures it is readily calculated that brown fat would be consuming 110–120 ml of O_2/100 g/min during summit metabolism. This would account for the whole of the proportion of summit metabolism attributable to nonshivering thermogenesis (40% or 70 ml of O_2/min in a 4-kg lamb) (62).

In the newborn rabbit, excision of 60–85% of the brown adipose tissue reduced the increase in oxygen consumption due to cold or noradrenaline infusion by about 80%, much higher than the 20–30% associated with major surgery alone (243); this is consistent with an important thermogenic role for brown fat in this animal.

In contrast with these experiments on newborn animals, those on the cold-adapted rat have been controversial, and it has been questioned whether a tissue that comprises only 1% of body weight could account for a large proportion of nonshivering thermogenesis (244). Janský and Hart (245) calculated that the contribution of brown fat to cold thermogenesis in these animals was unlikely to represent more than 6% of the total increase in metabolism. This estimate was based on the cold-induced increase in blood flow, estimated from the regional distribution of radioactive rubidium chloride (246) and assuming that oxygen was completely removed from the blood by the tissues both at 30°C and 9°C. Similar low values were also obtained in rats and arousing ground squirrels by other authors (247, 248) who used Sapirstein's method. However, Sapirstein's method appears to grossly underestimate blood flow in adipose tissue (see under "Circulatory Adjustments During Cold Thermogenesis"). Experiments based on the use of radioactive microspheres for blood flow measurements, and on the measurement of arteriovenous difference in oxygen content of blood flowing through the interscapular brown fat of cold-acclimated rats (249), indicate that the major masses of brown fat could account for 60% of the calorigenic response to noradrenaline.

Attempts to estimate the contribution of brown fat to nonshivering thermogenesis induced by noradrenaline in cold-adapted rats and mice have also been made by excision of the interscapular or other depots. The results have been extraordinarily variable, ranging from little or no immediate effect (250), through no immediate effect but a marked subsequent reduction of 40% (251), to an immediate reduction of 40% (252); reductions as high as 60% a week after surgery have also been reported (253). Attempts to explain these discrepancies in terms of nutrition, anesthesia, and acclimation temperatures have failed (250). However, age may be implicated, since young, cold-acclimated rats are more affected by brown fat removal than older rats (254), and the acclimation response of brown fat declines with age (223).

These observed declines in the response to noradrenaline are usually regarded as too large to be accounted for by the removal of the thermogenic potential in the form of brown fat (250), but the capacity of this tissue has clearly been underestimated due to unreliable estimates of blood flow in stimulated brown fat (see under "Circulatory Adjustments During Cold Thermogenesis"). The large reduction in thermogenesis, together with its sometimes delayed appearance, prompted a number of investigators to postulate that the interscapular depot has an endocrine-like function, producing a substance that facilitates nonshivering thermogenesis elsewhere in the body (251, 252, 254, 255), but stringent evidence is lacking. It now seems that the large decline simply represents removal of active thermogenic tissue; reasons for the delayed decline remain to be investigated, but temporary compensation by other brown fat depots may be involved (249).

Estimation of heat production of brown fat from in vitro measures of oxygen consumption of tissue slices, incubated under optimal conditions and stimulated with noradrenaline, have given much lower values than more physiological methods. In the newborn rabbit, for example, the maximum in vitro oxygen consumption of brown fat was about 20 ml of O_2/100 g/min compared with 60 ml of O_2/100 g/min, as measured in vivo (90, 256). A similar in vitro value was derived for the cold-acclimated rat (24 ml/100 g/min) (190) and warm-acclimated hamster (also 24 ml/100 g/min) (257), but this would account for less than 10% of the estimated level of nonshivering thermogenesis (about 8 ml of O_2/min) in cold-exposed, cold-adapted rats

(245) and only a small fraction of the heat required for arousal of the hibernating hamster (257). Likewise, an attempt to obtain realistic values in vitro in a hibernating bat was unsuccessful (258).

Thus, physiological methods show that brown adipose tissue, which represents only a small proportion of body weight, supplies the major part of nonshivering thermogenesis in newborn mammals and cold-acclimated rodents and will doubtless prove to do so in arousing hibernators when suitable blood flow methods are used. The special role in warming vital regions of the body (114, 190, 207, 208) also endows the tissue with an importance disproportionate to its small size.

Mechanisms of Thermogenesis in Brown Fat

The histology of brown fat, in contrast to white fat, is that of a tissue well suited to the oxidation of lipids, substrates that have a high energy content per carbon atom (259). The tissue is highly vascular (188), the disposition of the lipid is usually multilocular, so there is a high total surface area presented to the cytoplasm (260), and there are numerous mitochondria, the main sites of oxidative metabolism in tissue (261), in contact with the lipid locules (202). In contrast to white fat, brown fat rapidly loses lipid during exposure of the animal to cold (198), and there is a large amount of direct biochemical evidence that rapid utilization of lipid by oxidation of fatty acids is responsible for thermogenesis in brown fat (259, 262–265). Indeed, fatty acids appear to be essential for thermogenesis in brown fat (266).

A fully documented account of the biochemical mechanisms that are responsible for the mitochondrial oxidation of fatty acids and control of the rate of oxidation is beyond the scope of this chapter; for detailed documentation the reader is referred elsewhere (259, 265, 267–75).

The utilization of triglycerides of brown fat largely follows a conventional pathway (Figure 11). They are first hydrolyzed by a lipase that is activated by cyclic AMP in a process that requires ATP and activated adenyl cyclase. The resultant fatty acids enter the mitochondria by one of two pathways. They may be first activated by a thiokinase in the presence of ATP to fatty acyl CoA esters that cannot pass through the mitochondrial membrane without first reacting with carnitine to form fatty acyl carnitine; reconversion to fatty acyl CoA esters occurs within the mitochondrion. This carnitine-dependent pathway is the major of the two. Fatty acids may also enter the mitochondria unchanged, via the slower, carnitine-independent pathway, and be activated by thiokinase within the mitochondrion. The activated fatty acids are then degraded by β oxidation, a cyclic process that also requires ATP, into acetyl CoA (two-carbon units). Acetyl CoA enters the tricarboxylic acid cycle to yield CO_2, ATP or GTP, and energy-rich protons in the form of NADH; entry of acetyl CoA into the citric acid cycle appears to be regulated by the level of NADH (276). However, the citric acid cycle is unable to function at low temperatures, and in hypothermic hibernators the acetate pathway, which is temperature insensitive, may function in place of

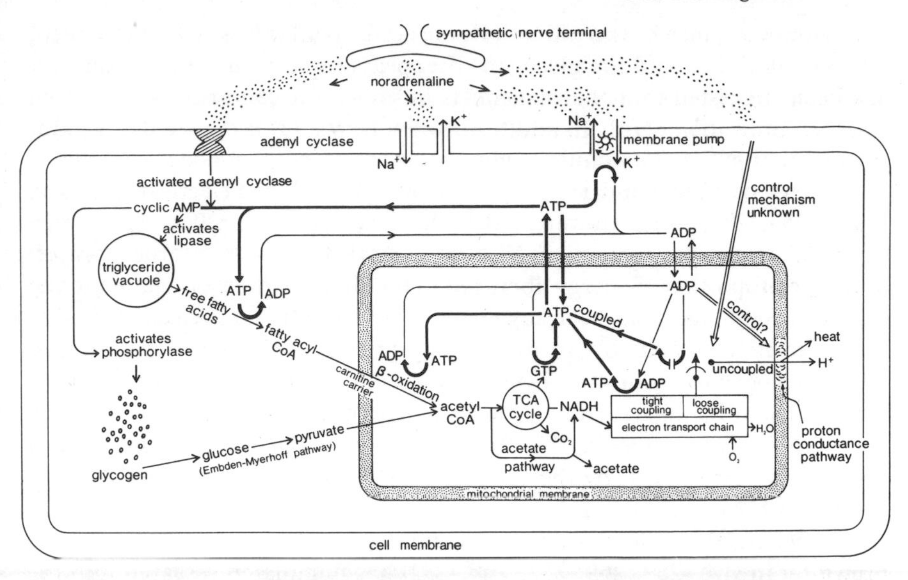

Figure 11. Schematic representation of likely mechanisms of thermogenesis in brown adipocytes (see text). Noradrenaline released from the sympathetic nerve terminals affects properties of the cell membranes. First, the activity of adenyl cyclase is increased, stimulating lipolysis and availability of free fatty acids for oxidation. Second, the permeability of the membrane to small ions is increased, and, third, the activity of the ouabain-sensitive Na^+/K^+ pump is increased and acts to maintain normal ionic concentrations. Thus, the utilization of energy (ATP) by the cell is increased, so substrate oxidation accelerates. Another possibility is that energy-rich proteins from the electron transport chain are short circuited through a specific protein conductance pathway, liberating heat in the process, but the control mechanisms are poorly understood. The electron transport chain is associated with the inner mitochondrial membrane, but for ease of illustration the chain is shown within the mitochondrial matrix.

the citric acid cycle to allow fatty acid oxidation to proceed (277). The energy-rich protons finally enter the electron transport chain in the inner mitochondrial membrane, and in normal mitochondria, from liver or muscle, for example, the final products are water and energy mainly in the form of ATP but also as heat (259, 272). The utilization of substrate is largely governed by the rate of production of ATP in the electron transport chain. The ATP can readily leave the mitochondria and is available for reactions that require energy, such as hormone-induced lipolysis and fatty acid activation in the cytoplasm. Normally about 90% of the ATP produced by animal cells arises from oxidation in the electron transport pathway. Additional high energy bonds are available in the form of GTP produced in the tricarboxylic acid cycle (substrate level phosphorylation) within the mitochondria, and these bonds are available for activation of fatty acids that enter by the carnitine-independent pathway. ATP is also produced in the cytoplasm by glycolysis through the Embden-Meyerhof pathway from glycogen, which is a characteristic component of brown fat, but the energy produced by this source is usually small in comparison to that produced by fatty acid oxidation. However, since brown adipose tissue of neonatal rats is rich in glycogen, but poor in lipids, and the glycogen disappears rapidly on cold exposure, it has been suggested that thermogenesis in brown fat can arise largely from glycogen utilization (278). In addition, when brown fat is depleted of lipid it can utilize significant amounts of glucose drawn from the bloodstream (279).

Brown fat mitochondria also possess a well developed electron transport chain (259, 280, 281). However, the chain appears to be less tightly tied (coupled) to the production of ATP from ADP than that in other tissues. Since the proportion of energy liberated in the chain is inversely related to the degree of coupling, it has been suggested that brown fat is especially suited to the production of heat by an uncoupling process such as can be induced pharmacologically by dinitrophenol (267). Under normal resting conditions the mitochondria appear to be at least partly coupled, especially at the first site of ATP production in the chain (267), and the rate of oxidation is limited, at least in part, by the supply of ADP, and also by the low ATPase content of brown fat mitochondria.

A debate about the degree of uncoupling in various states of activity of brown fat in vivo (259, 269, 272, 268, 282) has dominated research into the biochemical mechanisms of brown fat thermogenesis for 10 years. In numerous in vitro studies with isolated mitochondria from brown fat, the electron transport chain has been shown to be uncoupled by high concentrations of free fatty acids, such as might be produced when brown adipose tissue is stimulated with noradrenaline (271); this property is not shown by mitochondria from other tissues.

A likely mechanism for uncoupling in vivo has been described recently (275, 283). The inner mitochondrial membrane contains a specific protein with a proton conductance pathway that allows protons to leave the electron transport chain, pass through the membrane, and dissipate their energy as heat (Figure 11). The concentration of the protein shows adaptive changes that correlate with the ability of the brown fat to produce heat. The pathway is inhibited in vitro by purine nucleotides (ADP and GDP), and their inhibition leads to restoration of the protein electrochemical gradient across the membrane; recoupling and inhibition of thermogenesis result. However, neither the in vivo method of inhibiting the pathway nor the link between noradrenaline receptors on the cell membrane and purine nucleotide receptors in the inner mitochondrial membrane are understood.

Other schools have tended to regard the partially uncoupled state of isolated mitochondria as a result of the preparative procedure, and explain the thermogenesis in terms of accelerated ATP utilization (270). Two possible heat production cycles involving ATP are a triglyceride re-esterification cycle and a cycle involving fatty acid breakdown and resynthesis, but these do not appear to be involved to any significant extent (259). On the other hand the mechanism could involve increased activity of the Na^+/K^+ membrane pump (270, 272) associated with an increased permeability of the adipocytes to small ions (Figure 11). The increased membrane permeability allows ions such as Na^+ and K^+ to move down their electrochemical gradients and increase or decrease their intercellular concentration, so an increased use of energy to restore and maintain the original concentration is required in the form of increased pump activity, which necessitates increased use of ATP. The increased production of phosphate acceptor (ADP) could then play a regulatory role via the electron transport chain in the accelerated fatty acid oxidation, which is the essential feature of brown fat thermogenesis.

In vitro studies, with the pump activity of brown adipocytes blocked with ouabain, indicate that perhaps half of the noradrenaline-induced thermogenesis derives from the Na^+/K^+ pump activity (270), but the assumptions on which this evidence is based have been challenged (272).

Although the mechanism of initiation of thermogenesis is still unsettled, sympathetic nerve stimulation or treatment with noradrenaline depolarizes adipocyte membranes, increases their permeability to small ions, appears to stimulate the Na^+/K^+ pump (270), and activates membrane-bound adenyl cyclase (284). The activation of adenyl cyclase stimulates cyclic AMP synthesis and hence lypolysis, thus increasing the amount of free fatty acids available for oxidation. Protection of cyclic AMP from enzymic degradation by means of theophylline has a similar effect without depolarizing the cell membrane.

The activating effects on adenyl cyclase (Figure 11) are characteristic of those associated with a β-adrenergic response; but both α and β receptors appear to be associated with the membrane depolarization (285, 286) so it is possible that different adrenergic receptors activate different parts of the thermogenic mechanism.

Significant heat production in peroxisomes has also been suggested (287, 288). These organelles are smaller than mitochondria, are found in a variety of tissues, including liver and brown fat, and their numbers increase during cold acclimation. It has been proposed that they oxidize substrate with the direct liberation of heat (H. T. Hammel, cited in ref. 288), but their mechanisms and quantitative contribution to brown fat metabolism are not known.

OTHER SITES OF NONSHIVERING THERMOGENESIS

Ever since Depocas in 1958 (157) showed that the curarized cold-acclimated rat with an occluded blood supply to the viscera could respond to cold by a substantial increase in metabolic rate, thermal physiologists who work with cold-acclimated rats have been preoccupied with skeletal muscle as a source of nonshivering thermogenesis (272). Interest in muscle waned with the subsequent recognition of the thermogenic potential of brown fat (see under "Historical"), but waxed when estimates of blood flow through brown fat by Sapirstein's method (see under "Neural Control of Cold Thermogenesis") indicated that brown fat could account for only 10% of the potential for nonshivering thermogenesis in cold-adapted rats (245, 247) and when experiments with excision of brown fat led various workers to postulate an indirect endocrine role for brown fat in nonshivering thermogenesis (251) (see under, "Contribution to Nonshivering Thermogenesis"). However, recently Foster and Frydman (249) showed that earlier estimates of blood flow through brown fat, and its oxygen consumption calculated from blood flow and (a − v) oxygen differences, were grossly low, thereby reinstating brown fat as the site of most (>60%) nonshivering thermogenesis in cold-adapted rats; their flow measurements also indicated that rat skeletal muscle accounted for less than 12% of the calorigenic response to noradrenaline, even if muscle used all the oxygen in the blood flowing through it. Nevertheless, a number of experiments have indicated involvement of tissues other than brown fat in nonshivering thermogenesis.

Skeletal muscle was implicated in experiments with cold-adapted rats in which leg muscle, denervated or curarized to prevent contraction, was prefused with blood of known substrate composition at a constant rate (289–291); muscle oxygen consumption was doubled by noradrenaline or adrenaline infusion. The effects could not have been due to altered blood flow or exogenous substrate availability. A similar effect, due to cold exposure, was obtained with leg muscle of cold-acclimated dogs, but not with warm-acclimated dogs (92). In addition, noradrenaline and adrenaline stimulate oxygen consumption of dog hind limb (292) and human forearm (293), both in situ, and α-adrenergic agonists such as phenylephrine induce marked thermogenesis in striated muscle of the Pietrain breed of pig (294).

There are also indirect indications of nonshivering thermogenesis in skeletal muscle. For example, administration of noradrenaline to hamsters is followed by depolarization of cell membranes of skeletal muscle in much the same way that occurs with activation of thermogenesis in brown fat (295). There is also an increased turnover of mitochondrial proteins and major alterations in mitochondrial protein metabolism (296) in skeletal muscle as well as in brown fat when rats are cold acclimated (297).

Ultrastructural and biochemical studies of muscle mitochondria in rats have indicated changes after cold acclimation that could be associated with an enhanced calorigenic action of noradrenaline (298), but convincing proof was not obtained. In harp seals, ultrastructural studies of skeletal muscle and caval-sphincter muscle, which show mitochondrial aggregations and lipid droplets in dark fibers both in adults and pups, have been taken to indicate an adaptation for thermogenesis, comparable to that of brown fat (299, 300), but again functional proof is lacking. Similarly, experiments with hamsters affected with muscular dystrophy and showing a subnormal calorigenic response to β-adrenergic stimulation with isoproterenol (301) cannot be regarded as definitive, because no account was taken of the calorigenic response by brown fat.

There appears to be no evidence about the relative ability of different muscle types (red versus white) to increase oxygen consumption without contraction. On the other hand, it is well established that oxygen consumption of cardiac muscle is substantially increased by noradrenaline, but this is attributed to alterations in the intrinsic speed of contraction rather than to biochemical uncoupling of energy at the mitochondrial level (302).

Nonshivering thermogenesis has been shown to occur in liver by a variety of experimental methods, including liver perfusion (303), estimation of liver blood flow from clearance rate of injected radioactive chromic phosphate in curarized cold-exposed rats (304), or dilution of infused para-amino hippuric acid in cold-exposed sheep (305); also, a heated thermocouple technique was used to measure liver blood flow and metabolic heat production in rats at various ambient temperatures (93). In sheep, for example, the liver contributes about 14% of resting metabolism, but when the oxygen consumption of the animal is doubled by cold exposure, the oxygen consumption of the liver increases by only 36% and the portal-drained viscera (gut, spleen, pancreas, and omental fat) show no increase (305). The increased heat production in the liver of rats and sheep may be an indirect result of increased blood flow and oxygen and substrate supply in the blood (79, 305, 306). Noradrenaline has no effect on the heat production in the isolated liver (303), and the inhibitory effects of β-sympatholytic blockade (93) could result from a reduction in substrate supply, especially from adipose tissue (see under "Substrates for Cold Thermogenesis"). In addition, cold acclimation of rats is not accompanied by an increased turnover of mitochondrial proteins in liver, as it is in brown fat and skeletal muscle (297). There is no evidence that the liver is directly involved in cold acclimation (307), although this organ is indirectly involved in metabolic responses to chronic, as well as acute, cold exposure (308).

The contribution of increased heat production in liver to thermogenesis during severe cold exposure is uncertain, especially in view of the halving of portal flow, measured by microspheres, that occurs in young lambs exposed to conditions that produce summit metabolism; significant changes in liver arterial flow are not seen (48).

The kidneys appear to have been eliminated as contributors to nonshivering thermogenesis by experiments with cold-exposed, cold-acclimated rats with a kidney perfused extracorporally (289); as with liver there is no increased turnover of mitochondrial proteins during cold acclimation (297).

Implication of the alimentary canal in nonshivering thermogenesis (79) cannot be accepted, because the calculations were based on intestinal blood flow measured by Sapirstein's method; blood flow in these organs, and also in the kidney, as measured with microspheres, declines markedly in lambs during summit metabolism (48).

There is little evidence about whether or not white fat has any potential for nonshivering thermogenesis. The heat production of white fat increases severalfold in vitro after administration of adrenaline (309–311); and there have been suggestions that invite investigation into whether or not white fat could make a significant contribution to nonshivering thermogenesis (312, 313); in particular, the number of white adipocytes and their responsiveness to noradrenaline increase during cold acclimation in young rats (314).

Estimates of the contribution of a variety of tissues to nonshivering thermogenesis have also been made on the assumption that the potential for oxidative metabolism of a tissue is reflected by the total activity of its cytochrome oxidase, the terminal oxidative enzyme (79, 307, 315). These findings focused attention on the carcass (skeletal muscle) as the major contributer (63%) to nonshivering thermogenesis, with liver next in importance (23%). The finding that the activity in brown fat could account for up to three-quarters of the total nonshivering thermogenesis was discarded, presumably because it did not conform with estimates based on blood flow measurements (79). Despite the evidence for the presence of nonshivering thermogenesis in organs other than brown fat, such as the muscle and liver of animals that display a marked thermogenic response to noradrenaline, the magnitude of their contribution in a normal intact unanesthetized animal under noradrenaline stimulation or cold stress, particularly severe cold stress, remains to be defined.

The evidence for existence of nonshivering thermogenesis in warm- or cold-acclimated animals that lack brown fat, such as piglets (113), adult sheep (153), or chickens (110) (see under "Occurrence of Nonshivering Thermogenesis") is unsettled and appears to rest either on a visually or electromyographically assessed absence or decline in shivering during acclimation while metabolic rate increases (113, 153, 316, 317). Curarization would seem desirable to eliminate shivering thermogenesis. These animals show poor, if any, thermogenic response to noradrenaline (113, 153, 154, 318), although a calorigenic effect of adrenaline potentiated by thyroxine (see under "Thyroid Hormones") has been demonstrated in piglets (318), and thermogenesis in both sheep (153) and chickens (154), but not in piglets (113), is reduced by β-sympatholytic blockade. No discrete site for the production of thermoregulatory heat has yet been described in any of these species, but the magnitude, indeed existence, of nonshivering thermogenesis in them remains in question. However, an increase in resting metabolic rate during cold acclimatization is established for domestic ruminants (319).

Suggested mechanisms of nonshivering thermogenesis in tissues other than brown fat include substrate oxidation in nonmitochondrial organelles, the peroxisomes (see under "Mechanisms of Thermogenesis in Brown Fat"), and the operation of energy-wasting or futile cycles (272) involving pathways operating in opposite directions, e.g., pyruvate → oxaloacetate → phosphoenolpyruvate → pyruvate, triglycerides → fatty acids + glycerol → triglycerides, fructose 6-phosphate → fructose 1,6-diphosphate → fructose 6-phosphate (320–322). While the fructose phosphate cycle serves to prewarm flight muscles in insects (323), futile cycles have not been shown to have thermogenic significance in mammals or birds.

SYMPATHETIC CONTROL OF NONSHIVERING THERMOGENESIS

Brown Fat

The innervation of brown fat appears to carry both sympathetic and somatic components (188, 324, 325). There is no evidence of involvement of the parasympathetic system, although this is involved with other aspects of cold adaptation (326). The neuroanatomy of brown fat, especially of depots other than the interscapular and cervical, has received little attention, and very little is known about the supply in species such as ruminants, in which the interscapular pad is absent and the major depot is perirenal-abdominal. In the lamb, the perirenal brown fat receives many nerve fibers from a large ganglion in the sympathetic chain near the 3rd lumbar vertebra (Armati-Gulson and Alexander, unpublished data). In the rat and hibernating species, the cervical and interscapular depots are supplied by cervical spinal nerves, cranial nerves X (vagus) and XII (hypoglossal), and nerves from the sympathetic chain; fibers innervating the depots in the thorax and abdominal cavity are likely to be found in intercostal and visceral nerves. In the rat, the major sympathetic supply to the interscapular fat lies in a mixed nerve that also sends branches to the skin above the fat pad, and may include sensory fibers from both the cervical and interscapular fat (327). A similar distribution has been described for young rabbits (328). Light and electron microscopy show that nerve fibers run along both the major and minor brown fat vessels and between the brown fat cells (194, 329) and that these fibers are anatomically heterogenous, with no distinct fiber type being associated with brown adipocytes or their blood vessels (330). However, much of the evidence now available about brown fat innervation is based on visualization of the sympathetic structures by means of fluorescent microscopy of rat adipose tissue; histological sections are treated to convert the noradrenaline, associated with the sympathetic nerves, into quinoidal structures that fluoresce strongly under ultraviolet light (331). Fluorescent fibers are found along the arterial blood vessels, but not capillaries, and as a fine network about the fat cells, with the catecholamines apparently confined to the nerves (324, 325, 332, 333). Typically, the strands contain strongly fluorescent varicosities, presumably areas of catecholamine synthesis or release. White adipocytes do not have the same degree of innervation as brown adipocytes (308), although the arterial blood supply of the tissue is well innervated with fibers that fluoresce (200). Fluorescence is much brighter in the tissue of cold-acclimated rats than it is in warm-acclimated rats (324), and the degree of fluorescence increases in neonatal rats along with their increasing capacity for nonshivering thermogenesis (334). The same pattern of fluorescence is seen in brown fat of species other than the rat, such as ground squirrels (335) and lambs (refs. 211 and 336 and G. Alexander and D. Stevens, unpublished data) (Figure 12).

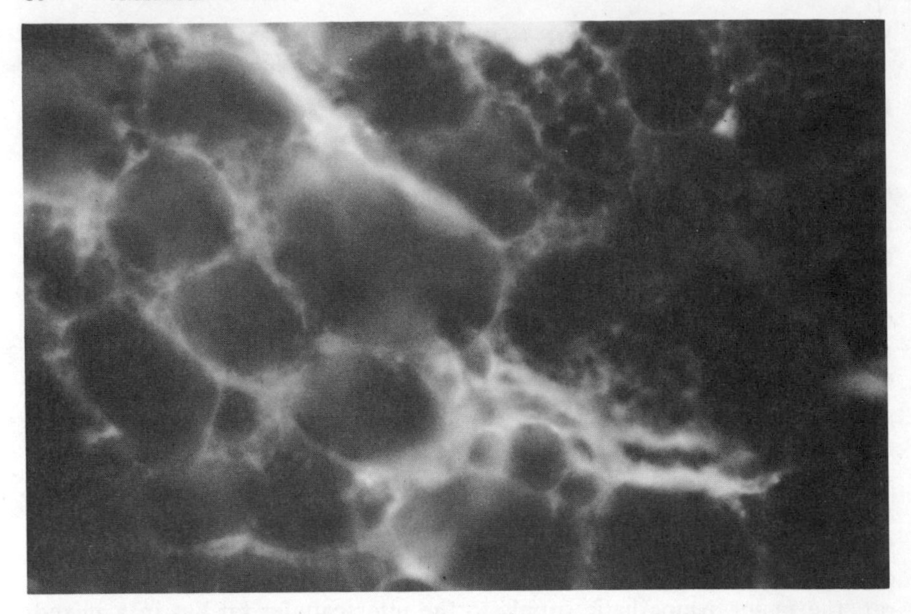

Figure 12. Catecholamine-containing nerves in perirenal adipose tissue of a 144-day fetal lamb. The catecholamines have been converted into quinoidal structures that fluoresce strongly under ultraviolet light. These nerves are found along the arterial blood vessels and between fat cells. × 600. Photograph supplied by D. Stevens.

Surgical denervation and immunosympathectomy indicate that the sympathetic innervation of the blood vessels and of the parenchyma tissue of brown fat is of two types (332). These treatments result in massive loss of fluorescence from the blood vessels, but not from the parenchyma. In addition, fluorescent structures with the appearance of ganglia within the brown fat are resistant to sympathectomy (332). Since destruction of ganglia causes postganglionic nerve fluorescence to vanish (337), it is possible that the parenchymal nerve supply is derived from ganglia within the brown adipose tissue by way of short adrenergic nerves, such as those seen in some reproductive organs (325), and that blood vessel innervation is of the conventional type, derived from the sympathetic chain by way of long adrenergic neurones. The proposal that sympathetic ganglia are present in rat brown fat is also supported by functional evidence, which is based on the fact that, in vitro, both noradrenaline and the nicotinic ganglionic stimulant dimethylphenyl piperazinium iodide invoke similar biochemical responses (338). However, this scheme for the innervation of brown fat parenchyma has been recently challenged, both on the grounds that brightly fluorescent structures in the parenchyma are not ganglia or ganglia-like cell groups, and that the nicotinic ganglion stimulant is not as specific as supposed (339).

Specialized cell contacts between fine naked axons and brown fat cells have been identified by electron microscopy (340), and most of the noradrenaline of sympathetic nerves appears to be in the nerve ending and is associated with dense cored or granular vesicles (storage granules) (324). Close junctions between adjacent adipocytes have also been observed; their function may be to ensure an even distribution of adrenergic stimuli throughout the adipose tissue (340). Brown fat of lambs is unusually rich in dopamine-containing mast cells, but the functional significance of this is unknown (336).

In addition to anatomical evidence and evidence based on the high noradrenaline content of brown fat, there is considerable functional evidence that activity of brown fat is under control of the sympathetic nervous system (324, 325, 341). Noradrenaline turnover in brown fat increases during cold acclimation of rats (342), and administration of noradrenaline to newborn rabbits, for example, stimulates thermogenesis and blood flow in brown fat and release of free fatty acids and glycerol into the circulation from this tissue (see under "Tissue Substrate Stores") (241, 279). Section of the cervical sympathetic trunk is followed by retention of lipid in brown fat of young rabbits during cold exposure (343). In addition, chemical sympathectomy with 6-hydroxydopamine, which destroys the nerve terminals both on the blood vessels and parenchyma of brown fat (339), in contrast with the differential effect of immunosympathectomy, drastically reduces nonshivering thermogenesis in cold-adapted rats (344) and in newborn lambs treated prenatally with 6-hydroxydopamine (G. Alexander and D. Stevens, unpublished data). Evidence of a more direct nature is provided by the increase in temperature of brown fat that follows electrical stimulation of its sympathetic nerve supply (343). Stimulation of the mixed nerve to interscapular brown fat of cold-acclimated rats leads to a membrane depolarization of brown adipocytes, similar to the depolarization resulting from noradrenaline injection (see under "Mechanisms of Thermogenesis in Brown Fat") (345). In addition, ganglionic blocking drugs, such as hexamethonium, and the so-called β-sympatholytic drugs, such as propranalol and pronethalol, which block the action of catecholamine transmitters at the sympathetic nerve terminals, inhibit the increased blood flow and thermogenesis in brown fat, as indicated by failure of the temperature or oxygen consumption of brown fat to rise either during cold exposure or under stimulation with noradrenaline (104, 256, 346).

The existence of a direct action of catecholamines or of the sympathetic nerves on blood flow through brown fat has been questioned (220, 256). However, the increased flow associated with noradrenaline administration does not appear to be an indirect result of increased oxygen demand by the tissue, since it still occurs without thermogenesis in older rabbits in which the adipose depots that were originally brown no longer have a thermogenic function. Nor does the increased flow appear to be an indirect effect of increased fatty acid release (see under "Substrates in Blood"), because it still occurs when the release is prevented (347).

The finding that nonshivering thermogenesis in brown fat of newborn rabbits was blocked by β-sympatholytic drugs (256) was substantiated by studies on other species, for example, in cold-exposed young guinea pigs (104), lambs (62), and mice (348). This conflicted with the scheme originally proposed for the classification of catecholamine receptors (349). Alpha receptors were defined as those stimulated principally by noradrenaline and were typified by the vasoconstriction effects; β-receptors were those stimulated by adrenaline. Alpha receptors were inhibited by a group of compounds referred to as α-blocking agents, for example, phentolamine, and β-receptors were inhibited by β-blockers, for example, propranolol. The identification of a response as being due to stimulation of α- or β-receptors depended on the relative potency of these catecholamines in producing the response, and upon the effects of α- and β-blockers on the response. However, the stimulating effects on blood flow and calorigenic action of noradrenaline (α-agonist) and isoprenaline (β-agonist) on brown fat of young rabbits were very similar (256); these catecholamines also had a similar effect on thermogenesis in kittens (107) and lambs (G. Alexander and D. Stevens, unpublished data) and cold-adapted rats (170) (Table 4 and Figure 13). In addition, adrenaline has an extremely variable effect on nonshivering thermogenesis. For example, adrenaline and noradrenaline given as a single injection intraperitoneally to newborn guinea pigs were equally potent in the stimulation of thermogenesis (64, 137); noradrenaline was about 50% more potent than adrenaline when infused into the cold-acclimated rat (350); noradrenaline was about twice as potent as adrenaline when infused into newborn rabbits (64, 256); and adrenaline had little or no effect compared with noradrenaline when injected subcutaneously into very young kittens, rats, mice, and rabbits (85). While some of these apparent species dif-

Table 4. Stimulation of nonshivering thermogenesis by noradrenaline (α-agonist) and isoprenaline (β-agonist)

	Noradrenaline		Isoprenaline	
Animal	Dose	Percent increase in basal metabolism	Dose	Percent increase in basal metabolism
Cold-adapted rats (253)	300 μg/kg (subcutaneous)	≃50[a]	0.8 μg/kg/min (I.V. infusion)	73
Week-old kittens (107)	400 μg/kg (subcutaneous)	260	400 μg/kg (subcutaneous)	230
Day-old lambs (Alexander and Stevens, unpublished data)	10 μg/kg/min (I.V. infusion)	86	10 μg/kg/min (I.V. infusion)	103

[a]None of the differences between the responses to the two drugs can be regarded as significant.

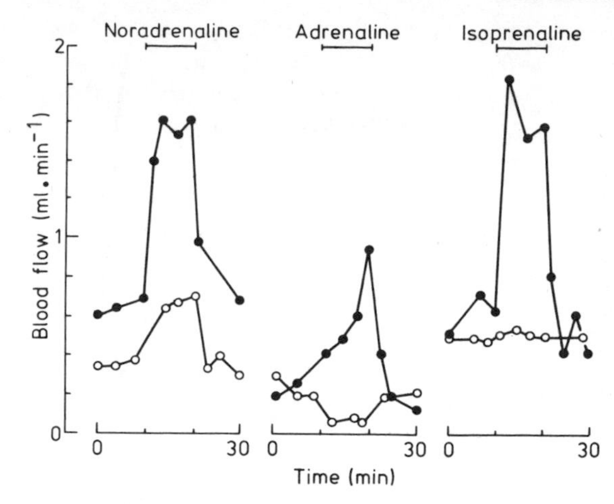

Figure 13. Blood flow from part of the cervical brown fat of 3–5-day-old rabbits during infusion of noradrenaline, adrenaline, and isoprenaline at the rate of 2 μg/kg/min, at air temperatures of 35 °C. ●—●, before propranolol; ○—○, after propranolol (1 mg/kg). Reproduced from Heim and Hull (256) with permission of the authors and publisher.

ferences could be due to methodological differences, such as route of administration of the catecholamines, this could not be so with apparent variations in potency of infused adrenaline relative to noradrenaline ranging from 0% to 100% in young lambs (351) (Figure 14).

The difficulty in classification of catecholamine receptors extends also to the effects of the catecholamines on blood flow through brown fat, because in young rabbits the blood flow increases in response to infusions of noradrenaline and to isoprenaline to a much greater extent than to adrenaline (256) (Figure 13). It is possible that these inconsistencies and apparent variability are at least partly explicable on the basis that a catecholamine can both stimulate and reduce blood flow, depending on dose, and that reduced blood flow would reduce metabolic rate in brown fat due to limitations of oxygen supply. This is illustrated by a study on chemical sympathectomy of lambs by treatment with 6-hydroxydopamine during gestation (G. Alexander and D. Stevens, unpublished data). In treated animals, at birth, noradrenaline in the infusion dose that gives a maximum thermogenic response in control lambs (10 μg/kg/min) (62) frequently produces an actual fall in oxygen consumption (Figure 15) under basal resting conditions, whereas a dose of only 1 μg/kg/min produces the expected increase in metabolic rate, indicating that the lamb is in a state of denervation hypersensitivity. Moreover, the expected increase can also be obtained if a dose of 10 μg/kg/min is preceded by α-blockade (phentolamine, 20 μg/kg/min). Premature lambs also appear particularly sensitive to noradrenaline, and a dose of 10 μg/kg/min depresses resting metabolism, whereas one-third of this dose stimulates metabolism (144). Similarly, kittens that normally show little or no thermogenic response to injected adrenaline give a response, after α-blockade

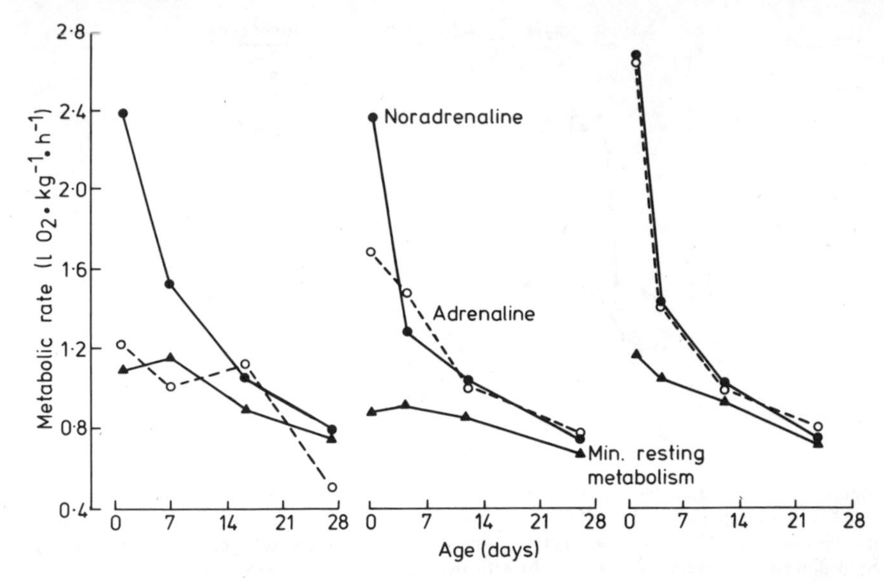

Figure 14. Effect of adrenaline and noradrenaline on metabolic rate in three young lambs. Infusion rates of catecholamines varied between 2.5 and 10 μg/kg/min, but the results appeared independent of the dose level. Reproduced from Alexander (351) with permission of the publisher (S. Karger AG, Basel).

with phenoxybenzamine, that is similar to that given by noradrenaline (352). In addition, adrenaline or noradrenaline, especially in high doses, has sometimes substantially reduced the metabolic response to cold in newborn rabbits, kittens, and lambs (64, 351, 353, 354) (Figure 16). The catecholamine levels induced by exogenous doses would have added to the levels induced endogenously by cold exposure. This metabolic depression has been associated with an increase in blood pressure (64, 351), and these results can be interpreted as indicating a baroreflex inhibition of thermogenesis (355) or that the catecholamine has induced vasoconstriction, thereby reducing blood supply to thermogenic tissues under summit metabolism.

It is suggested that the α-receptors in brown fat are relatively insensitive to catecholamines, but once stimulated they override the vasodilatory effects of β-receptors, and that α-receptors in brown fat are, in general, more sensitive to adrenaline than to noradrenaline. Indeed, the potency of noradrenaline in producing vasoconstriction in white adipose tissue is low, lower than in muscle, for example, although the constriction effect in white adipose tissue becomes effective after β-blockade (356). It would be illuminating to construct dose-response curves relating catecholamine dose to blood flow in brown fat. A study with canine white adipose tissue, denervated and autoperfused, showed that blood flow tended to increase with increasing perfusion rates of noradrenaline up to 2 μg/kg/min, but very high doses were not used (357); some preparations did show vasoconstriction, but this was

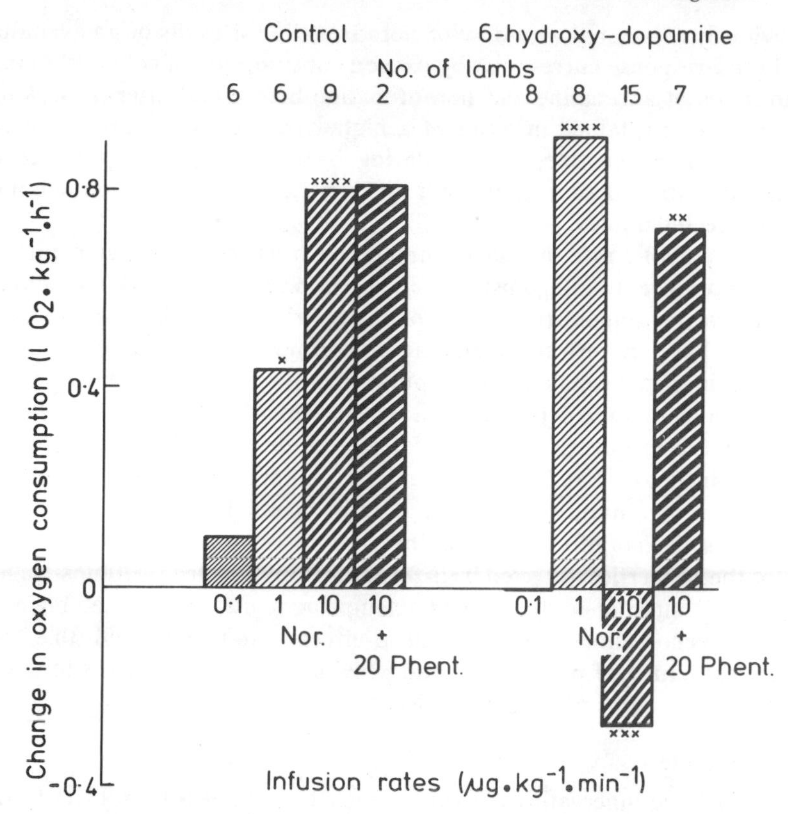

Figure 15. Effect of various doses of noradrenaline (Nor.) on oxygen consumption of newborn lambs under thermoneutral conditions. Lambs that were injected with 6-hydroxydopamine (100 mg/kg at fortnightly intervals) during fetal life show denervation hypersensitivity; the depression of oxygen consumption by the highest dose of noradrenaline in these animals is relieved by α blockade with phentolamine (Phent.). The statistical significance of the changes in oxygen consumption is indicated in the conventional manner. From G. Alexander and D. Stevens (unpublished data).

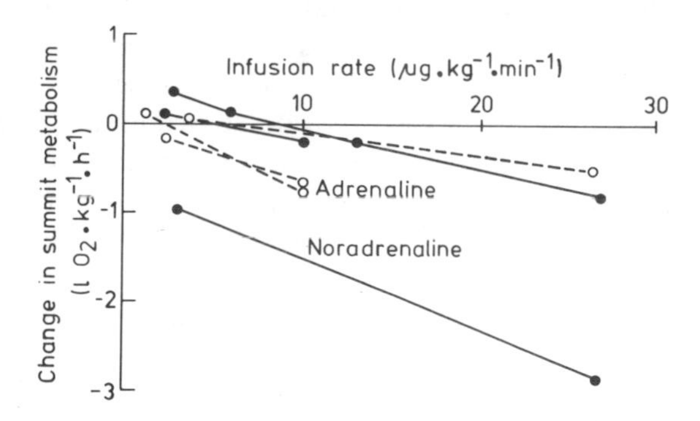

Figure 16. Depression of summit metabolism in six lambs less than 5 days old by infusion of catecholamines. Adapted from Alexander (351) with permission of the publisher (S. Karger AG, Basel).

not obviously associated with dose of noradrenaline. Results of an examination of dose-response curves relating oxygen consumption of cold-acclimated rats to doses of adrenaline and noradrenaline before and after β-blockade (350) could be explained in terms of a higher potency of adrenaline, compared with noradrenaline, in stimulating α-receptors in the presence of β-blockade, rather than in terms of a different metabolic mode of action of the two catecholamines.

The metabolic rate of intact animals, such as cold-acclimated rats, is clearly responsive to β-agonists, such as isoprenaline (Table 4), and is unresponsive to α-agonists such as phenylephrine (358); however, α- and β-agonists have an effect of similar magnitude in depolarizing brown adipocytes in partly isolated interscapular brown fat of cold-acclimated rats (286). Since membrane depolarization is regarded as an initial step in the activation of thermogenesis in brown fat (see under "Mechanisms of Thermogenesis in Brown Fat"), both α- and β-receptors may be implicated.

However, the nature of the receptors in brown fat appears to be very complex, as the noradrenaline binding sites in rat brown fat do not have many of the properties expected by β-receptors (359). Indeed, Himms-Hagen in 1967 (341) questioned the value of attempting to classify the metabolic effect of catecholamines into α- and β-effects, and suggested that the biochemical nature of receptors should provide a more useful basis for comparison. There seems no reason to dispute this suggestion.

Skeletal Muscle

The sympathetic innervation of skeletal muscle has not been subject to the scrutiny given to sympathetic innervation of adipose tissue (see under "Brown Fat"), and little is known about possible autonomic control of any nonshivering thermogenesis in skeletal muscle. In isolated and perfused resting dog or cat muscle, reflex or direct stimulation of noradrenergic sympathetic nerves generally produced a reduction in blood flow and oxygen uptake (360); both are attributed to closure of precapillary sphincters (361). However, stimulation at low frequencies had a direct effect on muscle metabolism, since oxygen uptake increased even though blood flow was reduced. Similarly, infusion of catecholamines increased oxygen consumption of muscle independently of vascular effects (see under "Other Sites of Nonshivering Thermogenesis"). Receptor sites for these metabolic changes cannot be strictly classified as either α or β, but the α-response appears to predominate, by contrast with the β-response by brown fat (362).

Circulation in skeletal muscle has been regarded as largely under autoregulatory control (363), which in dog skeletal muscle is not abolished by acute or chronic denervation. Traditionally, the muscle vascular bed has also been regarded as being provided with both sympathetic vasoconstrictor and vasodilator fibers, the constrictor fibers being adrenergic and the dilator fibers cholinergic (364). However, both α- and β-adrenergic receptors have been identified in the vascular bed of skeletal muscle of dogs (365, 366). It has also been shown that noradrenaline can exert both α- and β-effects on vascular smooth muscle as demonstrated by the use of α- and β-blocking agents (366), and there is some evidence, based on electrostimulation of nerves, that the same types of nerve fibers mediate both α- and β-effects. The vasodilatory effects produced by the cholinergic innervation can be blocked by atropine to reveal the adrenergic effects (367). The role of this vascular innervation in any nonshivering thermogenesis is unknown. However, during cold acclimation of rats there is a diminution in sensitivity of α-receptors to the vasoconstrictive effects of noradrenaline (368).

SHIVERING

Role in Homeothermy

There is no species of homeotherm, whether eutherian mammal or bird, that has been recorded as being unable to shiver in response to cold, although the mechanism may be poorly developed or absent at birth in relatively undeveloped young, such as laboratory rodents or domestic carnivores (see under "Magnitude of Shivering Thermogenesis"). The potential for thermogenesis by shivering has been infrequently measured, but is usually several times resting metabolic rate and is of the same order or even greater than the potential for nonshivering thermogenesis (see under "Quantitation of Shivering and Nonshivering Thermogenesis") yet its importance in facilitating homeothermy appears to have been underrated. Thus, shivering has been regarded as an emergency mechanism that functions only in severe cold (63, 129, 176). This is unrealistic, especially for species like birds, which have little or no nonshivering thermogenesis; as extreme examples, the willow ptarmigan, an arctic bird, shivers for much of its life (369) and the tiny chickadee (12 g) uses shivering both to maintain homeothermy and to control the depth of nocturnal torpor (370). Indeed, shivering may be an important component of metabolism at thermoneutrality in all birds (370). In addition, it has been frequently stated (95, 371) that shivering is replaced by nonshivering thermogenesis during cold acclimation, implying, perhaps unwittingly, that the potential for shivering wanes while that for nonshivering waxes. Such statements are not based on measurement of shivering potential (see under "Quantitation of Shivering and Nonshivering Thermogenesis"), but on the observation that, under conditions of moderate heat loss, shivering declines and may disappear. Had the conditions been colder, shivering would have persisted, because it is activated only when heat loss approaches the capacity for nonshivering thermogenesis (372) (see under "Neural Control of Cold Thermogenesis"). Indeed, it is likely that the capacity for shivering remains unchanged (178) during cold acclimation or actually increases, as appears to be the case in rats (373) and miniature pigs (113). In addition, it is possible that the ultrastructural and enzymic changes and the increased ion pumping in skeletal muscle that occur during acclimation (374–376) (see under "Other Sites of Nonshivering Thermogenesis") indicate increased efficiency of shivering, rather than improved capacity for nonshivering thermogenesis in skeletal muscle. Cold acclimation may be akin to exercise training, in which the capacity of both red and white muscle for mitochondrial respiration is increased (377); the capacity of rats for shivering thermogenesis also appears to be increased by exercise training (378).

Shivering has also been discounted on the grounds that the tremor may reduce external thermal insulation and interfere with voluntary muscular activity (81), but some degree of shivering thermogenesis can occur before the tremor becomes appreciable or strong enough to interfere with fine movement. In addition, shivering can be intermittent in an animal that is required to produce heat by this means for much of its life; the ptarmigan shivers at −10°C for only 16% of exposure time, and so for 84% of the time there is no tremor (369). In contrast, the chickadee shivers for about 60% of the time at 2°C (370). In man with minimum external insulation, thermal conductivity of tissue and surrounding air may increase during vigorous shivering, resulting in a low (11%) efficiency of shivering (16, 379). However, in animals with a significant pelage the evidence is conflicting; in anesthetized dogs up to three-quarters of the heat produced by shivering is lost due to increased convection (380), but in conscious goats and oxen shivering appears to be 100% efficient (381, 382). Whether or not the difference is due to differences in the effectiveness of the hair as an insulator or is associated with the effects of anesthesia on the type of shivering is not clear.

Magnitude of Shivering Thermogenesis

The thermogenic potential of shivering has been seldom estimated because of the measurement difficulties already outlined (see under "Quantitation of Shivering and Nonshivering Thermogenesis"); however, it is usually quoted as 2–3 times resting metabolism (383). The validity of the estimation depends on two major criteria: that any nonshivering is excluded and that a maximum shivering response is induced. The latter can be satisfied by adjusting the severity of the cold exposure so that body temperature begins to fall (see under "Definitions and Measurement"), but the first criterion may be difficult to satisfy. For example, it is unlikely that all forms of nonshivering, such as that in the liver, can be blocked by β-sympathetic blockade, and the control mechanisms of any nonshivering thermogenesis in muscle are quite unknown. However, any contribution from these sources appears to be small (see under "Components of Cold Thermogenesis," "Occurrence of Nonshivering Thermogenesis," and "Other Sites of Nonshivering Thermogenesis") so it may be adequate to use sympathetic blocking agents.

With these reservations, the two criteria have been satisfied in experiments with young lambs treated with propranolol (Table 2) (62). Maximum shivering thermogenesis measured soon after birth was twice the minimum resting metabolic rate; a month later it was still the same, but since resting metabolism had fallen this increase was now 3 times resting metabolism. Experiments indicating a very low shivering potential in young guinea pigs (104) do not satisfy the second criterion and probably grossly underestimate the shivering potential. On the other hand, experiments with young piglets indicating a thermogenic potential of 3 times basal fail to satisfy the first criterion and may overestimate the shivering potential; however, significant nonshivering thermogenesis has not been demonstrated in piglets (29, 113). Shivering probably does not occur in the least developed newborns (mice, lemmings, and hamsters), which show little or no thermogenic potential, but the potential for both shivering and nonshivering thermogenesis does appear some days later (129), and the former increases as nonshivering potential declines (21) (see under "Newborn Animals"). In birds, the shivering response appears to be weak at hatching but rapidly improves during the first few days of life (384). The conditions necessary for the elicitation and estimation of shivering in neonates that are imperfect homeotherms and allow body temperature to fall while thermogenesis is still increasing have not been adequately assessed.

Similarly, in adult animals, there are few, if any, experiments that provide a valid estimation of the thermogenic potential of shivering, although estimates of summit metabolism in animals that lack brown fat provide an upper limit to shivering thermogenesis, up to 10 times basal in the dog (385) and 8–10 times basal in the sheep (39). The estimate is 5–6 times basal in man (58), although it is unresolved whether adult man has functional brown fat (see under "Brown Fat"). Quantitative estimation of the contribution of various muscles to shivering in young cattle at twice resting metabolism has been made by direct methods by using blood flow and a − v oxygen differences (386); the muscles of one hind leg appear to contribute 15% of total cold-induced thermogenesis.

Nature of Shivering

The term "shivering," in a thermogenic sense, is used to denote involuntary, periodic contractions of voluntary (skeletal) muscle, and in this sense the term includes all grades of contraction from an increase in muscle tone (preshivering tone (383)) through a barely perceptible tremor to vigorous overt shivering. Shivering has been largely studied in laboratory animals and man, species that are not adapted to shivering; the patterns of shivering may be quite different in species, such as arctic birds, adapted to shivering (369, 370). Shivering, as distinct from shivering thermogenesis, is usually measured by mechanical devices sensitive to tremor (63, 387) or by electrical recording of action potentials of muscle groups, with electrodes fastened to the muscle or to the skin overlying the muscle (63, 81, 105, 388). Mechanical recordings give no indication of preshivering tone, but these two methods give similar tremor rates due to significant synchrony of the action potentials from different motor units in the same muscle, and because there is a predominant frequency at which the greatest number of fibers contract; however, an analysis of the electromyogram indicates that units are contracting at a wide range of frequencies (65, 389). With preshivering tone, contraction by the various muscle units is completely out of phase, and periodic bursts of synchronized activity that may be detected electrically, but not mechanically or visually, are also a feature of the initial stages of shivering. These bursts of electrical activity occur more often and become apparent as cold exposure continues, until shivering becomes a continuous, moderately regular sequence of bursts (16, 389). However, the pattern may be more intermittent in species adapted to shivering (369), and there appear to be differences in the overt nature of shivering that depend on the method of acclimation of animals such as sheep to cold (390).

The amplitude of electrical activity associated with shivering increases during inspiration and decreases during expiration, presumably due to a spinal reflex related to activity of the muscles of the diaphragm (16), but the frequency of the bursts remains unchanged (383). Shivering also tends to be suppressed when a muscle moves voluntarily (16), but apparently not if deep body temperature is depressed (391). Pressure on the skin can also inhibit shivering (392), and shivering in man can be suppressed psychosomatically (393). Unexplained periodic fluctuations in amplitude of shivering have also been described (16).

Both preshivering tone and shivering first appear in extensor muscles and proximal muscles on the upper limbs or trunk, rather than in the extremities, and the shivering intensity is always greatest in these groups (65, 379, 383, 394). For example, in cattle shivering first appears in the femoral region of the hind leg above the quadriceps femoris, an extensor, the only muscle that has shown consistent increases in blood flow during experimental cold exposure in this species (395). As shivering increases, most muscles, both red and white, participate, at least in small animals such as mouse and rabbit; they include the muscles of mastication, neck and trunk, pectoral and pelvic girdles, the limbs, and the small muscles of the extremities (105, 383, 396). Muscles that do not participate, such as those of the ear, face, and perineum, appear to be mostly those that lack muscle spindles (stretch receptors). In addition, the diaphragm and rhythmic units of the intercostal muscles do not shiver, although other intercostal units do (397). Panniculus muscle may be involved since it shows a small increase in blood flow on exposure to cold (48, 398) (see under "Circulatory Adjustments During Cold Thermogenesis"). Large animals, such as the horse, have been said to shiver with fewer muscles than small animals use (396), but this may depend on the relative severity of cold exposure. It is not known whether the proportion of different muscle fiber types (fast and slow twitch) determines the potential for shivering, in the same way that certain types of athletic performance depend on the proportion of fast twitch fibers (397a).

The average integrated electrical activity increases with decreasing environmental temperatures in all muscles, provided the animals do not become hypothermic, but the frequency of muscle contraction appears to be independent of environmental temperature (105).

The range of frequency of contractile oscillation among different muscles of an animal is similar (105, 383), but the frequency appears to be characteristic of the species depending upon body size. The frequency decreased with increasing body size in a series that included mice, rats, guinea pigs, rabbits, cats, and dogs (186). However, the range of discharge frequencies that occurs in an animal during shivering may be much larger than the 10–30 cycles/sec usually quoted (383, 389). When recording instruments with a much wider range of frequency response than the usual 100 Hz were used, a frequency range of 10–350 Hz was recorded in shivering rabbits (105), indicating that a mere 40% of electrical activity would be recorded by the conventional instruments (Figure 17). Similarly in birds, shivering frequencies in the range of 6–600 Hz have been recorded with appropriate instrumentation (399).

Neural Control

The stimuli for the onset of shivering and nonshivering thermogenesis are discussed elsewhere under "Neural Control of Cold Thermogenesis." Although shivering occurs in the voluntary muscles, it is an autonomic or involuntary function, and its control is mediated by spinal pathways other than the pyramidal tract, which mediates voluntary movement (16).

Shivering appears to be an adaptation of the rhythmic asynchronous variation in excursion of muscle fibers (physiological tremor) characteristic of normal voluntary or reflex muscular contraction of homeotherms,

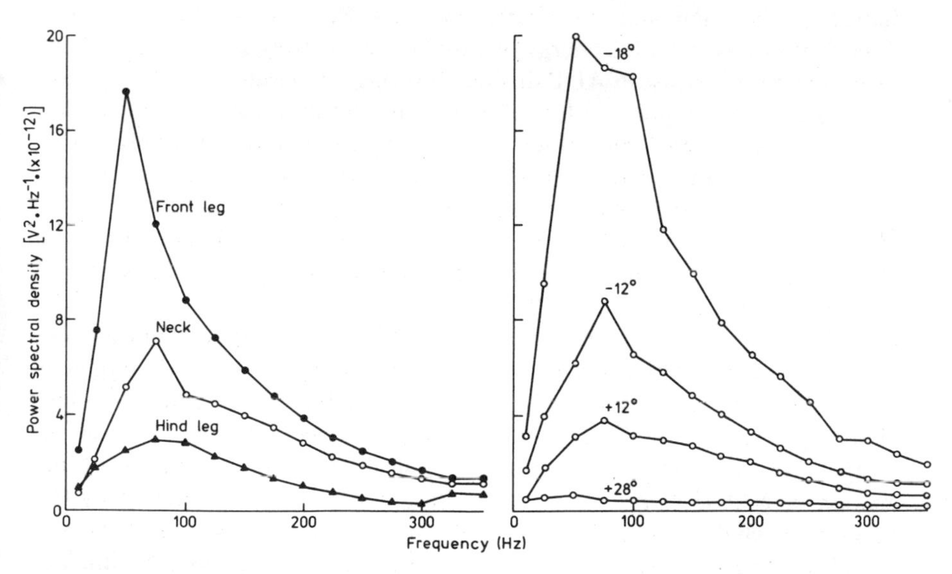

Figure 17. Power spectral density of shivering electrical activity in shaved rabbits. *Left,* at different locations of the body at an air temperature of −3°C; side and back were very similar to hind leg. *Right,* neck muscle at different air temperatures. Adapted from Janský et al. (105). Reproduced by permission.

although not of poikilotherms (383); the frequencies of this tremor and of shivering are similar. The neural control of this rapid oscillation of muscle contraction that is characteristic of shivering has received much attention, but is still not resolved (389); current knowledge was reviewed in a recent volume of this series (308).

Heat Production and Substrate Use

The shivering mechanism is an evolutionary development well suited to the production of heat by muscle without the need for movements of wide amplitude by the animal and the associated convectional heat loss; if heat-producing contractions of opposing muscles were not synchronized, limb excursions would have to be very wide to provide thermogenesis equivalent to 2–3-fold resting metabolism (383). In addition, heat production by a simple increase in rigidity (isometric contraction) would be much less effective than the isotonic contractions of shivering (400). The shivering mechanism probably produces minimum tension for maximum energy and heat production and interferes minimally with voluntary muscular activity (16).

The biochemical basis of the heat production of contracting muscle is well understood, although mechanisms of contraction are still debated. The heat-producing mechanism involves stimulation by motor nerves, depolarization of muscle, and the release of intracellularly bound calcium ions, resulting in activation of the contractile protein myosin, which is also an ATPase; ATP, produced largely from the electron transport chain (see under "Mechanisms of Thermogenesis in Brown Fat"), is hydrolyzed to ADP and inorganic phosphate and the muscle contracts. Because no external work is done during shivering, the energy released by the hydrolysis appears solely as heat. The rapid release of ADP during shivering accelerates substrate oxidations and the accompanying loss of energy as heat in mitochondria; the rate of reaction at the biochemical level is largely governed by the supply of this phosphate acceptor. In addition, minor contributions to heat production are made by ion pumping involved in restoration of the normal polarized state of the muscle membrane by Na^+/K^+-ATPase and return of calcium ions to the circumscribed stores, mainly by the pumping action of Ca^{2+}-ATPase of the sarcoplasmic reticulum (272, 401).

The high requirement of contracting muscle for ATP results in rapid acceleration of glycolysis, a process with a net yield of ATP. If oxygen is limiting, as is likely during very high rates of metabolism, the rate of production of pyruvate, the end product of glycolysis, exceeds its rate of oxidation by the citric acid cycle in the contracting muscle. In addition, the rate of formation of NADH in glycolysis would be greater than its oxidation by the respiratory chain, and so lack of NAD^+ would limit the rate of glycolysis. This is overcome by reduction of pyruvate by NADH to form NAD^+ and lactate, which explains the tendency for lactate to accumulate during summit metabolism (402) and intense exercise (403). Lactate can also be produced when white fat is metabolically stimulated while vasoconstricted (404).

Energy Source for Shivering

There is no longer any doubt that oxidation of fat plays a vital role in providing energy for sustained muscular activity (405), whereas not long ago it was generally considered that carbohydrate was the major fuel for working muscle. It has been shown, for example, that mitochondria from both red and white muscle are able to oxidize fat and carbohydrate (406). Fat is available to the muscle either from endogenous stores or in the form of free fatty acids or very low density lipoproteins, the usual form of triglyceride in the blood (407).

However, the proportion of fat and carbohydrate used during shivering appears to vary widely (405). Respiratory quotients in shivering man (about 0.80) indicate that fat is used rather than carbohydrate (393). Studies with ruminants, animals that rely almost entirely upon endogenous formation of glucose for metabolism, illustrate the apparent variable nature of substrates used. In sheep, energy balance studies indicated that the increase in metabolism due to cold exposure is entirely due to fat catabolism (408). However, studies with radioactive glucose (409) indicate that glucose utilization by the whole animal increases 2–3-fold in recently fed sheep, either exposed to chronic cold for several weeks or acutely exposed for several hours, and could account for up to 17% of the total CO_2 production. This could represent a large use of glucose by shivering muscle if this increase were confined to that tissue. Indeed, estimates of a more direct nature, in which blood flow and differences in arteriovenous concentration of substrates were examined in the shivering hind leg of young cattle that had been recently fed, indicate that glucose may supply about 70% of the leg metabolism during cold exposure; acetate, a major product of fermentation in the rumen, appears to supply most of the remaining energy (410) (Table 5). The glucose contribution was substantially reduced during fasting, and it seems possible that data indicating fat as the sole substrate (408) may have been associated with a poor nutritional state of the experimental sheep or with the mild degree of cold exposure.

Considerably more is known about substrate utilization by exercising than by shivering muscle (406), and there is no reason to believe that the two sets of mechanisms are basically different (272, 385). Exercise training, for example, increases the capacity of muscle to use fat and probably lactate (411), but carbohydrate appears to be essential for continued heavy exercise (412). The availability of fatty acids and glucose appears to be a major factor that influences their relative rate of oxidation (405), but it is doubtful whether the so-called glucose-fatty acid cycle (413) applies at high levels of energy expenditure. The cycle proposes that if the availability of one substrate predominates, the utilization of the other tends to be inhibited; fatty acids inhibit glucose transport and associated enzyme reactions, whereas glucose inhibits fatty acid oxidation through increased insulin secretion, suppressing lipolysis and accentuating lipogenesis. The cycle does not appear to apply in strenuous exercise in dogs (414), during severe cold ex-

Table 5. Effects of cold and feeding on the potential contribution of different substrates to leg metabolism in cattle (steers)[a]

	Thermoneutral (≃15°C, still air)		Cold (≃2°C, wind 7 km/hr, clipped and wetted)	
Substrate	Fasted (20 hr)	Fed (3 hr since feeding)	Fasted (20 hr)	Fed (3 hr since feeding)
Free fatty acids	net release	net release	0.22	0.04
Acetate	0.44	0.68	0.16	0.32
Net CHO[b]	0.29	0.52	0.57	0.74
Total	0.73	1.20	0.95	1.10

From Bell and Thompson (410). Reproduced by permission of the author.

[a]The net uptake by the leg of different metabolites has been converted to oxygen equivalents, assuming complete oxidation, and is expressed as a fraction of leg oxygen uptake.

[b]Glucose plus lactate.

posure in dogs (415), or even to skeletal muscle in vitro, although it does apply to the heart and diaphragm (416). Similarly, in newborn piglets the limited capacity for hepatic gluconeogenesis appears uninfluenced by increased fatty acid oxidation (417). The lack of mutual inhibition between fatty acid oxidation and glucose utilization would favor maximum substrate availability during cold thermogenesis in muscle.

Variates such as the depth and duration of cold stress (418), the degree of acclimation and endocrine status of the animal, and the nutritive status could also lead to wide differences in the proportion of fat and carbohydrate oxidized, and possibly to utilization of potential substrates such as lactate (419) and ketone bodies (420). Versatility of substrate use is indicated by experiments with rats in which antilipolytic agents did not impair thermogenesis nor did inhibition of glycogenolysis by adrenodemedullation, but if both lipolysis and glycogenolysis were inhibited together thermogenesis was impaired (405).

In young birds, carbohydrate appears to be the preferred substrate for shivering (421), but the role of lipid thermogenesis is still unresolved (422).

CIRCULATORY ADJUSTMENTS DURING COLD THERMOGENESIS

Estimation of the regional distribution of cardiac output during climatic stress became possible with the development of indicator fractionation techniques. These are based on the Fick principle: in essence, an indicator substance, mixed with blood leaving the heart, is distributed to all the tissues of the body in proportion to the fraction of the cardiac output reaching them.

It is assumed that all the indicator or a constant proportion of it is retained by the tissue until the animal is killed and the indicator content of the tissues is determined. Sapirstein's method, first described in 1958, employed $^{86}Rb^+$ as the marker and assumed that the indicator extraction ratio for each tissue was essentially the same, for about 1 min immediately after injection of the marker, as the mean for the whole body. However, a diffusion barrier to Rb was known to exist in the brain (246). Thermal physiologists, desperate for a simpler physical principles was introduced by Rudolph and Heymann in 1968 (424) and was exhaustively tested and used by Hales (425); it involves the injection of nuclide-labeled plastic spheres of a sufficiently large diameter (15 and 50 microns) to ensure that they do not pass through the capillaries. The blood flow should then be proportional to the radioactivity of the tissue, and the flow rate can be calculated if the cardiac output or blood flow in any individual organ is known.

A method that was more attractive because it appeared to rely on simpler physical principles was introduced by Rudolph and Heyman in 1968 (424) and was exhaustively tested and used by Hales (425); it involves the injection of nuclide-labeled plastic spheres of a sufficiently large diameter (15 and 50 microns) to ensure that they do not pass through the capillaries. The blood flow should then be proportional to the radioactivity of the tissue, and the flow rate can be calculated if the cardiac output or blood flow in any individual organ is known.

When this method was applied to the anesthetized cold-exposed newborn rabbit (426), brown fat was the only tissue showing a major increase in flow due to cold exposure, the increase being from 36 to 180 ml/100 g/min. Direct collection of blood from the venous outflow of part of the cervical brown fat of newborn rabbits stimulated with noradrenaline (241) had also demonstrated an increase, from 90 to 360 ml/100 g/min.

The microsphere method was also applied (48) to conscious newborn lambs during summit metabolism, with results that, in some respects, were strikingly different from those obtained with Sapirstein's method (Figure 18, Table 6). This was particularly marked with brown adipose tissue, for which the microsphere method gave blood flow values that were more than twice those obtained with the $^{86}Rb^+$ technique under thermoneutral conditions. Exposure to cold caused a 6-fold increase in the values obtained with microspheres, but only a 2-fold increase in $^{86}Rb^+$ values. Independent evidence leads to the conclusion that the microsphere values for blood flow to brown fat are correct and the $^{86}Rb^+$ values incorrect. For example, use of the microsphere value, together with the estimated oxygen consumption by brown fat of lambs, leads to the calculation of a value for the arteriovenous difference in blood oxygen content that conforms with the available data, whereas if the rubidium flow value is used, an untenable value for the arteriovenous difference in blood oxygen content is obtained (212) (see under "Contribution to Nonshivering Thermogenesis"). These results indicate that

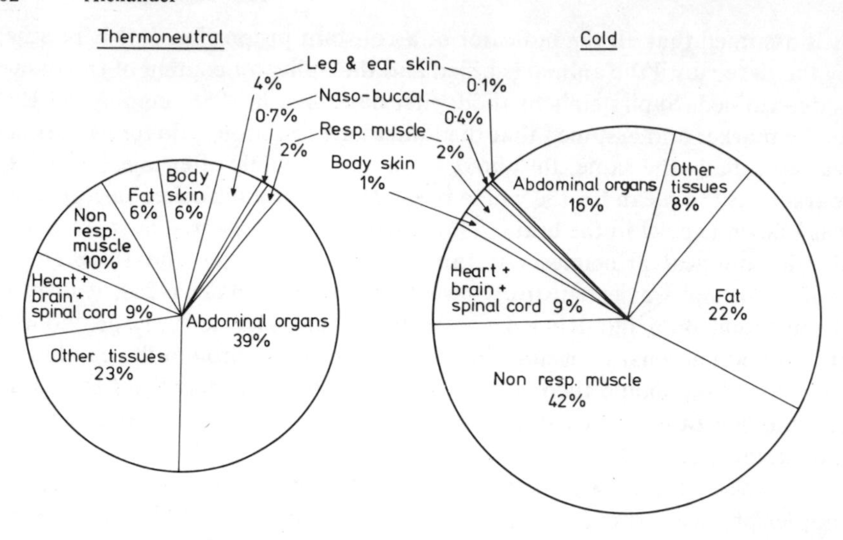

Figure 18. Redistribution of cardiac output when lambs are exposed to cold conditions that induce summit metabolism. The total areas of the circles represent cardiac output, and the sectional areas are proportional to the fraction of cardiac output received by the tissues. Reproduced from Alexander (21) with permission of the publisher.

Table 6. Mean blood flow (ml/100 g of tissue/min) to selected tissues of lambs in thermoneutral and summit conditions: comparison of microsphere and $^{36}Rb^+$ methods

	Mean blood flow (ml/100 g tissue/min)			
	Microsphere method		$^{86}Rb^+$ method	
	Thermo-neutral	Summit	Thermo-neutral	Summit
Perirenal brown fat	188	1,125	71	145
Skeletal muscle (biceps femoris)	13	111	21	92
Skin, ear	49	6	27	8
Skin, midside	23	6	25	19
Panniculus muscle	13	28		
Diaphragm	63	152		
Liver, arterial	37	43	62	94
Liver, portal	679	379	230	346
Liver, total	716	422	292	440

From Alexander et al. (48). Reproduced by permission

brown fat is a tissue, in addition to brain and testis (427), for which there is a barrier to rubidium uptake; they confirm previous indications that the two methods do not give identical results (425).

Subsequently, the two methods were compared in a detailed study by Foster and Frydman (428) in laboratory rats during stimulation with noradrenaline; the $^{86}Rb^{+}$ method gave values for flow in brown fat that were only one-tenth of those by the microsphere method, and the microsphere values were corroborated by direct measurments of the venous drainage from tissue.

In view of the unacceptability of the $^{86}Rb^{+}$ method, there are few acceptable studies on the distribution of cardiac output relevant to cold thermogenesis. Cold exposure in young rabbits (426) was accompanied by a 4–5-fold increase in blood flow to cervical and interscapular brown fat, but flow in the perirenal, inguinal, pectoral, and axillary depots was lower than that in these depots and changed less in response to cold (Table 7). By contrast, all brown fat depots in lambs showed a high flow and a 5-fold response to severe cold exposure, and cardiac output increased between 50% and 100% (48, 242). In the cold-adapted rat, stimulation with noradrenalinc doubled cardiac output and increased flow through all brown fat depots examined; the increases were even greater than in lambs (249) (Table 7).

Simultaneous comparisons of blood flow in white and brown fat by using the microsphere technique are not available. In adult sheep, flow through white fat from an area that had been brown at birth (see under "Ontogenic Changes: Conversion to White Fat") showed a 6-fold increase due to cold exposure (68). Three-fold increases were recorded in white fat of the hind leg of cold-exposed steers (395), and 3–4-fold increases in white fat have been recorded in man by the ^{133}xenon washout technique during prolonged heavy muscular work (429). In addition, noradrenaline infusion causes an increase in flow, directly measured with a drop counter, in white fat of dogs (347, 357). This does not appear to be due to noradrenaline-induced lypolysis and the vasodilatory effect of substances released with the fatty acids, since increased blood flow still results after inhibition of lipolysis with nicotinic acid (347). In adult rabbits with flow measured directly, the resting flow and the increased flow due to noradrenaline injection are much greater in white fat that was originally brown than in white fat that was white at birth (219); clearly the physiology of white adipose tissue may vary from site to site. Increased flow in white fat could facilitate lipid transport to muscle or other tissues for oxidation during cold exposure (see under "Tissue Substrate Stores").

Skeletal muscle blood flow, measured by microspheres, showed very little increase in the newborn rabbit during cold exposure (426), but estimation of muscle flows in cold stress of varying severity would be illuminating in this animal in which the role of shivering has not been fully explored (see under "Magnitude of Shivering Thermogenesis"). In lambs, which have a well

Table 7. Stimulation of blood flow through brown fat by cold or noradrenaline in three species

Brown fat depot	Blood flow (ml/100 g of tissue/min): Unstimulated (thermo-neutrality)	Blood flow (ml/100 g of tissue/min): Stimulated (cold or nor-adrenaline infusion)	Ratio of unstimulated to stimulated
Newborn rabbit (426)		(cold)	
Cervical/interscapular	44	193	4.4
Perirenal[a]	7	10	1.4
Inguinal[a]	8	9	1.1
Pectoral[a]	12	36	3.0
Axillary[a]	17	57	3.4
Newborn lamb (48)		(cold)	
Perirenal	188	1,125	6.0
Inguinal	144	681	4.7
Pericardial	136	749	5.5
Prescapular	129	731	5.7
Cervical	134	691	5.2
Cold-adapted rats (249)		(noradrenaline)	
Dorsalocervical	55	2,050	37.3
Periaortic	45	1,410	31.3
Interscapular	39	1,300	33.3
Perirenal-iliac-inguinal	47	750	16.0
Axillary	41	990	24.1

[a]In the newborn rabbit these depots may contain a high proportion of white fat.

developed shivering mechanism (62), muscle flow increased markedly during summit metabolism, 8-fold in longissimus dorsi and biceps femoris, 6-fold in trapezius, and 5-fold in gastrocnemius, representing an overall increase from 10% to nearly 50% of cardiac output (48) (Figure 18). Two- to four-fold increases also occurred in adult sheep under subsummit conditions (398). Likewise, in young cattle blood flow measured by dye detection methods in the muscle of the hind leg increased at least 3-fold (386) during moderate cold exposure. There appear to be wide differences in the magnitude and consistency of response of different muscles, but the total flow appears to depend on the number of muscles involved in shivering (see under "Shivering").

Among small animals, smaller increases in muscle flow have been recorded in cold- or warm-adapted rats exposed to cold (245) and in bats arousing from hibernation (430), but these estimates with the use of Sapirstein's method may be unreliable. Microspheres have been used in laboratory rodents, but only to examine the effects of noradrenaline-induced thermogenesis; changes in muscle flow under these conditions are barely perceptible (249).

Increased flow also occurred through the diaphragm, intercostal, and cardiac muscles in lambs (Table 6), adult sheep, and rats during cold exposure; these changes, no doubt, facilitate the increased respiratory effort and increased cardiac output that occur at high metabolic rates and which are necessary to maintain tissue oxygen supply and help buffer changes in acid-base balance (431). There was also an increase in blood flow through the panniculus muscle of the lambs (Table 6), so this muscle may be involved in shivering. In tissues not directly concerned with thermogenesis, blood flow during summit metabolism was either unchanged or reduced below that at thermoneutrality (Figure 18).

In lambs, arterial blood pressure during summit metabolism was only 20% higher than under thermoneutral conditions (48) so major changes in blood flow to brown fat, skeletal muscle, and other organs would largely reflect changes in local vascular activity. Differential outflow of sympathetic activity to various body regions is likely to be involved in the control of these changes in regional blood flow (389), but local regulatory mechanisms activated by changes in the metabolic rate of tissue may also be involved.

Increases in cardiac output also occur during cold exposure, as indicated above, but the role of increased heart rate and increased stroke volume are not clear, perhaps because of species differences or differences in the degree of effective cold stress applied or failure of the heart to fill at high heart rates. In lambs under summit conditions, the increase of 50–100% in cardiac output was largely accomplished by an increase in heart rate from 200–300 per min (242), but in man, during moderate cold exposure, increases in cardiac output of up to 95% were accomplished by increases in stroke volume without much change in heart rate (about 70 beats per min) (432). In this respect, cold exposure in man may be very different from maximum exercise in which cardiac output increases 5-fold, from 5–25 ml/100 g/min) (433, 434), largely due to an increase in heart rate to about 170 beats/min, stroke volume having plateaued at a work level of 35% of maximum capacity (435). However, cardiovascular changes in man during summit metabolism have yet to be examined.

With the microsphere technique the only changes in the distribution of cardiac output that appear to accompany acclimation are in the brown fat, where there is a 3-fold increase in blood flow under resting conditions and a 70-fold increase under noradrenaline stimulation (249). This large increase is accommodated by an increase in cardiac output. It is not known whether cold acclimation induces changes in the capillary supply to muscle fibers, as appears to occur in man during endurance training (436).

Cold acclimation appears to have no effect on cardiac output in rats at rest, but the increase that follows noradrenaline injection is much greater in cold- than in warm-acclimated rats (100% versus 30%) (249).

NEURAL CONTROL OF COLD THERMOGENESIS

Thermoreceptors are distributed throughout the body, particularly in the skin, hypothalamus, spinal cord, and viscera (389, 437–439). The mode of stimulation of these receptors, the relative importance of sensory inputs from different areas, and the nature and location of their integration in thermoregulation are still debated (437, 440–449). In particular, little is known about changes in the regulatory system that give rise to increasing thermal stability after birth (21, 450) or about congenital differences in thermal stability even within the same species of newborn (25). Similarly, it is not known whether the thermolability often seen in mammals during the first hours of life (29, 75) is due to characteristics of the central thermostat or thermoreceptors, excessive heat loss, inadequate oxygenation of tissues, or the metabolic consequence of birth. There is also much to learn about the mechanisms that result in thermolability, such as the habituation in certain types of cold adaptation in sheep and man (451), in daily cycles of torpidity in some small animals (see under "Hibernators"), or in hibernation (452–455), states in which the set-point for thermogenesis is lowered.

The relative importance of some receptor sites in controlling the onset of nonshivering thermogenesis and of shivering has been elegantly studied by Brück and colleagues (114, 456) by using young guinea pigs. Nonshivering thermogenesis is stimulated by the integrated effects of skin temperature and hypothalamic temperature (Figure 19). The lower the skin temperature, the higher must be the hypothalamic temperature to suppress nonshivering thermogenesis, and when hypothalamic temperature is low, a small decline in skin temperature evokes a large increase in nonshivering thermogenesis. This type of relationship between hypothalamic and skin temperature could explain the apparent independence of thermogenesis from rectal temperature of human infants less than 4 hr old (457) and the apparent dependence on central temperature of newborn rabbits and guinea pigs (458). Brück also showed that shivering in young guinea pigs was stimulated by the integrated effects of skin temperature and the temperature of the cervical part of the vertebral canal (Figure 19); hypothalamic temperature had little effect. Spinal cord temperature also appears to contribute to the control of shivering in birds (459, 460). Since stimulation of nonshivering thermogenesis in guinea pigs results in warming of the cervical vertebral canal, owing to the metabolism of strategically placed brown fat (see under "Distribution"), shivering is inhibited until nonshivering thermogenesis can no longer maintain the temperature of the canal. A similar mechanism appears to operate in the rat, in which shivering is activated only when heat loss approaches the capacity for nonshivering thermogenesis (372). However, in many species the location of temperature-sensitive areas in relation to the brown fat depots is unknown; interscapular brown fat is not found in all species (see under "Distribution"). Hypothalamic temperatures play a significant role in the control of shivering in the pig, which has no brown fat at all (114).

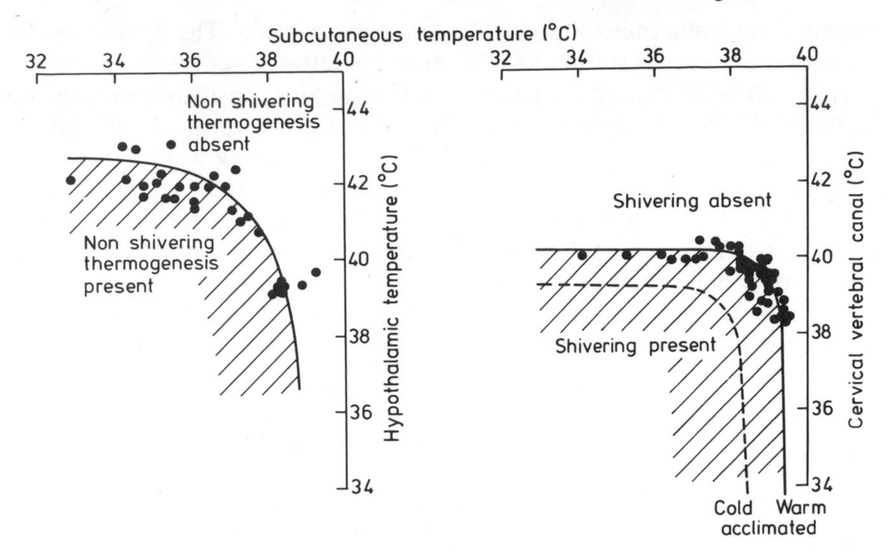

Figure 19. Temperature thresholds for nonshivering and shivering thermogenesis in the guinea pig. *Left*, temperature of receptors in the skin and hypothalamus of 2–7-day-old guinea pigs reared at 5°C. *Right*, temperature of receptors in the skin and cervical vertebral canal of 4-8-week old guinea pigs adapted to 30°C. The *dotted line* is for similar animals reared at 3°C. Cold thermogenesis is absent above and to the right of the lines and is present below and to the left of the lines (*shaded area*). Adapted from Brück and Wünnenberg (456) by permission of the authors and Charles C Thomas, Publisher, Springfield, Illinois.

Temperature of the spinal cord and hypothalamus are also likely to be involved in the control of regional distribution of cardiac output, since local manipulation of their temperatures produces a redistribution similar to that seen during manipulation of environmental temperatures (461, 462).

The mechanisms controlling thermogenesis appear to be affected by the temperature to which the animal was recently exposed, the duration and frequency of exposure (451), and the rate of change of ambient temperature (463, 464). Results may, therefore, be partly influenced by experimental design when animals are exposed consecutively to a number of different temperatures for short periods. The neural mechanisms of mutual interaction of exercise with cold thermogenesis also require examination to resolve the questions of whether or not there is a decrease in the thermoregulatory set point at the onset of exercise and whether or not the heat produced by exercise abolishes or interferes with concurrent shivering (391).

SUBSTRATES FOR COLD THERMOGENESIS

Tissue Substrate Stores

During the initial stages of cold exposure, substrates for thermogenesis in the principal thermogenic tissues, brown fat and skeletal muscle, are drawn largely from endogenous stores of fat and carbohydrate. The lipid stores in brown fat are usually substantial, for example, 30–40% of the tissue weight (wet) in lambs (465), and at times brown fat contains considerable glycogen, as revealed by electron-microscopy (188). When these stores are exhausted, for example, in brown fat of young rabbits, some degree of thermogenesis continues by oxidation of fat and glucose drawn from the circulation (266, 279, 466–468). Indeed, glucose from the circulation may be utilized by brown fat prior to significant lipid depletion (468). However, brown fat of newborn rabbits does not appear to take up plasma free fatty acids until its own lipid stores have been well mobilized (469). In brown fat, the activity of lipoprotein lipase that is essential for uptake of lipid from lipoproteins in plasma is increased by cold exposure, at least in rats after a mere 1 day at 4°C (470).

Unlike lipid stores in brown fat, the endogenous stores of lipid within muscle cells are meager, at least when expressed on a percentage basis; in man, red and white muscle fiber contains only about 2% and 0.7% triglyceride, respectively (471). However, muscle represents a much larger percentage of body weight than brown fat (25% and 1.5%, respectively, in lambs (48)). The lipid in muscle may take the form of small droplets, as revealed by electronmicroscopy of dog muscle (472). In young lambs the extractable neutral lipid from whole muscle that contains no adipocytes is less than 3%, as indicated by the level of chloroform-extractable lipids. In older animals, muscular tissue contains significant amounts of fat in the form of adipocytes between the bundles of muscle fibers; this fat, commonly referred to as marbling, can account for more than 25% of muscle weight on a dry matter basis (473). Skeletal muscle triglyceride is depleted during acute cold exposure in rats (474), but whether muscle can preferentially use this lipid is not clear; presumably, it would enter the venous circulation after sympathetically mediated lipolysis (405) and be returned to the muscle in arterial blood. Skeletal muscle lipoprotein lipase that is essential for lipolysis of lipoprotein from blood was unaffected by acute cold exposure in rats (470) but increased by cold acclimation (475).

Likewise, the carbohydrate stores in the form of glycogen in muscle are low, amounting to less than 1% wet weight in both red and white fibers in guinea pigs (476), rats (477), and man (471), although somewhat higher in the young lambs 2–3% (478). However, blood glucose, probably derived from liver glycogen (6% glycogen in lamb's liver (478)), would be available as a fuel for shivering.

Although the percentages of triglyceride and glycogen stores in muscle are low, it can be calculated that in a 4-kg newborn lamb, for example, these stores contain sufficient energy (approximately 180 and 120 Kcal, respectively) to maintain maximum shivering for about 10 hr if all muscles were equally involved. The stores in brown fat would last for a similar period if the tissue were working to capacity.

Substrates in Blood

Much more is known about changes in the concentration of substrates than about the more important aspect, substrate turnover. The concentration of free fatty acids, glucose, and other metabolites in the plasma usually increase markedly during cold exposure. For example, in neonatal lambs, plasma glucose more than doubled during summit metabolism, increasing from about 100 mg/100 ml to 300 mg/100 ml or higher, and lactate levels increased 2–3-fold from about 20 mg/100 ml to about 60 mg/100 ml; plasma free fatty acid levels also tended to increase, but any increases were small until the lambs were 2–3 weeks old, when variable increases, up to 6-fold, occurred (402).

Free Fatty Acids Substantial cold-induced increases in free fatty acid concentrations in plasma have been recorded in a variety of other animals including adult sheep (479, 480), newborn calves and adult oxen (481), piglets (112, 482), rats (483), and young rabbits (484). Concomitantly, the glycerol concentration also rises (431, 484).

The magnitude of change in plasma free fatty acids due to cold exposure depends, of course, on the balance between lipogenesis, lipolysis, and utilization, which is greatly influenced by the degree of cold exposure and whether or not the animal is shivering. The increase results from lipolysis, either in brown fat or in white fat, but the net release from brown fat appears to depend to some extent on the tissues' own intracellular requirements. For example, in cold-exposed young rabbits, the release, measured directly, is much lower if they are being reared in the cold than if they are undernourished and being reared in the warmth (218). In newborn lambs, which lack significant white fat, the lack of a consistent increase in free fatty acids under summit conditions is probably due to rapid utilization of lipid, both in brown fat and in shivering muscle; inhibition of shivering by muscular paralysis induced by suxamethonium chloride in cold-exposed lambs results in a rise in plasma free fatty acids (FFA) (402). The very large increases seen in older lambs probably reflect the decreased thermogenic capacity of the adipose tissue and its conversion to a storage organ (136); transport of lipid to muscle is then required for thermogenesis. On the other hand, the marked response observed in young rabbits with the poorest metabolic response to cold (484) could be due to failure of their muscle to use fat.

The relative roles of brown and white fat in contributing to total body lipolysis have been indicated by a number of experiments with starved and cold-exposed newborn rabbits, animals that have both brown and white fat. In starving newborn rabbits, white fat is rapidly depleted of lipid regardless of the degree of cold stress; in brown fat, on the other hand, the rate of depletion is much greater in the cold than at thermoneutrality although it increases as starvation progresses (279, 468, 485–488). Autopsy findings conforming with these observations have been made in human infants (489).

In addition, experimental conditions such as fasting and cold exposure or treatments with lipolytic agents increase the concentration of free fatty acids in white fat, but do not change the level of FFA in brown fat, whereas agents such as nicotinic acid that block lipolysis lower FFA in white fat but not in brown fat (490). Thus, the level of FFA in plasma depends more on the lipolysis and lipogenesis occurring in white fat than in brown fat. Inherent differences, probably genetic in origin, in the lipolytic capacity of white fat have been recorded in piglets (491).

Lipolysis and the release of fatty acids into the circulation by both brown and white fat appear to be under the influence of the sympathetic and splanchnic nerves with control exerted centrally, probably by the hypothalamic region, as indicated by intraventricular injection of neurotransmitters (244, 341, 405, 492). This region is sensitive to glycopenia, but is also activated when a rapid mobilization of free fatty acid is called for (493). The rise in plasma free fatty acids is mimicked, for example, by infusions of catecholamines into young lambs, rabbits, or chickens, and is blocked by sympatholytic drugs (279, 402, 494) (see under "Brown Fat"). However, white adipocytes appear poorly innervated by sympathetic nerves (200) (see under "Brown Fat"), and white fat appears unresponsive to a dose of the specific β-agonist, isoproterenol, which depletes the brown fat of lipid (495), indicating that the control of lipolysis in white fat may need further evaluation.

The form of lipid absorbed from the blood stream by thermogenic or other tissues is unclear. While some of the FFA produced by lipolysis appears to be utilized by peripheral tissues such as shivering muscle (496), a large fraction is absorbed by the liver, esterified, and released in the form of low density lipoprotein, a form of lipid that appears to constitute an important source of energy for thermogenic tissues (497).

Glucose The interpretation and significance of the increase in concentration of glucose in the plasma of acutely cold-exposed animals (112, 402, 419, 484) are less clear than that of the increased fatty acid concentrations. The rise is not simply a reflection of increased hepatic glycogenolysis under the influence of adrenaline released from the adrenal medulla in response to cold stress, as is often assumed, because adrenaline in physiological doses, as opposed to pharmacological doses, has little if any direct action on glycogen breakdown in liver (341).

The increase in plasma glucose concentration must arise from relative changes in rate of glucose utilization and rate of glucose production. There can be little doubt that glucose uptake and utilization by skeletal muscle increase during shivering. However, there have been suggestions that uptake in muscle is inhibited by the catecholamines, especially adrenaline, and that glucose is spared for use by the central nervous system (497a); other reports indicate that glucose utilization is increased by adrenaline. These differences appear to be due to experimental conditions (498).

Glucose production increases during cold exposure, but the mechanism and the role of sympathetic stimulation and changes in the secretion of adrenaline and the pancreatic hormones, glucagon and insulin, are unclear. Increased hepatic production of glucose and hepatic uptake of glucose precursors have been recorded in sheep (499) during cold exposure, and increased hepatic gluconeogenesis has been observed in rats especially after prolonged cold exposure (500). Adrenaline, released as a result of cold stress, appears to increase blood glucose in the short term by inhibiting insulin secretion and stimulating an increase in glucagon secretion and hence an increase in glycogenolysis in liver, and in the longer term by increasing the availability of gluconeogenic precursors such as alanine (498, 501). Indeed, increases in α-amino nitrogen have been observed in plasma of rats during prolonged cold exposure, supporting the view that protein metabolism of muscle may participate in thermogenesis (502). However, insulin secretion during cold exposure can be inhibited to some degree, without the mediation of adrenaline, presumably by stimulation of pancreatic α-sympathetics (503); β stimulation may stimulate glucagon release (504, 505). In addition, sympathetic stimulation via the splanchnic nerves stimulates glycogenolysis in hepatic cells (506).

Lactate Changes in plasma lactate during cold exposure are also variable (402, 507) and do not necessarily reflect the production or utilization rates. For example, there is a high oxidation rate of lactate during shivering in dogs (507), and human muscle possesses a pronounced capacity to oxidize lactate (508). Lactate produced by contracting muscle (see under "Heat Production and Substrate Use") also contributes to the increased blood glucose concentration observed during cold exposure, through hepatic resynthesis of glucose and hepatic uptake of lactate increases during cold exposure in sheep (509). This cycle, the Cori cycle, is active even in very young animals (417). The Cori cycle is strongly activated by adrenaline (498) and in this respect a dual role for adrenaline was proposed by Kusaka and Ui (498) on the basis of tracer studies. First, adrenaline promotes conversion of glucose, taken from the blood by the tissues, to lactate; increased levels of free fatty acids appear to facilitate this pathway (510). Second, adrenaline inhibits the secretion of insulin, thereby promoting glycolysis in muscle and complete oxidation of the glucose with the release of energy without the production of lactate. Whether there are indeed two glycolytic pathways, separated from each other in muscle, as suggested by Kusaka and Ui, remains to be confirmed.

Glycerol The rise in plasma glycerol concentration that occurs during cold exposure is indicative of increased lipolysis, since glycerol released by lipolysis is scarcely utilized by white fat and incompletely utilized in brown fat (120, 511). Reutilization of glycerol largely depends on the degree of re-esterification with free fatty acids, a process that requires the presence of glycerol kinase to convert glycerol into glycerol α-phosphate. This enzyme is absent from white fat (512) but is present in brown fat, at least in rats (511), and activity of the enzyme is increased during cold acclimation (513). Its concentration appears to be low in newborn rabbits (120). Adipose tissue that lacks glycerol kinase depends on blood glucose for resynthesis of triglycerides. Glycerol can also enter the glycolytic pathway after phosphorylation by glycerol kinase, but its contribution to thermogenesis in brown fat must be small. Glycerol is also a substrate for hepatic gluconeogenesis, and its uptake by liver has been demonstrated in cold-exposed sheep (305).

Ketone Bodies There have been several reports of elevated ketone body concentration in plasma during prolonged cold exposure in man (514, 515), and in fasted rats the level was greater in the cold than at thermoneutrality (516). On the other hand, there were no significant changes during cold-induced lipolysis in day-old chicks (517). These products of fatty acid metabolism (acetoacetate, β-hydroxybutyrate, and acetone) accumulate in various physiological conditions, as well as in abnormal states such as diabetes (518). Biochemically, they are produced from acetyl CoA, if its production from free fatty acid oxidation is disproportionately high in relation to the state of the pathways of removal, for example, if the availability of oxaloacetate in the citric acid cycle is reduced, as occurs in fasting and diabetes (519). Ketone bodies can be oxidized by tissue and can supply a large proportion (75%) of the energy requirements of organs such as heart muscle, brain, and skeletal muscle (420, 518). They have also been shown to be as important an energy source as free fatty acids in noradrenaline-induced nonshivering thermogenesis in the cold-adapted, nonfasted rat, even though there were no changes in blood levels (518). The site of this utilization is not clear, but the main production site appears to be the liver (518).

Dietary Substrate

The preponderance of evidence favors high fat diets as the most suitable to sustain thermogenesis in the cold (519, 520), but agreement is by no means complete. However, there is a rise in voluntary food intake that persists for weeks during prolonged cold exposure in pigs (521) and newly shorn sheep (522). In addition, cold-exposed ruminants on constant feed intakes show both an increased visceral release and peripheral utilization of volatile fatty acids (499, 523).

POSSIBLE CONSTRAINTS ON COLD THERMOGENESIS

Effect of Substrate Limitation

The capacity for cold thermogenesis, at least in young animals, declines during periods of food deprivation (see under "Effect of Prolonged Cold Exposure and Fasting"), for example, in lambs (37), newborn rabbits (467, 524), neonatal rats (525–527) and human infants (528). This decline is clearly associated with declining availability of substrate (467, 529). However, there are indications that central nervous control supervenes, via the sympathetic system, to reduce thermogenesis in the face of declining reserves and thereby conserve energy (527); for example, depression of thermogenesis in fasting infants is relieved by central nervous lesions (526), and young rabbits fasted at thermoneutrality have a poor metabolic response to cold, despite a high lipid content of their brown adipose tissue (467). The stimulus for the sympathetically mediated decline in thermogenesis may be partly, but not completely, a decline in blood glucose level; administration of glucose alone did not restore thermogenesis to normal (527) and in newborn rabbits initially depleted of substrate by prolonged cold exposure and then fed, thermogenesis was normal but blood glucose was low (467).

Although the depression of thermogenic capacity due to depletion of substrate can be reversed by supplying substrate (467, 528), little information is available about whether substrate availability normally limits the thermogenic capacity (summit metabolism) except in special circumstances. Thermogenesis in brown fat of the newborn rat, for example, appears to be retarded by lack of lipid (see under "Newborn Animals"), and newborn piglets, which have no brown fat and very little white fat at birth, show little mobilization of lipid for shivering thermogenesis during short-term cold exposure until their lipid status has been improved by suckling (482). Whether this is due to activation of adipose tissue lipase or to substrate lack is not clear; in piglets of different origins differences in adipose tissue lipase are associated with differences in thermoregulatory capabilities (491). In newborn piglets, the supply of carbohydrate on which thermogenesis largely depends could also limit the thermogenic capacity or at least the maintenance of thermogenesis; the muscle glycogen is rapidly depleted, the hepatic gluconeogenic capacity is relatively undeveloped (112, 417), and the animals are prone to hypoglycemia (530). However, it has been suggested that piglets are unable to utilize carbohydrate efficiently (531).

There are few reports of attempts to increase thermogenic capacity by increasing the amount of substrate available. In newborn lambs, sucking for the first time does not increase summit metabolism (37). However, infusion of an emulsion of soya bean oil into lambs resulted in a small but significant increase in summit metabolism (532). In newborn rabbits whose mothers had been fasted in late gestation, both the fat content of brown adipose tissue and the thermogenic capacity were increased (237), presumably because of fat mobilization by the mother and placental transfer of lipids.

Drugs, Hormones, and Other Substances

Many hormones and other biologically active substances have been reported to stimulate thermogenesis, especially nonshivering thermogenesis, or to stimulate activity of white and brown adipose tissue in vitro (533 and reviews in refs. 79, 272, and 534). For the most part, their mode of action is obscure, and it is not clear whether they have a normal role in the mechanisms of induction and maintenance of cold thermogenesis (see under "Mechanisms of Thermogenesis in Brown Fat" and "Heat Production and Substrate Use").

Nor is it clear whether their presence in physiological levels provides constraints on summit metabolism that would be relieved by pharmacological levels.

Catecholamines Turnover and excretion of catecholamines are almost invariably increased during cold exposure, and the role of the increased production of noradrenaline in stimulating nonshivering thermogenesis, lipid turnover, membrane depolarization, substrate transport, and blood flow through adipose tissue (see under "Brown Fat" and "Circulatory Adjustments During Cold Thermogenesis") are as well understood as the effects of any hormone on thermogenesis. Despite the many known metabolic and physiological effects of adrenaline (341), adrenomedullary secretion of this hormone appears to have only a minor role in both nonshivering and shivering thermogenesis. For example, in curarized rats adrenodemedullation leads to only a small decrease in cold thermogenesis (82), and in adrenalectomized lambs supported with adrenal steroids, summit metabolism, which is contributed to by both shivering and nonshivering thermogenesis, is only about 10% below that of controls and is restored by adrenaline infusion (G. Alexander and A. W. Bell, unpublished data). It is not clear whether the catecholamines have a significant role in shivering or any nonshivering thermogenesis in piglets, which lack brown fat, but cold exposure increases their excretion of catecholamines and the plasma concentration of adrenaline (112, 535). Adrenaline also has a small thermogenic effect in newborn piglets (111, 536). In the young fowl, which also lacks brown fat, there is an unidentified form of nonshivering thermogenesis, but the catecholamines appear to have little if any role in this process (422).

Exhaustion of noradrenaline stores with consequent failure of nonshivering thermogenesis during prolonged cold exposure has been suggested as a cause of the neonatal cold injury syndrome in premature human infants (537).

Attempts to increase summit metabolism by exogenous supplementation of endogenous catecholamines have been successful in very young rats (538), but not in older neonatal rats or in lambs (see under "Brown Fat").

Serotonin Serotonin (5-hydroxytryptamine), a ubiquitous neurotransmitter substance with physiological effects similar to those of noradrenaline, is present in high concentrations in brown fat (539). This compound can increase cyclic AMP levels, stimulate lipolysis, and promote thermogenesis in this tissue (533, 540, 541), but it appears to exert these effects by releasing noradrenaline from the neuronal endings in brown fat (541); therefore, its administration would be unlikely to increase summit metabolism.

Hypophyseal Peptide Hormones The hypophyseal peptide hormones, adrenocorticotropin (ACTH), thyroid-stimulating hormone (TSH), and growth hormone, are adipokinetic, at least in certain species (79, 534, 542, 543). In vitro ACTH and TSH induce lipolysis in brown fat (262), and ACTH appears to stimulate thermogenesis in brown fat in vivo in newborn rabbits (256), but neither ACTH nor TSH, when given as short-term infusions into lambs, significantly affected summit metabolism (532). However, significant increases in summit metabolism were obtained with longer-term administration of TSH over 3 days; presumably, this resulted from increased thyroxine secretion. ACTH also failed to stimulate thermogenesis in lambs under thermoneutrality (532). These results appear to represent species differences, so frequently encountered with peptide hormone action (542). Although growth hormone stimulates lipolysis in white fat in vitro, it has no effect on brown fat under similar conditions (544), and there appear to be no reports on its effect on cold thermogenesis.

Adenyl Cyclase–Cyclic AMP System The adenyl cyclase–cyclic AMP system (see under "Mechanisms of Thermogenesis in Brown Fat"), is regarded as the mediator of β-adrenergic metabolic effects of hormones, such as catecholamines, ACTH, serotonin, and TSH, in brown adipose tissue, although the addition of cyclic AMP either in vitro or in vivo does not appear to mimic the hormonal effects completely (533, 545, 546). The role of cyclic AMP can also be examined by addition of theophylline or caffeine, inhibitors of phosphodiesterase, an enzyme that degrades cyclic AMP. Both cyclic AMP and theophylline are calorigenic in brown fat in vitro (533, 547, 548), and theophylline has a large calorigenic effect in cold-adapted rats, but not if maximum thermogenesis is already induced by noradrenaline (545, 549). In addition, intravenous caffeine has no effect on summit metabolism in lambs (532), so there is no evidence that cold thermogenesis is normally limited by the level of activity of the adenyl cyclase–cyclic AMP system.

Carnitine Carnitine and carnitine acyltransferases facilitate transport of fatty acids across the mitochondrial membrane (see under "Mechanisms of Thermogenesis in Brown Fat"), and optimal rates of fatty acid oxidation can occur in brown fat only in their presence (550). Although carnitine, injected intramuscularly into young lambs for 4 days, had no effect on summit metabolism (532), carnitine injected into adult rats and mice enhanced noradrenaline-induced thermogenesis (551) and carnitine injected into newborn rats and rabbits enhanced cold thermogenesis in the absence of injected noradrenaline (552, 553). It appears, therefore, that carnitine status could limit cold thermogenesis not only in suckling rats and rabbits whose supply depends upon placental transfer or absorption from milk, but also in adult rats and mice. The milk yield of carnitine is much reduced by cold exposure of the lactating animal, at least in goats (509).

Glucagon Glucagon, the hormone secreted by the α cells of the islets of Langerhans in the pancreas, plays an essential role in glucose homeostasis through its stimulatory effect on hepatic glycogenolysis and gluconeogenesis (554) and effects on insulin secretion (560). However, glucagon also has a lipolytic effect on brown fat of rats (555) and is also lipolytic in chickens that lack brown fat (556). Under thermoneutral conditions, this hormone is calorigenic in adult rats (557) and newborn rabbits (256), but not in the newly hatched domestic chicken (558); indeed, glucagon inhibits shivering in some birds (559). The mechanism of the calorigenic action is not clear. It is widely believed to act via the adenyl cyclase–cyclic AMP system, but there is some evidence to the contrary; namely, calorigenesis is not prevented by β blockade with propranalol (256), and, although glucagon mediates the short-term glycogenolytic response to adrenaline (501), this would not appear to be important in the thermogenic response to cold since adrenaline plays only a minor role in this response. The effect of cold on glucagon dynamics does not appear to have been studied, but summit metabolism, at least in lambs, does not appear to be limited by the rate of glucagon secretion, since the response to cold is uninfluenced by concurrent infusion of the hormone (532).

Insulin Likewise, there is no unified hypothesis about the role of the other islet hormone, insulin, in cold thermogenesis, although there are many reports on its effect in cold-exposed animals and thermogenic tissues. Many studies deal with the effect of added insulin (79, 561, 562), and insulin has been reported to have a marked calorigenic effect in liver, muscles, and brain of rats and monkeys (79). However, changes in insulin turnover during cold exposure are unknown; blood levels in lambs decline during acute cold exposure and also in rats chronically cold exposed (563). Insulin secretion in man and young lambs is partially inhibited by catecholamines, which are likely to be present in elevated concentrations during cold exposure (564, 565). In dogs there may be a direct sympathetic inhibition of insulin secretion in the cold (503). Insulin stimulates glucose uptake and metabolism and decreases fatty acid release due to inhibition of lipolysis and/or accelerated lipogenesis (534, 544). In many tissues insulin appears to inhibit transport of fatty acids across cell membranes (561) and to inhibit catecholamine-induced lipolysis by lowering cyclic AMP levels (534). It also inhibits depolarization induced in brown fat cells by noradrenaline (566) and appears to inhibit the increased lipoprotein lipase activity observed in rat tissues during cold exposure (470); this enzyme regulates the uptake of circulating triglyceride. These various effects of insulin are antithermogenic and indicate that thermogenesis from fatty acid oxidation is promoted when insulin levels are low, a condition that would be unfavorable to increased utilization of glucose for thermogenesis, but would conserve glucose for use by central nervous tissue. This mutual exclusiveness of fat and carbohydrate utilization, exemplified in the so-called glucose-fatty acid cycle, does not appear to apply at high levels of energy (see under "Heat Production and Substrate Use"), but the mechanism of disruption of this mutual exclusion is unknown.

High plasma concentrations of insulin are, however, not necessarily inhibitory to thermogenesis from fatty acid oxidation; neonatal rabbits are capable of marked thermogenesis (see under "Newborn Animals"), although the concentration of circulating insulin is high (50 μU/ml) and slow to fall after birth (468). This observation led Hardman et al. (468) to suggest that the high circulating level of insulin in the neonatal rabbit may be responsible for inhibition of the release of fatty acids from brown fat for metabolism elsewhere. Conditions that induce a decline in insulin levels, such as starvation and cold exposure, would promote release of fatty acid from the tissue (534).

In view of the very high circulating levels of glucose and lactate in the blood of lambs under summit conditions (see under "Substrates in Blood"), it would appear that the availability of these substrates could not limit summit metabolism. Indeed, there is an inverse correlation between summit metabolism and blood glucose level (567). Nevertheless, the hyperglycemia could be associated with relative insulin insufficiency, especially in view of the inhibitory effects of catecholamines. However, when insulin was given to lambs, summit metabolism was slightly depressed rather than elevated (532).

It may prove profitable to consider the function of the islet hormones in terms of the insulin to glucagon ratio, which may carry out fine tuning of substrate availability, coarse tuning being provided by other inputs.

Corticosteroids Secretions of the adrenal cortex are essential for cold thermogenesis as demonstrated by adrenalectomy and replacement with cortical extracts (568, 569). In addition, marked increases in plasma corticosteroid levels occur during cold exposure; for example, in man (570), neonatal lambs (565), rats (571), dogs (572), and rabbits (573), suggesting an increased cortisol secretion rate, which has indeed been reported for sheep (574). The role of these steroids in cold thermogenesis is likely to be complex, since they interact metabolically with other hormones, such as thyroid hormone, and act directly on the thermoregulatory centers of the brain (575).

Although high doses of corticosteroids in intact rats and sheep improve survival under prolonged cold conditions (576, 577) the metabolic response of lambs to cold was not affected by exogenous cortisol, either infused as the soluble form, cortisol sodium succinate, during summit metabolism or administered previously in the form of cortisol acetate, injected intramuscularly over 3 days (532). Experiments on adrenalectomized lambs indicate that corticosteroids play a largely permissive role in the summit response, but there are species, such as the hedgehog, in which corticosteroids have marked thermogenic effects in at least some brown fat depots (213). High doses of cortisol can be toxic (578), which could account for the decrease in fatty acid oxidation observed in brown adipose tissue of young rats (579, 580).

Thyroid Hormones The necessity of the thyroid for survival in the cold and the calorigenic effects of thyroid hormones, thyroxine (T_4) and the more potent triiodothyronine (T_3), have been recognized for a long time (79, 534, 581). For example, thyroid hormones given to rats or lambs over periods of weeks more than double resting metabolism (138, 582) and can increase thermogenesis in lambs by a small amount even when given twice daily for only 3 days (532). However, the effect of thyroid hormone on the increment of heat production that is due to cold exposure appears to be much smaller than the effect of the hormone on resting metabolism (138).

Thyroid hormones have been given to animals in an attempt to mimic the effects of cold adaptation; the effects on hypertrophy of brown fat, enzyme activity, substrate content, mitochondrial function and the activity of the Na^+/K^+ pump in brown fat and skeletal muscle have also been observed (226, 583–590). However, there is now enough evidence to indicate that chronic treatment with thyroxine is not a satisfactory model for cold adaptation (176, 591). In brown fat of young rats, cold induced an increase in phospholipid and altered the fatty acid pattern of phospholipids, whereas thyroxine prevented this increase in phospholipid and had small and different effects on its composition, but promoted an accumulation of triglycerides (592). Rats thryoidectomized after cold adaptation can be maintained on low doses of thyroxine, but still display increased nonshivering thermogenesis (593). In addition, animals can be cold acclimated without evidence of increased thyroid activity (594, 595), and the hyperthyroidism observed, on occasions, appears to be the result of increased food intake that accompanies cold exposure without being essential to the process of cold acclimation (176, 595). Nevertheless, the presence of thyroid hormones is necessary for thermogenesis due to cold or catecholamine treatment; this feature of cold resistance is inhibited in hypothyroidism, whether it be natural or surgically or pharmacologically induced (138, 581, 596–598).

The exact role of thyroid hormones in the control of cold thermogenesis remains to be clarified. There is recent evidence that interactions with the catecholamines and sympathetic nervous system are involved (590, 599), although at present they are largely regarded as playing a permissive role in the response of thermogenesis to other hormones (591). In acute cold exposure there is little if any change in the thyroxine turnover or blood levels of thyroxine or thyroid-stimulating hormone in adult rabbits (573, 594) or man (570, 600). However, there is a prompt increase in TSH in infants and children (600), and in the young pig, cold exposure for several hours increases thyroxine level, and the calorigenic action of adrenaline is potentiated by thyroxine (535, 536). Although a thermogenic effect of thyroxine is apparent within an hour or two of subcutaneous injection, the increase in oxygen consumption with or without adrenaline is less than 40% of the resting level (536, 601), and it is not clear whether this reflects a sort of nonshivering thermogenesis that is independent of brown fat, a pharmacological artifact, or whether these hormones have a role in shivering, the only identified form of thermogenesis in normal piglets. Neither TSH nor T_3 given to lambs has any effect on summit metabolism measured about an hour later (532).

Respiratory Gases

Reduction of the partial pressure of oxygen in the inspired air, either by changes in barometric pressure or alterations in the percentage of oxygen, tends to reduce the oxygen consumption and body temperature of animals, particularly if they are exposed to low temperatures (602). The degree of reduction of thermogenesis varies with body size and age. For example, the effect occurs at thermoneutrality in newborn rats breathing 18% oxygen, but in rats a few days old no reduction occurs until the oxygen content of air is reduced below 15% (98); in kittens and adult guinea pigs at thermoneutrality there is no reduction unless the air contains <10% oxygen, but as ambient temperature is decreased the hypoxic effect appears and becomes increasingly severe (97, 603). The same phenomenon has been recorded in the newborn of a variety of other species (man, monkey, dog, rabbit, and lamb) (26, 604) and in adults (man, dog, miniature pigs, and rats) (605–608). The mechanisms of hypoxic depression of cold thermogenesis are still debated. There is supporting evidence for the suggestion that it is due to selective inhibition of nonshivering thermogenesis (606). For example, 12–13% oxygen in inspired gas reduced summit metabolism of newborn lambs by 25%, but there was no effect in week-old lambs at which stage nonshivering thermogenesis was minimal (604). In addition, in adult miniature pigs, which lack brown fat, exposure to 10% oxygen at an ambient temperature of 7°C has no effect either on the intensity of visible shivering or oxygen consumption. Heat production in periaortic brown fat of rats has also been shown to be abolished by hypoxia (609). On the other hand, cold thermogenesis in the adult guinea pig, which also lacks brown fat (137), is reduced by hypoxia (97); therefore, shivering may be suppressed by hypoxia as well. It is obvious that, if the hypoxia is severe enough, the insufficiency of oxygen supply to thermogenic tissues must limit the rate of oxidation and heat production. However, the sensitivity of the tissues to oxygen partial pressure and the balance of tissue demands against the supply of oxygen do not appear to have been considered in studies (97, 610) that have led to the suggestion that oxygen lack at tissue level is directly responsible for the thermogenic depression, and hence for the decline in body temperature. The very low level of oxygen in the venous blood draining from noradrenaline-stimulated brown fat of newborn rabbits (241) and in the mixed venous blood of lambs during summit metabolism (242) indicates that there may be very little reserve in oxygen supply under conditions of maximum thermogenic stimulation. On the other hand, there are several findings indicating that hypoxia cannot always act in this way (582). Hypoxic effects in rats can be abolished by central nervous lesions that do not affect thermogenesis; the uncoupling agent dinitrophenol (see under "Mechanisms of Thermogenesis in Brown Fat") can restore oxygen consumption to normal during depression by hypoxia (611), and in pyrogen-treated rats hypoxia increases oxygen consumption but decreases body temperature. In addition, hypoxia appears to have independent effects on heat production and on body temperature in newborn rats and adult guinea pigs (603, 612). Largely on the basis of these observations, Donhoffer's group concluded that hypoxia acts primarily on the central thermoregulatory mechanisms that control and coordinate thermoregulatory heat production and heat conservation mechanisms (603). Whether this applies under conditions of maximum thermogenic stimulation remains to be shown. In dogs, the mechanism of hypoxia depression does not seem to involve a diminished mobilization of substrate (605), and in lambs it is not due to a reduction in cardiac output (604). There are also indications that hypoxia during the birth process could impair the thermogenic potential at birth and contribute to neonatal mortality (613, 614).

In contrast to hypoxia, increases in the oxygen content of inspired gas (hyperoxia) to around 40% can increase maximum oxygen consumption, at least in neonatal rats (98) and lambs (604); in the latter, the maintenance of summit metabolism and cardiac output are also improved. However, the increase is small (10–20%) and the mechanisms are unclear.

Increases in the carbon dioxide content of inspired air (hypercapnia) to 5% or even higher also reduce cold thermogenesis, for example, in adult rats and guinea pigs (615, 616), but the effect appears to be small or absent in newborn animals such as rats and rabbits (26, 617). Hypercapnia, like hypoxia, suppresses nonshivering thermogenesis in cold-adapted rats (616) and is believed to act, at least in part, via the central nervous system, but differences between the effects of hypoxia and hypercapnia indicate that the mechanisms are not identical (615). Central involvement is also indicated by findings that severe hypercapnia (7.5% and 15% CO_2), continued for 1 hr in the adult guinea pig, decreases by 4°C the skin and cervical spinal cord temperatures, at which shivering is initiated (618). However, because of metabolic adaptation over about 3 days the effects of hypercapnia in the guinea pig appear to be transient (618).

Acidosis and Pulmonary Ventilation

During severe cold exposure of lambs there is a pronounced acid-base disturbance (431). The pH of venous blood declines from about 7.3 to 7.1 during summit metabolism, and the pH of arterial blood likewise declines from 7.3 to 7.2. Since there is a clear decrease in base excess and a small increase in arterial P_{CO_2} (5 mm Hg), the acidosis is of mixed respiratory and metabolic origin. The incompleteness of respiratory compensation raises the question of whether or not pulmonary ventilation reaches its physiological limit under summit metabolism. However, this does not appear to be so because arterial P_{O_2} is maintained at about the normal level. Adequacy of pulmonary ventilation during maximum oxygen uptake has also been indicated in experiments with small rodents (52). Restoration of blood pH to control levels of infusion of base (bicarbonate or Tris) had no effect on summit metabolism (G. Alexander, A. W. Bell, and J. R. S. Hales, unpublished data).

Blood Pressure

Cold thermogenesis is often accompanied by a small increase in systemic blood pressure, for example, about 10 mm Hg, in young lambs during summit metabolism (242) and in squirrel monkeys at 10°C (355), but further marked increases induced by sympathetic agonists are associated with a marked depression in oxygen consumption (351, 355) (see under "Sympathetic Control of Nonshivering Thermogenesis"). This depression does not appear to be due to a direct effect of the agonists, since it is abolished by antagonists, but it appears to be partly due to reflex inhibition of central controlling mechanisms; it is reduced in animals subjected to bilateral sinoaortic denervation (355). The depression could also be partly due to reduced blood flow to thermogenic tissues.

Trauma

Trauma transiently impairs thermoregulation (272), and injured animals often die in hypothermia (619, 620). Experimental trauma is usually applied in the form of bilateral restriction of blood flow (ischemia) to the hind limbs. During the period of limb ischemia in rats, both shivering and heat loss mechanisms are inhibited; the threshold ambient temperatures at which shivering appears are lowered, and those for the opening of arteriovenous anastomoses are raised. When the restriction in blood flow is removed, the pathways for stimulation of nonshivering thermogenesis are also inhibited (621, 622). Inhibition of nonshivering thermogenesis likewise occurs in neonatal rabbits given the same treatment (620). Heat production in brown fat in traumatized rats appears normally responsive to catecholamines (622), so it appears that central control of nonshivering is inhibited by trauma. In animals that survive severe trauma, the transient suppression of thermogenesis is replaced by a period of elevated heat production of a nonshivering nature, but the significance, origin, and control of this temperature rise are poorly understood (272).

Endotoxins

Thermogenesis can be induced by injections of bacterial endotoxins (pyrogens) as well as by cold exposure, and the response to endotoxins (fever) has much in common with cold-induced thermogenesis, for example, the mobilization of substrate (623) and involvement of both shivering and nonshivering thermogenesis in brown fat (624, 625). Pyrogen-induced thermogenesis may be mediated by prostaglandins in the central nervous system, and its central control appears to be distinctly from that of cold thermogenesis. Cold thermogenesis can be abolished by lesions in the anterior hypothalamic-preoptic area of the brain, but lesions do not inhibit pyrogen-induced thermogenesis (626).

Although fever is dealt with elsewhere in this volume (Chapter 4) it is relevant to consider whether summit metabolism is influenced by bacterial endotoxin. The only available evidence indicates that summit metabolism in young lambs is significantly reduced by endotoxin (*Escherichia coli,* Bactolipopolysaccharide) (532), but a different result might have been obtained had the lambs been first sensitized to the toxin (627).

CAN RESTRAINTS BE LIFTED?

We have seen that summit metabolism is readily reduced by a variety of environmental agents, such as hypoxia, hypercapnia, starvation, prolonged exposure to very severe cold, trauma, endotoxins, pharmacological blocking agents, and high doses of catecholamines, but there are few agents that increase summit metabolism and the ability to maintain high metabolic rates. Indeed, progress in this area has been much slower than in the parallel and rapidly expanding field of exercise and athletic performance. This may be partly because the potential for thermogenesis is rarely assessed adequately. The most spectacular of these facilitatory agents is cold acclimation, at least among rodents and other species that show substantial activation and hypertrophy of brown fat during prolonged exposure to moderate cold, and there are less well substantiated or defined increases in shivering and in nonshivering thermogenesis in tissues other than brown fat. Hyperoxia, hyperthyroidism, and hyperlipidemia also have small effects, exercise training may improve the capacity for shivering (378), and noradrenaline administration may be effective in relatively undeveloped newborns (538). It is, however, far from clear what physiological or biochemical factors set the ceiling on the thermogenic response of which an animal is capable, and it is likely that these factors will vary with species, age, and physiological factors such as physical fitness, nutritional state, and previous thermal history. Theoretically, there is a number of possible limiting factors.

First, the capacity of the thermogenic tissue for production and utilization of ATP, or production of heat by an uncoupling process, must have a ceiling that could limit thermogenesis. This may indeed operate, because cold acclimation in some species increases the activities of a variety of enzymes in brown fat, especially cytochrome oxidase, (224, 225, 628), the mitochondria enlarge, and the phospholipid composition of their membranes change (589). Changes indicative of increased metabolism are also seen in muscle of cold-acclimated animals.

The supply of substrate to the thermogenic machinery is also clearly important, but there is a little evidence (532) that this is normally limiting except after periods of deprivation or excessive utilization. The poor thermogenic response to cold in some newborn animals, such as rabbits, rats, and piglets, could be limited by substrate availability; in newborn rabbits cold thermogenesis is correlated with the fat content of brown fat (343). Experimental supplementation of substrate supply in these species would be illuminating. In lambs, the level of summit metabolism is inversely correlated with the blood glucose level (correlation coefficient is approximately -0.9) and the level of free fatty acids in blood (567), so substrate utilization rate, rather than supply, appears to be limiting. Likewise, there is little evidence that normal circulating levels of hormones that affect mobilization and utilization of substrate are suboptimal.

The proportion of brown fat in the body appears to be correlated with the thermogenic potential in at least some animals, such as newborn rabbits (243) and rats (226), and changes in the weight of brown fat achieved through cold acclimation or manipulation of maternal diet (237, 240) are associated with changes in cold tolerance. However, the correlation is low (correlation coefficient is only 0.4 in newborn rabbits) and is not apparent in newborn lambs, animals in which shivering also makes a substantial contribution to summit metabolism (567). Nor was the proportion of skeletal muscle in lambs correlated with summit metabolism (567), so the major limitations of summit metabolism appear to lie elsewhere.

Pulmonary gas exchange and the delivery of oxygen to the thermogenic tissues could also place a ceiling on summit metabolism (538) and merit further investigation. There has, however, been considerable study of the limitations placed on capacity for muscular exercise by the capacity of the cardiovascular system in man. A ventilatory limit in man has been ruled out, because voluntary maximum ventilation exceeds the ventilation at maximum work-induced oxygen consumption (629). Cardiac output appears to play a role in limiting oxygen consumption. For example, in man the oxygen supply to one large group of exercising muscles appears to be limited when another large group of muscles is exercised at the same time (51), and maximum oxygen uptake in exercising man and in lambs during summit metabolism is correlated with cardiac output (55, 242). In addition, cardiac output in exercising man and dog reaches a plateau before maximum aerobic oxygen consumption is achieved (630). However, a large part of the variability in oxygen consumption, in exercising man and cold-exposed lambs, is associated with variations in arteriovenous difference in oxygen content of blood (242, 631), suggesting that the amount or efficiency of thermogenic tissue plays an important limiting role. Thus, at the present, the biochemical mechanisms in thermogenic tissues and their oxygen supply appear to place the major limitations on summit metabolism, except perhaps in relatively undeveloped newborn animals, in which substrate limitations may play a major role.

CONCLUSION

The level of cold thermogenesis that birds or mammals can achieve and maintain is of vital importance for survival during severely inclement weather unless the animal can escape from the constraints of strict homeothermy and regulate body temperature at a new low level in torpidity or hibernation (454, 455).

The major sources of cold thermogenesis have been clearly identified with accelerated substrate utilization in brown fat mitochondria and with shivering, but there appear to be minor sites of nonshivering thermogenesis in liver, skeletal muscle, and other tissues, particularly after cold acclimation; these are not well understood or even well substantiated. Definition and precise measurement of the thermogenic potential of these various sources of thermoregulatory heat production in individuals present certain difficulties, particularly in regard to criteria that the measured oxygen consumption is indeed maximum and that only one source of thermogenesis is being measured. If the thermal conditions do not evoke a maximum response, blockade of one source of thermogenesis may result in enhancement of another. The first criterion can be satisfied in animals that maintain strict homeothermy if body temperature is induced to fall slowly during the measurement, but uncertainties will remain in animals that can exhibit thermolability in the guise of cold-induced neonatal poikilothermy, habituation, torpidity, or hibernation. However, crude average estimates can be obtained by constructing graphs relating metabolism to ambient temperature.

The second criterion can be at least partially achieved by measuring summit metabolism before and after pharmacological blockade. However, there are no available methods for blocking nonshivering thermogenesis in tissues other than brown fat, and attempts to measure this as residual thermogenesis in intact animals after pharmacological blockade of both brown fat and shivering would be subject to serious error and artifact. We are left with cumbersome indirect methods based on surgical intervention.

Failure to observe these criteria will result in incorrect estimates of the various classes of thermogenesis and misinterpretation of the effects of various agents on them. Unfortunately, available methods for estimating maximum thermogenesis have been infrequently used in this field of thermal physiology.

Summit metabolism represents a very intense physiological effort. Much remains to be learned about the nature and integration of respiratory, cardiovascular, neural, hormonal, and biochemical mechanisms upon which it depends, but it is not surprising that it is easy to depress and difficult to enhance.

ACKNOWLEDGMENTS

I am grateful to librarians Moya Frost and Sandra Drake for assistance in obtaining articles and checking references, and I appreciate the tenacity of typist Helen Williams in deciphering the handwritten draft. The constructive criticisms of various colleagues—Alan Bell, John Bennett, Kevin Ward, Denise Stevens, and Bob Hales—were appreciated.

REFERENCES

1. Whittow, G. C. (ed.). (1970–73). Comparative Physiology of Thermoregulation, Vols. I, II, and III. Academic Press, Inc., New York.
2. Stevens, E. D. (1973). The evolution of endothermy. J. Theor. Biol. 38:597.
3. Vernberg, F. J., and Vernberg, W. B. (1970). Aquatic invertebrates. *In* G. C. Whittow (ed.), Comparative Physiology of Thermoregulation. Vol. 1, pp. 1–14. Academic Press, Inc., New York.
4. Somero, G. N. (1972). Molecular mechanisms of temperature compensation in aquatic poikilotherms. *In* F. E. South et al. (eds.), Hibernation and Hypothermia, Perspectives and Challenges, pp. 55–80. Proceedings of a symposium, January, 1971, Aspen, Colorado. Elsevier-North Holland Publishing Company, New York.
5. Fry, F. E. J., and Hochachka, P. W. (1970). Fish. *In* G. C. Whittow (ed.), Comparative Physiology of Thermoregulation, Vol. I, pp. 79–134. Academic Press, Inc., New York.

6. Harri, M., and Hedenstam, R. (1972). Calorigenic effect of adrenaline and noradrenaline in the frog *Rana temporaria.* Comp. Biochem. Physiol. (A) 41:409.
7. Cloudsley-Thompson, J. L. (1970). Terrestrial invertebrates. *In* G. C. Whittow (ed.), Comparative Physiology of Thermoregulation. I. Invertebrates and Nonmammalian Vertebrates, pp. 15–77. Academic Press, Inc., New York.
8. Ishay, J. (1972). Thermoregulatory pheromones in wasps. Experentia 28:1185.
9. Southwick, E. E., and Mugaas, J. N. (1971). A hypothetical homeotherm: the honeybee hive. Comp. Biochem. Physiol. (A) 40:935.
10. Aleksiuk, M. (1976). Metabolic and behavioural adjustments to temperature change in the red-sided garter snake (*Thamnophis sirtalis parietalis*): an integrated approach. J. Therm. Biol. 1:153.
11. Tromp, W. I., and Avery, R. A. (1977). A temperature-dependent shift in the metabolism of the lizard *Lacerta vivipara.* J. Therm. Biol. 2:53.
12. Asplund, K. K. (1970). Metabolic scope and body temperatures of whiptail lizards (*Cnemidophorus*). Herpetologica 26:403.
13. Vinegar, A., Hutchison, V. H., and Dowling, H. G. (1970). Metabolism, energetics and thermoregulation during brooding of snakes of the genus Python (*Reptilia, Boidae*). Zoologica (N.Y.) 55(2):19.
14. Louw, G., Young, B. A., and Bligh, J. (1976). Effect of thyroxine and noradrenaline on thermoregulation, cardiac rate and oxygen consumption in the monitor lizard *Varanus albigularis albigularis.* J. Therm. Biol. 1:189.
15. Scholander, P. F., Hock, R., Walters, V., Johnson, F., and Irving, L. (1950). Heat regulation in some Arctic and tropical mammals and birds. Biol. Bull. 99:237.
16. Burton, A. C., Edholm, O. G., (1955). Man in a Cold Environment, pp. 148–161. Edward Arnold, London.
17. Mitchell, D. (1974). Physical basis of thermoregulation. *In* D. Robertshaw (ed.), MTP International Review of Physiology, Vol. 7, Environmental Physiology I, pp. 1–32. University Park Press, Baltimore.
18. Webster, A. J. F. (1974). Physiological effects of cold exposure. *In* D. Robertshaw (ed.), MTP International Review of Physiology, Vol. 7, Environmental Physiology I, pp. 33–69. University Park Press, Baltimore.
19. Hart, J. S. (1957). Climatic and temperature induced changes in the energetics of homeotherms. Rev. Can. Biol. 16:133.
20. Alexander, G. (1961). Temperature regulation in the new-born lamb. III. Effect of environmental temperature on metabolic rate, body temperatures, and respiratory quotient. Aust. J. Agric. Res. 12:1152.
21. Alexander, G. (1975). Body temperature control in mammalian young. Br. Med. Bull. 31:62.
22. Rosen, R. C. (1975). Ontogeny of homeothermy in *Microtus pennsylvanicus* and *Octodon degus.* Comp. Biochem. Physiol. (A) 52:675.
23. Holyoak, G. W., and Stones, R. C. (1971). Temperature regulation of the little brown bat *Myotis lucifugus* after acclimation at various ambient temperatures. Comp. Biochem. Physiol. (A) 39:413.
24. Scopes, J. W. (1966). Metabolic rate and temperature control in the human baby. Br. Med. Bull. 22:88.
25. Vârnai, I., Farkas, M., and Donhoffer, S. (1970). Thermoregulatory effects of hypoxia in new-born rats and their correlation to body temperature and heat production prior to and after exposure to hypoxia. Acta Physiol. Acad. Sci. Hung. 38:19.
26. Vârnai, I., Farkas, M., and Donhoffer, S. (1971). Thermoregulatory responses to hypercapnia in the newborn rabbit: comparison with the effect of hypoxia. Acta Physiol. Acad. Sci. Hung. 40:145.

27. Hey, E. N., and Mount, L. E. (1967). Heat loss from babies in incubators. Arch. Dis. Child. 42:75.
28. Alexander, G. (1974). Heat loss from sheep. *In* J. L. Monteith and L. E. Mount (eds.), Heat Loss from Animals and Man: Assessment and Control, pp. 173–203. Proceedings of the Twentieth Easter School in Agricultural Science, University of Nottingham. Butterworths, London.
29. Mount, L. E. (1968). The Climatic Physiology of the Pig. Monographs of the Physiological Society, No. 18. p. 271. Arnold, London.
30. Hart, J. S., Héroux, O., Cottle, W. H., and Mills, C. A. (1961). The influence of climate on metabolic and thermal responses of infant caribou. Can. J. Zool. 39:845.
31. Alexander, G. (1962). Temperature regulation in the new-born lamb. IV. The effect of wind and evaporation of water from the coat on metabolic rate and body temperature. Aust. J. Agric. Sci. 13:82.
32. Holmes, C. W., and McLean, N. A. (1975). Effects of air temperature and air movement on the heat produced by young Friesian and Jersey calves, with some measurements on the effects of artificial rain. N. Z. J. Agric. Res. 18:277.
33. Lentz, C. P., and Hart, J. S. (1960). The effect of wind and moisture on heat loss through the fur of newborn caribou. Can. J. Zool. 38:679.
34. Frisch, J., Øritsland, N. A., and Krog, J. (1974). Insulation of furs in water. Comp. Biochem. Physiol. (A) 47:403.
35. Giaja, J. (1925). Le métabolisme de sommet et la quotient métabolique. Ann Physiol. Physiochim. Biol. 1:596.
36. Gelineo, M. S. (1934). Influence du milieu thermique d'adaptation sur la thermogènése de homéothermes. Ann. Physiol. Physiochim. Biol. 10:1083.
37. Alexander, G. (1962). Temperature regulation in the new-born lamb. V. Summit metabolism. Aust. J. Agric. Res. 13:100.
38. Mount, L. E., and Stephens, D. B. (1970). The relation between body size and maximum and minimum metabolic rates in the new-born pig. J. Physiol. (Lond.) 207:417.
39. Bennett, J. W. (1972). The maximum metabolic response of sheep to cold: effects of rectal temperature, shearing, feed consumption, body posture, and body weight. Aust. J. Agric. Res. 23:1045.
40. Depocas, F. J., Hart, S., and Héroux, O. (1957). Energy metabolism of the white rat after acclimation to warm and cold environments. J. Appl. Physiol. 10:393.
41. Rosenmann, M., and Morrison, P. (1974). Maximum oxygen consumption and heat loss facilitation in small homeotherms by He-O_2. Am. J. Physiol. 226:490.
42. Musacchia, X. J. (1976). Helium-cold hypothermia, an approach to depressed metabolism and thermoregulation. *In* L. Janský and X. J. Musacchia (eds.), Regulation of Depressed Metabolism and Thermogenesis, pp. 137–156. Proceedings of a symposium, October, 1974, Prague. Charles C Thomas Publisher, Springfield, Illinois.
43. Lindsay, A. L., and Bromley, L. A. (1950). Thermal conductivity of gas mixtures. Ind. Eng. Chem. 42:1508.
44. Alexander, G. (1961). Temperature regulation in the new-born lamb. II. A climatic respiration chamber for the study of thermoregulation with an appendix on a correction for leaks and imperfect measurement of temperature and pressure in closed circuit respiration chambers. Aust. J. Agric. Res. 12:1139.
45. Gosselin, R. F. (1949). Acute hypothermia in guinea pigs. Am. J. Physiol. 157:103.
46. Adolph, E. F. (1950). Oxygen consumptions of hypothermic rats and acclimatization to cold. Am. J. Physiol. 161:359.

47. Alexander, G., and Williams, D. (1966). Teat-seeking activity in new-born lambs: the effects of cold. J. Agric. Sci. 67:181.
48. Alexander, G., Bell, A. W., and Hales, J. R. S. (1973). Effects of cold exposure on tissue blood flow in the new-born lamb. J. Physiol. (Lond.) 234:65.
49. Hutchinson, K. J., and McRae, B. H. (1969). Some factors associated with the behavior and survival of newly shorn sheep. Aust. J. Agric. Res. 20:513.
50. Denckla, W. D. (1971). A new interpretation of $V_{O_2 max}$. J. Appl. Physiol. 31:168.
51. Secher, N. H., Clausen, J. P., Klausen, K., Noer, I., and Trap-Jensen, J. (1977). Central and regional circulatory effects of adding arm exercise to leg exercise. Acta Physiol. Scand. 100:288.
52. Pasquis, P., Lacaisse, A., and DeJours, P. (1970). Maximal oxygen uptake in four species of small mammals. Respir. Physiol. 9:298.
53. Hart, J. S., and Janský, L. (1963). Thermogenesis due to exercise and cold in warm- and cold-acclimated rats. Can. J. Biochem. Physiol. 41:629.
54. Popovic, V., Kent, K., Mojovic, N., Mojovic, B., and Hart, J. S. (1969). Effect of exercise and cold on cardiac output in warm- and cold-acclimated rats. Fed. Proc. 28:1138.
55. Mitchell. J. H., Sproule, B. J., and Chapman, C. B. (1958). The physiological meaning of the maximal oxygen intake test. J. Clin. Invest. 37:538.
56. Macnab, R. B. J., Conger, P. R., and Taylor, P. S. (1969). Differences in maximal and submaximal work capacity in men and women. J. Appl. Physiol. 27:644.
57. Bergh, U., Kanstrup, I.-L., and Ekblom, B. (1976). Maximal oxygen uptake during exercise with various combinations of arm and leg work. J. Appl. Physiol. 41:191.
58. Behnke, A. R., and Yaglou, C. P. (1951). Physiological responses of men to chilling in ice water and to slow and fast rewarming. J. Appl. Physiol. 3:591.
59. Hayward, J. S., Eckerson, J. D., and Collis, M. L. (1975). Thermal balance and survival time prediction of man in cold water. Can. J. Physiol. Pharmacol. 53:21.
60. Lefevre, J., and Auguet, A. (1934). Les courbes thermoregulatrices et les rendements de la machine vivante dans les grandes puissances de travail. Ann. Physiol. Physicochim. Biol. 10:1116.
61. Nadel, E. R., Holmer, I., Bergh, U., Åstrand, P. O., and Stolwijk, J. A. J. (1974). Energy exchanges of swimming man. J. Appl. Physiol. 36:465.
62. Alexander, G., and Williams, D. (1968). Shivering and non-shivering thermogenesis during summit metabolism in young lambs. J. Physiol. (Lond.) 198:251.
63. Hemingway, A. (1963). Shivering. Physiol. Rev. 43:397.
64. Dawes, G. S., and Mestyán, G. (1963). Changes in the oxygen consumption of newborn guinea-pigs and rabbits in exposure to cold. J. Physiol. (Lond.) 168:22.
65. Vapaatalo, H., Mäkeläinen, A., and Nikki, P. (1975). Effects of some inhalation anaesthetics on thermoregulation and metabolic processes. *In* P. Lomax et al. (eds), Temperature Regulation and Drug Action, pp. 319–324. Proceedings of a symposium, April, 1974, Paris. S. Karger, Basel.
66. Jeddi, E., and Chatonnet, J. (1969). Influence de la narcose sur la potentialisation de l'action thermogène de la noradrénaline chez le rat adapté au froid. C. R. Soc. Biol. 163:168.
67. Mejsnar, J., and Janský, L. (1971). Nonshivering thermogenesis and calorigenic action of catecholamines in the white mouse. Physiol. Bohemoslov. 20:157.

68. Hales, J. R. S., Fawcett, A. A., and Bennett, J. W. (1978). Radioactive microsphcre partitioning of blood flow between capillaries and anteriovenous anastomoses in skin of conscious sheep. Pfluegers Arch. 376:87.
69. Ogilvie, D. M. (1967). Adaptation of white mice (*Mus musculus*) to repeated cooling. Can. J. Zool. 45:321.
70. Schmidt-Neilsen, K. (1970). Energy metabolism, body size, and problems of scaling. Fed. Proc. 29:1524.
71. Graham, N. McC., Searle, T. W., and Griffiths, D. A. (1974). Basal metabolic rate in lambs and young sheep. Aust. J. Agric. Res. 25:957.
72. Alexander, G. (1962). Summit metabolism in young lambs. J. Physiol. (Lond.) 162:31.
73. Alexander, G. (1974). Birth weight of lambs: influences and consequences. Ciba Found. Symp. 27:215.
74. Peirce, A. W. (1934). The basal (standard) metabolism of the Australian Merino sheep—II. Bulletin No. 84, pp. 22. Council for Scientific and Industrial Research, Melbourne.
75. Hey, E. N. (1972). Thermal regulation in the newborn. Br. J. Hosp. Med. 8:51.
76. Watts, A. J. (1971). Hypothermia in the aged: a study of the role of cold sensitivity. Environ. Res. 5:119.
77. Stine, R. J. (1977). Accidental hypothermia. J. Am. Coll. Emergency Physicians 6(9):1413.
78. Donhoffer, S. (1971). Non-shivering thermogenesis: past and present. *In* L. Janský (ed.), Nonshivering Thermogenesis, pp. 11–26. Proceedings of a symposium, April, 1970, Prague. Swets and Zeitlinger, Amsterdam.
79. Janský, L. (1973). Non-shivering thermogenesis and its thermoregulatory significance. Biol. Rev. 48:85.
80. Davis, T. R. A., and Mayer, J. (1955). Demonstration and quantitative determination of the contributions of physical and chemical thermogenesis on acute cold exposure. Am. J. Physiol. 181:675.
81. Hart, J. S., Héroux, O., and Depocas, F. (1956). Cold acclimation and the electromyogram of unanesthetized rats. J. Appl. Physiol. 9:404.
82. Cottle, W. H., and Carlson, L. D. (1956). Regulation of heat production in cold-adapted rats. Proc. Soc. Exp. Biol. Med. 92:845.
83. Hsieh, A. C. L., and Carlson, L. D. (1957). Role of adrenaline and noradrenaline in chemical regulation of heat production. Am. J. Physiol. 190:243.
84. Hsieh, A. C. L., Carlson, L. D., and Gray, G. (1957). Role of the sympathetic nervous system in the control of chemical regulation of heat production. Am. J. Physiol. 190:247.
85. Moore, R. E., and Underwood, M. C. (1960). Possible role of noradrenaline in control of heat production in the newborn mammal. Lancet 278:1277.
86. Depocas, F. (1960). The calorigenic response of cold-acclimated white rats to infused noradrenaline. Can. J. Biochem. Physiol. 38:107.
87. Smith, R. E. (1961). Thermogenic activity of the hibernating gland in the cold-acclimated rat. Physiologist 4:113.
88. Smith, R. E., and Hock, R. J. (1963). Brown fat: thermogenic factor of arousal in hibernators. Science 140:199.
89. Smith, R. E., and Hock, R. J. (1963). Thermogenesis of brown fat in the hibernator. Fed. Proc. 22:341.
90. Dawkins, M. J. R., and Hull, D. (1964). Brown adipose tissue and the response of new-born rabbits to cold. J. Physiol. (Lond.) 172:216.
91. Janský, L. (1971). Participation of body organs during nonshivering heat production. *In* L. Janský (ed.), Nonshivering Thermogenesis, pp. 159–169. Pro-

ceedings of a symposium, April, 1970, Prague. Swets and Zeitlinger, Amsterdam.
92. Davis, T. R. A. (1967). Contribution of skeletal muscle to nonshivering thermogenesis in the dog. Am. J. Physiol. 213:1423.
93. Stoner, H. B. (1973). The role of the liver in non-shivering thermogenesis in the rat. J. Physiol. 232:285.
94. Brück, K., and Wünnenberg, B. (1965). Blockade der chemischen Thermogenese und Auslösung von Muskelzittern durch Adrenolytica und Ganglienblockade beim newgeborenen Meerschweinchen. Pflüegers Arch. 282:376.
95. Héroux, O., Depocas, F., and Hart, J. S. (1959). Comparison between seasonal and thermal acclimation in white rats. I. Metabolic and insulative changes. Can. J. Biochem. Physiol. 37:473.
96. Jensen, C., and Ederstrom, H. E. (1955). Development of temperature regulation in the dog. Am. J. Physiol. 183:340.
97. Hill, J. R. (1959). The oxygen consumption of new-born and adult mammals: its dependence on the oxygen tension in the inspired air and on the environmental temperature. J. Physiol. (Lond.) 149:346.
98. Taylor, P. M. (1960). Oxygen consumption in new-born rats. J. Physiol. (Lond.) 154:153.
99. Brück, K. (1961). Temperature regulation in the newborn infant. Biol. Neonate 3:65.
100. Hull, D. (1965). Oxygen consumption and body temperature of new-born rabbits and kittens exposed to cold. J. Physiol. (Lond.) 177:192.
101. Brück, K., and Wünnenberg, B. (1965). Über die Modi der Thermogenese beim neugeborenen Warmbluter: Untersuchungen am Meerschweinchen. Pfluegers Arch. 282:362.
102. Mejsnar, J., and Janský, L. (1971). Methods for estimating nonshivering thermogenesis. *In* L. Janský (ed.), Non-Shivering Thermogenesis, pp. 27–36. Proceedings of a symposium, April 1–2, 1970, Prague. Swets and Zeitlinger, Amsterdam.
103. Zeisberger, E., Brück, K., Wünnenberg, W., and Wietasch, C. (1967). Das Ausmass der zitterfreien Thermogenese des Meerschweinchens in Abhangigkeit vom Lebensalter. Pfluegers Arch. 269:276.
104. Brück, K., and Wünnenberg, B. (1966). Alterations in the thermogenic mechanism during postnatal development: the dependence on environmental temperature conditions. Helgol. Wiss. Meeresunters 14:514.
105. Janský, L., Funke, E., and Hart, J. S. (1970). Analysis of electromyograms in shivering rabbits. Physiol. Bohemoslov. 19:397.
106. Dawson, R. W., Bennett, A. F., and Hudson, J. W. (1976). Metabolism and thermoregulation in hatchling ringbilled gulls. Condor 78:49.
107. Moore, R. E. (1963). Control of heat production in newborn mammals: role of noradrenaline and mode of action. Fed. Proc. 22:920.
108. Rowlatt, V., Mrosovsky, N., and English, A. (1971). A comparative survey of brown fat in the neck and axilla of mammals at birth. Biol. Neonate 17:53.
109. Freeman, B. M. (1967). Some effects of cold on the metabolism of the fowl during the perinatal period. Comp. Biochem. Physiol. 20:179.
110. Johnston, D. W. (1971). The absence of brown adipose tissue in birds. Comp. Biochem. Physiol. (A) 40:1107.
111. LeBlanc, J., and Mount, L. E. (1968). Effects of noradrenaline and adrenaline on oxygen consumption rate and arterial blood pressure in the newborn pig. Nature (Lond.) 217:77.
112. Stanton, H. C., and Mueller, R. L. (1973). Metabolic responses to cold and catecholamines as a function of age in swine (*Sus domesticus*). Comp. Biochem. Physiol. (A) 45:215.

113. Brück, K., Wünnenberg, W., and Zeisberger, E. (1969). Comparison of cold-adaptive metabolic modifications in different species, with special reference to the miniature pig. Fed. Proc. 28:1035.
114. Brück, K. (1970). Non-shivering thermogenesis and brown adipose tissue in relation to age, and their integration in the thermoregulatory system. *In* O. Lindberg (ed.), Brown Fat, p. 117–154. Elsevier-North Holland Publishing Company, New York.
115. Jenkinson, D. M., Noble, R. C., and Thompson, G. E. (1968). Adipose tissue and heat production in the newborn ox (*Bos taurus*). J. Physiol. (Lond.) 195:639.
116. Alexander, G., Bennett, J. W., and Gemmell, R. T. (1975). Brown adipose tissue in the new-born calf (*Bos taurus*). J. Physiol. (Lond.) 244:223.
117. Thompson, G. E., and Bell, A. W. (1976). Heat production in the newborn ox during noradrenaline infusion. Biol. Neonate 28:375.
118. ter Meulen, V., Duchanowa, H., Kahles, H., Nordbeck, H., Preusse, C. J., and Molnar, S. (1976). Untersuchungen zur zitterfrieren Wärmebildung beim neugeborenen Kalb unter Kältebelastung. Z. Tierphysiol. 36:283.
119. Rink, R. D. (1969). Oxygen consumption, body temperature and brown adipose tissue in the postnatal golden hamster (*Mesocritetus auratus*). J. Exp. Zool. 170:117.
120. Hull, D. (1973). Thermoregulation in young mammals. *In* G. C. Whittow (ed.), Comparative Physiology of Thermoregulation, Vol. III, pp. 167–200. Academic Press, Inc., New York.
121. Heim, T., and Szelényi, Z. (1965). Temperature regulation in rats, semistarved since birth. Acta Physiol. Acad. Sci. Hung. 27:247.
122. Hill, J. R., and Rahimtulla, K. A. (1965). Heat balance and the metabolic rate of newborn babies in relation to environmental temperature; and the effect of age and weight on basal metabolic rate. J. Physiol. (Lond.) 180:239.
123. Stern, L., Lees, M. H., and Leduc, J. (1965). Environmental temperature, oxygen consumption and catecholamine excretion in newborn infants. Pediatrics 36:367.
124. Hey, E. N. (1969). The relation between environmental temperature and oxygen consumption in the new-born baby. J. Physiol. (Lond.) 200:589.
125. Heaton, J. M. (1972). The distribution of brown adipose tissue in the human. J. Anat. 112:35.
126. Merklin, R. J. (1974). Growth and distribution of human fetal brown fat. Anat. Rec. 178:637.
127. Thompson, G. E., and Jenkinson, D. McE. (1970). Adipose tissue in the newborn goat. Res. Vet. Sci. 11:102.
128. Grav, H. J., and Blix, A. S. (1976). Brown adipose tissue—a factor in the survival of harp seal pups. Can. J. Physiol. Pharmacol. 54:409.
129. Hissa, R. (1968). Postnatal development of thermoregulation in the Norwegian lemming and the golden hamster. Ann. Zool. Fenn. 5:345.
130. Pshennikov, A. E., and Morodosov, I. I. (1972). Content of brown fat in some mammals of Yakutiya. Sov. J. Ecol. 3:177.
131. Dawkins, M. J. R., and Hull, D. (1965). The production of heat by fat. Sci. Am. 213(2):62.
132. Chaffee, R. R. J., Allen, J. R., Arine, R. M., Fineg, J., Rochelle, R. H., and Rosander, J. (1975). Studies on thermogenesis in brown adipose tissue in temperature-acclimated *Macaca mulatta*. Comp. Biochem. Physiol. (A) 50:303.
133. Rovetto, M. J., and Ferguson, J. H. (1971). Effects of acclimation temperature on brown adipose tissue in the red squirrel (*Tamiasciurus hudsonicus*). Comp. Biochem. Physiol. (A) 39:39.

134. Mason, E. B., and Prychodko, W. (1975). Interscapular brown adipose tissue mass in two species of *Peromyscus* raised in different thermal environments. J. Mammal. 56:683.
135. Heldmaier, G. (1971). Zitterfreie Warmebildung und Körpergrösse bei Säugetieren. Z. Vergl. Physiol. 73:222.
136. Gemmell, R. T., Bell, A. W., and Alexander, G. (1972). Morphology of adipose cells in lambs at birth and during subsequent transition of brown to white adipose tissue in cold and in warm conditions. Am. J. Anat. 133:143.
137. Brück, K., and Wünnenberg, B. (1965). Untersuchungen über die Bedeutung des multilokularen Fettgewebes für die Thermogenese des neugeborenen Meerschweinchens. Pfluegers Arch. 283:1.
138. Alexander, G., Bell, A. W., and Williams, D. (1970). Metabolic response of lambs to cold: effects of prolonged treatment with thyroxine and acclimation to low temperatures. Biol. Neonate 15:198.
139. Moore, R. E., and Simmonds, M. A. (1966). Decline with age in the thermogenic response of the young rat to *l*-noradrenaline. Fed. Proc. 25:1329.
140. Hsieh, A. C. L., Emery, N., and Carlson, L. D. (1971). Calorigenic effect of norepinephrine in newborn rats. Am. J. Physiol. 221:1568.
141. Tarkkonen, H., and Julku, H. (1968). Brown adipose tissue in young mice: activity and role in thermoregulation. Experientia 24:798.
142. Janský, L., Bartůňková, R., Kočkova, J., Mejsnar, J., and Zeisberger, E. (1969). Interspecies differences in cold adaptation and non-shivering thermogenesis. Fed. Proc. 28:1053.
143. Gemmell, R. T., and Alexander, G. (1978). Ultrastructural development of adipose tissue in foetal sheep. Aust. J. Biol. Sci. 31:505.
144. Alexander, G., Nicol, D., and Thorburn, G. (1973). Thermogenesis in prematurely delivered lambs, Foetal and Neonatal Physiology, pp. 410–417. Proceedings of the Sir Joseph Barcroft Centenary Symposium, July, 1972, Cambridge. Cambridge University Press, Cambridge.
145. Hart, J. S. (1958). Metabolic alterations during chronic exposure to cold. Fed. Proc. 17:1045.
146. Heldmaier, G. (1974). Cold adaptation by short daily cold exposures in the young pig. J. Appl. Physiol. 36:163.
147. Kočková, J., and Janský, L. (1968). Cold acclimation in the rabbit. Physiol. Bohemoslov. 17:309.
148. Vybíral, S., and Janský, L. (1974). Non-shivering thermogenesis in the golden hamster. Physiol. Bohemoslov. 23:235.
149. Joy, R. J. T. (1963). Responses of cold-acclimated men to infused noradrenaline. J. Appl. Physiol. 18:1209.
150. Budd, G. M., and Warhaft, N. (1966). Cardiovascular and metabolic responses to noradrenaline in man, before and after acclimatization to cold in Antarctica. J. Physiol. (Lond.) 186:233.
151. Hong, S. K. (1973). Pattern of cold adaptation in women divers of Korea (Ama). Fed. Proc. 32:1614.
152. Chaffee, R. R. J., and Allen, J. R. (1973). Effects of ambient temperature on the resting metabolic rate of cold- and heat-acclimated *Macaca mulatta*. Comp. Biochem. Physiol. (A) 44:1215.
153. Webster, A. J. F., Heitman, J. H., Hays, F. L., and Olynyk, G. P. (1969). Catecholamines and cold thermogenesis in sheep. Can. J. Physiol. Pharmacol. 94:719.
154. Freeman, B. M. (1970). Thermoregulatory mechanisms of the neonate fowl. Comp. Biochem. Physiol. 33:219.
155. Nagasaka, T., and Carlson, L. D. (1965). Responses of cold- and warm-

adapted dogs to infused norepinephrine and active body cooling. Am. J. Physiol. 209:227.

156. LeBlanc, J., Robinson, D., Sharman, D. F., and Tousignant, P. (1967). Catecholamines and short-term adaptation to cold in mice. Am. J. Physiol. 213:1419.
157. Depocas, F. (1958). Chemical thermogenesis in the functionally eviscerated cold-acclimated rat. Can. J. Biochem. Physiol. 36:691.
158. Héroux, O. (1961). Comparison between seasonal and thermal acclimation in white rats. V. Metabolic and cardiovascular response to noradrenaline. Can. J. Biochem. Physiol. 39:1829.
159. Janský, L., Bartůňková, R., and Zeisberger, E. (1967). Acclimation of the white rat to cold: noradrenaline thermogenesis. Physiol. Bohemoslov. 16:366.
160. Kreider, M. B. (1972). Stimulus for metabolic acclimation to cold-intensity versus duration. *In* R. E. Smith (ed.), Bioenergetics, pp. 88–89. Proceedings of an International Symposium on Environmental Physiology and Bioenergetics, 1971, Dublin. Federation of American Society of Experimental Biology, Maryland.
161. Heldmaier, G. (1975). The effect of short daily cold exposures on development of brown adipose tissue in mice. J. Comp. Physiol. 98:161.
162. Tarkkonen, H. (1972). Effect of repeated short-term cold exposure on brown adipose tissue of mice. Ann. Zool. Fenn. 8:434.
163. LeBlanc, J. (1967). Adaptation to cold in three hours. Am. J. Physiol. 212:530.
164. LeBlanc, J., Roberge, C., Vallières, J., and Oakson, G. (1971). The sympathetic nervous system in short-term adaptation to cold. Can. J. Physiol. Pharmacol. 49:96.
165. Brück, K., Baum, E., and Schwennicke, H. P. (1976). Cold-adaptation modifications in man induced by repeated short-term cold-exposures and during a 10-day and -night cold-exposure. Pfluegers Arch. 363:125.
166. Banet, M., and Hensel, H. (1976). Nonshivering thermogenesis induced by repetitive hypothalamic cooling in the rat. Am. J. Physiol. 230:522.
167. Banet, M., and Hensel, H. (1976). Nonshivering thermogenesis induced by repetitive cooling of spinal cord in the rat. Am. J. Physiol. 230:720.
168. Hsieh, A. C. L., and Wang, J. C. C. (1971). Calorigenic responses to cold of rats after prolonged infusion of norepinephrine. Am. J. Physiol. 221:335.
169. LeBlanc, J., and Pouliot, M. (1964). Importance of noradrenaline in cold adaptation. Am. J. Physiol. 207:853.
170. LeBlanc, J., Vallières, J., and Vachon, C. (1972). Beta-receptor sensitization by repeated injections of isoproterenol and by cold adaptation. Am. J. Physiol. 222:1043.
171. Suter, E. R. (1969). The fine structure of brown adipose tissue. I. Cold-induced changes in the rat. J. Ultrastruct. Res. 26:216.
172. Héroux, O. (1962). Seasonal adjustments in captured Norway rats. II. Survival time, pelt insulation, shivering, and metabolic and tissue responses to noradrenaline. Can. J. Biochem. Physiol. 40:537.
173. Bartholomew, G. A. (1972). Aspects of timing and periodicity of heterothermy. *In* F. E. South et al. (eds.), Hibernation and Hypothermia, Perspectives and Challenges, pp. 663–680. Proceedings of a Symposium held at Aspen, Colorado, June, 1971. Elsevier-North Holland Publishing Company, New York.
174. Hudson, J. W. (1973). Torpidity in mammals. *In* G. C. Whittow (ed.), Comparative Physiology of Thermoregulation. III. Special Aspects of Thermoregulation, pp. 97–165. Academic Press, Inc., New York.
175. Pivorun, E. (1977). Mammalian hibernation. Comp. Biochem. Physiol. (A) 58:125.

176. Chaffee, R. R. J., and Roberts, J. C. (1971). Temperature acclimation in birds and mammals. Annu. Rev. Physiol. 33:155.
177. Cassuto, Y., and Amit, Y. (1968). Thyroxine and norepinephrine effects on the metabolic rates of heat-acclimated hamsters. Endocrinology 82:17.
178. Pohl, H., and Hart, J. S. (1965). Thermoregulation and cold acclimation in a hibernator, *Citellus tridecemlineatus*. J. Appl. Physiol. 20:398.
179. Hissa, R. (1970). Calorigenic effect of noradrenaline in the Norwegian lemming, *Lemmus lemmus* (*L*). Experentia 26:266.
180. Hayward, J. S. (1968). The magnitude of noradrenaline-induced thermogenesis in the bat (*Myotis lucifugus*) and its relation to arousal from hibernation. Can. J. Physiol. Pharmacol. 46:713.
181. Mejsnar, J., and Janský, L. (1970). Shivering and non-shivering thermogenesis in the bat (*Myotis myotis* Borkh.) during arousal from hibernation. Can. J. Physiol. Pharmacol. 48:102.
182. Hammel, H. T., Dawson, T. J., Abrams, R. M., and Andersen, H. T. (1968). Total calorimetric measurements on *Citellus lateralis* in hibernation. Physiol. Zool. 41:341.
183. Hayward, J. S. (1969). Non-shivering thermogenesis in hibernating animals. *In* L. Janský (ed.), Non-shivering Thermogenesis, pp. 119–137. Proceedings of a symposium, April, 1970, Prague. Swets and Zeitlinger, Amsterdam.
184. Hayward, J. S., and Lyman, C. P. (1967). Non-shivering heat production during arousal from hibernation and evidence for the contribution of brown fat. *In* K. C. Fisher ct al. (eds.), Mammalian Hibernation. III. Proceedings of an International Symposium, September, 1965, Toronto. Oliver and Boyd, Edinburgh.
185. Pagels, J. F. (1972). The effect of short and prolonged cold exposure on arousal in thc free-tailed bat *Tadarida brasilionsis cyanocephala* (Le Conte). Comp. Biochem. Physiol. (A) 42:559.
186. Spaan, G., and Klussmann, F. W. (1970). Die Frequenz des Kältezitterns bei Tierarten Verschiedener Grösse. Pfluegers Arch. 320:318.
187. Smalley, R. L., and Dryer, R. L. (1967). Brown fat in hibernation. *In* K. L. Fisher et al. (eds.), Mammalian Hibernation III, pp. 325–345. Proceedings of a symposium, September, 1965, Toronto. Oliver and Boyd, Edinburgh.
188. Smith, R. E., and Horwitz, B. A. (1969). Brown fat and thermogenesis. Physiol. Rev. 49:330.
189. Afzelius, B. A. (1970). Brown adipose tissue: its gross anatomy, histology and cytology. *In* O. Lindberg (cd.), Brown Adipose Tissue, pp. 1–31. Elsevier-North Holland Publishing Company, New York.
190. Smith, R. E., and Roberts, J. C. (1964). Thermogenesis of brown adipose tissue in cold-acclimated rats. Am. J. Physiol. 206:143.
191. Dawkins, M. J. R., and Hull, D. (1963). Brown fat and the response of the newborn rabbit to cold. J. Physiol. (Lond.) 169:101P.
192. Cockburn, F., Hull, D., and Walton, I. (1968). The effect of lipolytic hormones and theophylline on heat production in brown adipose tissue *in vivo*. Br. J. Pharmacol. 31:568.
193. Hull, D., and Hardman, M. J. (1970). Brown adipose tissue in newborn animals. *In* O. Lindberg (ed.), Brown Adipose Tissue, pp. 97–115. Elsevier-North Holland Publishing Company, New York.
194. Johansson, B. (1959). Brown fat: a review. Metab. Clin. Exp. 8:221.
195. Lindberg, O. (ed). (1970). Brown Adipose Tissue, pp. 337. Elsevier-North Holland Publishing Company, New York.
196. Rasmussen, A. T. (1923). The so-called hibernating gland. J. Morphol. 38:147.
197. Sheldon, E. F. (1924). The so-called hibernating gland in mammals: a form of adipose tissue. Anat. Rec. 28:331.

198. Hull, D., and Segall, M. M. (1966). Distinction of brown from white adipose tissue. Nature (Lond.) 212:469.
199. Hull, D. (1966). The structure and function of brown adipose tissue. Br. Med. Bull. 22:92.
200. Daniel, H., and Derry, D. M. (1969). Criteria for differentiation of brown and white fat in the rat. Can. J. Physiol. Pharmacol. 47:941.
201. Napolitano, L. (1963). The differentiation of white adipose cells: an electron microscope study. J. Cell Biol. 18:663.
202. Lever, J. D. (1957). The fine structure of brown adipose tissue in the rat with observations on the cytological changes following starvation and adrenalectomy. Anat. Rec. 128:361.
203. Napolitano, L., and Fawcett, D. (1958). The fine structure of brown adipose tissue in the newborn mouse and rat. J. Biophys. Biochem. Cytol. 4:685.
204. Itoh, S., and Hiroshige, T. (1967). Presence of brown adipose tissue in monkeys. J. Physiol. Soc. Jpn. 29:322.
205. Heaton, J. M. (1973). A study of brown adipose tissue in hypothermia. J. Pathol. 110:105.
206. Tanuma, Y., Ohata, M., Ito, T., and Yokochi, C. (1976). Possible function of human brown adipose tissue as suggested by observation on perirenal brown fats from necropsy cases of variable age groups. Arch. Histol. Jpn. 39:117.
207. Rauch, J. C., and Hayward, J. S. (1969). Topography and vascularization of brown fat in a small non-hibernator (deer mouse, *Peromyscus maniculatus*). Can. J. Zool. 47:1301.
208. Rauch, J. C., and Hayward, J. S. (1969). Topography and vascularization of brown fat in a hibernator (little brown bat, *Myotis lucifugus*). Can. J. Zool. 47:1315.
209. Rechardt, L., and Hervonen, H. (1976). Electron microscopic localization of adenylate cyclase activity of white and brown adipose tissue of the rat and chicken. Histochemistry 50:57.
210. Fawcett, D. W. (1947). Differences in physiological activity in brown and white fat as revealed by histochemical reactions. Science 105:123.
211. Cannon, B., Romert, L., Sundin, U., and Barnard, T. (1977). Morphological and biochemical properties of perirenal adipose tissue from lamb (*Ovis aries*): a comparison with brown adipose tissue. Comp. Biochem. Physiol. (B) 56:87.
212. Alexander, G., and Bell, A. W. (1975). Quantity and calculated oxygen consumption during summit metabolism of brown adipose tissue in new-born lambs. Biol. Neonate 26:214.
213. Wünnenberg, W., Merker, G., and Brück, K. (1974). Do corticosteroids control heat production in hibernators? Pfluegers Arch. 352:11.
214. Hassi, J. (1971). Histochemical investigations on brown adipose tissue (BAT) in man at different ages. Scand. J. Clin. Lab. Invest. (Suppl. 116) 27:4.
215. Sidman, R. L. (1956). Histogenesis of brown adipose tissue in vivo and in organ culture. Anat. Rec. 124:581.
216. Smalley, R. L., and Smalley, K. N. (1967). Brown and white fats: development in the hamster. Science 157:1449.
217. Vernon, R. G. (1977). Development of perirenal adipose tissue in the neonatal lamb: effects of dietary safflower oil. Biol. Neonate 32:15.
218. Hardman, M. J., and Hull, D. (1971). The effect of environmental conditions on the growth and function of brown adipose tissue. J. Physiol. (Lond.) 214:191.
219. Hull, D., and Hardman, M. (1973). Active fat. *In* Foetal and Neonatal Physiology, pp. 418–419. Proceedings of Sir Joseph Barcroft Centenary Symposium, July, 1972, Cambridge. Cambridge University Press, Cambridge.

220. Hardman, M. J., and Hull, D. (1973). Blood flow and fatty acid release by cervical adipose tissue of rabbits. J. Physiol. (Lond.) 235:1.
221. Ballard, K., and Rosell, S. (1969). The unresponsiveness of lipid metabolism in canine mesenteric adipose tissue to biogenic amines and to sympathetic nerve stimulation. Acta Physiol. Scand. 77:442.
222. Weiss, A. K. (1960). Factors which affect cold acclimation. Fed. Proc. (Suppl. 5) 19:137.
223. Cameron, I. L. (1975). Age-dependent changes in the morphology of brown adipose tissue in mice. Tex. Rep. Biol. Med. 33:391.
224. Pospíšilová, D., and Janský, L. (1976). Effect of various adaptational temperatures on oxidative capacity of the brown adipose tissue. Physiol. Bohemoslov. 25:519.
225. Chaffee, R. R. J., Roberts, J. C., Conaway, C. H., Sorenson, M. W., and Kaufman, W. C. (1970). Comparative effects of temperature exposure on mass and oxidative enzyme activity of brown fat in insectivores, tupaiads and primates. Lipids 5:23.
226. LeBlanc, J., and Villemaire, A. (1970). Thyroxine and noradrenaline on noradrenaline sensitivity, cold resistance, and brown fat. Am. J. Physiol. 218:1742.
227. Steiner, G., Loveland, M., and Schonbaum, E. (1970). Effect of denervation on brown adipose tissue metabolism. Am. J. Physiol. 218:566.
228. Heldmaier, G., and Hoffmann, K. (1974). Melatonin stimulates growth of brown adipose tissue. Nature (Lond.) 247:224.
229. Lynch, R. G., and Epstein, A. L. (1976). Melatonin induced changes in gonads, pelage and thermogenic characters in the white-footed mouse, *Peromyscus leucopus*. Comp. Biochem. Physiol. (C) 53:67.
230. Hunt, T. E., and Hunt, E. A. (1967). A radioautographic study of proliferation in brown fat of the rat after exposure to cold. Anat. Rec. 157:537.
231. Cameron, I. L., and Smith, R. E. (1964). Cytological responses of brown fat tissue in cold-exposed rats. J. Cell Biol. 23:89.
232. Barnard, T. (1969). The ultrastructural differentiation of brown adipose tissue in the rat. J. Ultrastruct. Res. 29:311.
233. Suter, E. R. (1969). The fine structure of brown adipose tissue. III. The effect of cold exposure and its mediation in newborn rats. Lab. Invest. 21:259.
234. Lindgren, G., and Barnard, T. (1972). Changes in interscapular brown adipose tissue of rat during perinatal and early postnatal development and after cold acclimation. IV. Morphometric investigation of mitochondrial membrane alterations. Exp. Cell Res. 70:81.
235. Thomson, J. F., Habeck, D. A., Nance, S. L., and Beetham, K. L. (1969). Ultrastructural and biochemical changes in brown fat in cold-exposed rats. J. Cell Biol. 41:312.
236. Pellet, H., and Lheritier, M. (1975). Mise en évidence de deux types cellulaires dans le tissu adipeux brun du rat. Étude en microscopie électromique. C. R. Soc. Biol. 169:1220.
237. Edson, J. L., and Hull, D. (1977). The effect of maternal starvation on the metabolic response to cold of the newborn rabbit. Pediatr. Res. 11:793.
238. Hyvärinen, H., Pasanen, S., Heikura, H., Heinineva, R., and Laru, H. (1976). Effects of a cold environment on energy-related enzyme activities in the postnatal rat. Growth 40:41.
239. Cresteil, T. (1977). Effect of linoleic acid concentration of mother's diet on adenyl cyclase activity of fetal and neonatal rat brown adipose tissue. Can. J. Physiol. Pharmacol. 55:1242.
240. Cogneville, A.-M., Cividino, N., and Tordet-Caridroit, C. (1975). Lipid com-

position of brown adipose tissue as related to nutrition during the neonatal period in hypotrophic rats. J. Nutr. 105:982.

241. Heim, T., and Hull, D. (1966). The blood flow and oxygen consumption of brown adipose tissue in the new-born rabbit. J. Physiol. (Lond.) 186:42.
242. Alexander, G., and Williams, D. (1970). Cardiovascular function in young lambs during summit metabolism. J. Physiol. (Lond.) 208:65.
243. Hull, D., and Segall, M. M. (1965). The contribution of brown adipose tissue to heat production in the new-born rabbit. J. Physiol. (Lond.) 181:449.
244. Himms-Hagen, J. (1972). Effects of catecholamines on metabolism. *In* H. Blaschko and E. Muscholl (eds.), Catecholamines, pp. 363–462. Handbook of Experimental Pharmacology, New Series XXXIII. Springer-Verlag, New York.
245. Janský, L., and Hart, J. S. (1968). Cardiac output and organ blood flow in warm- and cold-acclimated rats exposed to cold. Can. J. Physiol. Pharmacol. 46:653.
246. Sapirstein, L. A. (1958). Regional blood flow by fractional distribution of indicators. Am. J. Physiol. 193:161.
247. Imai, Y., Horwitz, B. A., and Smith, R. E. (1968). Calorigenesis of brown adipose tissue in cold-exposed rats. Proc. Soc. Exp. Biol. Med. 127:717.
248. Horwitz, B. A., Smith, R. E., and Pengelley, E. T. (1968). Estimated heat contribution of brown fat in arousing ground squirrels. (*Citellus lateralis*). Am. J. Physiol. 214:115.
249. Foster, D. O., and Frydman, M. L. (1978). Nonshivering thermogenesis in the rat. II. Measurements of blood flow with microspheres point to brown adipose tissue as the dominant site of the calorigenesis induced by noradrenaline. Can. J. Physiol. Pharmacol. 56:110.
250. Foster, D. O. (1974). Evidence against a mediatory role of brown adipose tissue in the calorigenic response of cold-acclimated rats to noradrenaline. Can. J. Physiol. Pharmacol. 52:1051.
251. Himms-Hagen, J. (1974). Interscapular location of brown adipose tissue: role in noradrenaline-induced calorigenesis in cold-acclimated rats. Can. J. Physiol. Pharmacol. 52:225.
252. Hayward, J. S., and Davies, P. F. (1972). Evidence for the mediatory role of brown adipose tissue during non-shivering thermogenesis in the cold-acclimated mouse. Can. J. Physiol. Pharmacol. 50:168.
253. LeBlanc, J., LaFrance, L., Villemaire, C., Roberge, C., Vallieres, J., and Rousseau, S. (1972). Catecholamines and cold adaptation. *In* R. E. Smith (ed.), Bioenergetics, pp. 71–76. Proceedings of a Symposium on Environmental Physiology, July, 1971, Dublin. Federation of American Society for Experimental Biology, Bethesda, Maryland.
254. Leduc, J., and Rivest, P. (1969). Effets de l'ablation de la graisse brune interscapulaire sur l'acclimatation au froid chez le rat. Rev. Can. Biol. 28:49.
255. Himms-Hagen, J. (1969). The role of brown adipose tissue in the calorigenic effect of adrenaline and noradrenaline in cold-acclimated rats. J. Physiol. (Lond.) 205:393.
256. Heim, T., and Hull, D. (1966). The effect of propranolol on the calorigenic response in brown adipose tissue of new-born rabbits to catecholamines, glucagon, corticotrophin and cold exposure. J. Physiol. (Lond.) 187:271.
257. Nedergaard, J., Cannon, B., and Lindberg, O. (1977). Microcalorimetry of isolated mammalian cells. Nature (Lond.) 267:518.
258. Hayward, J. S., and Ball, E. G. (1966). Quantitative aspects of brown adipose tissue thermogenesis during arousal from hibernation. Biol. Bull. 313:94.
259. Prusiner, S., and Poe, M. (1970). Thermodynamic considerations of mammalian heat production. *In* O. Lindberg (ed.), Brown Adipose Tissue, pp. 263–282. Elsevier-North Holland Publishing Company, New York.

260. Ahlabo, I., and Barnard, T. (1974). A quantitative analysis of triglyceride droplet structural changes in hamster brown adipose tissue during cold exposure and starvation. J. Ultrastruct. Res. 48:361.
261. Hogeboom, G. H., Schneider, W. C., and Pallade, G. H. (1948). Cytochemical studies of mammalian tissues. I. Isolation of intact mitochondria from rat liver: some biochemical properties of mitochondria and submicroscopic particulate material. J. Biol. Chem. 172:619.
262. Joel, C. D. (1966). Stimulation of metabolism of rat brown adipose tissue by addition of lipolytic hormones *in vitro*. J. Biol. Chem. 241:814.
263. Kornacker, M. S., and Ball, E. G. (1968). Respiratory processes in brown adipose tissue. J. Biol. Chem. 243:1638.
264. Giaja, J. (1929). Le métabolisme de sommet. C. R. Soc. Biol. 101:3.
265. Flatmark, T., and Pedersen, J. I. (1975). Brown adipose tissue mitochondria. Biochim. Biophys. Acta 416:53.
266. Knight, B. L. (1972). The effects of glucose, free fatty acids and lipid depletion on the metabolism *in vitro* of brown fat from newborn rabbits. Biochem. J. 129:1175.
267. Drahota, Z. (1970). Fatty acid oxidation by brown adipose tissue mitochondria. *In* O. Lindberg (ed.), Brown Fat, pp. 225–244. Elsevier-North Holland Publishing Company, New York.
268. Horwitz, B. A. (1971). Brown fat thermogenesis: physiological control and metabolic basis. *In* L. Janský (ed.), Non-shivering Thermogenesis, pp. 221–240. Proceedings of a symposium, April, 1970, Prague. Swets and Zeitlinger, Amsterdam.
269. Horwitz, B. A., and Smith, R. E. (1972). Function and control of brown fat thermogenesis during cold exposure. *In* R. E. Smith (ed), Bioenergetics, pp. 134–140. Proceedings of the International Symposium on Environmental Physiology, July, 1971, Dublin. Federation of the American Society for Experimental Biology, Bethesda, Maryland.
270. Horwitz, B. A. (1975). Physiological and biochemical characteristics of adrenergic receptors and pathways in brown adipocytes. *In* P. Lomax et al. (eds.), Temperature Regulation and Drug Action, pp. 150–158. Proceedings of a symposium, 1974, Paris. S. Karger, Basel.
271. Lindberg, O., Bieber, L. L., and Houštěk, J. (1976). Brown adipose tissue metabolism: an attempt to apply results from *in vitro* experiments on tissue *in vivo*. *In* L. Janský and X. J. Musacchia (eds.), Regulation of Depressed Metabolism and Thermogenesis, pp. 117–136. Proceedings of a symposium, October, 1974, Prague. Charles C Thomas Publisher, Springfield, Illinois.
272. Himms-Hagen, J. (1976). Cellular thermogenesis. Annu. Rev. Physiol. 38:315.
273. Horwitz, B. A., and Eaton, M. (1977). Ouabain-sensitive liver and diaphragm respiration in cold-acclimated hamsters. J. Appl. Physiol. 42:150.
274. Drahota, Z., and Houštěk, J. (1976). Biochemical aspects of non-shivering thermogenesis in brown adipose tissue. *In* L. Janský and X. L. Musacchia (eds.), Regulation of Depressed Metabolism, pp. 213–224. Proceedings of a symposium, October, 1974, Prague. Charles C Thomas Publisher, Springfield, Illinois.
275. Nicholls, D. G. (1976). Hamster brown-adipose-tissue mitochondria: purine nucleotide control of the ion conductance of the inner membrane, the nature of the nucleotide binding site. Eur. J. Biochem. 62:223.
276. Cannon, B. (1971). Control of fatty-acid oxidation in brown-adipose-tissue mitochondria. Eur. J. Biochem. 23:125.
277. Bernson, V. S. M., and Nicholls, D. G. (1974). Acetate, a major end product of fatty-acid oxidation in hamster brown-adipose-tissue mitochondria. Eur. J. Biochem. 47:517.

278. Langer, H., Rautenberg, W., and Zinkler, D. (1974). Thermogenese durch Glycogenumsatz im braunen Fettgewebe der neugeborenen Ratte. Verh. Dtsch. Zool. Ges. 67:242.
279. Hardman, M. J., and Hull, D. (1970). Fat metabolism in brown adipose tissue *in vivo*. J. Physiol. (Lond.) 206:263.
280. Horwitz, B. A., Herd, P. A., and Smith, R. E. (1970). Bioenergetics of brown adipose tissue. Lipids 5:30.
281. Prusiner, S. B., Cannon, B., and Lindberg, O. (1968). Oxidative metabolism in cells isolated from brown adipose tissue. 1. Catecholamine and fatty acid stimulation of respiration. Eur. J. Biochem. 6:15.
282. Thureson-Klein, Å, Mill-Hyde, B., Barnard, T., and Lagercrantz, H. (1976). Ultrastructural effects of chemical sympathectomy on brown adipose tissue. J. Neurocytol. 5:677.
283. Heaton, G. M., Wagenvoord, R. J., Kemp, A., and Nicholls, D. G. (1978). Brown-adipose-tissue mitochondria: photoaffinity labelling of the regulatory site of energy dissipation. Eur. J. Biochem. 82:515.
284. Fain, J. N., Jacobs, M. D., and Clement-Cormier, Y. C. (1973). Interrelationship of cyclic AMP, lipolysis and respiration in brown fat cells. Am. J. Physiol. 224:346.
285. Fink, S. A., and Williams, J. A. (1976). Adrenergic receptors mediating depolarization in brown adipose tissue. Am. J. Physiol. 231:700.
286. Flaim, K. E., Horwitz, B. A., and Horowitz, J. M. (1977). Coupling of signals to brown fat: α- and β-adrenergic responses in intact rats. Am. J. Physiol. 232:R101.
287. Ahlabo, I., and Barnard, T. (1971). Observations on peroxisomes in brown adipose tissue of the rat. J. Histochem. Cytochem. 19:670.
288. Horwitz, B. A. (1976). The effect of cold exposure on liver mitochondrial and peroxisomal distribution in the rat, hamster and bat. Comp. Biochem. Physiol. (A) 54:45.
289. Janský, L., and Hart, J. S. (1963). Participation of skeletal muscle and kidney during nonshivering thermogenesis in cold-acclimated rats. Can. J. Biochem. Physiol. 41:953.
290. Mejsnar, J., and Mejsnarová, B. (1971). Calorigenic action of noradrenaline, measured in gracilis anticus muscle of cold-adapted rats *in vivo*. Physiol. Bohemoslov. 20:389.
291. Grubb, B., and Folk, G. E. (1976). Effect of cold acclimation on norepinephrine stimulated oxygen consumption in muscle. J. Comp. Physiol. 110:217.
292. Schmitt, M., Meunier, P., Rochas, A., and Chatonnet, J. (1973). Catecholamines and oxygen uptake in dog skeletal muscle in situ. Pfluegers Arch. 345:145.
293. Lundholm, L., and Svedmyr, N. (1965). Influence of adrenaline on blood flow and metabolism in the human forearm. Acta Physiol. Scand. 65:344.
294. Hall, G. M., Lucke, J. N., and Lister, D. (1977). Porcine malignant hyperthermia. V. Fatal hyperthermia in the Pietrain pig associated with the infusion of α-adrenergic agonists. Br. J. Anaesth. 49:855.
295. Teskey, N., Horwitz, B., and Horowitz, J. (1975). Norepinephrine-induced depolarization of skeletal muscle cells. Eur. J. Pharmacol. 30:352.
296. Bukowiecki, L., and Himms-Hagen, J. (1976). Alteration of mitochondrial protein metabolism in liver, brown adipose tissue and skeletal muscle during cold-acclimation. Biochim. Biophys. Acta 428:591.
297. Himms-Hagen, J., Bukowiecki, L., Behrens, W., and Bonin, M. (1972). Mechanisms of non-shivering thermogenesis in rats. *In* R. E. Smith (ed.), Bioenergetics, pp. 127–133. Proceedings of the International Symposium on

Environmental Physiology, July, 1971, Dublin. Federation of American Societies for Experimental Biology, Bethesda, Maryland.
298. Himms-Hagen, J., Behrens, W., Hbous, A., and Greenway, D. (1976). Altered mitochondria in skeletal muscle of cold acclimated rats and the adaptation for nonshivering thermogenesis. *In* L. Jansky and X. J. Musacchia (eds.), Regulation of Depressed Metabolism and Thermogenesis, pp. 243–260. Proceedings of a symposium, October, 1974, Prague. Charles C Thomas Publisher, Springfield, Illinois.
299. George, J. C., and Ronald, K. (1973). The harp seal, *Pagophilus groenlandicus* (Erxleben, 1777). XXV. Ultrastructure and metabolic adaptation of skeletal muscle. Can. J. Zool. 51:833.
300. George, J. C., and Ronald, K. (1975). The harp seal, *Pagophilus groenlandicus* (Erxleben, 1777). XVII. Structure and metabolic adaption of the caval sphincter muscle with some observations on the diaphragm. Acta Anat. 93:88.
301. Horwitz, B. A., and Hanes, G. E. (1974). Isoproterenol-induced calorigenesis of dystrophic and normal hamsters. Proc. Soc. Exp. Biol. Med. 147:392.
302. Coleman, H. N., Sonnenblick, E. H., and Braunwald, E. (1971). Mechanism of norepinephrine-induced stimulation of myocardial oxygen consumption. Am. J. Physiol. 221:778.
303. Janský, L., Zeisberger, E., and Doležal, V. (1964). Effects of oxygen supply and noradrenaline infusion on liver metabolism of rats acclimatized to cold. Nature (Lond.) 202:397.
304. Kawahata, A., and Carlson, L. D. (1959). Role of rat liver in nonshivering thermogenesis. Proc. Soc. Exp. Biol. Med. 101:303.
305. Thompson, G. E., Gardner, W., and Bell, A. W. (1975). The oxygen consumption, fatty acid and glycerol uptake of the liver in fed and fasted sheep during cold exposure. Q. J. Exp. Physiol. 60:107.
306. Mejsnar, J., and Janský, L. (1976). Mode of catecholaminic action during organ regulation of non-shivering thermogenesis. *In* L. Janský and X. J. Musacchia (eds.), Regulation of Depressed Metabolism and Thermogenesis, pp. 225–242. Proceedings of a symposium, October, 1974, Prague. Charles C Thomas Publisher, Springfield, Illinois.
307. Janský, L. (1966). Body organ thermogenesis of the rat during exposure to cold and at maximal metabolic rate. Fed. Proc. 25:1297.
308. Thompson, G. E. (1977). Physiological effects of cold exposure. *In* D. Robertshaw (ed.), MTP International Review of Physiology, Vol. 15, Environmental Physiology II, pp. 29–69. University Park Press, Baltimore.
309. Orth, R. D., Odell, W. D., and Williams, R. H. (1960). Some hormonal effects on the metabolism of acetate-1-C^{14} by rat adipose tissue. Am. J. Physiol. 198:640.
310. Flatt, J. P., and Ball, E. G. (1963). Studies on the metabolism of adipose tissue. XIV. The manometric determination of total CO_2 production and oxygen consumption in bicarbonate buffer. Biochem. Z. 338:73.
311. Himms-Hagen, J., and Ball, E. G. (1961). Studies in the metabolism of adipose tissue: the effect of adrenaline on oxygen consumption and glucose utilization. Endocrinology 69:752.
312. Novak, M., Penn-Walker, D., and Monkus, E. F. (1975). Oxidation of fatty acids by mitochondria obtained from newborn subcutaneous (white) adipose tissue. Biol. Neonate 25:95.
313. Heim, T., Schenk, H., Varga, F., and Goetze, E. (1977). White adipose tissue (WAT) and heat production in the newborn rabbit. Pediatr. Res. 11:406.
314. Therriault, D. G., and Mellin, D. B. (1971). Cellularity of adipose tissue in cold-exposed rats and the calorigenic effect of norepinephrine. Lipids 6:486.
315. Janský, L., Votápková, Z., and Feiglová, E. (1969). Total cytochrome-oxidase

activity of the brown fat and its thermogenetic significance. Physiol. Bohemoslov. 18:443.

316. Sykes, A. R., and Slee, J. (1968). Acclimatization of Scottish Blackface sheep to cold. 2. Skin temperature, heart rate, respiration rate, shivering intensity and skinfold thickness. Anim. Prod. 10:17.
317. Freeman, B. M. (1966). The effects of cold, noradrenaline and adrenaline upon the oxygen consumption and carbohydrate metabolism of the young fowl (*Gallus domesticus*). Comp. Biochem. Physiol. 18:369.
318. Kaciŭba-Uscilko, H., and Poczopko, P. (1975). Role of catecholamines in cold-induced thermogenesis in the newborn pig. *In* P. Lomax et al. (eds.), Temperature Regulation and Drug Action, pp. 202–208. Proceedings of a symposium, April, 1974, Paris. S. Karger, Basel.
319. Young, B. A. (1975). Effects of winter acclimatization on resting metabolism in beef cows. Can. J. Anim. Sci. 55:619.
320. Clark, D., Lee, D., Rognstad, R., and Katz, J. (1975). Futile cycles in isolated perfused rat liver and in isolated rat liver parenchymal cells. Biochem. Biophys. Res. Commun. 67:212.
321. Bloxham, D. P., Clark, M. G., Holland, P. C., and Lardy, H. A. (1973). A model study of the fructose-diphosphatase-phosphofructokinase substrate cycle. Biochem. J. 134:581.
322. Debeer, L. J., Mannaerts, G., and de Schepper, P. J. (1974). Effects of octanoate and oleate on energy metabolism in the perfused rat liver. Eur. J. Biochem. 47:591.
323. Newsholme, E. A., Crabtree, B., Higgins, S. J., Thornton, S. D., and Start, C. (1972). The activities of fructose diphosphatase in flight muscles from the bumble-bee and the role of this enzyme in heat generation. Biochem. J. 128:89.
324. Cottle, W. H. (1970). The innervation of brown adipose tissue. *In* O. Lindberg (ed.), Brown Adipose Tissue, pp. 155–178. Elsevier-North Holland Publishing Company, New York.
325. Schönbaum, E., Steiner, G., and Sellers, E. A. (1970). Brown adipose tissue and norepinephrine. *In* O. Lindberg (ed.), Brown Adipose Tissue, pp. 179–196. Elsevier-North Holland Publishing Company, New York.
326. LeBlanc, J., Dulac, S., Côte, J., and Girard, B. (1975). Autonomic nervous system and adaptation to cold in man. J. Appl. Physiol. 39:181.
327. Horowitz, J. M. (1971). Neural control of thermogenesis in brown adipose tissue. *In* R. E. Smith (ed.), Bioenergetics, pp. 115–121. Proceedings of an International Symposium on Environmental Physiology, July, 1971, Dublin. Federation of the American Society for Experimental Biology, Bethesda, Maryland.
328. Nouri, T. N. (1972). Innervation of brown adipose tissue in new-born rabbits. J. Physiol. (Lond.) 227:42.
329. Sidman, R. L., and Fawcett, D. W. (1954). The effect of peripheral nerve section on some metabolic responses of brown adipose tissue in mice. Anat. Rec. 118:487.
330. Flaim, K. E., Horowitz, J. M., and Horwitz, B. A. (1976). Functional and anatomical characteristics of the nerve-brown adipose interaction in the rat. Pfluegers Arch. 365:9.
331. Corrodi, H., and Jonsson, G. (1967). The formaldehyde fluorescence method for the histochemical demonstration of biogenic monoamines: a review on methodology. J. Histochem. Cytochem. 15:65.
332. Derry, D. M., Schönbaum, E., and Steiner, G. (1969). Two sympathetic nerve supplies to brown adipose tissue of the rat. Can. J. Physiol. Pharmacol. 47:57.
333. Ochi, J., Konishi, M., and Yoshikawa, H. (1969). Morphologischen Nachweis

der sympathischen Innervation des braunen Fettgewebes bei der Ratte. Z. Anat. Entwicklungsgesch. 129:259.
334. Derry, D. M., and Daniel, H. (1970). Sympathetic nerve development in the brown adipose tissue of the rat. Can. J. Physiol. Pharmacol. 48:160.
335. Cottle, M. K. W., Cottle, W. H., and Nash, C. W. (1974). Adrenergic innervation of brown adipose tissue from the Ground Squirrel (*Citellus richardsonii*). Can. J. Physiol. Pharmacol. 52:70.
336. Thompson, G. E., Clough, D. F., Scobie, A., and Kenney, J. D. R. (1971). The localization of noradrenaline and dopamine in brown adipose tissue of the newborn lamb. Biol. Neonate 17:394.
337. Falck, B. (1962). Observations on the possibilities of the cellular localization of monoamines by a fluorescence method. Acta Physiol. Scand. (suppl. 197) 56:6.
338. Steiner, G., and Evans, S. (1972). Sympathetic ganglia in brown adipose tissue: a new tool to study ganglionic stimulants. Am. J. Physiol. 222:111.
339. Thureson-Klein, Å, Lagercrantz, H., and Barnard, T. (1976). Chemical sympathectomy of interscapular brown adipose tissue. Acta Physiol. Scand. 98:8.
340. Linck, G., Stoeckel, M. E., Porte, A., and Petrovic, A. (1973). An electron microscope study of the specialized cell contents and innervation of adipocytes in the brown fat of the European Hamster (*Cricetus cricetus*). Cytobiologie 7:431.
341. Himms-Hagen, J. (1967). Sympathetic regulation of metabolism. Pharmacol. Rev. 19:367.
342. Cottle, W. H., Nash, C. W., Veress, A. T., and Ferguson, B. A. (1967). Release of noradrenaline from brown fat of cold-acclimated rats. Life Sci. 6:2267.
343. Hull, D., and Segall, M. M. (1965). Sympathetic nervous control of brown adipose tissue and heat production in the new-born rabbit. J. Physiol. (Lond.) 181:458.
344. Cottle, W. H., Cottle, M. K. W., and Nash, C. W. (1975). Thermogenesis of cold-acclimated rats following 6-hydroxydopamine treatment. *In* P. Lomax, E. Schönbaum, and J. Jacob (eds.), Temperature Regulation and Drug Action. Proceedings of a Symposium on Temperature Regulation and Drug Action, April, 1974, Paris. S. Karger, Basel.
345. Smith, R. E., and Imai, Y. (1969). Electrical activity of brown fat cells *in vivo.* Fed. Proc. 28:721.
346. Smith, R. E. (1964). Thermoregulatory and adaptive behaviour of brown adipose tissue. Science 146:1686.
347. Mjös, O. D., and Akre, S. (1971). Effect of catecholamines on blood flow, oxygen consumption, and release/uptake of free fatty acids in adipose tissue. Scand. J. Clin. Lab. Invest. 27:221.
348. Estler, C.-J., and Ammon, H. P. T. (1969). The importance of the adrenergic beta-receptors for thermogenesis and survival of acutely cold-exposed mice. Can. J. Physiol. Pharmacol. 47:427.
349. Ahlquist, R. P. (1948). A study of the adrenotropic receptors. Am. J. Physiol. 153:586.
350. Mejsnar, J., Červinka, and Janský, L. (1976). Substitution of calorigenic effects of noradrenaline and adrenaline and differences in their inhibition by propranolol. Physiol. Bohemoslov. 25:201.
351. Alexander, G. (1969). The effect of adrenaline and noradrenaline on metabolic rate in young lambs. Biol. Neonate 14:97.
352. Baum, H., Moore, R. E., and Underwood, M. C. (1960). Stimulation of heat production in new-born cats and adult rats by adrenaline and noradrenaline. J. Physiol. (Lond.) 154:49.

353. Scopes, J. W., and Tizard, J. P. M. (1963). The effect of intravenous noradrenaline on the oxygen consumption of newborn mammals. J. Physiol. (Lond.) 165:305.
354. Alexander, G., and Mills, S. C. (1968). Free fatty acids and glucose in plasma of newly born lambs: effects of environmental temperature. Biol. Neonate 13:53.
355. Wasserstrum, N., and Herd, J. A. (1977). Baroreflexive depression of oxygen consumption in the squirrel monkey at 10°C. Am. J. Physiol. 232:451.
356. Hjemdahl, P., and Fredholm, B. B. (1976). Influence of acidosis on noradrenaline-induced vasoconstriction in adipose tissue and skeletal muscle. Acta Physiol. Scand. 97:319.
357. Ballard, K. (1973). Blood flow in canine adipose tissue during intravenous infusion of norepinephrine. Am. J. Physiol. 225:1026.
358. Fregly, M. J., Field, F. P., Nelson, E. L., Tyler, P. J., and Dasler, R. (1977). Effect of chronic exposure to cold on some responses to catecholamines. J. Appl. Physiol. 42:349.
359. Girardier, L., Seydoux, J., Giacobino, J. P., and Chenet, A. (1976). Catecholamine binding and modulation of the thermogenic effect in brown adipose tissue of the rat. *In* L. Janský and X. J. Musacchia (eds.), Regulation of Depressed Metabolism and Thermogenesis. Proceedings of a symposium, October, 1974, Prague. Charles C Thomas Publisher, Springfield, Illinois.
360. Renkin, E. M., and Rosell, S. (1962). The influence of sympathetic adrenergic vasoconstrictor nerves on transport of diffusible solutes from blood to tissue in skeletal muscle. Acta Physiol. Scand. 54:223.
361. Duran, W. N., and Renkin, E. M. (1976). Influence of sympathetic nerves on oxygen uptake of resting mammalian skeletal muscle. Am. J. Physiol. 231:529.
362. Grubb, B., and Folk, G. E. (1977). The role of adrenoceptors in norepinephrine-stimulated $\dot{V}O_2$ in muscle. Eur. J. Pharmacol. 43:217.
363. Jones, R. D., and Berne, R. M. (1964). Intrinsic regulation of skeletal muscle blood flow. Circ. Res. 14:126.
364. Roddie, I. C., and Shepherd, J. T. (1963). Nervous control of the circulation in skeletal muscle. Br. Med. J. 19:115.
365. Brungardt, J. M., Swan, K. G., and Reynolds, D. G. (1974). Adrenergic mechanisms in canine hindlimb circulation. Cardiovasc. Res. 8:423.
366. Viveros, O. H., Garlick, D. G., and Renkin, E. M. (1968). Sympathetic beta adrenergic vasodilatation in skeletal muscle of dog. Am. J. Physiol. 215:1218.
367. Mauskopf, J. M., Gray, S. D., and Renkin, E. M. (1969). Transient and persistent components of sympathetic cholinergic vasodilatation. Am. J. Physiol. 216:92.
368. Koo, A., and Liang, I. Y. S. (1978). Microvascular responses to norepinephrine in skeletal muscle of cold acclimated rats. J. Appl. Physiol. 44:190.
369. Aulie, A. (1976). The shivering pattern in an Arctic (willow ptarmigan) and a tropical bird (bantam hen). Comp. Biochem. Physiol. (A) 53:347.
370. Chaplin, S. B. (1976). The physiology of hypothermia in the black-capped chickadee, *Parus atricapillus.* J. Comp. Physiol. (B) 112:335.
371. Davis, T. R. A., Johnston, D. R., Bell, F. C., and Cremer, B. J. (1960). Regulation of shivering and non-shivering heat production during acclimation of rats. Am. J. Physiol. 198:471.
372. Banet, M., and Hensel, H. (1977). The control of shivering and non-shivering thermogenesis in the rat. J. Physiol. (Lond.) 269:669.
373. Sobolev, V. I. (1974). Physiological mechanisms of heat production during adaptation to cold. Fiziol. Zh. S.S.S.R. 60:1267.
374. Depocas, F. (1966). Concentration and turnover of cytochrome C in skeletal

muscles of warm- and cold-acclimated rats. Can. J. Physiol. Pharmacol. 44:875.

375. Himms-Hagen, J., Behrens, W., Muirhead, M., and Hbous, A. (1975). Adaptive changes in the calorigenic effect of catecholamines: role of changes in the adenylcyclase system and of changes in the mitochondria. Mol. Cell. Biochem. 6:15.
376. Stevens, E. D., and Kido, M. (1974). Active sodium transport: a source of metabolic heat during cold adaptation in mammals. Comp. Biochem. Physiol. (A) 47:395.
377. Holloszy, J. O., and Booth, F. W. (1976). Biochemical adaptations to endurance exercise in muscle. Annu. Rev. Physiol. 38:273.
378. Strømme, S. B., and Hammel, H. T. (1967). Effects of physical training on tolerance to cold in rats. J. Appl. Physiol. 23:815.
379. Horvath, S. M., Spurr, G. B., Hutt, B. K., and Hamilton, L. H. (1956). Metabolic cost of shivering. J. Appl. Physiol. 8:595.
380. Behmann, F. W., and Bontke, E. (1956). Intravasale Kühlung: Eine Methode zur Erzielung steuerbarer Hypothermia. Pfluegers Arch. 263:145.
381. Mercer, J. B., and Jessen, C. (1978). Effects of total body core cooling on heat production of conscious goats. Pfluegers Arch. 373:259.
382. McLean, J. A., Hales, J. R. S., Jessen, C., and Calvert, D. T. (1970). Influences of spinal cord temperature on heat exchange of the ox. Proc. Aust. Physiol. Pharmacol. Soc. 1(2):32.
383. Stuart, D. G., Eldred, E., Hemmingway, A., and Kawamura, Y. (1963). Neural regulation of the rhythm of shivering. *In* C. M. Herzfeld and J. D. Hardy (eds.), Temperature, Its Measurement and Control in Science and Industry, Vol. 3, pp. 545–557. Van Nostrand Reinhold, Company, New York.
384. Aulie, A. (1976). The pectoral muscles and the development of thermoregulation in chicks of willow ptarmigan (*Lagopus lagopus*). Comp. Biochem. Physiol. (A) 53:343.
385. Chatonnet, J., and Minaire, Y. (1966). Comparison of energy expenditure during exercise and cold exposure in the dog. Fed. Proc. 25:1348.
386. Bell, A. W., Thompson, G. E., and Findlay, J. D. (1974). The contribution of the shivering hind leg to the metabolic response to cold in the young ox (*Bos taurus*). Pfluegers Arch. 346:341.
387. Frens, J. (1973). A device for quantitative measuring of shivering in goats. Lab. Anim. 7:287.
388. Burton, A. C., and Bronk, D. W. (1937). The motor mechanism of shivering and thermal muscular tone. Am. J. Physiol. 119:284.
389. Simon, E. (1974). Temperature regulation: the spinal cord as a site of extrahypothalamic thermoregulatory functions. Rev. Physiol. Biochem. Pharmacol. 71:1.
390. Webster, A. J. F., Hicks, A. M., and Hays, F. L. (1969). Cold climate and cold temperature induced changes in the heat production and thermal insulation of sheep. Can. J. Physiol. Pharmacol. 47:553.
391. Nadel, E. R., Holmér, I., Bergh, U., Ästrand, P. O., and Stolwijk, J. A. J. (1973). Thermoregulatory shivering during exercise. Life Sci. 13:983.
392. Kosaka, M., Takagi, K., and Satoh, T. (1975). Inhibitory effect of electrical stimulation of the spinal cord on cold shivering. Nagoya J. Med. Sci. 20:41.
393. Glickman, N., Mitchell, H. H., Keeton, R. W., and Lambert, E. H. (1967). Shivering and heat production in men exposed to intense cold. J. Appl. Physiol. 22:1.
394. Petajan, J. H., and Williams, D. D. (1972). Behaviour of single motor units during pre-shivering tone and shivering tremor. Am. J. Phys. Med. 51:16.
395. Bell, A. W., Hilditch, T. E., Horton, P. W., and Thompson, G. E. (1976). The

distribution of blood flow between individual muscles and non-muscular tissues in the hind limb of the young ox (*Bos taurus*): values at thermoneutrality and during exposure to cold. J. Physiol. (Lond.) 257:229.

396. Slonim, A. D. (1969). Neural mechanisms of thermal regulation under normal living conditions. *In* E. Bajusz (ed.), Physiology and Pathology of Adaptation Mechanisms: Neural-Neuroendocrine-Humoral, pp. 410–435. Pergamon Press, Inc., New York.
397. Lupandin, Y. V., and Kuzimina, G. I. (1976). Interaction of regulation systems for contractile thermogenesis and external respiration. Fiziol. Zh. S.S.S.R. 62:1848.

397a. Komi, P. V., Rusko, H., Vos, J., and Vihko, V. (1977). Anaerobic performance capacity in athletes. Acta Physiol. Scand. 100:107.

398. Hales, J. R. S., Bennett, J. W., and Fawcett, A. A. (1976). Effects of acute cold exposure on the distribution of cardiac output in the sheep. Pfluegers Arch. 366:153.
399. West, G. C., Funke, E. R. R., and Hart, J. S. (1968). Power spectral density and probability analysis of electromyograms in shivering birds. Can. J. Physiol. Pharmacol. 46:703.
400. Fenn, W. O. (1923). The relation between the work performed and the energy liberated in muscular contraction. J. Physiol. (Lond.) 58:373.
401. Chinet, A., Clausen, T., and Girardier, L. (1977). Microcalorimetric determination of energy expenditure due to active sodium-potassium transport in the soleus muscle and brown adipose tissue of the rat. J. Physiol. 265:43.
402. Alexander, G., Mills, S. C., and Scott, T. W. (1968). Changes in plasma glucose, lactate and free fatty acids in lambs during summit metabolism and treatment with catecholamines. J. Physiol. (Lond.) 198:277.
403. Margaria, R., Aghemo, P., and Sassi, G. (1971). Lactic acid production in supramaximal exercise. Pfluegers Arch. 326:152.
404. Fredholm, B. B., and Karlsson, J. (1970). Metabolic effects of prolonged sympathetic nerve stimulation in canine subcutaneous adipose tissue. Acta Physiol. Scand. 80:567.
405. Himms-Hagen, J. (1972). Lipid metabolism during cold-exposure and during cold-acclimation. Lipids 7:310.
406. Pande, S. V., and Blanchaer, M. C. (1971). Carbohydrate and fat in energy metabolism of red and white muscle. Am. J. Physiol. 220:549.
407. Newsholme, E. A., and Start, C. (1973). Regulation in Metabolism, pp. 238–243. John Wiley and Sons, Inc., New York.
408. Graham, N. McC., Wainman, F. W., Blaxter, K. L., and Armstrong, D. G. (1959). Environmental temperature, energy metabolism and heat regulation in sheep. I. Energy metabolism in closely clipped sheep. J. Agric. Sci. 52:13.
409. McKay, D. G., Young, B. A., and Milligan, L. P. (1974). Energy substrates for cold thermogenesis. *In* K. H. Menke, H. J. Lantzsch, and J. R. Reichl (eds.), Energy Metabolism in Farm Animals. Proceedings of Sixth Symposium, September, 1973, Hohenheim. University of Hohenheim, Hohenheim.
410. Bell, A. W., and Thompson, G. E. (1976). The energy metabolism of shivering muscle measured *in vivo* during cold exposure of young cattle. *In* M. Vermorel (ed.), Energy Metabolism in Farm Animals, pp. 41–44. Proceedings of a symposium held in Vichy (France), September, 1976. G. de Bussac, Clermont-Ferrand.
411. Henriksson, J. (1977). Training induced adaptation of skeletal muscle and metabolism during submaximal exercise. J. Physiol. (Lond.) 270:661.
412. Costill, D. L., Coyle, E., Dalsky, G., Evans, W., Fink, W., and Hoopes, D. (1977). Effects of elevated plasma FFA and insulin on muscle glycogen usage during exercise. J. Appl. Physiol. 43:695.

413. Randle, P. J., Garland, P. B., Newsholme, E. A., and Hales, C. N. (1965). The glucose fatty acid cycle in obesity and maturity onset diabetes mellitus. Ann. N.Y. Acad. Sci. 131:324.
414. Issekutz, B., Issekutz, A. C., and Nash, D. (1970). Mobilization of energy sources in exercising dogs. J. Appl. Physiol. 29:691.
415. Minaire, Y., Vincent-Falquet, J. C., Pernod, A., and Chatonnet, J. (1973). Energy supply in acute cold-exposed dogs. J. Appl. Physiol. 35:51.
416. Beatty, C. H., and Bocek, R. M. (1971). Interrelations of carbohydrate and palmitate metabolism in skeletal muscle. Am. J. Physiol. 220:1928.
417. Helmrath, T. A., and Bieber, L. L. (1974). Development of gluconeogenesis in neonatal pig liver. Am. J. Physiol. 227:1306.
418. Bell, A. W., Clarke, P. L., and Thompson, G. E. (1975). Changes in the metabolism of the shivering hind leg of the young ox during several days of continuous cold exposure. Q. J. Exp. Physiol. 60:267.
419. Bell, A. W., Gardner, J. W., Manson, W., and Thompson, G. E. (1975). Acute cold exposure and metabolism of blood glucose, lactate and pyruvate and plasma amino acids in the hind leg of the fed and fasted young ox. Br. J. Nutr. 33:207.
420. Rudderman, N. B., and Goodman, M. N. (1973). Regulation of ketone body metabolism in skeletal muscle. Am. J. Physiol. 224:1391.
421. Freeman, B. M. (1976). Thermoregulation in the young fowl (*Gallus domesticus*). Comp. Biochem. Physiol. (A) 54:141.
422. Freeman, B. M. (1977). Lipolysis and its significance in the response to cold of the neonatal fowl, *Gallus domesticus*. J. Therm. Biol. 2:145.
423. Alexander, G., Bell, A. W., and Setchell, B. P. (1972). Regional distribution of cardiac output in young lambs: effect of cold exposure and treatment with catecholamines. J. Physiol. (Lond.) 220:511.
424. Rudolph, A. M., and Heymann, M. A. (1968). The circulation of the fetus in utero: methods for studying distribution of blood flow, cardiac output and organ blood flow. Circ. Res. 21:163.
425. Hales, J. R. S. (1974). Radioactive microsphere techniques for studies of the circulation. Clin. Exp. Pharmacol. Physiol. (suppl.) 1:31.
426. Járai, I. (1969). The redistribution of cardiac output on cold exposure in newborn rabbits. J. Physiol. (Lond.) 202:559.
427. Waites, G. M. H., and Setchell, B. P. (1966). Changes in blood flow and vascular permeability of the testis, epididymis and accessory reproductive organs of the rat after the administration of cadmium chloride. J. Endocrinol. 34:329.
428. Foster, D. O., and Frydman, M. L. (1978). Comparison of microspheres and $^{86}Rb^{+}$ as tracers of the distribution of cardiac output in rats indicates invalidity of $^{86}Rb^{+}$ based-measurements. Can. J. Physiol. Pharmacol. 56:97.
429. Bülow, J., and Madsen, J. (1976). Adipose tissue blood flow during prolonged, heavy exercise. Pfluegers Arch. 363:231.
430. Rauch, J. C., and Hayward, J. S. (1970). Regional distribution of blood flow in the bat (*Myotis lucifugus*) during arousal from hibernation. Can. J. Physiol. Pharmacol. 48:269.
431. Alexander, G., Bell, A. W., and Hales, J. R. S. (1972). The effect of cold exposure on the plasma level of glucose, lactate, free fatty acids and glycerol, and on the blood gas and acid-base status in young lambs. Biol. Neonate 20:9.
432. Raven, P. B., Niki, I., Dahms, T. E., and Horvath, S. M. (1970). Compensatory cardiovascular responses during an environmental cold stress, 5°C. J. Appl. Physiol. 29:417.
433. Hamilton, W. F. (1962). Measurement of cardiac output. *In* W. F. Hamilton

(ed.), Handbook of Physiology, Section 2, Circulation, pp. 551–584. American Physiological Society, Washington, D. C.

434. Douglas, F. G. V., and Becklake, M. R. (1968). Effect of seasonal training on maximal cardiac output. J. Appl. Physiol. 25:600.

435. Faulkner, J. A., Roberts, D. E., Elk, R. L., and Conway, J. (1971). Cardiovascular responses to submaximum and maximum effort of cycling and running. J. Appl. Physiol. 30:457.

436. Brodal, P., Ingjer, F., and Hermansen, L. (1977). Capillary supply of skeletal fibres in untrained and endurance-trained men. Am. J. Physiol. 232:H705.

437. Hensel, H. (1974). Thermoreceptors. Annu. Rev. Physiol. 36:233.

437a. Herd, P. A., Hammond, R. P., and Hamolsky, M. W. (1973). Sodium pump activity during norepinephrine-stimulated respiration in brown adipocytes. Am. J. Physiol. 224:1300.

438. Klussmann, F. W., and Pierau, F. K. (1972). Extrahypothalamic deep body thermosensitivity. *In* J. Bligh and R. Moore (eds.), Essays on Temperature Regulation. Elsevier-North Holland Publishing Company, New York.

439. Kuzmina, G. I., and Lupandin, Y. V. (1977). Role of vascular thermoreception in mechanisms regulating shivering. Fiziol. Zh. S.S.S.R. 63:573.

440. Horowitz, J. M., Fuller, C. A., and Horwitz, B. A. (1976). Central neural pathways and the control of non-shivering thermogenesis. *In* L. Janský and X. J. Musacchia (eds.), Regulation of Depressed Metabolism and Thermogenesis, pp. 3–25. Charles C Thomas Publisher, Springfield, Illinois.

441. Banet, M., and Hensel, H. (1976). The interaction between cutaneous and spinal thermal inputs in the control of oxygen consumption in the rat. J. Physiol. (Lond.) 260:461.

442. Bacon, M., and Bligh, J. (1976). Interaction between the effects of spinal heating and cooling and of injections into a lateral cerebral ventricle of noradrenaline, 5-hydroxytryptamine and carbachol on thermoregulation in sheep. J. Physiol. (Lond.) 254:213.

443. Cabanac, M., and Massonnet, B. (1977). Thermoregulatory responses as a function of core temperature in humans. J. Physiol. (Lond.) 265:587.

444. Buguet, A. G. C., Livingstone, S. D., Reed, L. D., and Limmer, R. E. (1976). Cold-induced shivering in men with thermoneutral skin temperatures. J. Appl. Physiol. 41:142.

445. Hayward, J. S., Eckerson, J. D., and Collis, M. L. (1977). Thermoregulatory heat production in man: prediction equation based on skin and core temperatures. J. Appl. Physiol. 43:377.

446. Timbal, J., Boutelier, C., Loncle, M., and Bougues, L. (1976). Comparison of shivering in man exposed to cold in water and in air. Pfluegers Arch. 365:243.

447. Wünnenberg, W. (1976). Thermointegrative functions of the hypothalamus. *In* L. Janský and X. J. Musacchia (eds.), Regulation of Depressed Metabolism, pp. 26–40. Proceedings of a symposium, October, 1974, Prague. Charles C Thomas Publisher, Springfield, Illinois.

448. Zeisberger, E., and Brück, K. (1976). Alteration of shivering threshold in cold- and warm-adapted guinea pigs following intrahypothalamic injections of noradrenaline and of an adrenergic alpha-receptor blocking agent. Pfluegers Arch. 362:113.

449. Hensel, H., Andres, K. H., and von During, M. (1974). Structure and function of cold receptors. Pfluegers Arch. 352:1.

450. Cooper, K. E., Pittman, Q. J., and Veale, W. L. (1976). The effect of noradrenaline and 5-hydroxytryptamine injected into a lateral cerebral ventricle on thermoregulation in the new-born lamb. J. Physiol. (Lond.) 261:223.

451. Slee, J. (1972). Habituation and acclimatization of sheep to cold following exposures of varying length and severity. J. Physiol. 227:51.

452. Janský, L., and Novotná, R. (1976). The role of central aminergic transmission, in thermoregulation and hibernation. *In* L. Janský and X. J. Musacchia (eds.), Regulation of Depressed Metabolism and Thermogenesis, pp. 64–78. Proceedings of a symposium, October, 1974, Prague. Charles C Thomas Publisher, Springfield, Illinois.
453. Heller, H. C., Colliver, G. W., and Beard, J. (1977). Thermoregulation during entrance into hibernation. Pfluegers Arch. 369:55.
454. Lyman, C. P., and O'Brien, R. C. (1972). Sensitivity to low temperature in hibernating rodents. Am. J. Physiol. 222:864.
455. Florant, G. L., and Heller, H. C. (1977). CNS regulation of body temperature in euthermic and hibernating marmots (*Marmota flaviventris*). Am. J. Physiol. 232:203.
456. Brück, K., and Wünnenberg, W. (1970). 'Meshed' control of two effector systems: non-shivering and shivering thermogenesis. *In* J. D. Hardy, A. P. Gagge, and J. A. J. Stolwijk (eds.), Physiological and Behavioural Temperature Regulation, pp. 562–580. Charles C Thomas Publisher, Springfield, Illinois.
457. Adamsons, K., Gandy, G. M., and James, L. S. (1965). The influence of thermal factors upon oxygen consumption of the newborn human infant. J. Pediatr. 66:495.
458. Farkas, M., Varnai, I., and Donhoffer, S. (1972). Fallacies in the interpretation of body temperature changes in the newly born: the "inability" to increase heat production and the "unfavourable" body mass-body surface ratio. Acta Physiol. Acad. Sci. Hung. 42:31.
459. Rautenberg, W., Necker, R., and May, B. (1972). Thermoregulatory responses of the pigeon to changes of the brain and the spinal cord temperatures. Pfluegers Arch. 338:31.
460. Hammel, H. T., Maggert, J., Kaul, R., Simon, E., and Simon-Opperman, C. (1976). Effects of altering spinal cord temperature on temperature regulation in the Adelie Penguin *Pygoscelis Adeliae*. Pfluegers Arch. 362:1.
461. Hales, J. R. S., and Iriki, M. (1975). Integrated changes in regional circulatory activity evoked by spinal cord and peripheral thermoreceptor stimulation. Brain Res. 87:267.
462. Hales, J. R. S., Bennett, J. W., and Fawcett, A. A. (1977). Integrated changes in regional circulatory activity evoked by thermal stimulation of the hypothalamus. Pfluegers Arch. 372:157.
463. Grausz, J. P. (1968). The effects of environmental temperature changes on the metabolic rate of newborn babies. Acta. Paediatr. Scand. 57:98.
464. Adair, E. R., and Rawson, R. O. (1973). Step-wise changes in thermoregulatory responses to slowly changing thermal stimuli. Pfluegers Arch. 339:241.
465. Alexander, G. (1978). Quantitative development of adipose tissue in foetal sheep. Aust. J. Biol. Sci. 31:489.
466. Hardman, M. J., Hey, E. W., and Hull, D. (1969). Fat metabolism and heat production in young rabbits. J. Physiol. (Lond.) 205:51.
467. Hardman, M. J., Hey, E. W., and Hull, D. (1969). The effect of prolonged cold exposure on heat production in newborn rabbits. J. Physiol. (Lond.) 205:39.
468. Hardman, M. J., Hull, D., and Milner, A. D. (1971). Brown adipose tissue metabolism *in vivo* and serum insulin concentrations in rabbits soon after birth. J. Physiol. (Lond.) 213:175.
469. Schenk, H., Heim, T., Mende, T., Varga, F., and Goetze, E. (1975). Studies on plasma free-fatty-acid metabolism and triglyceride synthesis of brown adipose tissue *in vivo* during cold-induced thermogenesis of the newborn rabbit. Eur. J. Biochem. 58:15.

470. Radomski, M. U., and Orme, T. (1971). Response of lipoprotein lipase in various tissues to cold exposure. Am. J. Physiol. 220:1852.
471. Essen, B., Jansson, E., Hendriksson, J., Taylor, A. W., and Saltin, B. (1975). Metabolic characteristics of fibre types in human skeletal muscle. Acta Physiol. Scand. 95:153.
472. Frey, H. M. M., and Skjörten, F. (1967). Peripheral circulatory and metabolic consequences of thyrotoxicosis in the dog. IX. Ultrastructural changes in skeletal muscle during experimental thyrotoxicosis in the dog. Scand. J. Clin. Lab. Invest. 19:351.
473. Pryor, W. J., and Warren, G. H. (1973). Chemical fat in musculature of the sheep carcass. J. Agric. Sci. 80:219.
474. Therriault, D. G., and Poe, R. H. (1965). The effect of acute and chronic cold exposure on tissue lipids in the rat. Can. J. Biochem. 43:1427.
475. Bégin-Heick, N., and Heick, H. M. C. (1977). Increased lipoprotein lipase activity of skeletal muscle in cold-acclimated rats. Can. J. Biochem. 55:1241.
476. Gillespie, C. A., Simpson, D. R., and Edgerton, V. R. (1970). High glycogen content of red as opposed to white skeletal muscle fibres in guinea pigs. J. Histochem. Cytochem. 18:552.
477. Short, F. A., Cobb, L. A., Kawabori, I., and Goodner, C. J. (1969). Influence of exercise training on red and white rat skeletal muscle. Am. J. Physiol. 217:327.
478. Alexander, G. (1962). Energy metabolism in the starved newborn lamb. Aust. J. Agric. Res. 13:144.
479. Bost, J., and Dorleac, E. (1965). Action du froid sur le taux plasmatique des acids gras non esterifies (FFA) chez le mouton. C. R. Soc. Biol. 159:2209.
480. Slee, J., and Halliday, R. (1968). Some effects of cold exposure, nutrition and experimental handling on serum free fatty levels in sheep. Anim. Prod. 10:67.
481. Thompson, G. E., and Clough, D. P. (1972). The effect of cold exposure on plasma lipids of the new-born and adult ox. Q. J. Exp. Physiol. 57:192.
482. Curtis, S. E., and Rogler, J. C. (1970). Thermoregulatory ontogeny in piglets: sympathetic and adipokinetic responses to cold. Am. J. Physiol. 218:149.
483. Himms-Hagen, J. (1965). Lipid metabolism in warm-acclimated and cold-acclimated rats exposed to cold. Can. J. Physiol. Pharmacol. 43:379.
484. Hardman, M. J., and Hull, D. (1969). The effects of age and environmental temperature on the blood concentrations of glucose, free fatty acids and glycerol in new-born rabbits. J. Physiol. (Lond.) 201:685.
485. Heim, T., and Kellermayer, M. (1967). The effect of environmental temperature in brown and white adipose tissue in the starving newborn rabbit. Acta Physiol. Acad. Sci. Hung. 31:339.
486. Hajós, F., Heim, T., and Kerpel-Fronius, S. (1970). Electron-microscopic observations in various functional stages of the brown and white adipose tissue in the newborn rabbit. Biol. Neonate 15:94.
487. Heim, T., Schenk, H., Wagner, H., Winkler, L., Goetze, E., and Varga, F. (1974). *In vivo* metabolism of ^{14}C-labelled palmitic acid in serum and in brown and white adipose tissues of well-fed and starved newborn rabbits. I. Trace kinetic analysis of free fatty acid metabolism in serum. Biol. Neonate 24:244.
488. Rakow, L., Beneke, G., Mohr, W., and Brauchle, I. (1970). Vergleichende morphologische und chemische Untersuchungen am weissen und braunen Fettgewebe der Maus nach chronischem Hunger und Wiederauffutterung. Beitr. Pathol. 141:349.
489. Heim, T., Kellermayer, M., and Dani, M. (1968). Thermal conditions and the mobilization of lipids from brown and white adipose tissue in the human neonate. Acta Paediatr. Acad. Sci. Hung. 9:109.

490. Bizzi, A., Codegoni, A. M., Lietti, A., and Garattini, S. (1968). Different responses of white and brown adipose tissue to drugs affecting lipolysis. Biochem. Pharmacol. 17:2407.
491. Horn, G. W., Foley, C. W., Seerley, R. W., and Munnell, J. F. (1973). Role of adipose tissue lipolysis in postnatal amelioration of thermogenesis in domestic and wild piglets. J. Anim. Sci. 37:1356.
492. Clough, D. P., and Thompson, G. E. (1972). Effect of intraventricular noradrenaline and acetylcholine on plasma unesterified fatty acid concentration in sheep. Neuroendocrinology 9:365.
493. Gross, J. L., and Migliorini, R. H. (1977). Further evidence for a central regulation of free fatty acid mobilization in the rat. Am. J. Physiol. 232:E165.
494. Freeman, B. M. (1969). Effect of noradrenaline on the plasma free fatty acid and glucose levels in *Gallus domesticus.* Comp. Biochem. Physiol. 30:993.
495. Derry, D. M., Ranson, J., and Morrow, E. (1971). Effect of isoproterenol on brown and white fat of the adult rat. Can. J. Physiol. Pharmacol. 49:8.
496. Bell, A. W., and Thompson, G. E. (1974). Effects of cold exposure and feeding on net exchange of plasma free fatty acids and glycerol across the hind leg of the young ox. Res. Vet. Sci. 17:265.
497. McBurney, L. J., and Radomski, M. W. (1969). Metabolism of serum free fatty acids and low-density lipoproteins in the cold-acclimated rat. Am. J. Physiol. 217:19.
497a. Chatonnet, J., Minaire, Y., Pernod, A., and Vincent-Falquet, J.-C. (1972). Inhibitor of glucose uptake by epinephrine in dogs during cold exposure. J. Appl. Physiol. 32:170.
498. Kusaka, M., and Ui, M. (1977). Activation of the Cori cycle by epinephrine. Am. J. Physiol. 232:145.
499. Thompson, G. E., Manson, W., Clarke, P. L., and Bell, A. W. (1978). Acute cold exposure and the metabolism of glucose and some of its precursors in the liver of the fed and fasted sheep. Q. J. Exp. Physiol. 63:189.
500. Klain, G. J., and Hannon, J. P. (1969). Gluconeogenesis in cold-exposed rats. Fed. Proc. 28:965.
501. Chideckel, E. W., Goodner, C. J., Koerker, D. J., Johnson, D. G., and Ensinck, J. W. (1977). Role of glucagon in mediating metabolic effects of epinephrine. Am. J. Physiol. 232:E464.
502. Smith, O. K. (1976). Rise in plasma α-amino nitrogen concentration in rats eviscerated after cold exposure. Am. J. Physiol. 231:174.
503. Forichon, J., Jomain, M. J., Dallevet, G., and Minaire, Y. (1977). Effect of cold and epinephrine on glucose kinetics in dogs. J. Appl. Physiol. 43:230.
504. Iversen, J. (1973). Adrenergic receptors and the secretion of glucagon and insulin from the isolated perfused canine pancreas. J. Clin. Invest. 52:2102.
505. Luyckx, A. S., Dresse, A., Cession-Fossion, A., and Lefebvre, D. J. (1975). Catecholamines and exercise-induced glucagen and fatty acid mobilization in the rat. Am. J. Physiol. 229:376.
506. Edwards, A. V., and Silver, M. (1970). The glycogenolytic response to stimulation of the splanchnic nerves in adrenalectomized calves. J. Physiol. (Lond.) 211:109.
507. Minaire, Y., Pernod, A., Jomain, M.-J., and Mottaz, M. (1971). Lactate turnover and oxidation in normal and adrenal-demedullated dogs during cold exposure. Can. J. Physiol. Pharmacol. 49:1063.
508. Hermansen, L., and Stensvold, I. (1972). Production and removal of lactate during exercise in man. Acta Physiol. Scand. 86:191.
509. Thomson, E. M., Snoswell, A. M., Clarke, P. C., and Thompson, G. E. Effect of cold exposure on mammary gland uptake of fat precursors, and secretion of milk fat and carnitine in the goat. Q. J. Exp. Physiol. In press.

510. Weil, R., Ho, P.-P., and Altszuler, N. (1965). Effect of free fatty acids on metabolism of pyruvic and lactic acids. Am. J. Physiol. 208:887.
511. Treble, D. H., and Ball, E. G. (1963). The occurrence of glycerolkinase in rat brown adipose tissue. Fed. Proc. 22:357.
512. Margolis, S., and Vaughan, M. (1962). α-Glycerophosphate synthesis and breakdown in homogenates of adipose tissue. J. Biol. Chem. 237:44.
513. Bertin, R. (1976). Glycerokinase activity and lipolysis regulation in brown adipose tissue of cold acclimated rats. Biochimie 58:431.
514. Sargent, F., Johnson, R. E., Robbins, E., and Sawyer, L. (1958). The effect of environment and other factors on nutritional ketosis. Q. J. Exp. Physiol. 43:345.
515. Itoh, S., and Kuroshima, A. (1972). Lipid metabolism of cold-adapted man. *In* S. Itoh et al. (eds.), Advances in Climatic Physiology, pp. 260–277. Igaku-Shoin, Tokyo.
516. Scott, J. L., and Engel, F. C. (1953). The influence of the adrenal cortex and cold stress on fasting ketosis in the rat. Endocrinology 53:410.
517. Davison, T. F. (1973). Metabolite changes in the neonate fowl in response to cold stress. Comp. Biochem. Physiol. (A) 44:979.
518. Maekubo, H., Moriya, K., and Hiroshige, T. (1977). Role of ketone bodies in nonshivering thermogenesis in cold-acclimated rats. J. Appl. Physiol. 42:159.
519. Masoro, E. J. (1966). Effect of cold on metabolic use of lipids. Physiol. Rev. 46:67.
520. Kuroshima, A., Doi, K., Yahata, T., and Ohno, T. (1977). Improved cold tolerance and its mechanism in cold-acclimated rats by high fat diet feeding. Can. J. Physiol. Pharmacol. 55:943.
521. Ingram, D. L., and Legge, K. F. (1974). Effects of environmental temperature on food intake in growing pigs. Comp. Biochem. Physiol. (A) 48:573.
522. Weston, R. H. (1970). Voluntary consumption of low quality roughage by sheep during cold exposure. Aust. J. Exp. Agric. Anim. Husb. 10:679.
523. Bell, A. W., Gardner, J. W., and Thompson, G. E. (1974). The effects of acute cold exposure and feeding on volatile fatty acid metabolism in the hind leg of the young ox. Br. J. Nutr. 32:471.
524. Cardasis, C. A., Blanc, W. A., and Sinclair, J. C. (1972). The effects of ambient temperature on the fasted newborn rabbit. II. Gross and microscopic changes in cervical and interscapular brown adipose tissue. Biol. Neonate 21:347.
525. Bignall, K. E., Heggeness, F. W., and Palmer, J. E. (1974). Effects of acute starvation in cold induced thermogenesis in the preweanling rat. Am. J. Physiol. 227:1088.
526. Bignall, K. E., Heggeness, F. W., and Palmer, J. E. (1975). Effect of neonatal decerebration on thermogenesis during starvation and cold exposure in the rat. Exp. Neurol. 49:174.
527. Bignall, K. E., Heggeness, F. W., and Palmer, J. E. (1977). Sympathetic inhibition of thermogenesis in the infant rat: possible glucostatic control. Am. J. Physiol. 233:23.
528. Brooke, O. G., Harris, M., and Salvosa, C. B. (1973). The response of malnourished babies to cold. J. Physiol. 233:75.
529. Cardasis, C. A., and Sinclair, J. C. (1972). The effects of ambient temperature on the fasted newborn rabbit. I. Survival time, weight loss, body temperature and oxygen consumption. Biol. Neonate 21:330.
530. Morrill, C. C. (1952). Studies on baby pig mortality. VIII. Chemical observations on the newborn pig, with special reference to hypoglycemia. Am. J. Vet. Res. 13:164.

531. Curtis, S. E. (1970). Environmental thermoregulatory interactions and neonatal piglet survival. J. Anim. Sci. 31:576.
532. Alexander, G. (1970). Summit metabolism in young lambs: effect of hormones and drugs that affect mobilization and utilization of substrate for cold-induced thermogenesis. Biol. Neonate 15:37.
533. Beviz, A., Lundholm, L., and Mohme-Lundholm, E. (1971). Cyclic AMP as a mediator of hormonal metabolic effects in brown adipose tissue. Acta Physiol. Scand. 81:145.
534. Steiner, G. (1975). Neural and humoral regulation of brown adipose tissue metabolism. *In* P. Lomax et al. (eds.), Temperature Regulation and Drug Action, pp. 159–171. Proceedings of a Symposium, April, 1974, Paris. S. Karger, Basel.
535. Kaciŭba-Uścilko, H. (1972). Hormonal regulation of thermogenesis in the newborn pig. Biol. Neonate 21:245.
536. Kaciŭba-Uścilko, H. (1971). The effect of previous thyroxine administration on the metabolic response to adrenaline in new-born pigs. Biol. Neonate 19:220.
537. Anagnostakis, D., Economoŭ-Mavrou, C., Agathopoulos, A., and Matsaniotis, N. (1974). Neonatal cold injury: evidence of defective thermogenesis due to impaired norepinephrine release. Pediatrics 53:24.
538. Thompson, G. E., and Moore, R. E. (1968). A study of newborn rats exposed to cold. Can. J. Physiol. Pharmacol. 46:865.
539. Stock, K., and Westermann, E. O. (1963). Concentration of norepinephrine serotonin and histamine, and of amine metabolizing enzymes in mammalian adipose tissue. J. Lipid Res. 4:297.
540. Yoshimura, K., Hiroshige, T., and Itoh, S. (1969). Lipolytic action of serotonin in brown adipose tissue *in vitro.* Jpn. J. Physiol. 19:176.
541. Steiner, G., and Evans, S. (1976). Effect of serotonin on brown adipose tissue and on its sympathetic neurones. Am. J. Physiol. 231:34.
542. Rudman, D., Girolamo, M. D., Malkin, M. F., and Garcia, L. A. (1965). The adipokinetic property of hypophyseal peptides and catecholamines: a problem in comparative endocrinology. *In* A. E. Renold and G. F. Cahill (eds.), Handbook of Physiology, Section 5, pp. 533–539. American Physiological Society, Washington, D. C.
543. Lebovitz, H. E., and Engel, F. L. (1965). *In* A. E. Renold and G. F. Cahill (eds.), Handbook of Physiology, Section 5, pp. 541–547. American Physiological Society, Washington, D. C.
544. Fain, J. N., Reed, N., Sapirstein, R. (1967). The isolation and metabolism of brown fat cells. J. Biol. Chem. 242:1887.
545. Foster, D. O., Frydman, M. L., and Usher, J. R. (1977). Nonshivering thermogenesis in the rat. I. The relation between drug-induced changes in thermogenesis and changes in the concentration of plasma cyclic AMP. Can. J. Physiol. Pharmacol. 55:52.
546. Portet, R., Laury, M. C., Bertin, R., Senault, C., Hluszko, M. T., and Chevillard, L. (1974). Hormonal stimulation of substrate utilization in brown adipose tissue of cold acclimated rats. Proc. Soc. Exp. Biol. Med. 147:807.
547. Reed, N., and Fain, J. N. (1968). Stimulation of respiration in brown fat cells by epinephrine, dibutyryl-3′,5′-adenosine monophosphate and *m*-chloro(carbonyl cyanide)phenylhydrazone. J. Biol. Chem. 243:2843.
548. Fain, J. N., and Reed, N. (1970). A mechanism for hormonal activation of lipolysis and respiration in free brown fat cells. Lipids 5:210.
549. Leduc, J. (1975). Potentiation of the calorigenic effect of noradrenaline in cold acclimated rats by theophylline. *In* L. Janský (ed.), Depressed Metabolism and Cold Thermogenesis, pp. 133–138. Proceedings of a symposium, October, 1974, Prague. Charles University, Prague.

550. Fritz, I. B., and Marquis, N. R. (1965). The role of acylcarnitine esters and carnitine palmityltransferase in the transport of fatty acyl groups across mitochondrial membranes. Proc. Natl. Acad. Sci. U.S.A. 54:1226.
551. Hahn, P., Skala, J., and Davies, P. (1971). Carnitine enhances the effect of norepinephrine on oxygen consumption in rats and mice. Can. J. Physiol. Pharmacol. 49:853.
552. Stave, U., Novak, M., Wieser, P. B., and Buch, M. (1976). Heat producing effect of carnitine in newborns. Pediatr. Res. 10:326.
553. Hahn, P., and Skala, J. P. (1975). The role of carnitine in brown adipose tissue of suckling rats. Comp. Biochem. Physiol. (B) 51:507.
554. Sokal, J. E. (1966). Glucagon—an essential hormone. Am. J. Med. 41:331.
555. Kuroshima, A., Ohno, T., and Doi, K. (1977). In vivo lipolytic action of glucagon in brown adipose tissue of warm-acclimatized and cold-acclimatized rats. Experentia 33:240.
556. Palokangas, R., Vihko, V., and Nuuja, I. (1973). The effect of cold and glucagon on lipolysis, glycogenolysis and oxygen consumption in young chicks. Comp. Biochem. Physiol. (A) 45:489.
557. Davidson, I. W. F., Salter, J. M., and Best, C. H. (1957). Calorigenic action of glucagon. Nature (Lond.) 180:1124.
558. Freeman, B. M. (1975). On a possible role for glucagon in the thermogenic response of the neonate fowl, *Gallus domesticus*. J. Therm. Biol. 1:59.
559. Hohtola, E., Hissa, R., and Saarela, S. (1977). Effect of glucagon on thermogenesis in the pigeon. Am. J. Physiol. 232:451.
560. Bassett, J. M. (1971). The effects of glucagon on plasma concentrations of insulin, growth hormone, glucose, and free fatty acids in sheep: comparison with the effects of catecholamines. Aust. J. Biol. Sci. 24:311.
561. Hardman, M. J., and Hull, D. (1972). The action of insulin on brown adipose tissue *in vivo*. J. Physiol. (Lond.) 221:85.
562. Steiner, G., and Cahill, G. F. (1966). Brown and white adipose tissue metabolism response to norepinephrine and insulin. Am. J. Physiol. 211:1325.
563. Beck, L. V., Zaharko, D. S., and Kalser, S. C. (1967). Variation in serum insulin and glucose of rats with chronic cold exposure. Life Sci. 6:1501.
564. Porte, D., Graber, A. L., Kuzuya, T., and Williams, R. H. (1966). The effect of epinephrine on immunoreactive insulin levels in man. J. Clin. Invest. 45:228.
565. Bassett, J. M., and Alexander, G. (1971). Insulin, growth hormone and corticosteroids in neonatal lambs: normal concentrations and the effects of cold. Biol. Neonate 17:112.
566. Krishna, G., Moskowitz, J., Dempsey, P., and Brodie, B. B. (1970). The effect of norepinephrine and insulin on brown fat cell membrane potentials. Life Sci. 9:1353.
567. Alexander, G., and Bell, A. W. (1975). Maximum thermogenic response to cold in relation to the proportion of brown adipose tissue and skeletal muscle in the body and to other parameters. Biol. Neonate 26:182.
568. Artundo, A. (1927). Le metabolisme de sommet des rats prives de surrenales. C. R. Soc. Biol. 97:409.
569. Widström, G. (1935). On the biological assay of adrenal cortical preparations using white rats and mice. Acta Med. Scand. 87:1.
570. Wilson, O., Hedner, P., Laurell, S., Nosslin, B., Rerup, C., and Rosengren, E. (1970). Thyroid and adrenal response to acute cold exposure in man. J. Appl. Physiol. 28:543.
571. Levin, L. (1945). The effect of several varieties of stress on the cholesterol content of the adrenal glands and of the serum of rats. Endocrinology 37:34.
572. Edgahl, R. H., and Richards, J. B. (1956). Effect of extreme cold exposure on adrenocortical function in the unanaesthetised dog. Am. J. Physiol. 185:239.

573. Leppäluoto, J., Lybeck, H., Ranta, T., and Virkkunen, P. (1973). Effect of acute exposure to cold on blood thyrotrophin (TSH) and corticosterone concentrations in the rabbit. Acta Physiol. Scand. 89:423.
574. Panaretto, B. A., and Vickery, M. R. (1970). The rates of plasma cortisol entry and clearance in sheep before and during their exposure to a cold, wet environment. J. Endocrinol. 47:273.
575. Wilkerson, J. E., Raven, P. B., Bolduan, N. W., and Horvath, S. M. (1974). Adaptations in man's adrenal function in response to acute cold stress. J. Appl. Physiol. 36:183.
576. Kirsteins, A. (1956). Survival of cortisone and corticotropin-treated rats during exposure to cold. Surgery 40:337.
577. Panaretto, B. A., and Ferguson, K. A. (1969). Pituitary adrenal interactions in shorn sheep exposed to cold, wet conditions. Aust. J. Agric. Res. 20:99.
578. Ramey, E. R., and Goldstein, M. S. (1957). The adrenal cortex and the sympathetic nervous system. Physiol. Rev. 37:155.
579. Hahn, P., Drahota, Z., Skala, J., Kazola, S., and Towell, M. E. (1969). The effect of cortisone on brown adipose tissue of young rats. Can. J. Physiol. Pharmacol. 47:975.
580. Skala, J., and Hahn, P. (1971). Effects of single cortisone injections on brown adipose tissue of developing rats. Can. J. Physiol. Pharmacol. 49:501.
581. Hemon, P. (1976). Some aspects of rat metabolism in the brown adipose tissue of normal and hypothyroid rats during early postnatal development. Biol. Neonate 28:241.
582. Székely, M. (1970). Effects of thyroxine treatment of different duration on oxygen consumption and body temperature at different ambient temperatures in the rat. Acta Physiol. Acad. Sci. Hung. 37:51.
583. Heick, H. M. C., Vachon, C., Kallai, M. A., Bégin-Heich, N., and LeBlanc, J. (1973). The effect of thyroxine and isopropylnoradrenaline on cytochrome oxidase activity in brown adipose tissue. Can. J. Physiol. Pharmacol. 51:751.
584. Patton, J. F., and Platner, W. S. (1970). Cold acclimation and thyroxine effects on liver and liver mitochondrial fatty acids. Am. J. Physiol. 218:1417.
585. Rabi, T., and Cassuto, Y. (1976). Metabolic activity of brown adipose tissue in T_3-treated hamsters. Am. J. Physiol. 231:161.
586. Winder, W. W., and Holloszy, J. O. (1977). Response of mitochondria of different types of skeletal muscle to thyrotoxicosis. Am. J. Physiol. 232:180.
587. Ismail-Beigi, F., and Edelman, I. S. (1970). Mechanism of thyroid calorigenesis: role of active sodium transport. Proc. Natl. Acad. Sci. U.S.A. 67:1071.
588. Asano, Y., Liberman, U. A., and Edelman, I. S. (1976). Thyroid thermogenesis: relationships between Na^+-dependent respiration and $Na^+ + K^+$-adenosine triphosphatase activity in rat skeletal muscle. J. Clin. Invest. 57:368.
589. Simon, R. G., Eybel, C. E., Galster, W., and Morrison, P. (1971). Mitochondrial involvement in cold acclimation. Comp. Biochem. Physiol. (B) 40:601.
590. Kennedy, D. R., Hammond, R. P., and Hamolsky, M. W. (1977). Thyroid cold acclimation influences on norepinephrine metabolism in brown fat. Am. J. Physiol. 232:565.
591. Hemon, P., Ricquier, D., and Mory, G. (1976). A role for thyroid hormones in the response of brown adipose tissue to chronic cold. *In* L. Janský and X. J. Musacchia (eds.), Regulation of Depressed Metabolism and Thermogenesis, pp. 174–195. Proceedings of a symposium, October, 1974, Prague. Charles C Thomas Publisher, Springfield, Illinois.
592. Ricquier, D., Mory, G., and Hemon, P. (1976). Effects of chronic treatments

upon the brown adipose tissue of young rats. I. Cold exposure and hyperthyroidism. Pfluegers Arch. 362:241.

593. Hsieh, A. C. L. (1966). Thyroid hormone requirement in rats exposed to cold. Gunma Symp. Endocrinol. 3:239.

594. Carlson, L. D., Roohk, H. V., and Wilson, O. (1973). Thyroid function studies in normal and cold-exposed rabbits using ^{125}I. Acta Physiol. Scand. 89:359.

595. Evans, S. E., and Ingram, D. L. (1977). The effect of ambient temperature upon the secretion of thyroxine in the young pig. J. Physiol. (Lond.) 264:511.

596. Thompson, R. H., Buskirk, E. R., and Whedon, G. D. (1971). Temperature regulation against cold: effect of induced hyperthyroidism in men and women. J. Appl. Physiol. 31:740.

597. Therminarias, A., Chirpaz, M. F., and Tanche, M. (1975). Influence des hormones thyroidiennes sur l'effet calorigénique de l'adrénaline. Rev. Can. Biol. 34:101.

598. Ikemoto, H., Hiroshige, T., and Itoh, S. (1967). Oxygen consumption of brown adipose tissue in normal and hypothyroid mice. Jpn. J. Physiol. 17:516.

599. Roy, M. L., Sellers, E. M., Flattery, K. V., and Sellers, E. A. (1977). Influence of cold exposure and thyroid hormones on regulation of adrenal catecholamines. Can. J. Physiol. Pharmacol. 55:804.

600. Fisher, D. A., and Odell, W. D. (1971). Effect of cold on TSH secretion in man. J. Clin. Endocrinol. Metab. 33:859.

601. Kaciǔba-Uścilko, H., Legge, K. F., and Mount, L. E. (1970). The development of the metabolic response to thyroxine in the new-born pig. J. Physiol. (Lond.) 206:229.

602. Lintzel, W. (1931). Über die Wirkung der Luftverdunnung auf Tier. V. Mitteilung. Gaswechsel weisser Ratten. Pfluegers Arch. 227:693.

603. Farkas, M., and Donhoffer, S. (1973). Interrelationships between the effects of hypoxia and ambient temperature on heat production and body temperature in the adult guinea-pig: cause and effect in the response of body temperature and heat production to hypoxia. Acta Physiol. Acad. Sci. Hung. 43:301.

604. Alexander, G., and Williams, D. (1970). Summit metabolism and cardiovascular function in young lambs during hyperoxia and hypoxia. J. Physiol. (Lond.) 208:85.

605. Blatteis, C. M., and Lutherer, L. O. (1973). Cold-induced thermogenesis in dogs: its reduction by moderate hypoxia. J. Appl. Physiol. 35:608.

606. Blatteis, C. M., and Lutherer, L. O. (1976). Effect of altitude exposure on thermoregulatory response of man to cold. J. Appl. Physiol. 41:848.

607. Blatteis, C. M., and Gilbert, T. M. (1974). Hypoxia and shivering thermogenesis in cold acclimatized miniature pigs. J. Appl. Physiol. 36:453.

608. Bhatia, B., George, S., and Rao, T. L. (1969). Hypoxic poikilothermia in rats. J. Appl. Physiol. 27:583.

609. Székely, M., Kellermayer, M., and Donhoffer, S. (1971). The effect of hypoxia, hypercapnia, β-adrenergic blockade and anaesthesia on heat production by periaortic brown adipose tissue. Acta Physiol. Acad. Sci. Hung. 40:261.

610. Blatteis, C. M. (1972). Shivering and nonshivering thermogenesis during hypoxia. *In* R. E. Smith (ed.), Bioenergetics, pp. 151–160. Proceedings of the International Symposium in Environmental Physiology, Dublin, July, 1971. Federation of the American Society for Experimental Biology, Bethesda, Maryland.

611. Járai, I., and Lendvay, B. (1957). The action of alpha-dinitrophenol on heat production and body temperature in hypoxic hypoxia. Acta Physiol. Acad. Sci. Hung. 13:147.

612. Farkas, M., and Donhoffer, S. (1974). The effect of hypoxia on ther-

moregulatory heat production and body temperature in the new-born and young guinea pig. Acta Physiol. Acad. Sci. Hung. 45:181.
613. Stanton, H. C., Brown, L. J., and Mueller, R. L. (1973). Interrelationships between maternal and neonatal factors and thermoregulation in fasted neonatal swine (*Sus domesticus*). Comp. Biochem. Physiol. (A) 44:97.
614. Harding, P. G. R., and Ralph, E. D. (1970). Effects of chronic hypoxia on lipolysis in brown adipose tissue in the fetal rabbit. Am. J. Obstet. Gynecol. 106:907.
615. Farkas, M. (1973). The effect of hypercapnia in heat production and body temperature in the adult guinea pig. Acta Physiol. Acad. Sci. Hung. 43:309.
616. Pepelko, W. E., and Dixon, G. A. (1974). Elimination of cold-induced non-shivering thermogenesis by hypercapnia. Am. J. Physiol. 227:264.
617. Várnai, I., Farkas, M., and Donhoffer, S. (1970). Thermoregulatory effects of hypercapnia in the new-born rat: comparison with the effect of hypoxia. Acta Physiol. Acad. Sci. Hung. 38:225.
618. Schaefer, K. E., and Wünnenberg, W. (1976). Threshold temperatures for shivering in acute and chronic hypercapnia. J. Appl. Physiol. 41:67.
619. Stoner, H. B. (1954). Studies on the mechanism of shock: the effect of limb ischaemia on tissue temperature and blood flow. Br. J. Exp. Pathol. 35:487.
620. Little, R. A. (1974). The impairment of thermoregulation by trauma during the first days of life in the rabbit. Biol. Neonate 24:363.
621. Stoner, H. B. (1971). Effect of injury on shivering thermogenesis in the rat. J. Physiol. (Lond.) 214:599.
622. Stoner, H. B. (1974). Inhibition of thermoregulatory non-shivering thermogenesis by trauma in cold-acclimated rats. J. Physiol. (Lond.) 238:657.
623. Matsaniotis, N., Pastelis, V., Agathopoulos, A., and Constantsas, N. (1971). Fever and biochemical thermogenesis. Pediatrics 47:571.
624. Blatteis, C. M. (1976). Fever: exchange of shivering by nonshivering pyrogenesis in cold acclimated guinea-pigs. J. Appl. Physiol. 40:29.
625. Székely, M., Szelényi, Z., and Sümegi, I. (1973). Brown adipose tissue as a source of heat during pyrogen-induced fever. Acta Physiol. Acad. Sci. Hung. 43:85.
626. Veale, W. L., Cooper, K. E., and Pittman, Q. J. (1976). Ontogenesis of thermoregulation and sensitivity to pyrogens. *In* M. Brazier and F. Coceani (eds.), Brain Dysfunction in Infantile Febrile Convulsions, pp. 153–160. Proceedings of an International Brain Research Organization Symposium, October, 1975, Toronto. Raven Press, New York.
627. Pittman, Q. J., Cooper, K. E., Veale, W. L., and van Petten, G. R. (1974). Observations on the development of the febrile response to pyrogens in sheep. Clin. Sci. Mol. Med. 46:591.
628. Hahn, P., and Skala, J. (1972). Changes in interscapular brown adipose tissue of the rat during perinatal and early postnatal development and after cold acclimation. III. Some cytoplasmic enzymes. Comp. Biochem. Physiol. (B) 41:147.
629. Asmussen, E. (1965). Muscular exercise. Handbook of Physiology: Respiration, Section 3, Vol. II, pp. 939–978. American Physiological Society, Washington, D. C.
630. Ouellet, Y., Poh, S. C., and Becklake, M. R. (1969). Circulatory factors limiting maximal aerobic exercise capacity. J. Appl. Physiol. 27:874.
631. Cureton, K. J., Boileau, R. A., and Massey, B. H. (1977). Sources of variance in maximal oxygen uptake in children. Med. Sci. Sports 9:54.
632. Blaxter, K. L., Graham, N. McC., and Wainman, F. W. (1959). Environmental temperature, energy metabolism and heat regulation in sheep. III. The metabolism and thermal exchanges of sheep with fleeces. J. Agric. Sci. 52:41.

International Review of Physiology
Environmental Physiology III, Volume 20
Edited by D. Robertshaw

3 Hyperthermia and Exercise

J. E. GREENLEAF

Laboratory of Human Environmental Physiology, Biomedical Research Division, NASA-Ames Research Center, Moffett Field, California

---

LIMITS OF BODY TEMPERATURE 159

SEQUENCE OF THERMAL RESPONSES DURING EXERCISE IN MAN 160

HYPOTHESES FOR THE CONTROL OF BODY TEMPERATURE DURING EXERCISE 162

THE GENERAL FLUID-ELECTROLYTE HYPOTHESIS 177
Historical Perspective (1899–1967) 179
Recent Research (1968–1978) 181
Intracerebral Infusion 181
Rest 181
Exercise 185
Intravenous Infusion 189
Ingestion of Fluid and Electrolytes 191
Normal Environmental and Exercise-induced Responses 197

CONCLUSIONS 198

ACKNOWLEDGMENTS 199

Of the mammals that regulate their internal temperature essentially independently of environmental temperature (homeotherms), man is unique because of his large potential capacity for heat dissipation by evaporation of sweat. Because a resting man's cooling capacity is greater than his capacity for heating, it has been suggested that he was and continues to be, basically, a tropical animal (1). This is fortunate because most of our thermal problems arise from overheating. Protection from excessive cooling is mainly achieved through reducing heat loss by increasing body insulation via peripheral vasoconstriction, wearing clothing, and seeking shelter; protection from overheating relies mainly upon physiological responses for heat dissipation, i.e., vasodilation with increased blood flow and sweating.

Metabolic heat is both a byproduct and a necessary constituent in the process of energy transfer within the human body. For effective temperature regulation, heat loss must ultimately equal heat production. Hyperthermia during physical exercise is one example in which heat production is greater than heat loss until equilibrium conditions are attained, but this latter state is only temporary. Heat is transferred to, within, and from the body in accordance with the physical laws of conduction, convection, radiation, and evaporation (2, 3); and evaporation of sweat is the most efficient avenue for heat dissipation during exercise. In a temperate environment, with little or no solar radiation, the primary source of heat is from metabolism, and maximum physical exertion can increase the metabolic rate for a few minutes to 1,800 kcal/hr, about 30 times the basal metabolic rate.

The major factors that limit human physical performance are 1) insufficient metabolic substrates and oxygen delivery for the maintenance of muscular contraction, 2) impaired blood pressure control, and 3) hyperthermia. Of the three, hyperthermia is the most dangerous because fatigue or syncope or both usually intervene before heat stroke becomes manifest. Yet relatively little research has been directed toward understanding the function of changes in fluid volumes and electrolyte concentrations for the maintenance of thermoregulation. This is surprising since blood is the vehicle for delivery of oxygen to tissues, for removal of the end products of metabolism, and for transportation of metabolic heat to the periphery; it also provides fluid for sweat production. In addition, the final common pathway for thermoregulatory responses resides in the nervous system, and neuronal as well as general cellular membrane function depends upon water as well as upon Na^+, K^+, and Ca^{2+} fluxes.

Henderson (4) has delineated the various physical properties of water that make it the optimal liquid to facilitate survival of the organism. These properties are as follows.

1. Its power as a solvent.
2. Its dielectric constant, together with its related ionizing potential.
3. Its relatively high thermal conductivity, which favors equilibration of temperature within and between cells.
4. Its high latent heat of evaporation.
5. Its high specific heat, which allows for the transfer of relatively large quantities of heat with relatively small changes in body temperature, since an increase of 10 °C essentially doubles the velocity of a chemical reaction.

To quote Henderson in 1913: "It is therefore incontestable that the unusually high specific heat of water tends automatically and in most marked degree to regulate the temperature of the whole environment, of both air and water, land and sea, and that of the living organism itself. . . . Here is a striking instance of natural fitness which in like degree is unattainable with any other substance except ammonia" (4).

Intuitively, it seems that, with such an optimal and intimate relationship between body water and temperature, there should be an interaction between these two parameters to maintain their respective homeostases.

This chapter emphasizes fluid and electrolyte parameters that affect the hyperthermia of physical exercise (metabolic heat production). Bligh (5) has suggested that the control of mammalian thermoregulation may have two separate components: a broad-band control that operates when core temperature deviates widely from its normal range and reaches critical upper or lower levels, and a fine control that operates between these two limits. The major hypothesis discussed in this chapter is that fluid and electrolyte changes influence thermal regulation within the fine control boundaries. A second working hypothesis is that the elevation of core temperature during exercise is a regulated phenomenon that is beneficial for the organism in terms of efficiency and potential for survival and is not merely a failure of the thermoregulatory control system.

Excellent reviews of thermoregulatory mechanisms are available on resting animals (5–10), exercising animals (11, 12), resting man (1, 13), and exercising man (13, 14), as well as on the pharmacological aspects of thermoregulation (15–17). The effect of electrolytes on thermoregulation has been treated in reviews by Gale (7), Greenleaf (18), Hellon (10), Myers (19, 20), Myers et al. (21), and Nielsen and Greenleaf (22).

LIMITS OF BODY TEMPERATURE

Men have survived with core temperatures in the range of 24 °C and 45.6 °C (23). However, in clothed subjects at rest, death usually occurs during cold water immersion when body temperature falls below 27 °C and when it exceeds 42 °C (24). If we assume the average normal (mythical) core temperature to be 37.0 °C, then death ensues with a drop in core temperature of 10 °C but with an increase of only 5 °C. Thus, overheating appears to be more critical than overcooling. The lower limit of core temperature for the onset of heat stroke is between 41.1 °C and 42.0 °C (25, 26), but cases of classical heatstroke have been reported with core temperatures of 40.6 °C (25). Unacclimatized subjects resting in a hot, humid environment (42 °C, 90% *rh*) are near their limit of tolerance and consciousness with rectal temperatures of 38.5 °C (27).

On the other hand, during intensive isotonic exercise, rectal temperatures near that upper lethal level have been recorded and the subjects have exhibited no lasting adverse effects. Robinson (28) reported rectal temperatures of 40.0°C and 41.1°C in two champion runners after a 3-mile race, and Pugh et al. (29) observed rectal temperatures of 41.1°C, 40.5°C, and 40.2°C in the first, third, and fourth place finishers, respectively, after a marathon race. Clearly, the absolute level of core temperature must be used advisedly for the determination of physiological state.

SEQUENCE OF THERMAL RESPONSES DURING EXERCISE IN MAN

Before movement actually begins, there is a simultaneous increase in sympathetic activity and a decrease in parasympathetic activity in the autonomic nervous system. The result is an increase in pulmonary ventilation and in the force and frequency of the heartbeat.

When muscular contraction begins, the increased sympathetic adrenergic activity facilitates a major shift of blood to the active muscles (13, 30, 31). The bulk of this blood appears to come initially from the skin, renal, and splanchnic circulations (13, 31).

In the process of this early blood redistribution, cooler blood from the periphery is mixed with warmer blood from the core and the result is a slight drop in core temperature in the first 2–3 min of exercise (32). This decrease in core temperature is not a consistent finding, but appears more frequently during exercise in the supine than the upright position (33), and in a cool ambient temperature (34). The reduced hydrostatic pressure in the supine position probably facilitates blood redistribution. Venous constriction (35), the pressure of contracting muscles on the veins, and the more forceful movement of the diaphragm assist venous return to the heart. Systolic blood pressure, heart rate, and oxygen uptake increase immediately as movement begins and, with a constant work load, reach equilibrium within 3–5 min (36). Reflex sweating begins within 1.5 s after exercise begins (37, 38).

At the beginning of exercise there is a shift of blood plasma, presumably to the interstitial space (36, 39, 40). The rate of change of the percentage change in plasma volume is directly proportional to the work load (36, 41, 42). Plasma electrolytes, except potassium, also shift out with the plasma in quantities approximately proportional to their normal plasma concentrations at relative work loads below 43% of maximum aerobic capacity. At relative work loads above 62% there is a net increase in plasma osmolality because the influx of osmols (mainly potassium and probably lactate) into the plasma is greater than the combined loss. In the working muscles, increased blood pressure and elevated concentrations of lactate, carbon dioxide, and hydrogen and potassium ions facilitate dilation of arterioles not opened by sympathetic stimulation (43–46). As the exertion proceeds, core temperature continues to rise, the skin vessels dilate to facilitate heat transfer, regulatory sweating increases, and the falling skin temperature begins to reach equilibrium.

The time taken to reach equilibrium core temperature depends upon the site of measurement, environmental conditions, the body level of hydration, the subject's maximum oxygen uptake, and the work load. The absolute versus relative work load hypothesis for the control of core temperature is discussed in the following section. Temperatures measured in the working muscles (47–50), on the tympanic membrane (51–53), along the external auditory canal (54), and in the esophagus (47, 55, 56) have different time constants and all reach equilibrium earlier than rectal temperature.

At metabolic equilibrium the cardiac output is divided between providing blood for the metabolic demands of working muscles and for the "resting" muscles, organs, and tissues; it also functions as the major vehicle for heat transportation to the skin (13). At thermal equilibrium, heat loss is equal to heat production and the level of core temperature during exercise is directly proportional to work load and is independent of ambient conditions between +5°C and +29°C dry bulb temperature, 42–83% relative humidity, and 149 mm Hg (350 m altitude) to 87 mm Hg (4,000 m altitude) of oxygen partial pressure (14, 55, 57). At a constant but submaximal work load, acute exposure to differing altitudes has no effect on the equilibrium level of core temperature; however, a different strategy of heat loss is used at high altitude; sweat rate is increased whereas tissue conductance is proportionately decreased (55).

With various work loads performed at a constant, normal, ambient temperature, mean skin temperatures are relatively constant, but sweat rates and tissue heat conductances are directly proportional to net metabolic heat production and relatively independent of core temperature (58). These results suggest that heat dissipation responses react to heat flow and that core temperature, rather than acting as the stimulus for heat dissipation, is the passive result of imbalance between heat production and heat dissipation. At various ambient temperatures between +5°C and +29°C, a constant moderate submaximal work load results in the same equilibrium level of core temperatures, but sweating increases linearly with skin temperature and ambient temperature (59). All these observations serve to confirm Nielsen's original hypothesis that the control of core temperature during exercise involves the sequential coordination of heat transportation to the skin via regulated changes in peripheral blood flow (for conductive-convective and radiation heat exchange) and in the rate of sweating (for evaporative heat loss) (57).

Cerebral and local hypothalamic temperatures are determined mainly by cerebral arterial blood temperature and blood flow (60). Since brain tissue temperature is warmer than arterial blood temperature, increased blood flow cools the brain. The sparse evidence suggests that mean blood flow through the brain is essentially constant during exercise, but the findings of Olesen (61) indicate that there is a large increase in flow in some areas and a compensatory decrease in others. In the region of the temperature regulatory centers in the hypothalamus or in the vicinity of the choroid plexus, variations in blood flow to change the ionic composition of the cerebral spinal fluid could play a role in exercise temperature regulation.

When exercise ceases, heart rate, blood pressure, and sweating are lowered almost instantaneously, but there are no significant changes in skin or core temperatures (36, 57). Restitution of plasma volume and plasma sodium, potassium, chloride, total calcium, ionized calcium, total protein, and osmotic contents takes at least 40 min after maximum ergometer exercise with the subjects in the sitting or supine positions (41) and 1 hr after submaximal ergometer exercise (relative oxygen uptakes of 23, 43, and 62%) with subjects in the upright position (36). It may be significant that recovery of plasma volume and electrolytes is not complete until core temperature returns close to control levels.

HYPOTHESES FOR THE CONTROL OF BODY TEMPERATURE DURING EXERCISE

A multitude of hypotheses have been proposed to explain exercise thermoregulation. Research in this area has, classically, been divided into two groups. The first group, termed physical heat regulation, in which investigations center on the thermal responses of the body to both environmental and metabolic heating, utilize measurements of various body temperatures and partitional calorimetry. Most human exercise and some behavioral thermoregulatory studies fall into this group. The second group, termed chemical heat regulation, usually involves the mechanisms for increasing heat production and other body warming responses in response to cold exposure. Studies on hibernation and behavioral thermoregulation that utilize various animal models would be included in this group. Fortunately, this dichotomy is breaking down. It seems inconceivable that the true, complete mechanisms for thermoregulation in man will be uncovered without both physical and chemical knowledge and techniques. Consider the comment by Gale (7) to the effect that all hormones except the parathyroid participate in thermoregulation. Even if only one-quarter of the hormones participate in thermoregulation, the understanding of this mechanism would surely be incomplete, since exercise and/or heat exposure induces activation of many hormones (e.g., vasopressin and renin-angiotensin (27), epinephrine, norepinephrine, testosterone, and thyroid stimulating hormone (62)).

In the following section a rather brief summary of the many hypotheses suggested for the control of body temperature during exercise is presented to provide a background for the fluid-electrolyte hypothesis that is discussed in more detail in a subsequent section.

Christensen (32) has reviewed the literature between 1878 and 1926 on thermal regulation during exercise, with the following observations and conclusions.

1. At relatively low work loads, the equilibrium level of rectal temperature is reached only after 1 hr of exercise, and because of the difficulties in measuring rectal temperature in athletes during exertion, "bicycle riding

appears to offer the most ideal conditions for studying the influence of muscular activity on body-temperature" (63).

2. The increased rectal temperature after physical exercise is not a pathological phenomenon, but an entirely normal physiological response because normal body temperature is restored within 1 hr of recovery.
3. The increase in body temperature during work is an adaptive process that increases the velocity of chemical reactions in the working organs.
4. At ambient temperatures of −7°C to −27°C, moderate physical work elevates body temperature to 38.5–38.75°C.
5. Within certain limits, the magnitude of the increase in rectal temperature can be determined by varying the intensity and duration of the exercises.
6. At body temperatures of 40.5°C, the capacity for work is decreased considerably, and the danger of heat stroke becomes imminent.

To these findings Christensen (32) added the following results from his study on body temperature responses during and after heavy physical work.

1. The greater the work rate the greater the rate of rise in rectal temperature (Figure 1).
2. Simultaneous measurements of temperatures at widely spaced points within the rectum during rest and exercise give the same values, so rectal temperature measurements give an accurate estimate of general core temperature.
3. At the same O_2 uptake, both arm and leg work result in the same increase in core temperature.
4. Rectal temperature increases in proportion to the rate of metabolism, but the magnitude of the temperature rise differs in various test subjects.
5. Increased body temperature is one of the main factors limiting working capacity, and the increase in metabolism that occurs during fatigue may be partly due to an elevated body temperature.

It was this paper by Christensen (32) that laid the foundation for much of the subsequent work on exercise temperature regulation, particularly that from the Scandinavian workers.

The next giant step was reported in a paper by Nielsen (57), in which he sought to determine whether the increase in body temperature during muscular exercise was a regulated process or whether it was due to failure of the heat regulatory system. In addition to measurements of rectal temperature, he made, for the first time, continuous measurements of skin temperature on the upper trunk and of body weight (sweat) loss during exercise of various intensities and at various ambient temperatures in a controlled environment chamber. His major findings are reproduced in Figures 2 and 3. These results show that the equilibrium levels of rectal temperature and regulatory sweating are directly proportional to the absolute work load. At a constant work load, the equilibrium level of rectal temperature is indepen-

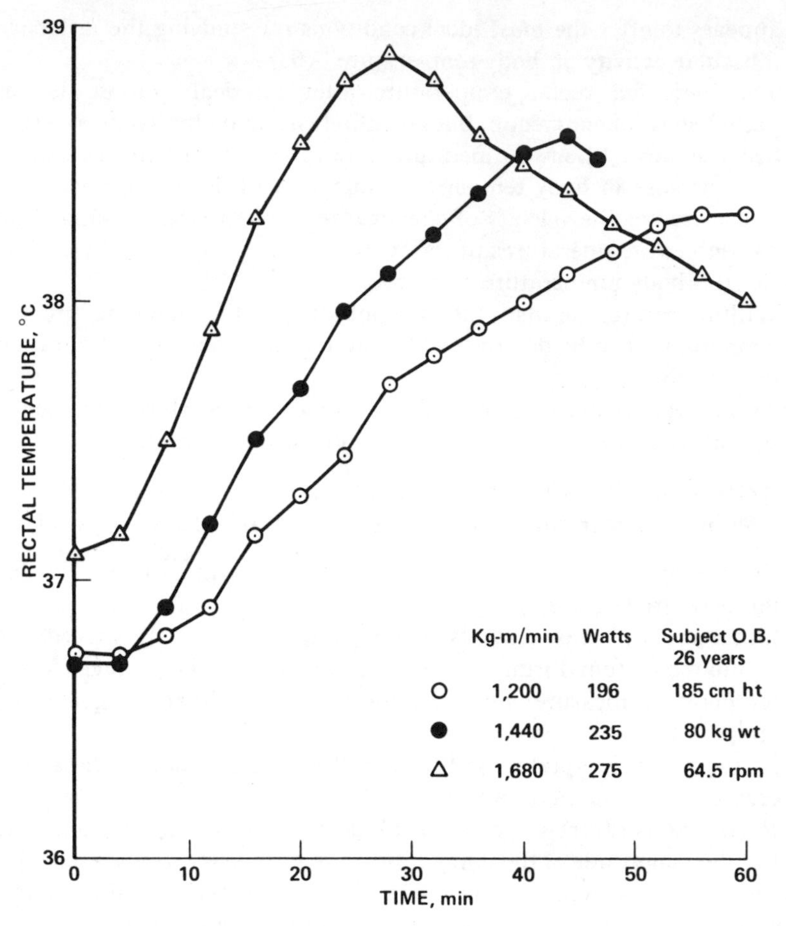

Figure 1. Rectal temperature responses to three different work loads. Reproduced from Christensen (32) with permission of Springer-Verlag.

dent of ambient temperature between 5°C and 23°C; but his subsequent experiments extended the upper limit to 29°C (57). The equilibrium level of skin temperature was 30–31.5°C and did not change with greater work loads, but followed changes in ambient temperature. To quote Nielsen,

> If the regulatory process during the work would tend mainly to maintain the normal rest temperature, the increase would be proportionally less at better heat dissipation conditions. However, if the body temperature during the work is adjusted to higher values, such increase would have to be independent of changes in heat dissipation conditions within certain limits. The experimental results show uniformly that the body temperature during work (900 kg-m/min) increased in the first 30–40 min; thereafter, an approximately constant value was reached which, in all experiments, was entirely independent of major changes in the ambient temperature. The changes made in the ambient temperature corresponded to very great changes in the heat dissipation conditions and were con-

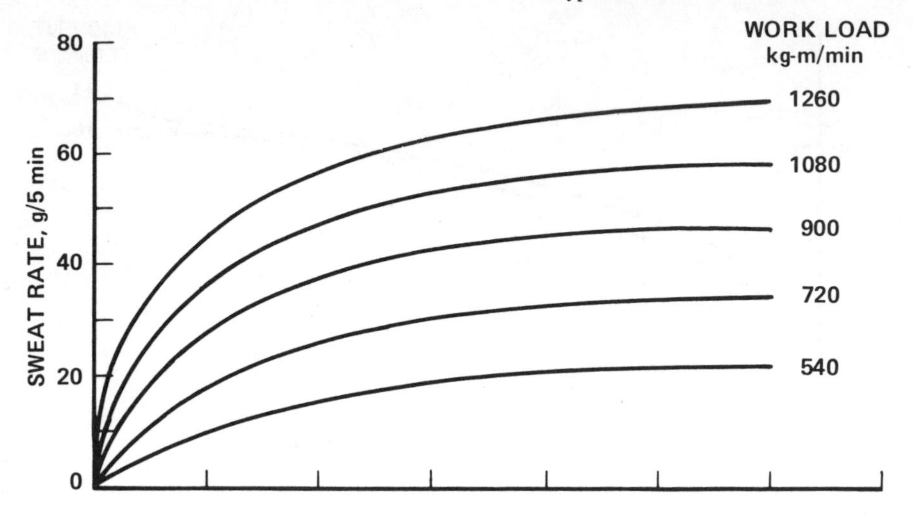

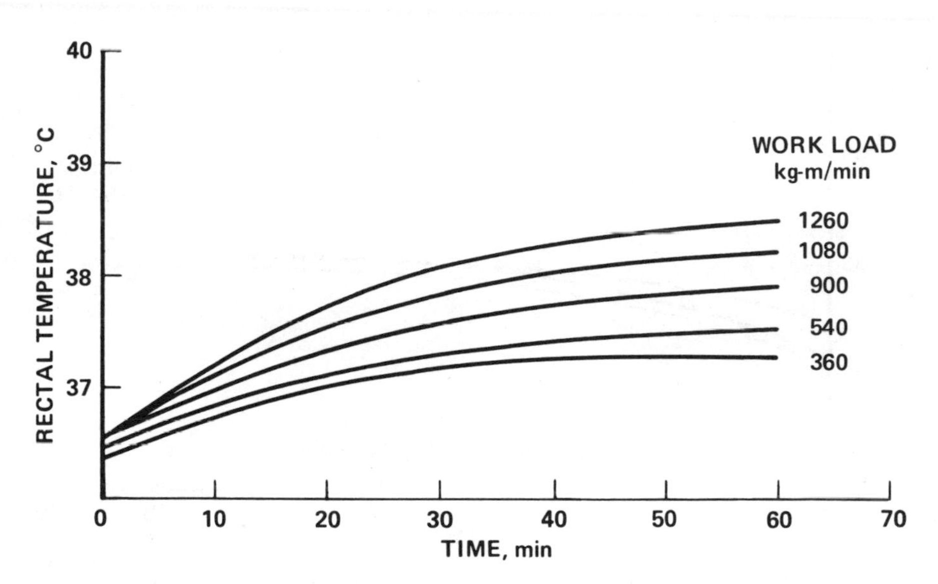

Figure 2. Time courses of sweat rates and rectal temperatures during various submaximal work loads. Reproduced from Nielsen (57) with permission of Acta Physiologica Scandinavica.

nected with a deep-seated change in the metabolic heat. At low ambient temperatures . . . the major portion of the heat dissipation took place by convection + radiation. On an increase in ambient temperature, the convection + radiation was regularly reduced until, at the maximum ambient temperature, negative values were reached. These extensive variations in the convection + radiation, however, were compensated so accurately by variations in the transpiration (sweating) that the body temperature, in all experiments, increased to practically the same value. From this, it can be concluded that the increase in

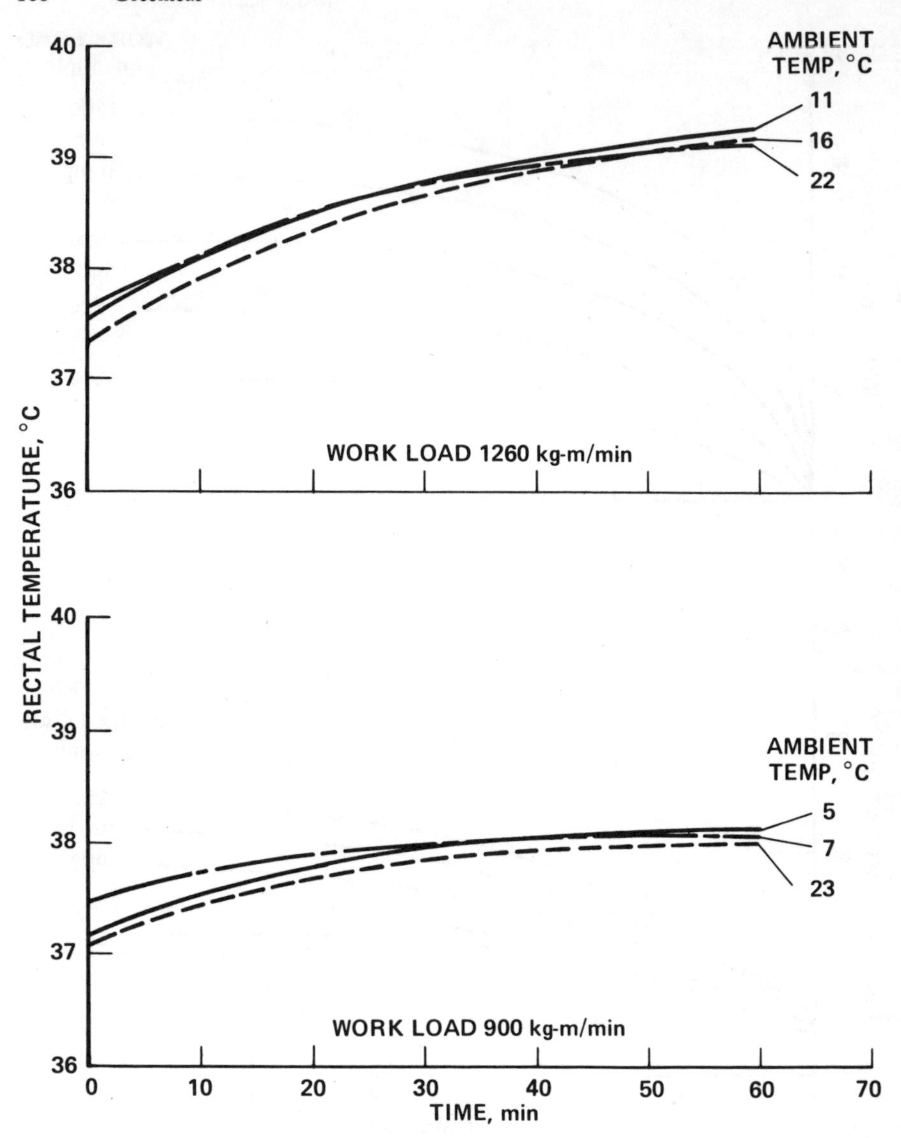

Figure 3. Time courses of rectal temperatures at two submaximal work loads at various ambient dry bulb temperatures. Reproduced from Nielsen (57) with permission of Acta Physiologica Scandinavica.

> body temperature during work is subject to regulatory processes and cannot be interpreted as a failure of the heat regulation. (57)

In his discussion of the close relationship between the increases in core temperature and sweat rate, Nielsen (57) suggested that the increased "sensitivity" of the sweating centers in the CNS could have been instigated by "cortical impulses" or "by changes in blood composition." It was not clear whether Nielsen's use of the word "sensitivity" meant a change in threshold or gain of the system. But "the most likely reason might be the fact that the excitability of the centers is indirectly influenced by changes in skin temperature" (57). The latter hypothesis, in conjunction with skin blood flow changes, has been investigated extensively and reviewed recently in Nadel's (64) symposium volume. The cortical impulse hypothesis is nearly impossible to investigate in human subjects, but the blood composition hypothesis is beginning to receive concerted attention. While earlier investigators (65, 66) understood the functions of sweating and peripheral blood flow in the genesis of exercise thermoregulation, Nielsen's work (57) provided the first and most complete analysis of the various avenues of heat exchange (partitional calorimetry) during exercise.

The next notable finding was published by Winslow and Gagge (67), who measured the effects of various thermal environmental conditions (T_{db} = 15.7–32.4°C, T_{ra} = 8.6–34.3°C, operative temperature = 12.4–33.2°C, rh = 40–50%, wind = 5–8 cm/s) in two men during moderate exercise (>300 kg-cal/hr) for 90 min in the upright position on a bicycle ergometer. Sweat rate was calculated from changes in body weight. They made the following conclusions.

1. The radiation area for a man exercising on an ergometer was about 70%, the same as for a resting subject.
2. The convection constant was essentially maximum at a pedaling frequency of 30 rpm; this increased cooling is equivalent to an air flow of 30–40 cm/s.
3. The increased sweat secretion and evaporative cooling balanced the increased heat production from the exercise, in confirmation of Nielsen's (57) findings. (Nielsen's paper was not included among Winslow and Gagge's (67) references.) In winter, active sweat secretion during exercise began at an operative temperature (T_{op}) of 20°C; in summer sweating began at an T_{op} between 16°C and 20°C. These were probably the first quantitative data on a seasonal increase in sweating sensitivity presumably due to heat acclimatization.
4. The regulation of evaporation in working subjects operated so efficiently that skin temperature was held remarkably constant; it was actually lower during work than at rest for any given operative temperature. On the other hand, rectal temperature rose appreciably during exercise. Therefore, "it would seem, under these circumstances, clear that internal temperature—not skin temperature—must control the sweat-secreting mechanism" (67). Nielsen (57) made these same observations but did not state this important, specific conclusion.

Asmussen and Nielsen (68) investigated the hypothesis that the supposed resetting (changed sensitivity) of the thermoregulatory center during work at a simulated altitude of 4,000 m might be influenced by chemical factors (lactate) in the blood or afferent proprioceptive stimuli (reflex) arising from moving limbs. Rectal and stomach temperatures were measured in two men who exercised with the arms and again with the legs at equal metabolic rates of 7.2 kcal/min and again at 14.3 kcal/min. Lower efficiency during arm work resulted in slightly greater heat productions. After 40 min the increases in rectal temperatures during leg exercise were greater than during arm exercise, by 0.22 °C after light work and by 0.36 °C after heavier work. Factors eliminated as the cause of the higher rectal temperature with leg exercise were as follows.

1. Arm work created a greater subjective feeling of strain because leg exercise at 4,000 m was perceived as more stressful than comparable exercise at sea level.
2. The level of anaerobic metabolism (blood lactate concentration) and ventilation were eliminated because both of these variables were greater at altitude, whereas the increases in rectal temperature were similar.
3. A changed blood distribution was eliminated because of blood pooling in the legs during arm work in the upright position. Results from Nielsen et al. (69) showed that during 45° head-up tilting with blood pooling in the legs, rectal temperature increased, i.e., gave the opposite change from the attenuated rectal temperature response to arm work. An alternative explanation for the greater rise in core temperature during leg work is that activity with a larger muscle mass forces a greater shift (loss) of plasma volume and possibly greater residual plasma sodium and osmotic concentrations, which have been found to be correlated with equilibrium levels of core temperature during exercise (70).
4. Errors due to placement of the thermocouple in the rectum were not the cause of a higher recorded temperature, since stomach temperature responded similarly to rectal temperature. This was the first time stomach temperatures were measured during exercise. It was concluded "that the difference in the 'setting' of the heat-dissipating centre in the two kinds of work is brought about either by irradiation of the efferent impulses from the motor-cortex, or by a summation of proprioceptive impulses arising in the working muscles" (68). The sequence of steps proposed by Asmusson and Nielsen (68) for the regulation of body temperature during exercise was as follows:

> The heat-dissipating centre is stimulated by the increased temperature of the blood, warmed in the working muscles. The centre sends out impulses by which the secretion of sweat and the blood flow through the skin is increased, until a balance between blood temperature and heat dissipation is reached. The temperature at which this balance is reached is, however, dependent on the sensitivity of the centre, and this in turn may among other factors be determined by the degree of the nervous activity including nervous impulses, reaching the centre from the periphery and, possibly, from the cortical motor areas. The peripheral impulses might include sensations from the skin as well as proprioceptive impulses from muscles and joints. The integration in the centre of these impulses might contribute to the final 'setting' of the body's "thermostat" (68).

To explain the observations of Nielsen (57) for the control of body temperature during exercise, Bazett (71) proposed an hypothesis based upon reflex controls to account for the two major responses: first, an adequate explanation of the afferent stimulus for control of sweating; second, an accounting of the regulation of rectal temperature by dermal sensory receptors that are prone to adaptation. Bazett postulated that reflex responses to both heat and cold exist, and dermal receptors that activate these reflexes are stimulated by two levels of dermal temperature receptors: superficial and deep heat and cold receptors found near venous and arterial plexuses. The superficial receptors would be stimulated by changes in environmental temperature and be concerned with vascular adjustments, whereas the deeper receptors located at the level of the subcutaneous junction would be stimulated by metabolic heat from working muscles and be concerned with the control of sweating. The relationship between sweating and skin temperature can be explained by these thermal gradients. Rather than emphasize the rise in core temperature during exercise, he emphasized the stability of the skin temperature, "*which depends on a failure of tissue conductance to be increased proportionally to the increased heat load.* The deficit is not one due to chance, but one which is regulated to cause a precise rise in deep temperature, including muscle temperature. . . . It is not likely that the contrasting reactions can be ascribed to differences in the mode of heating of hypothalamic centres, since no such differences could exist other than in secondary accompaniments, such as *blood acidity or other chemical changes*" (71).

Application of this theory to the sequence of thermal responses during exercise results in the following. Increased heat production increases the temperature of the blood leaving the active muscles and stimulates the deep thermal receptors, which were assumed to be the sole regulators for sweat secretion and to be uninvolved in vasomotor adjustments. Stimulation of the deep receptors would leave the peripheral circulation generally unaffected except for a generalized, temporary vasoconstriction that would lower skin temperature and cause heat retention. At the same time, heat loss to the skin surface would be accelerated through conduction from the warmed arterial blood and by direct transfer of heat to the superficial veins from the heated blood leaving the active muscles. All these heat exchanges would occur without any extra vascular response. The net result would be a rise in core temperature with an increase in subclavian arterial temperature to precede the rise in rectal temperature. The increased arterial temperature sensed by the hypothalamus would stimulate increased vasodilation until heat loss balanced heat production. Thus, the equilibrium level of rectal temperature would be determined by the rate of heat production (work), rather than by environmental conditions. The greater blood (plasma) volume, with a more efficient heat loss through the superficial veins, could explain the attenuated rise in rectal temperature after heat acclimatization. The explanation for the direct relationship between the rate of sweating and skin temperature, at a constant work load, depends upon stimulation of the deep receptors by the resultant blood temperature from the mixing of cooler cutaneous blood with blood of essentially constant but warmer temperature from the working muscles. This resultant blood temperature would be lower in comparison with the thermal difference between the core and skin surface, and the rate of sweating might be inversely proportional to this temperature difference. Thus, the rate of sweating could be increased by an increasing hypothalamic temperature, and, at the same time, sweating could be progressively decreased by a progressively increasing core-to-skin temperature gradient. "These two effects might tend to cancel one another and to create a fictitious dependence on surface temperature since skin temperature is merely rectal temperature minus the observed gradient" (71).

Because of the difficulty in verifying the data concerning the afferent stimuli for control of sweating as influenced by core temperature and/or skin temperature, Kerslake (72) proposed an extension of Bazett's hypothesis to account for it. Kerslake's hypothesis was that at equilibrium conditions, "the *heat flow* through the skin will bear a fairly constant relation to metabolic rate, and the temperature of those (deep epidermal) nerve endings will exceed the surface temperature of the skin by an amount proportional to the heat flow through the skin and the thermal resistance of the layers superficial to them" (72). If it were assumed that the thermal resistance superficial to these nerve endings was constant, and that sweat rate depends only on the temperature of these receptors, then it should be possible to relate sweat rate to skin temperature and to heat flow through the skin. The major difficulty with testing this hypothesis is in the technique for measuring heat flow. Kerslake estimated heat flow by subtracting the estimated heat gain of the skin from the air by convection and radiation from the evaporative heat loss, all difficult measurements and prone to error. The major problem with this hypothesis is that it is improbable that there is a constant skin temperature at all parts of the body as well as a constant subepidermal temperature. The latter would be controlled by the anatomical distribution of the superficial arteries and veins and the regulation of blood flow through them. The latter would depend, in part, on the intensity of the heat production of the local skeletal muscular activity, as well as on the summation of the factors controlling vascular smooth muscle activity in the vascular system, e.g., plasma K^+, pH, etc.

Nielsen (14) could not confirm the skin temperature receptor hypotheses of Bazett or Kerslake. She measured four temperatures at 1–6 mm in the skin during rest and exercise at various ambient temperatures. In addition to confirming many of Nielsen's (57) basic findings, she concluded the following.

1. Sweat secretion cannot be regulated solely by the stimulation of thermal receptors in the deep layers of the skin.
2. The amount of heat transported to the skin surface by the skin blood

flow must be exactly equal to the quantity of heat dissipated by the evaporation of sweat.

3. The changes in muscle temperature with changes in total heat production are different with negative and positive work, so muscle temperature cannot be the stimulus that connects sweat rate with heat production.
4. Since none of the current theories explained the correlation between heat production and sweat rate during equilibrium conditions, it was suggested that sweat secretion is regulated from changes in the thermal balance, which activates some thermal receptors, thereby instituting reflex adjustments of the sweat rate.

While it was generally accepted that control of heat dissipation via changes in blood flow through peripheral circulation was an extremely important factor for body thermoregulation, the heat flow hypothesis still remained tentative.

In his provocative paper, Benzinger (73) attempted the nearly impossible task of determining whether cutaneous temperature and internal temperature, either both of them or neither of them, act as stimuli for eliciting the known thermoregulatory reactions at rest and during ergometer exercise. He wrote, "The experimental approach was by functional not anatomical analysis of the apparatus, without blocking measures, with all parts of the system intact and in vigorous thermoregulatory action, by simultaneous and continuous measurements of the responses and the two possible stimuli: *internal* temperature and *skin* temperature." But it is very difficult in man to hold one temperature while determining causes for variations in the other. In addition, Benzinger, as well as many other investigators, eliminated consideration of any inputs to the physical thermoregulatory system from the so-called chemical factors. So, even though his findings were addressed to the mechanism of physical thermoregulation, his conclusions were, at best, tenuous. Benzinger reasoned that if the vasomotor and sudomotor thermoregulatory responses were completely insensitive to internal temperature, "responses plotted against skin temperature would fall upon one best line, with all deviations random and within experimental errors" (73). A similar, converse argument was made pertaining to internal temperature. A new, relatively unsubstantiated location to measure internal temperature, the tympanic membrane, was introduced. Physical exercise (360 kcal/min and 720 kcal/min) was used to vary internal temperature while skin temperature remained essentially constant, and various ambient temperatures between +10°C and +45°C were used to vary skin temperature.

One controversial finding was that sweating during work did not occur until skin temperature reached 32°C. Benzinger's major conclusion was that sweating is controlled only by changes in core (tympanic membrane) temperature; and, since the relation of skin temperature to sweat rate was inconsistent and coincidental, sweating is not controlled at all by changes in skin temperature. These results and conclusions have been challenged and much evidence argues against them (37, 38, 53). In fact, during their exercise studies, Robinson et al. (53) discontinued the use of tympanic temperature because it changed more slowly and did not rise as high as rectal temperature. For tympanic temperature to faithfully reflect core temperature, the ear must be insulated from ambient air, especially at or below normal room temperature (54).

In an extension of the work to locate deep body thermal sensing organs, Robinson et al. (53) designed an experiment (using treadmill exercise by four heat-acclimated men) to determine the relationships between sweat rates and corresponding changes in temperatures of the skin, rectum, tympanic membrane, working leg muscles (gastrocnemius), the deeper femoral vein draining blood from the leg muscles, and the more superficial long saphenous vein on the thigh and leg. The most significant finding was that the onset and rate of increase in sweating during work and the decline of sweating in recovery were more closely related to the corresponding changes in femoral vein temperature than to any of the other temperatures measured. Since the activity of no one thermal sensing organ could account for the varied sweating patterns in response to the independent variations of environmental temperatures and work intensities, it was concluded that control of sweating involved the integrated regulation of multiple stimuli that included the following.

a) reflexes originating from cutaneous thermoreceptors,
b) increased sensitivity and activity of the hypothalamic center as its temperature increases,
c) possible thermoreceptors located in the muscles themselves, or in the veins draining blood from the working muscle, and
d) neuromuscular influences on the hypothalamic center (53).

Probably no other hypothesis has stirred more controversy than the set-point concept Burton (74) proposed to help explain the mechanism of thermoregulation during rest and exercise. If body temperature is regulated, regulatory control theory requires the existence of a reference temperature that the system attempts to maintain, i.e., a thermostat that does not have to be constant under all conditions (75). The difference between this reference or set temperature and body temperature is called the load error. In theory then, the control mechanism attempts to minimize the load error by appropriate physiological responses: increased heat dissipation if body temperature is greater than the set-point, and heat production and/or heat conservation responses if body temperature is lower than the set-point. In man, the best regulatory model is that of a proportional control system in which there is a continuous, linear relationship between the load error and the magnitude of the stimuli (1).

The first reference to the change in the "setting of the thermoregulatory centre" as applied to exercise temperature regulation was made by Asmussen and Nielsen (68) in reference to Nielsen's previous work. Nielsen suggested that the close association between core temperature and sweat rate during exercise could be influenced by "cortical impulses, by changes in blood composition . . . or the fact that the excitability of the centers is indirectly influenced by changes in skin temperature" (57). Thus, Nielsen has postulated both thermal and/or nonthermal stimuli that influence the sensitivity of the thermal centers, whereas Asmussen and Nielsen (68) have suggested a changed setting or set-point. They did not say the setting was increased, as many others have inferred (28, 76–78). What they actually said was that the higher body temperature during work was "due to a new setting of the thermoregulatory centre" (68).

The concepts of altered sensitivity (gain) and set-point are not necessarily the same (79). The classical concept of the functioning of the set-point, as mentioned above, would require the set-point to be lowered during exercise (76, 78, 80). Body temperature must be higher than the set-point for heat dissipation responses to be initiated, which is certainly the response during exercise. If the set-point were raised, then heat production and heat conservation responses that would tend to raise body temperature until it reached the set-point would be activated; but there would be no signal to activate heat dissipation responses. Sweating, a heat dissipation response, begins when the skin is warm, within 1.5 s after exercise begins (37); this would be a proper response to a decreased set-point. Conversely, there is a generalized peripheral vasoconstriction, a heat conservation response, at the beginning of exercise (13, 31); this would be a proper response to an increased set-point. Clearly, the set-point cannot change in opposite directions simultaneously if it is assumed that the set-point is changed within the first few seconds after exercise commences. If preoptic hypothalamic temperature is the regulated temperature, then the response time for changing its temperature becomes critical. Because of the high metabolism of central nervous system tissue, the temperatures of deep areas in the brain are warmer than cerebral arterial blood temperature in direct proportion to their distance from the circle of Willis (81). The supraoptic and paraventricular nuclei are supplied by the most dense capillary networks in the brain. In resting mammals that have an internal carotid artery (monkey, rabbit), the blood temperature in the circle of Willis is the same as the blood temperature in the aortic arch (81). Therefore, thermoreceptors in the preoptic area probably receive a direct thermal stimulus from the carotid arterial blood. During exercise, carotid arterial blood temperature becomes an integrated temperature that reflects the net effect of all heat production and dissipation responses. Metabolic heat from exercising muscle provides the extra heat production that increases arterial blood temperature and is sensed by the temperature-sensitive neurons in the hypothalamus. With leg exercise of moderate intensity, vastus lateralis muscle temperature increases to one-half its equilibrium level of 39 °C about 3–5 min after work starts (48, 49). Deep venous (vena cava, femoral) and arterial (subclavian) temperatures rise faster than tympanic or rectal temperatures (53, 71). Since hypothalamic temperature is warmer than its blood temperature, an increased blood flow would cool hypothalamic tissue. Presumably, the change in hypothalamic temperature during exercise would be the net result of the hyperthermic effect of the elevated arterial temperature and the hypothermic effect of any increase in blood flow. There are no data concerning the rate of change or equilibrium levels of hypothalamic temperature during exercise in man.

Thus, the set-point thesis may be summarized with a quote from Cabanac: "If the student of temperature regulation chooses hypothalamic temperature as the regulated variable, then he is forced to accept the theory of the adjustable set-point" (82).

Alternative hypotheses would be as follows.

1. There is no change in set-point from thermal stimuli (82, 83).
2. There is a change in set-point from nonthermal stimuli, e.g., the catecholamines (84, 85).
    a. Norepinepherine can constrict hypothalamic vessels and decrease blood flow (86).
    b. The aminergic theory for thermoregulation (20).
    c. The ionic theory for thermoregulation (20, 87).
3. The set-point concept may not be applicable for exercise thermoregulation.

Perhaps other hypotheses, such as changes in sensitivity or gain (79) or simply invocation of changes in heat content and/or storage (88–91), rather than changes in temperature per se, should be considered.

The classical example of increase in thermoregulatory set-point was the increase during exercise when the resting body temperature was raised by fever (1, 84, 92); i.e., the exercise hyperthermic response was the same as in euthermia, but it started at a higher temperature. Maskrey and Bligh (93) have concluded that the set-point concept is unnecessary when explaining pyrogen-induced interference with normal body temperature; the only necessary factors are functional warm sensors and heat loss effectors. Mitchell et al. (79) have suggested that in all thermoregulatory behavior, especially fever, the classical hypothetical reference signal of the set-point concept is unnecessary; it is only necessary to have opposing warm and cold stimuli with a change in gain or sensitivity of the thermoregulatory system during fever. In human heat acclimation, which involves intermittent bouts of exercise in a hot environment, the progressive increase in sweat rate with each succeeding exposure that is accompanied by a progressive decrease in equilibrium level of core temperature has been described as an increase in sensitivity of the sweating mechanism; i.e., a greater output (sweating) from a lowered input (core temperature). Houdas and Guieu's (91) analysis also negates the classical set-point theory because the latter does not account for changes in body core temperature of nonthermal origin, e.g., circadian and estrous cyclic variations. They conclude that "the body does not seek to return its central temperature to a set point, either variable or nonvariable (temperature regulation). Instead it seeks to return its instantaneous heat storage rate to zero as soon as possible: this is an exchange servomechanism" (91).

Snellen (88, 89, 94) has proposed a similar heat storage or quantity theory, in which body heat sensors are sensitive to body mass; the lower the mass, e.g., with dehydration (95), the higher the body temperature. Thus, the input signal to the controller will return to normal when body mass or some derivative of body mass, such as plasma osmolality, returns to normal. The obvious problem with the body mass hypothesis per se is that a person can lose 20 kg of body mass and retain a normal plasma osmolality and body temperature. It is more likely that a derivative of body mass, perhaps total body water and/or its osmotic or sodium content, could be a more integral part of the hour by hour thermal control mechanism. This is discussed in the following section.

The hypothesis of I. Åstrand (96), confirmed by Saltin and Hermansen (50), that the core temperature during exercise was "set according to the relative load of the individual and not to the absolute work load performed," was based upon the change in variability of equilibrium levels of esophageal temperatures between various test subjects when core temperatures were plotted against the relative oxygen uptake (expressed as a percentage of the maximal oxygen uptake) as opposed to the absolute oxygen uptake (expressed in liters per minute). If the equilibrium level of core temperature were set according to the level of the relative oxygen uptake, then any treatment that changed the relative oxygen uptake but retained the same heat production should result in a proportional change in core temperature.

At high altitudes, the maximal oxygen uptake is reduced because of the reduced partial pressure of inspired oxygen, so at any given absolute work load the relative oxygen uptake and core temperature should be greater at higher altitudes than at sea level. With acute exposure to various altitudes, the equilibrium levels of core temperature during constant work load are the same as at sea level (55, 68) (Figure 4). The simplest explanation is that use of the relative load calculation provides a method of reducing interindividual variability, which equalizes the degree of "stress" applied to the system between subjects of different work capacities. However, Saltin (97) has subsequently measured rectal temperature in subjects during a standardized (750 kg-m/min) load before physical training ($\overline{V}O_{2max}$ = 2.4 liters/min, $\overline{V}O_{2rel}$ = 71%, T_{re} = 38.6°C), after 2 months of physical training ($\overline{V}O_{2max}$ = 3.6 liters/min, $\overline{V}O_{2rel}$ = 51%, T_{re} = 38.1°C), and again 1 week later, after 3 days acclimation at 43,000 m ($\overline{V}O_{2max}$ = 2.5 liters/min, $\overline{V}O_{2rel}$ = 73%, T_{re} = 38.6°C). Exercise training and altitude acclimation maintained the relationship between the relative oxygen uptake and core temperature, although it was not present during acute altitude exposure. With essentially similar absolute work loads in the acute altitude (55) and training (97) studies, the heat production would have been the same. So the reason for the disparity between core temperature and relative oxygen uptake in these two studies must

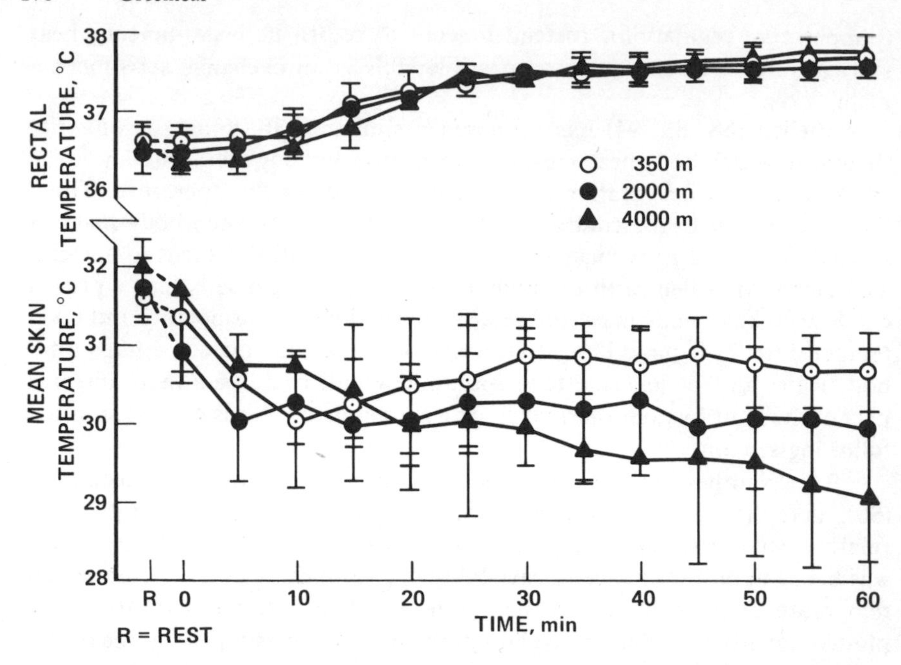

Figure 4. Mean (±S.D.) rectal and mean skin temperatures at rest (R) and during submaximal exercise at three altitudes. Reproduced from Greenleaf et al. (55) with permission of the American Physiological Society.

be found in the heat dissipation responses of sweating and/or blood flow that appear to be controlled by a nonthermal factor (or factors) proportional to the level of stress on the system. In the acute altitude experiments, the constant core temperatures were achieved by proportional increases in sweating and decreases in calculated tissue conductance.

When considering various hypotheses for the control of body temperature, perhaps insufficient attention has been given to the physical properties of water, the most plentiful substance in the body. Henderson's (4) thesis on this subject has been summarized in the introductory paragraphs. Calloway (98) has extended this thesis and further hypothesized that the optimal body temperature for homothermic animals (40 °C) is due to the peculiar thermodynamic properties of water. With chemical reactions in the biological (biokinetic) range of 0–100 °C, it has been found that the temperature at which the equilibrium constant of a reaction had half the value between 0 °C and 100 °C, the range of least thermodynamic stress, was 40 °C. In the Kelvin scale, the reaction rate at 40 °C, or 313.15 K, is half the difference between the value of the equilibrium constant at 0 °C (273.15 K) and 100 °C (373.15 K). Thus, a body temperature of 40 °C keeps the equilibrium of a chemical reaction at one-half the value of the difference between 0 °C and 100 °C. Since these findings are based upon logarithmic functions, then the values obtained are also related to some function of *e*, the base of the natural logarithms. Other equations used to calculate thermodynamic behavior of physical-chemical functions of aqueous phases (e.g., ionization constants, equilibrium constants, the potential at neutral membrane-liquid junctions, electrode potentials, surface tension-vapor pressure junctions, free energy, osmosis, viscosity and ionic influences) all contain the variable of temperature and the basic expression e to the $1/T$ power. Thus, the temperature function is such that the one-half value of this function between 0°C and 100°C is close to 40°C. In addition, the normal resting temperature of man is a close function of many thermodynamic qualities of water as related through e. For instance, the difference between the boiling point 373.15 K and the freezing point 273.15 K, divided by e (2.71828), equals 36.7879, the normal body temperature. Or, conversely, the difference of the boiling and freezing points of water (100), divided by body temperature (36.79), may be substituted for e, the natural base of numbers. Other functions of water, namely thermal expansivity and compressibility, which may be important for enzyme efficiency, have their least thermodynamic stress temperatures at 38.5°C and 40°C, respectively. Between 0°C and 100°C, the point of lowest specific heat of water (where the least amount of heat is expended to raise 1 g of water 1 degree) is basal body temperature, about 35–36°C. Thus, the cell has adapted its function to the inherent properties of water so as to carry out its energy-producing reactions in the physical state of least thermodynamic stress. This physical-chemical definition of homeostasis also provides an explanation of how body temperature is controlled at 37°C, without utilizing a variable set-point, and of how body temperature fits with the heat content or flow hypotheses. Calloway (98) concludes, "There is little evidence that temperature is fixed because of the nature of the cell contents other than water and perhaps some of its inorganic solutes." The nature of some of these inorganic solutes and their effects upon thermoregulation are discussed in the following section.

THE GENERAL FLUID-ELECTROLYTE HYPOTHESIS

The fluid-electrolyte composition and concentration of body cells are of prime importance for the maintenance of normal cellular function. As discussed above, there is an intimate association between the physical-chemical properties of water and body temperature. At constant ambient temperature and pressure, water is the next most important substance after oxygen for maintenance of life, and probably for the maintenance of physical and mental performance.

The critical function of sodium, potassium, and calcium ions for optimal cellular membrane activity is well known. Changes in these ionic concentrations or kinetics could influence the neurons involved with thermoregulation. Of the many hypotheses proposed for the control of body temperature during exercise, no satisfactory explanation has yet been found that takes into account a local and a central factor for controlling sweating that is integrated simultaneously with changes in peripheral blood flow. Sodium or calcium ions injected intracerebroventricularly or intravenously can change body temperature at rest (99, 100) and during exercise (101–103). Thermal equilibrium is elevated with dehydration (104–106) or after drinking saline solution (95, 107, 108), but is decreased after drinking plain water (95, 104, 106, 109). Increased plasma sodium and osmotic concentrations are associated with decreased sweating at rest (110) and during exercise (108, 111, 112). Increased plasma potassium and osmolality can alter peripheral blood flow (43–46, 113, 114). Fluid-electrolyte metabolism seems to play an important role in thermoregulation, especially under conditions of stress.

There has been confusion over use of terms to describe various states of body hydration. The terms "hyperhydration," "euhydration," and "hypohydration" refer here to the equilibrium states of an excess, normal (Δ1–2% of body H_2O), or deficit volume, respectively, of body water (Figure 5). The terms "overhydration" and "rehydration" refer to changes in body water from the three equilibrium states. The term "dehydration" is used for the process of fluid depletion when going from hyperhydration to euhydration and from euhydration to hypohydration.

A brief historical summary of the effects of changes in body fluid and electrolyte composition on body temperature is presented next, followed by an evaluation of the more recent findings. The latter is divided into four areas corresponding to various routes of introduction of fluid-electrolyte solutions into the organism: 1) direct injection or infusion into specific areas of brain tissue or cerebral ventricles; 2) injection or infusion into the systemic circulation, 3) absorption from the gastrointestinal tract from food or drink, and 4) "natural" changes induced by heat and/or exercise dehydration.

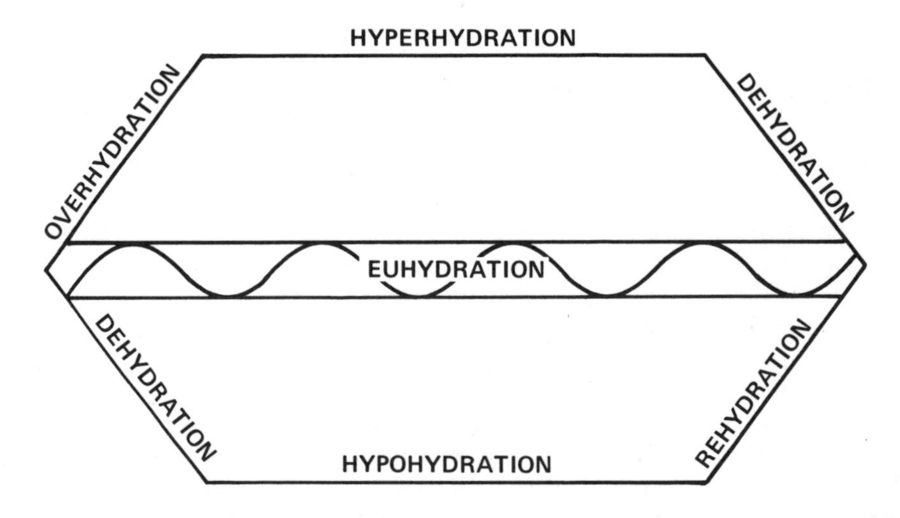

Figure 5. Depiction of terminology for various states of body hydration.

Historical Perspective (1899–1967)

The influence of fluid-electrolyte changes upon body temperature was observed many years ago. In 1899 Crandall suggested that the fever often observed in newborn babies was caused by dehydration (inanition fever), since it was relieved after administration of water (115, 116). Intravenous infusion of hypertonic sodium chloride increased core temperature in resting rabbits (117). Barbour (118) reported on work in 1911 by Montuori, who found that hyperthermia decreased blood osmotic pressure and increased sweating in cats; sweating was decreased after injection of hypertonic sodium chloride. Montuori suggested that changes in blood osmolality influenced the nerve centers that controlled sweating, a theory to be tested later by Dontas (119). Intravenous injections of hypertonic saline or 36% glucose solutions produced a diuresis that resulted in a large negative water balance accompanied by a 5–6 °C increase in body temperature; spinal cord section between C6 and C7 did not reduce the hyperpyrexia (120). Keith (121) utilized intravenous infusions of 50% solutions of either saccharose or glucose (8–9 g/kg/hr for 2 hr) and found no change in rectal temperature in response to diuretic weight losses of 7–10%; but hyperthermias of 1.8–3.4 °C occurred when these sugars were infused into dogs whose drinking water was withheld overnight. Forced ingestion of water warmed to body temperature resulted in a significant lowering of body temperature in resting dogs (122). Increased core temperatures, not caused by impairment of sweating, have been observed in resting men on restricted fluids (116). Rietschel and Beck (123) exercised five young test subjects for about 1 hr at step climbing, first in a euhydrated condition and again after a fluid restriction to 375 g of either red wine or water daily for 3–4 days. After exercise, mean rectal temperature was increased by 1.1 °C during euhydration and by 1.5 °C after hypohydration. To counter the objection that the excessive rise in rectal temperature after hypohydration was due to some local phenomenon from warm blood draining the leg muscles, they repeated these experiments in four additional subjects, who performed arm exercise, and found essentially the same results (euhydration ΔT_{re} = +0.5 °C, hypohydration ΔT_{re} = +1.4 °C). The equilibrium levels of T_{re} were uniformly lower after arm work compared with leg work. It was concluded that hypohydration results in increased body temperature during exercise. This so-called thirst fever, alimentary fever, or salt fever was not a true fever; it was a hyperthermia due to a physical-dynamic accumulation of heat that was caused by insufficient heat dissipation. The subjects' body temperature returned to normal when the fluid deficit was replaced. Rietschel and Beck speculated that part of the salt fever hyperthermia was due to the heat liberated by the osmotic reactions within the body, i.e., a dynamic salt effect. They did not put much credence in the inhibitory effect of sodium chloride on sweating and reduced heat dissipation.

Barbour and associates (118, 124–126) conducted a series of experiments to support the hypothesis that the water content of the blood was the essential factor for maintenance of a constant body temperature. Their major findings were as follows.

1. The nervous system must have been involved in the etiology of fever.
2. The onset of fever with pyretic agents was accompanied by concentration of the blood.
3. The hypothermia induced by antipyretic agents was accompanied by blood dilution.
4. An increased blood concentration reduced peripheral blood flow and water loss and thereby protected the body from adding or eliminating excessive amounts of heat; i.e., it had an insulative effect.
5. Thus, constancy of serum osmolality was sacrificed in the interest of thermoregulation.

Dontas (119), contrary to the prevailing opinion that fever was a symptom resulting from stimulation of the "heat centers," hypothesized that fever was instead due to a "paralysis" of the heat centers and that antipyretics increased the excitability of the "paralyzed" heat centers that resulted in a lowering of body temperature. Evidence for this hypothesis came from experiments in which the onset of polypnea in relation to the change in rectal temperature, i.e., the thermal width, was measured in dogs. The higher the rectal temperature (or the thermal width) at which polypnea began, the lower the excitability of the heat center. Dehydration induced by withholding water or intravenous injection of hypertonic sodium chloride raised body temperature from 39.4°C to 41.4°C, increased the thermal width, lowered excitability, and paralyzed the heat centers by "excessive drying of their tissues." Similar responses were obtained with human subjects. Turlejska-Stelmasiak (127) also concluded that a reduction in sensitivity of thermally responsive neurons, due to cellular dehydration, was the most likely explanation for inhibition of panting in response to preoptic heating in water-deprived rabbits. Parenthetically, Koep et al. (128) found at autopsy that total body dehydration prior to death resulted in marked cytoplasmic vacuolization and excessive increases in neuronal and nucleolar size in the supraoptic and paraventricular nuclei, perhaps in response to excessive antidiuretic hormone (ADH) activity. Intravenous injection of normal sea water into resting dogs also resulted in increases of 1.0–1.9°C in body temperature (129).

Injections of potassium chloride into the cerebral ventricles or the tuber cinereum region of the hypothalamus in rabbits (130) and in cats (131, 132) caused hyperpyrexia. In a more recent study intrahypothalamic injection of 10.6 μEq of potassium chloride resulted in an increase in rectal temperature of 1.3°C in rabbits; injection of normal sodium chloride had no effect (133).

While most of the early research on fluid and electrolyte aspects of thermoregulation dealt with sodium chloride, dehydration, and fluid replacement, a few investigators have reported on the effects of other ions, mainly calcium and magnesium. Schutz (134) showed clearly that intravenous injection of 10 ml of 10% calcium chloride or 4 ml of 20% magnesium chloride lowered normal core temperature in rabbits, and that calcium acted synergistically with magnesium. These findings were confirmed by Hasama (131), who also observed that injection of 2 mg of calcium chloride into the tuber cinereum of cats reduced body temperature by 0.6–1.1 °C; and a previous intravenous injection of calcium chloride or magnesium chloride suppressed the hyperthermia and sweating of subsequent injections of sodium chloride, potassium chloride, or barium chloride into the tuber cinereum. In addition, injection of magnesium or calcium chloride into the tuber cinereum of rabbits prevented or reduced pyrogen-induced fevers (130). Thus, the action and the antagonistic effects of the major hyperthermic and hypothermic ions were described by 1935. But there was some confusion, still unresolved, among those researchers concerning the effect of these ions on the direction and magnitude of changes in "excitability," or sensitivity, of the thermoregulatory center in the hypothalamus. Heagy and Burton (135) investigated the mechanism of magnesium-induced hypothermia in panting and shivering dogs. Intravenous magnesium caused, in all conditions, a decrease in rectal temperature by decreasing shivering and voluntary motor activity in a cold environment, and by increasing panting in a warm environment. The most likely explanation was a central, curare-like effect on the neuromuscular junction for the former response, and stimulation of the panting center for the latter response. There was no evidence that magnesium acted specifically through the temperature-regulating centers. Infusion of excess magnesium into the cerebral ventricles of resting hamsters failed to alter colonic temperature (136).

Recent Research (1968–1978)

Intracerebral Infusion

Rest The important problem with all experiments that involve injection or perfusion of ions into the brain is the difficulty in determining the ionic concentration at the site of action. The most likely response is for the infusion concentration to be diluted by the time it reaches the active site, but the degree of dilution is very difficult to determine. The concentration of a transmitter substance within a synapse might be extremely high and, to mimic the activity at that synapse, fairly high levels of that substance would have to be injected into the extracellular fluid.

Thermoregulatory effects from perfusion of various cations into cerebral ventricles or directly into the hypothalamus have been studied extensively by Robert Myers and his colleagues. Feldberg et al. (137) perfused various salt solutions from a lateral cerebral ventricle to the cisterna magna in unanesthetized resting cats while rectal temperature was recorded. They found that calcium must be present in the artificial cerebrospinal fluid (CSF) perfusion fluid to prevent body temperature from rising. Intense shivering, vasoconstriction, and hyperthermia, with a rate of rise in rectal temperature of 0.1 °C/min, occurred with an infusion of 0.9% sodium chloride. The final rectal temperature often reached 42°C. Perfusions with 0.9% sodium chloride plus potassium and calcium chloride in physiological concentrations did not affect body temperature. Perfusions with 0.9% sodium chloride plus calcium chloride resulted in the same level of hyperthermia as with isotonic saline alone. They (137) concluded, "This hyperthermia emphasizes the role of calcium in preventing temperature from rising to fever level and suggests a correct level of calcium or its permeability in the hypothalamus as the physiological basis of this set-point." But Clark and Cumby (138) found no significant change in retroperitoneal temperature of cats after one 0.1-ml injection of 0.9% saline into a lateral cerebral ventricle. This set-point hypothesis was taken one step further by Myers and Veale (87). They found that perfusion of either an excess or normal physiological concentration of sodium ions into the posterior hypothalamus resulted in a rise in rectal temperature of cats if calcium was not in the perfusate, whereas perfusion of an excess or normal concentration of calcium ions caused a fall in rectal temperature if sodium was not present. (Excess Na^+ or Ca^{2+} is that quantity above the normal resting CSF concentration, as seen in Table 1.) These temperature responses were found only with perfusions into the posterior hypothalamus and not into the anterior hypothalamus. Body temperature remained stable as long as the ratio between sodium and calcium was maintained within the posterior hypothalamus in spite of the fact that the concentrations of ions employed were well beyond the normal range for CSF—3–13 mM excess calcium and 14–34 mM excess sodium. Myers and Veale state, "We hypothesize here that the constancy in the concentration of extracellular ionic constituents maintains the firing rate of the neurons of the posterior hypothalamus. This steady-state discharge pattern thus keeps the set-point at or about 37°C" (87). These findings have been confirmed in resting

Table 1. Normal plasma and cerebrospinal fluid sodium and total calcium concentrations[a] in various animals

Species	CSF		Plasma	
	Na	Ca_{tot}	Na	Ca_{tot}
Man	141.2	2.4	140.0	5.0
Cat	162.0	2.6	143.0	4.1
Dog	156.3	2.6	147.0	5.3
Monkey	151.1*	2.6*	155.1	5.3
Rat	128.0†	2.6†	148.7	5.3
Sheep	154.0‡	2.3‡	163.0‡	4.8‡

Data from Altman and Dittmer (170) except for the following values: *, Turbyfill et al. (171); †, Gisolfi et al. (147); ‡, Pappenheimer et al. (172).

[a]Values in mEq/liter.

monkeys (100, 139), cats (99), rats (140), hamsters (136), and rabbits (141). Myers and Yaksh (100) suggested that monkeys, who had their body temperature changed, presumably by a shift in the sodium to calcium ratio, were thermoregulating around the new set-point because the animals exhibited the proper responses of shivering or panting when given a load of cold or warm water, respectively, through a stomach tube. One complication arises from the observations that body temperature responses to infused ions are modified to some extent by changes in ambient temperature. Dhumal and Gulati (142) found that, in a warm environment (28–35°C), injection into anesthetized dogs of artificial CSF containing 11–48 mM excess calcium produced a rise in rectal temperature, whereas 35 mM excess sodium caused a fall in rectal temperature. At 21°C ambient temperature, rectal temperature responded "properly" in confirmation of Myers et al. (139). Hanegan and Williams (143, 144) also confirmed the hyperthermic effect of excess (11 mM) sodium and the hypothermic effect of excess (6.7 mM) calcium with push-pull perfusions into the posterior hypothalamus of ground squirrels. But they also found that perfusion of excess (6.7–20.2 mM) calcium into the preoptic area of the anterior hypothalamus interfered with oxygen consumption and temperature regulation. Excess sodium had no effect there. The direction and magnitude of their responses also depended upon ambient temperature; the decrease in rectal temperature was greater at 12°C, smaller at 25°C, and at 33°C excess calcium increased rectal temperature, in agreement with Dhumal and Gulati (142). Results from the opposite procedure, in which CSF ionic composition was measured in cats during pyrogen-induced fever, indicated that CSF [Ca^{2+}] and [Na^{+}] changed in their predicted fashion (145). Smaller changes in ionic concentrations can evoke thermoregulatory responses, as shown in monkeys after citrate injections (100). Chelation of CSF calcium with EDTA infusion resulted in an elevated rectal temperature, vasoconstriction, and shivering in conscious dogs at 18–23°C ambient temperature (146). Rectal temperature reached a plateau after 40–60 min when both shivering and panting were present; no effect on rectal temperature was noted with injection of excess sodium up to 88 mM.

Myers et al. (21) have recently confirmed the reversal effect of ambient temperature on ionically induced body temperature responses. Exposure of cats to ambient cooling or warming at the same time the ionic perfusions began resulted in the typical calcium-induced hypothermia and sodium-induced hyperthermia. However, if ambient cooling or heating were started 30 min before the perfusions, the fall or rise in rectal temperature produced by calcium or sodium was attenuated or prevented. An increase in ambient temperature caused a retention of calcium in the posterior hypothalamus, and a decrease in ambient temperature enhanced the efflux of calcium. Local warming of the preoptic area reduced calcium efflux; i.e., it aided calcium retention in the posterior hypothalamus as a stimulus for reducing the temperature. Cooling the preoptic area caused a loss of calcium from the posterior hypothalamus. Injection of hyperthermic substances (5-HT, PGE_1, or *Salmonella typhosa*) into the anterior hypothalamus increased calcium loss from the posterior hypothalamus, whereas a similar injection of norepinephrine reduced calcium efflux. Myers et al. (21) have summarized their current hypothesis:

> . . . the change in cation flux could serve a dual purpose. By means of the shift in local kinetics of calcium or sodium or both ions, either the depolarization rate in the posterior hypothalamus, or synaptic transmission there, could be facilitated or suppressed transiently, without the cells in the posterior hypothalamus losing their intrinsic reference value, their set-point. Once the temporary shift in ion balance imposed by anterior-preoptic cells terminates, the capacious equilibrating property of extracellular fluid within the hypothalamus would re-establish instantaneously the activity of the set-point neurons as a result of the perpetual constancy in the interstitial concentration of these cations (21).

Thus, there should be little doubt that the cations, mainly calcium and probably sodium, can influence thermoregulation when introduced directly into the hypothalamic temperature-sensitive areas or into the cerebral ventricles in excessive concentrations. The unsettled question is whether these body temperature responses are activated with changes in ionic concentrations within the normal range for these ionic species.

In an attempt to determine the relationship between change in the equilibrium level of resting core temperature and the level of excess sodium and calcium concentrations infused into the hypothalamus or cerebral ventricles, results from experiments with various animals are presented in Figure 6. These data were taken from the cat (21, 87, 99), monkey (100, 139), rat (147), dog (142, 146), hamster (136), sheep (148), and ground squirrel (143) at normal ambient temperatures (T_a = 18–25°C). The hypothalamic perfusions were into the posterior area, except those of the two ground squirrels, which were into the anterior hypothalamus. These two latter points fell very close to the calcium regression line (Figure 6). Excess calcium perfusion always resulted in a decreased core temperature, and there was a significant linear correlation of 0.85 between the level of excess calcium and the decrease in core temperature (Figure 6, *top*). The y-intercept of −0.16°C suggests that there may be a range of brain temperature within which calcium has no effect. The slope of the regression line indicates a reduction in core temperature of about 0.1°C for each millimole of excess calcium perfused. Conversely, excess sodium perfusion caused either no change or an increase in core temperature; it never resulted in a decreased temperature in normal ambient conditions (Figure 6, *bottom*). The majority of the sodium data fall between a change in core temperature of +0.5°C and +1.3°C, regardless of the sodium concentration. So, at rest, the threshold concentration for sodium perfusion appears to be about 10 mM, much less sensitive than with excess calcium when compared on a millimolar basis, but much more sensitive when compared with their respective normal CSF concentrations. Normal CSF calcium concentration is 2.3–2.6 mEq/liter, and normal sodium concentration is 128–162 mEq/liter.

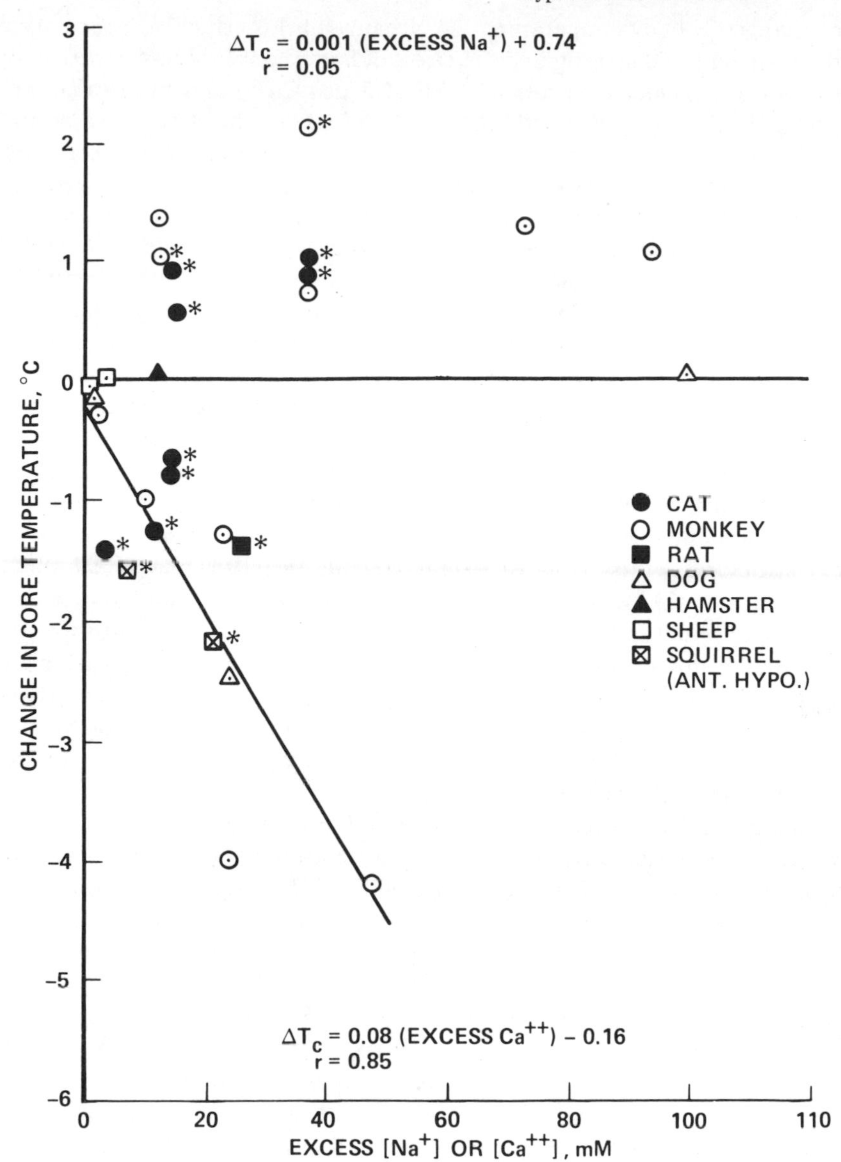

Figure 6. Change in resting core temperature (T_c), with infusion of excess sodium ($+T_c$) and excess calcium ($-T_c$) into the hypothalamus (*) or lateral cerebral ventricles, in various mammals at normal (18–25°C) ambient temperature. Reproduced from Greenleaf (173) with permission of Birkhäuser Verlag.

Exercise Since the rise in core temperature during exercise appears to be a regulated process, the fluid-ionic hypothesis has been suggested as a possible contributory factor in the control mechanism. As mentioned above, one major problem with most of the previous ionic perfusion experiments has been the excessively large concentrations of ions required to induce significant changes in body temperature. To circumvent this criticism, Sobocinska and Greenleaf (103) investigated the effect of lateral cerebral intraventricular perfusion of normocalcic artificial CSF (1.3 mM Ca^{2+}) and hypercalcic artificial CSF (2.6 mM Ca^{2+}) with normal osmolality or the effect of no perfusion on rectal temperature responses of dogs at rest and during moderate physical exercise for 1 hr on a treadmill (T_a = 22–24°C). There were no significant effects of normocalcic or hypercalcic perfusions on the gradual decline of rectal temperatures at rest (Figure 7). But the rise in rectal temperature during exercise was significantly attenuated during hypercalcic perfusion. This relative hypothermia was associated with a greater body weight loss (respiratory evaporative heat loss) compared with dogs perfused with normal artificial CSF or those who performed as nonperfused controls. No differences in plasma sodium, potassium, or osmotic concentrations were observed in these three experimental conditions. Lack of effect of the hypercalcic perfusion for the first 20 min of exercise was probably due to a buildup of CSF calcium concentration and transit time to reach the ion-sensitive sites, presumably the anterior area of the third cerebral ventricle, or perhaps other structures within the CSF compartment, including the spinal cord. Nielsen et al. (149) found no relationship between resting rectal temperature and lumbar spinal fluid sodium concentration, total calcium concentration, or the sodium-to-calcium ratio. These findings lead to two conclusions: near normal CSF calcium concentrations can significantly influence thermoregulation during exercise, but not at rest; and some factor or factors associated, activated, or released by exercise increase the sensitivity of the calcium-induced thermoregulatory mechanism.

A large excess concentration of calcium (26 mM) perfused into the third ventricle or posterior hypothalamus reduces core temperature during exercise in the rat (147) and in the monkey (102, 150, 151). The hypothermic response in the rat was accompanied by peripheral vasodilation in the tail. Chelation of calcium ions, with ethylene glycol tetraacetic acid within the hypothalamic region, resulted in an increase in colonic temperature, at rest, of 0.68°C and a greater increase, by 0.47°C ± S.D. 0.26°C, of the normal rise in colonic temperature during exercise of 1.21°C ± 0.18°C (152). With progressively increasing concentrations of calcium, infused intraventricularly, there was a progressive attenuation in the rise of colonic temperature during exercise in monkeys (Figure 8), followed by a progressively greater fall in body temperature during recovery that appeared to be mediated by increased peripheral vasodilation (102). During exercise, when the body temperature rose, there was a marked increase (up to 4-fold) in the concurrent efflux of ^{45}Ca ions from the third cerebral ventricle that remained elevated for 60–90 min during the recovery (150). They de-emphasized the importance of the increased heat loss mechanisms of panting (153) and vasodilation (114) that occur with excess calcium infusions. Instead, the importance of the central mechanism was emphasized with a postulated increase in set-point during exercise, induced by a lowering of the $Na^{+}:Ca^{2+}$

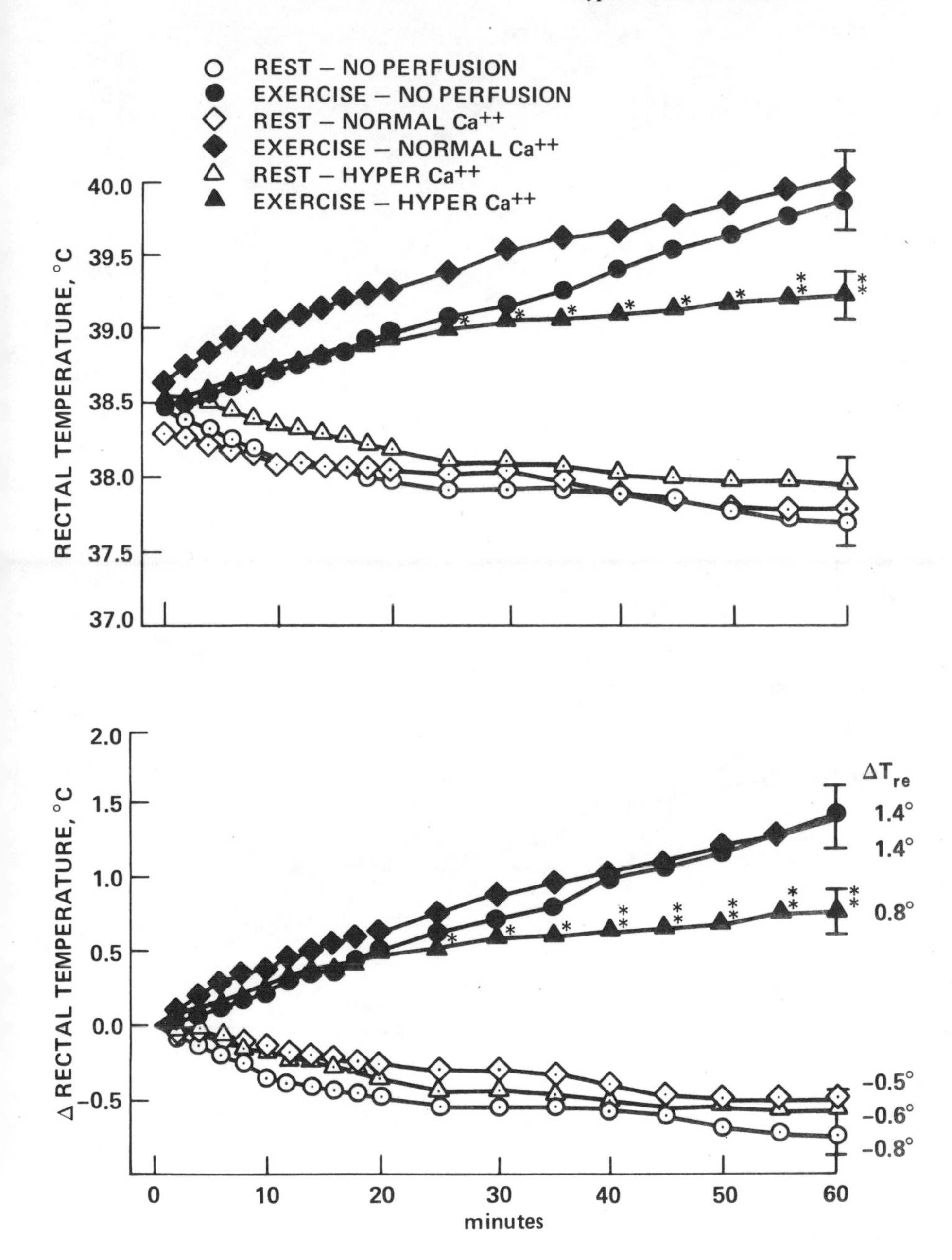

Figure 7. Mean (±S.E.) rectal temperatures of five dogs at rest and during 60 min of moderate treadmill exercise. $*P < 0.05$ or $**P < 0.01$ between hyper Ca^{2+} and normal Ca^{2+}. Reproduced from Sobocińska and Greenleaf (103) with permission of the American Physiological Society.

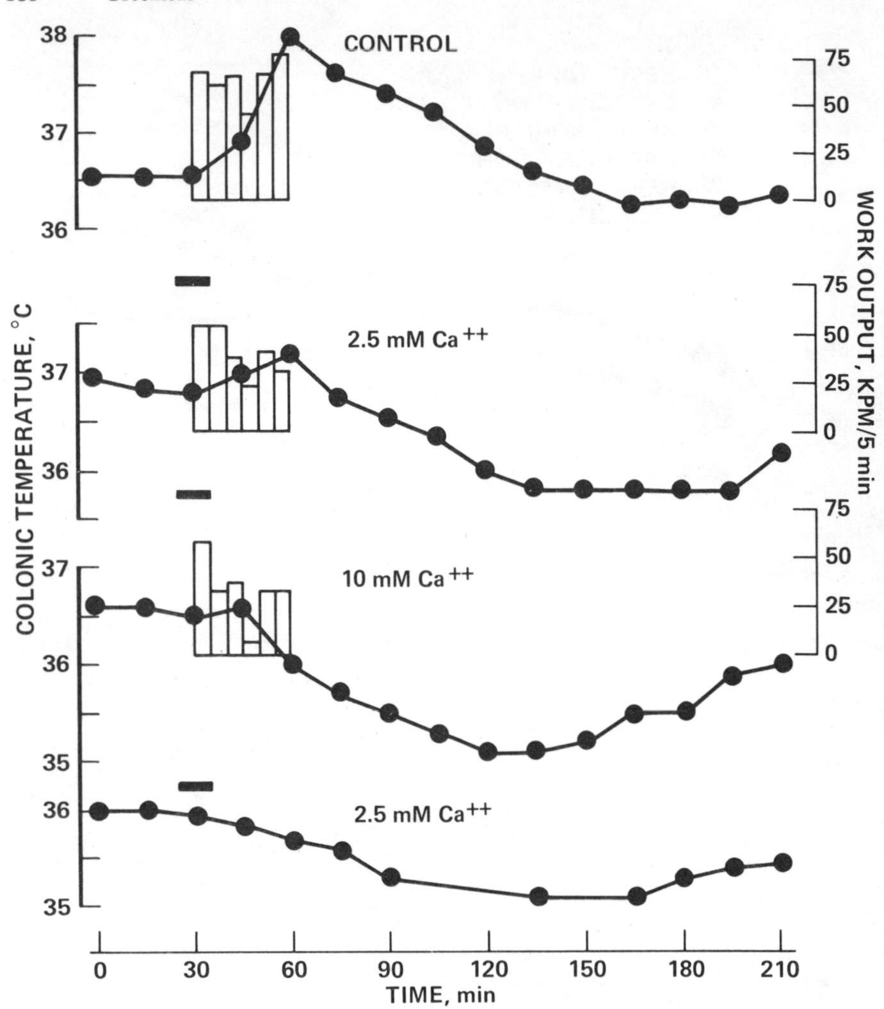

Figure 8. Colonic temperature responses in the monkey before, during, and after exercise (*vertical bars*). Calcium was infused at the *black bar*. Reproduced from Myers et al. (102) with permission of the American Physiological Society.

ratio, caused by the excess calcium perfused. They further postulated that the external addition of excess calcium ions into the hypothalamus stabilizes neuronal membranes and reduces neuronal firing rates of those cells responsible for heat conservation; the result is a decrease in body temperature (150).

As discussed previously, it is probably more correct to conceive of a lowering of the hypothetical set-point during exercise. A decrease in the set-point would result in activation of heat dissipation responses because body temperature would theoretically be higher than the set-point. Calcium infusion results in increased heat dissipation responses in conjunction with attenuation or actual blocking of the rise in core temperature with exercise and a frank decrease in core temperature at rest (Figure 8). So the calcium effect seems most consistent with the concept of a decrease in the set-point. But merely suggesting that calcium ions change the set-point will never explain the mechanism of exercise thermoregulation without an accurate accounting of the various components of the physical heat balance equation.

A second conceptual conflict involves the effect of sodium and calcium ions on the sensitivity of the hypothalamic temperature centers. Since normal brain temperature is higher than its blood temperature, increased vascular sodium and osmotic concentrations have been hypothesized to decrease sensitivity and to paralyze the heat loss centers in the hypothalamus, with a reduction in heat loss and increased core temperature (119, 127). That is, highly sensitive and active central nervous system (CNS) neurons must function positively to extrude heat actively and prevent brain temperature from rising. It is difficult to reconcile this hypothesis with that of Gisolfi et al. (150), who have suggested that excess calcium stabilized neuronal membranes, i.e., decreased sensitivity, and resulted in hypothermia.

Since water exchanges rapidly between the blood and cerebrospinal fluid (the half-life of deuterium oxide is between 2 min and 37 min (154)) minute fluid shifts from the systemic circulation to the CSF could possibly change CSF calcium concentration. Perhaps the hypovolemia and redistribution of other body fluid compartments, which accompany physical exercise, could influence the ionic composition and concentration of the CSF and thermoregulation.

Intravenous Infusion The thermal centers in and about the hypothalamus respond readily and are more sensitive than other centers within the central nervous system to infusion of calcium and sodium ions. This sensitivity appears to be increased during exercise. As the point of entry of the ions moves farther away from the brain, it would seem that larger concentrations of ions would be required to elicit thermoregulatory responses because more membranes would have to be crossed on their way to the hypothalamus if, indeed, the hypothalamus were the sole target area.

Intravenous injection of 20 ml of 3% solutions of calcium chloride or sodium chloride into resting Chinchilla rabbits caused a fall and rise, respectively, of rectal temperature (114). The changes in rectal temperature were aided by appropriate changes in blood flow in the ears, as reflected in earskin temperatures. Hypertonic sodium chloride solutions (6.7–10% at 91–229 ml/hr, 39°C) were infused into a hind leg vein of mongrel dogs before and during 60 min of moderate exercise (101, 111). The rates of infusion were proportional to their estimated extracellular fluid volumes and were designed to raise plasma osmolality to between 320 mosm/kg and 330 mosm/kg. In addition, one injection of 1.8 ml/kg of 3.8% sodium citrate was given just before exercise. The results (Figure 9) indicate greater increases in exercise rectal temperatures with pre-exercise infusion of hypertonic saline (prehypertonic) and with infusion of the same dose (hypertonic) during exercise com-

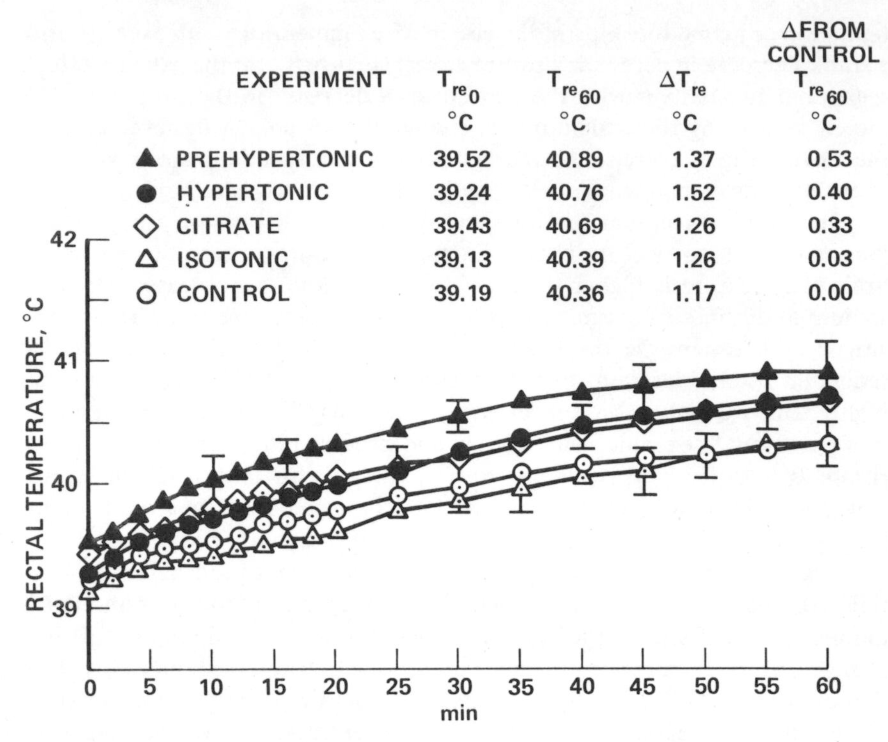

Figure 9. Mean (±S.E.) rectal temperatures of dogs at rest and during 60 min of moderate treadmill exercise during and after various ionic infusions. Reproduced from Greenleaf et al. (153) with permission of Birkhäuser Verlag.

pared with infusion and isotonic saline or no infusion (control). Rectal temperature response to citrate injection was between control and hypertonic values, which would be expected if the citrate bound some plasma calcium. The highest resting rectal temperature (39.52 °C) and the highest postexercise temperature (40.89 °C) were observed after prehypertonic infusion. Two additional prehypertonic experiments were conducted: a control experiment in which postexercise rectal temperature reached 41.0 °C and, after a few days for recovery, a second experiment in which 1 liter of tap water (39 °C) was offered to the dogs 10 min after exercise started (101). Each dog drank all of his water eagerly and within 5 min the rate of rise of rectal temperature decreased; by 15 min after drinking, rectal temperature reached an equilibrium level 0.5 °C lower than the control value. Water ingestion resulted in a progressive increase in plasma volume, as judged by the progressive decreases in plasma sodium (from 170 mEq/liter to 158 mEq/liter) and osmotic (from 340 mosm/kg to 312 mosm/kg) concentrations, by the end of exercise. Thus, intravenous infusion of sodium elevates rectal temperatures at rest and during moderate exercise. Dilution of hypernatremic-osmotic plasma results in an attenuation of the equilibrium level of rectal temperature during exercise, so peripheral vascular fluid shifts can influence thermoregulation during exercise.

Sodium and especially calcium are also important for sweat gland function. Intravenous injection of $^{24}Na^{+}$ appears in thermal sweat in about 1.5 min (155). Extracellular calcium, but not magnesium, is essential for the maintenance of sweat secretion (156). Intravenous injection of $^{47}Ca^{2+}$ is removed quickly from plasma but does not appear in thermal sweat (157). Thus, sodium is passed rapidly into the sweat, whereas calcium appears to be retained by the sweat glands.

Ingestion of Fluid and Electrolytes The classic paper of Pitts et al. (106) proved conclusively that the rise in rectal temperature during exercise in heat ($T_a = 37.7$°C) was directly proportional to the level of body dehydration (Figure 10). With no water consumption, rectal temperature rose steadily. With water consumption, the rise in body temperature was essentially inhibited for 3 hr of exercise; but with forced water ingestion equal to sweat loss, the increase in rectal temperature was minimal and performance was excellent. Senay (109) has confirmed those results and added the important conclusion that core temperatures in men resting in the heat were lower when hemodilution was present. Ekblom et al. (158) found that exercise rectal temperature was significantly elevated with no fluid replacement, compared with water consumption equal to sweat loss. This small hyperthermic response occurred with only a 1% loss in body weight, emphasizing the sen-

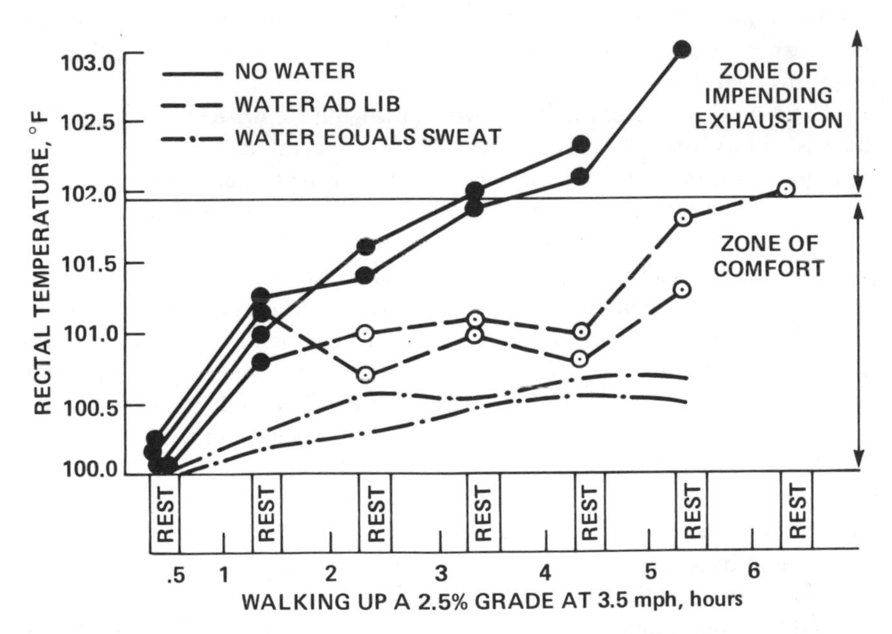

Figure 10. Effect of water consumption on rectal temperature while walking in the heat: subject J.S., $T_a = 100$°F, *rh*, 35–45%. Reproduced from Pitts et al. (106) with permission of the American Physiological Society.

sitivity of this mechanism. To quantify the effect of changes in sodium concentration on changes in core temperatures, eight men were exercised for 1 hr ($T_a = 23.6\,°C$) at three hydration levels: 1) hyperhydration (Δ body weight = 1.2%) induced by drinking 40 ml/kg (2.5–3.0 liters) of tap water (37°C) before exercise; 2) ad libitum (Δ body weight = −1.6%) where the subjects drank 100–200 ml of water before exercise; and 3) hypohydration (Δ body weight = −5.2%) induced by prior exercise in the heat (112). The equilibrium level of exercise rectal temperature was inversely related to the body fluid volume (Figure 11). Changes in body heat storage were accounted for by changes in sweat rate. Linear relationships were present between body weight (kg) and body temperature (°C).

$$\Delta T_{re} = -0.10\,(\Delta \text{ body weight}) + 1.14 \qquad (1)$$

$$\Delta \bar{X} \text{ body temperature} = -0.09\,(\Delta \text{ body weight}) + 0.47 \qquad (2)$$

The slopes of these equations indicated that equilibrium levels of rectal temperature were elevated 0.1°C for each 1% of body weight loss. Serum sodium and osmolality were both significantly correlated ($r = +0.71$) with equilibrium levels of rectal temperatures (Figure 12). These correlations imply that about half of the variability in core temperature is due to factors other than sodium or osmotic concentrations. There was essentially no relationship between sodium, osmolality or plasma volume and rectal temperature at rest. Nielsen (105) also concluded that exercise-induced shifts in equilibrium levels of esophageal temperature were not due to changes in plasma volume. With heavy exercise for 1.5–2.5 hr in the heat, rectal temperature rose 0.4°C for each 1% of body weight loss after a 2% body weight loss had been incurred (159).

Construction of prediction equations utilizing stepwise linear regression analysis shows that the equilibrium level of rectal temperature can be calculated with greater probability ($r = 0.92$) from serum sodium and pH, heart rate, and respiratory water loss (Figure 13, *top*). Similarly, equilibrium mean body temperature can be calculated from serum sodium, osmolality and pH, heart rate, and respiratory water loss. Adding a skin temperature component in the calculation of mean body temperature paralleled the addition of osmolality to the prediction equation. It may be important that serum pH appears in both of these equations even though intracerebroventricular injections of saline, with pH between 4.4 and 11.6, did not significantly change resting body temperature in the cat (138). Myers et al. (102) have suggested that blood P_{O_2} and P_{CO_2} levels may influence the sodium/calcium balance in the hypothalamus, and it is known that reduced venous P_{O_2} is related to the control of vasodilation during exercise (46). Further investigation of acid-base balance and thermoregulation may prove fruitful, particularly during exercise at altitude.

There is unanimous agreement that core temperature during exercise is elevated after procedures that induce plasma hemoconcentration, hypernatremia, and related hyperosmotemia, that is, dehydration from sweating

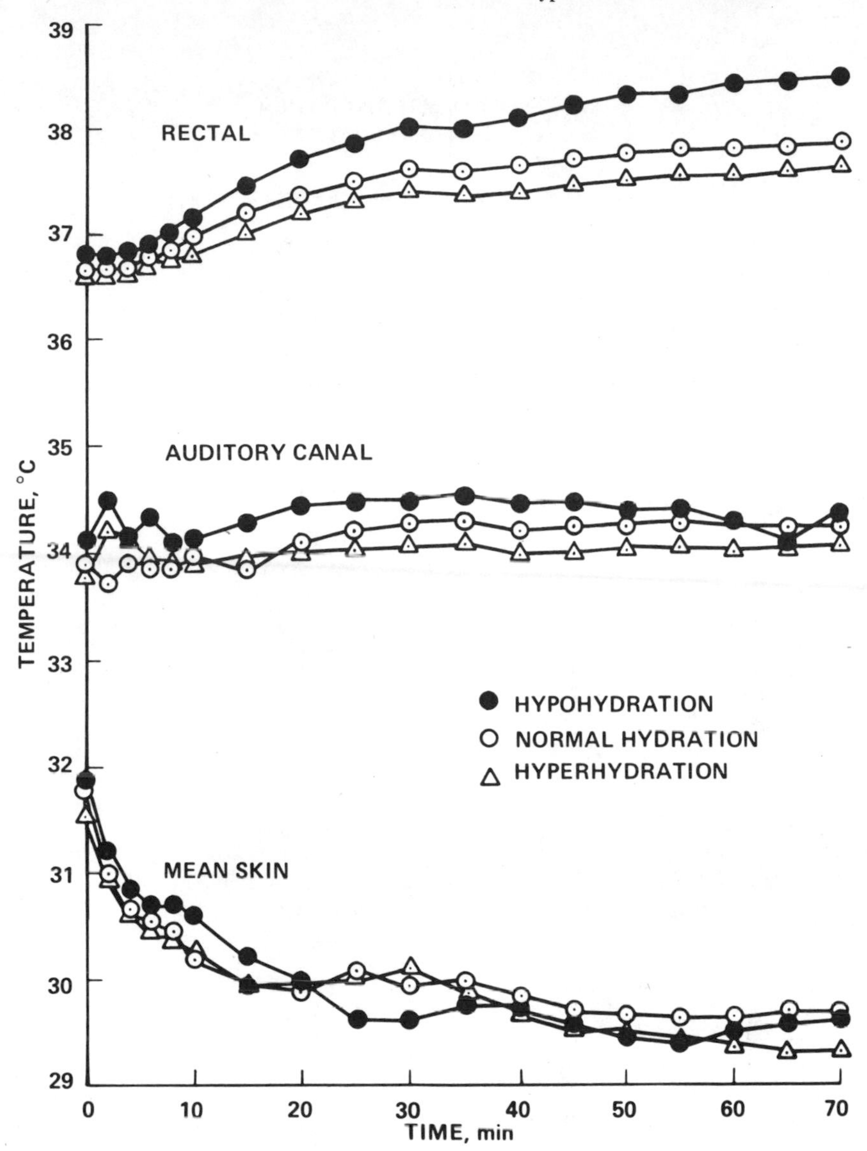

Figure 11. Mean rectal, auditory canal, and mean skin temperatures at rest and during moderate submaximal exercise under three hydration regimens. Hypohydration T_{re0}, T_{re70}, and ΔT_{re} were higher ($P < 0.01$) than corresponding hyperhydration values. Reproduced from Greenleaf and Castle (112) with permission of the American Physiological Society.

(105, 112, 159, 160) and saline ingestion (105, 107, 108, 160). Conversely, the rise in core temperature is attenuated after procedures that induce plasma hemodilution, hyponatremia, and hypo-osmotemia from water ingestion (105, 106, 109, 112, 158–160). Calcium ingestion, which increased plasma calcium concentration by 7–11%, had no effect on exercise rectal

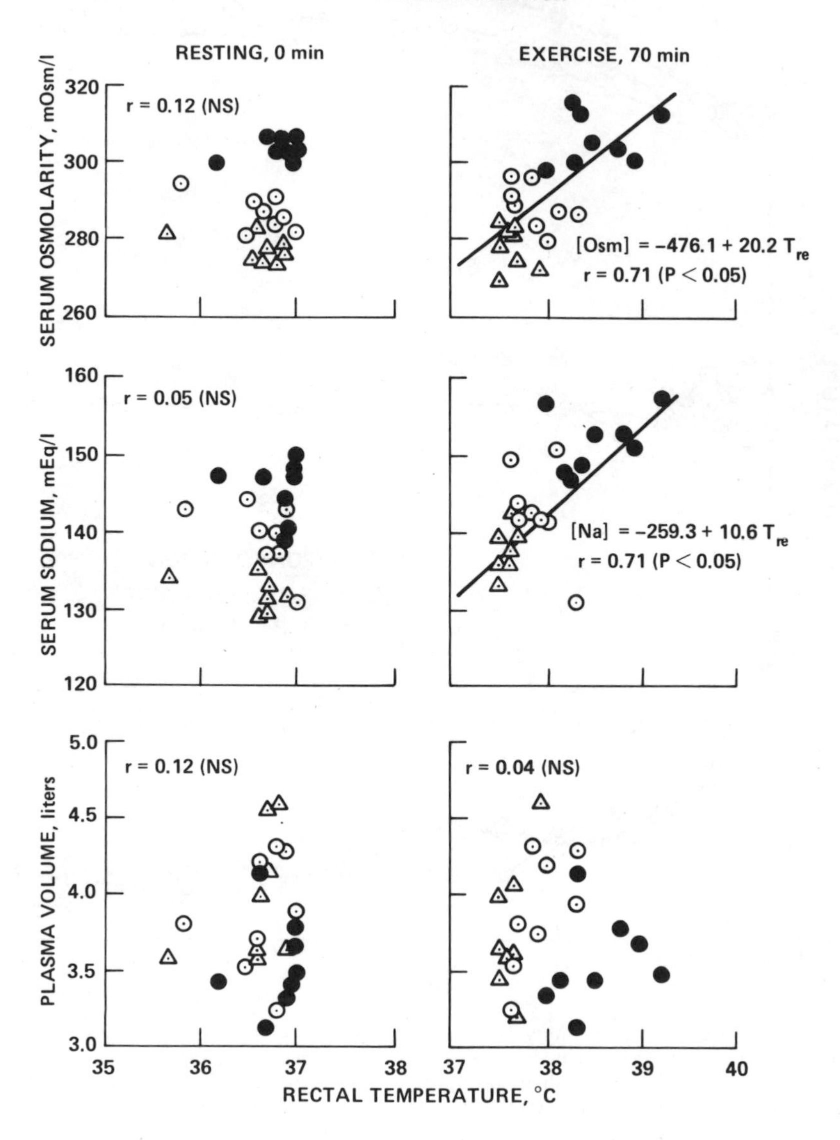

Figure 12. Correlations between rectal temperature and serum sodium, osmolarity and calculated plasma volume at rest and after exercise under three hydration regimens. Data from Greenleaf and Castle (112).

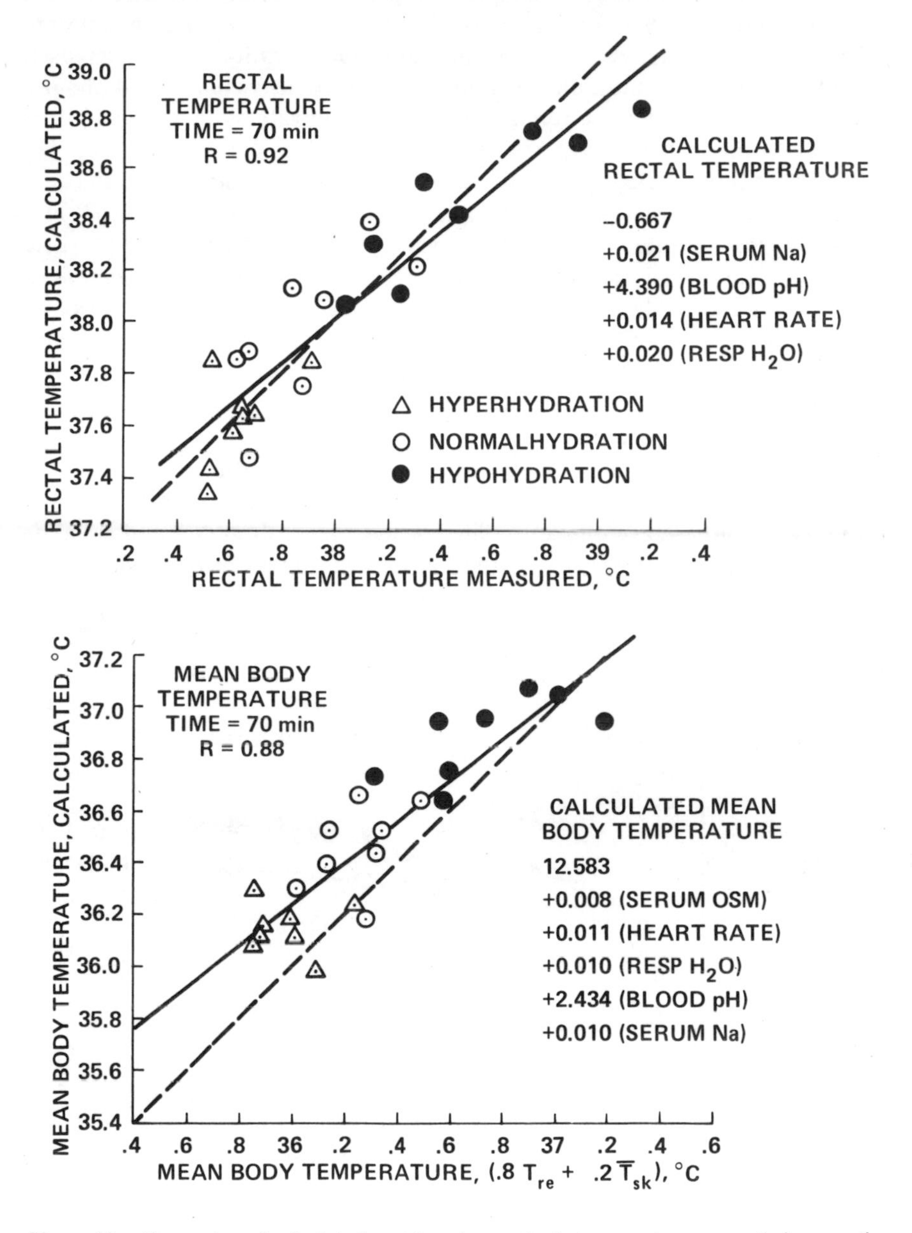

Figure 13. Regression of calculated rectal and mean body temperatures upon their respective measured body temperature after exercise under three hydration regimens. *Broken lines* are lines of equality. Data from Greenleaf and Castle (112).

temperatures in normal ($T_a = 26.5°C$) or hot ($T_a = 39.4°C$) environments (107). However, the rise in esophageal temperature during exercise has been reported to be attenuated after drinking 1–1.5 liters of 2% calcium chloride solution (Figure 14) when plasma calcium concentrations were increased 25–30% (108). Maximum plasma ionic concentrations were reached about 2 hr after ingestion. Hypercalcemia induces a greater rate of increase and final rate of sweating, whereas hypernatremia has the opposite response (107, 108, 112). Sweat rates during progressive dehydration at rest and work in the heat were negatively correlated with changes in serum sodium ($r = -0.63$) and osmolality ($r = -0.62$) (110). Resting rectal temperature and blood pressure levels were disturbed in rats fed diets deficient in calcium and magnesium (161). After the 28-day dietary period, the highest rectal temperatures and blood pressures were observed in animals which consumed the low calcium and high magnesium diet; the lowest values for body temperature and pressure occurred with the high calcium and low magnesium diet. The mean differences between these two dietary groups of animals were 1°C in rectal temperature and 40 mm Hg in blood pressure. The calcium concentration in brain tissue rose and fell with the blood levels. Blood, but not brain, levels of 5-HT were decreased significantly with the low magnesium diet and could be associated with accentuated vasodilation that occurs with magnesium deficiency.

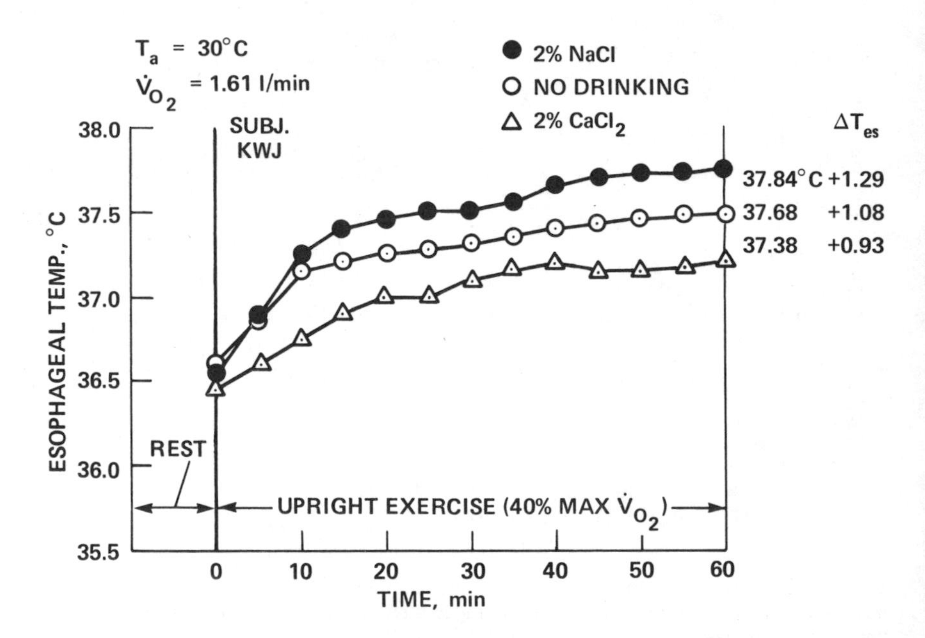

Figure 14. Esophageal temperature responses at rest and during 60 min of submaximal exercise after ingestion of hypertonic sodium and calcium solutions. Reproduced from Nielsen (108) with permission of Acta Physiologica Scandinavica.

The findings discussed previously suggest that induced changes in plasma electrolyte contents, similar to electrolyte changes occurring under natural conditions of dehydration and/or physical exercise, have an effect on temperature regulation. The specific hyperthermic response to sodium is well documented, but it has been more difficult to make the case for calcium as a hypothermic agent. A 25% increase in plasma calcium concentration is not within the normal range of variability for this ion and is not well tolerated in normal man. The evidence favors the conclusion that the actions of sodium and calcium ions are specific, and the hyperthermia associated with hyperosmotemia is due to sodium and not hyperosmolality per se.

Normal Environmental and Exercise-induced Responses For the sodium-calcium hypothesis to be accepted as a significant factor in the mechanism of thermoregulation, particularly in man, it must be shown that changes in plasma cation concentrations within the range of normal variability are in the proper direction and of sufficient magnitude to account for the change in body temperature. In addition, the influence of shifts in body fluid volumes, particularly the plasma volume, on exercise thermoregulation still remains an enigma. Most of the negative evidence of an effect of changes in plasma volume on body temperature responses has been obtained from changes in plasma volume calculated directly from changes in hematocrit and/or protein concentrations (105, 153, 160), which can be erroneous (39). Recognizing the importance of maintaining blood volume for optimal thermoregulation, it is interesting to speculate why plasma volume should be shifted to the interstitial space in direct proportion to the intensity of exercise (36).

The most important problem, however, is to determine whether fluid-electrolyte shifts control or even influence thermoregulation under normal exercise conditions without artificial introduction of excess ions. Intrahypothalamic perfusion obviously cannot be utilized in humans, so we must analyze spinal fluids and blood. Systematic measurements of ventricular or lumbar spinal fluid electrolyte concentrations have not been measured in man or in animals during exercise. If no significant changes in CSF sodium or calcium concentrations occur during exercise, any effect upon CSF electrolyte concentrations from fluid-electrolyte shifts in peripheral blood would be negated unless there was a direct connection, not involving the CSF, between electrolytes in peripheral blood and thermal-sensitive hypothalamic neurons. CSF composition is very closely controlled. The CSF to serum sodium and osmolar ratios are very close to 1.0 in normal man, with CSF sodium concentration about 141 mEq/liter (Table 1) and osmolality about 285 mosm/kg; their correlation is better than 0.98 (162). Severe neurological abnormalities occur when CSF osmolality exceeds serum osmolality by ≥ 7 mosm/kg (162). The half-time for equilibration of water between ventricular and lumbar spinal fluid is 5-50 min (154), compared with over 7 hr for sodium (163). Therefore, lumbar spinal fluid electrolyte concentrations would probably not reflect ventricular spinal fluid concentrations during relatively short-term exercise. The extracellular fluid of the brain communicates freely with the CSF, and the chemical composition of these two fluids is similar; but the composition of the extracellular fluid depends upon the site where the sample is taken (164). CSF concentrations of sodium, potassium, chloride, osmolality, and protein in resting dogs were essentially unchanged at five brain temperatures between 37°C and 10°C (with progressively decreasing cerebral metabolism) and during circulatory arrest in which only CSF potassium concentration increased in proportion to the body temperature (165). Maintenance of normal CSF constituent concentrations in the presence of these rather drastic treatments suggests that moderate to heavy physical exercise would probably not change them to any significant degree. It is inconceivable that fluid-electrolyte shifts could occur anywhere in the body to change the set-point within the first minute of exercises, essentially within the minimum circulation time, unless it were by a reflex stimulus. Thus, it is not surprising that no relationship has been found between normal variations in the calcium concentration or in the sodium-to-calcium ratio in peripheral blood and body temperature during exercise (36, 149, 166–168). However, Greenleaf et al. (36) found a significant correlation of +0.62 between total calcium and rectal temperature at rest; additionally, they found that equilibrium levels of rectal temperature at three levels of submaximal exercise correlated +0.78 with corresponding plasma sodium concentrations, in confirmation of previous results at three levels of hydration (112, 153). Also, a correlation of −0.46 was found between the change in plasma volume and rectal temperature. There were very low correlations between rectal temperature and plasma volume, total calcium and ionized calcium concentrations, and the sodium-to-calcium ratio. During stressful situations, e.g., acceleration or exercise, when plasma fluid shifted from the vascular space and presumably moved to the interstitial spaces, the change in plasma constituent (ionic) concentrations was the net result of their concomitant efflux and influx. At moderate exercise, plasma sodium, chloride, ionized calcium, and total osmols leave with the plasma in essentially isotonic quantities so that their residual concentrations are unchanged (36). With moderately heavy exercise (relative oxygen uptake of 63%), there was a net retention of chloride and the cations so that the plasma concentrations were increased. At all levels of exercise, from moderate to maximal, the influx of potassium greatly exceeded its efflux, so its plasma concentration increased by a much larger percentage than the other constituents. It was concluded that the shift (loss) of plasma fluid, with its accompanying ionic exchanges, accounted for the positive correlations between these ions and equilibrium levels of core temperature (36).

CONCLUSIONS

Introduction of sodium and calcium ions into the hypothalamus, cerebral ventricles, or the venous system influences temperature regulation at rest and during exercise. Excess sodium perfused into the brain or procedures that produce hypernatremia result in increased body temperature; excess calcium perfused into the brain or hypercalcemia results in decreased body temperature. As the evidence accumulates, particularly that showing significant thermoregulatory responses to sodium and calcium ion concentrations very near or within normal CSF or plasma ranges, it becomes increasingly difficult to ascribe changes in body temperature to nonspecific effects of these cations. The central thermoregulatory mechanism seems more responsive to the hypothermic effect of calcium than to the hyperthermic effect of sodium.

Thermoregulatory control of evaporative sweating during exercise is more closely related to changes in plasma sodium concentration than it is to plasma calcium concentration. The reversal of the hypothermic effect of intrahypothalamic or intraventricular perfusions of calcium in resting animals at high ambient temperatures emphasizes the importance of skin temperature and the basic level of core temperature for the ionic thermal response. It cannot be determined at this time whether changes in systemic calcium concentrations exert their influence mainly on peripheral effectors or centers for thermoregulation within the central nervous system.

Physical exercise, with its increased metabolism, activates, modifies, or sensitizes an additional factor or factors that attenuate the fall in core temperature associated with infusion of near normal CSF concentrations of calcium. Because of the widespread action of sodium and calcium within the body, it is unlikely that the ratio of these ionic concentrations within the hypothalamus would provide the physicochemical basis for a set-point that controls the level of core temperature at rest or during short-term exercise. Most of the observations concerning the effects of cations on thermoregulatory responses have been qualitative in nature. To determine the relative importance of the fluid-ionic hypothesis as a factor for thermoregulation, precise quantitative heat balance studies at rest and during exercise must be conducted. In view of the relatively large increase in plasma potassium concentration during exercise, which is known to facilitate peripheral vasodilation, additional studies should be done to determine if potassium has a direct thermoregulatory effect. The mechanisms controlling plasma fluid-electrolyte shifts, particularly during exercise and recovery from exercise, may play an important role for exercise thermoregulation.

Perhaps the essence of the thermoregulatory control mechanism during exercise has already been uncovered: " . . . there is no one human attribute of more importance than the ability to sweat skillfully" (169).

ACKNOWLEDGMENTS

The author is grateful to Carol Greenleaf and Deanna Sciaraffa for their very competent assistance in preparing this manuscript.

REFERENCES

1. Hardy, J. D. (1961). Physiology of temperature regulation. Physiol. Rev. 41:521.
2. Mitchell, D. (1974). Physical basis of thermoregulation. *In* D. Robertshaw (ed.), MTP International Review of Physiology, Vol. 7, Environmental Physiology I, pp. 1-32. University Park Press, Baltimore.
3. Mitchell, D. (1977). Physical basis of thermoregulation. *In* D. Robertshaw (ed.), MTP International Review of Physiology, Vol. 15, Environmental Physiology II, pp. 1-27. University Park Press, Baltimore.
4. Henderson, L. J. (1958). The Fitness of the Environment. Beacon Press, Boston.
5. Bligh, J. (1966). The thermosensitivity of the hypothalamus and thermoregulation in mammals. Biol. Rev. 41:317.
6. Chaffee, R. R. J., and Roberts, J. C. (1971). Temperature acclimation in birds and mammals. Annu. Rev. Physiol. 33:155.
7. Gale, C. C. (1973). Neuroendocrine aspects of thermoregulation. Annu. Rev. Physiol. 35:391.
8. Hales, J. R. S. (1974). Physiological responses to heat. *In* D. Robertshaw (ed.), MTP International Review of Physiology, Vol. 7, Environmental Physiology I, pp. 107-162. University Park Press, Baltimore.
9. Hammel, H. T. (1968). Regulation of internal body temperature. Annu. Rev. Physiol. 30:641.
10. Hellon, R. F. (1975). Monoamines, pyrogens and cations: their actions on central control of body temperature. Pharmacol. Rev. 26:289.
11. Taylor, C. R. (1974). Exercise and thermoregulation. *In* D. Robertshaw (ed.), MTP International Review of Physiology, Vol. 7, Environmental Physiology I, pp. 163-184. University Park Press, Baltimore.
12. Taylor, C. R. (1977). Exercise and environmental heat loads: different mechanisms for solving different problems. *In* D. Robertshaw (ed.), MTP International Review of Physiology, Vol. 15, Environmental Physiology II, pp. 119-146. University Park Press, Baltimore.
13. Rowell, L. B. (1974). Human cardiovascular adjustments to exercise and thermal stress. Physiol. Rev. 54:75.
14. Nielsen, B. (1969). Thermoregulation in rest and exercise. Acta Physiol. Scand. (suppl. 323):1.
15. Cooper, K. E., Lomax, P., and Schönbaum, E. (eds.). (1977). Drugs, Biogenic Amines and Body Temperature, p. 283. S. Karger, Basel.
16. Lomax, P., Schönbaum, E., and Jacob, J. (eds.). (1975). Temperature Regulation and Drug Action, p. 405. S. Karger, Basel.
17. Schönbaum, E., and Lomax, P. (eds). (1973). The Pharmacology of Thermoregulation, p. 583. S. Karger, Basel.
18. Greenleaf, J. E. (1973). Blood electrolytes and exercise in relation to temperature regulation in man. *In* E. Schönbaum and P. Lomax (eds.), The Pharmacology of Thermoregulation, pp. 72-84. S. Karger, Basel.
19. Myers, R. D. (1975). An integrative model of monamine and ionic mechanisms in the hypothalamic control of body temperature. *In* P. Lomax, E. Schönbaum, and J. Jacob (eds.), Temperature Regulation and Drug Action, pp. 32-42. S. Karger, Basel.
20. Myers, R. D. (1976). Chemical control of body temperature by the hypothalamus: a model and some mysteries. Proc. Aust. Physiol. Pharmacol. Soc. 7:15.
21. Myers, R. D., Simpson, C. W., Higgins, D., Natterman, R. A., Rice, J. C., Redgrave, P., and Metcalf, G. (1976). Hypothalamic Na^+ and Ca^{2+} ions and

temperature set-point: new mechanisms of action of a central or peripheral thermal challenge and intrahypothalamic 5-HT, NE, PGE_1 and pyrogen. Brain Res. Bull. 1:301.
22. Nielsen, B., and Greenleaf, J. E. (1977). Electrolytes and thermoregulation. *In* K. E. Cooper, P. Lomax, and E. Schönbaum (eds.), Drugs, Biogenic Amines and Body Temperature, pp. 39–47. S. Karger, Basel.
23. DuBois, E. F. (1951). The many different temperatures of human body and its parts. West. J. Surg. 59:476.
24. Gagge, A. P., and Herrington, L. P. (1947). Physiological effects of heat and cold. Annu. Rev. Physiol. 9:409.
25. Leithead, C. S., and Lind, A. R. (1964). Heat Stress and Heat Disorders, pp. 195–196. F. A. Davis Company, Philadelphia.
26. Shibolet, S., Lancaster, M. C., and Danon, Y. (1976). Heat stroke: a review. Aviat. Space Environ. Med. 47:280.
27. Convertino, V. A., Greenleaf, J. E., and Bernauer, E. M. (1977). Role of exercise in the mechanism of the chronic increase in plasma volume. Physiologist 20:18 (abstract).
28. Robinson, S. (1963). Temperature regulation in exercise. Pediatrics 32(part II):691.
29. Pugh, L. G. C. E., Corbett, J. L., and Johnson, R. H. (1967). Rectal temperatures, weight losses, and sweat rates in marathon running. J. Appl. Physiol. 23:347.
30. Bevegard, B. S., and Shepherd, J. T. (1965). Changes in tone of limb veins during supine exercise. J. Appl. Physiol. 20:1.
31. Christensen, E. H., Nielsen, M., and Hannisdahl, B. (1942). Investigations of the circulation in the skin at the beginning of muscular work. Acta Physiol. Scand. 4:162.
32. Christensen, E. H. (1931). Beiträge zur Physiologie schwerer körperlicher Arbeit. II. Die Körpertemperatur während und unmittelbar nach schwerer körperlicher Arbeit. Arbeitsphysiologie 4:154.
33. Greenleaf, J. E., Van Kessel, A. L., Ruff, W., Card, D. H., and Rapport, M. (1971). Exercise temperature regulation in man in the upright and supine positions. Med. Sci. Sports 3:175.
34. Kitzing, J., and Bleichert, A. (1965). Untersuchungen zur Arbeitshyperthermie des Menschen. Pfluegers Arch. 282:242.
35. Rowell, L. B., Brengelmann, G. L., Detry, J.-M. R., and Wyss, W. (1971). Venomotor responses to rapid changes in skin temperature in exercising man. J. Appl. Physiol. 30:64.
36. Greenleaf, J. E., Convertino, V. A., Stremel, R. W., Bernauer, E. M., Adams, W. C., Vignau, S. R., and Brock, P. J. (1977). Plasma [Na^+], [Ca^{2+}], and volume shifts and thermoregulation during exercise in man. J. Appl. Physiol.: Respir. Environ. Exercise Physiol. 43:1026.
37. Van Beaumont, W., and Bullard, R. W. (1963). Sweating: its rapid response to muscular work. Science 141:643.
38. Van Beaumont, W., and Bullard, R. W. (1966). Sweating exercise stimulation during circulatory arrest. Science 152:1521.
39. Van Beaumont, W., Greenleaf, J. E., and Juhos, L. (1972). Disproportional changes in hematocrit, plasma volume, and proteins during exercise and bed rest. J. Appl. Physiol. 33:55.
40. Van Beaumont, W., Strand, J. C., Petrofsky, J. S., Hipskind, S. G., and Greenleaf, J. E. (1973). Changes in total plasma content of electrolytes and proteins with maximal exercise. J. Appl. Physiol. 34:102.
41. Greenleaf, J. E., Van Beaumont, W., Brock, P. J., Morse, J. T., and Mangseth, G. R. (1979). Plasma volume and electrolyte shifts with heavy exer-

cise in sitting and supine positions. Am. J. Physiol.: Regulatory Integrative Comp. Physiol. 5:R206.
42. Wilkerson, J. E., Gutin, B., and Horvath, S. M. (1977). Exercise-induced changes in blood, red cell, and plasma volumes in man. Med. Sci. Sports 9:155.
43. Kjellmer, J. (1961). The role of potassium ions in exercise hyperaemia. Med. Exp. 5:56.
44. Lundvall, J., Mellander, S., and White, T. (1969). Hyperosmolality and vasodilatation in human skeletal muscle. Acta Physiol. Scand. 77:224.
45. Mellander, S., and Lundvall, J. (1971). Role of tissue hyperosmolality in exercise hyperemia. Circ. Res. (suppl. 1) 28–29:I-39.
46. Skinner, N. S., Jr., and Costin, J. C. (1970). Interactions of vaso-active substances in exercise hyperemia: O_2, K^+, and osmolality. Am. J. Physiol. 219:1386.
47. Aikas, E., Karvonen, M. J., Piironen, P., and Ruosteenoja, R. (1962). Intramuscular, rectal and oesophageal temperature during exercise. Acta Physiol. Scand. 54:366.
48. Asmussen, E., and Bøje, O. (1945). Body temperature and capacity for work. Acta Physiol. Scand. 10:1.
49. Saltin, B., Gagge, A. P., and Stolwijk, J. A. J. (1968). Muscle temperature during submaximal exercise in man. J. Appl. Physiol. 25:679.
50. Saltin, B., and Hermansen, L. (1966). Esophageal, rectal, and muscle temperature during exercise. J. Appl. Physiol. 21:1757.
51. Benzinger, T. H. (1969). Heat regulation: homeostasis of central temperature in man. Physiol. Rev. 49:671.
52. Nadel, E. R., and Horvath, S. M. (1970). Comparison of tympanic membrane and deep body temperature in man. Life Sci. 9:869.
53. Robinson, S., Meyer, F. R., Newton, J. L., Ts'ao, C. H., and Holgersen, L. O. (1965). Relations between sweating, cutaneous blood flow, and body temperature in work. J. Appl. Physiol. 20:575.
54. Greenleaf, J. E., and Castle, B. L. (1972). External auditory canal temperature as an estimate of core temperature. J. Appl. Physiol. 32:194.
55. Greenleaf, J. E., Greenleaf, C. J., Card, D. H., and Saltin, B. (1969). Exercise-temperature regulation in man during acute exposure to simulated altitude. J. Appl. Physiol. 26:290.
56. Nielsen, B., and Nielsen, M. (1962). Body temperature during work at different environmental temperatures. Acta Physiol. Scand. 56:120.
57. Nielsen, M. (1938). Die Regulation der Körpertemperatur bei Muskelarbeit. Skand. Arch. Physiol. 79:193.
58. Nielsen, B. (1966). Regulation of body temperature and heat dissipation at different levels of energy and heat production in man. Acta Physiol. Scand. 68:215.
59. Nielsen, B., and Nielsen, M. (1965). On the regulation of sweat secretion in exercise. Acta Physiol. Scand. 64:314.
60. Hayward, J. N., and Baker, M. A. (1968). Role of cerebral arterial blood in the regulation of brain temperature in the monkey. Am. J. Physiol. 215:389.
61. Olesen, J. (1971). Contralateral focal increase of cerebral blood flow in man during arm work. Brain 94:635.
62. Galbo, H., Hummer, L., Petersen, I. B., Christensen, N. J., and Bie, N. (1977). Thyroid and testicular hormone responses to graded and prolonged exercise in man. Eur. J. Appl. Physiol. 36:101.
63. Benedict, F. G., and Cathcart, E. P. (1913). Muscular work: a metabolic study with special reference to the efficiency of the human body as a machine.

Carnegie Institution of Washington Publication No. 187, pp. 1–176. The University Press, Cambridge, Massachusetts.
64. Nadel, E. R. (ed.). (1977). Problems with Temperature Regulation During Exercise. Academic Press, Inc., New York.
65. Dill, D. B., Edwards, H. T., Bauer, P. S., and Levenson, E. J. (1931). Physical performance in relation to external temperature. Arbeitsphysiologie 4:508.
66. MacKeith, N. W., Pembrey, M. S., Spurrell, W. R., Warner, E. C., and Westlake, H. J. W. J. (1923). Observations on the adjustment of the human body to muscular work. Proc. R. Soc. Lond. (Biol.) 95:413.
67. Winslow, C.-E. A., and Gagge, A. P. (1941). Influence of physical work on physiological reactions to the thermal environment. Am. J. Physiol. 134:664.
68. Asmussen, E., and Nielsen, M. (1947). The regulation of the body-temperature during work performed with the arms and with the legs. Acta Physiol. Scand. 14:373.
69. Nielsen, M., Herrington, L. P., and Winslow, C.-E. A. (1939). The effect of posture upon peripheral circulation. Am. J. Physiol. 127:573.
70. Greenleaf, J. E. (1973). Temperature regulation during isotonic exercise. Acta Physiol. Pol. 24:67.
71. Bazett, H. C. (1951). Theory of reflex controls to explain regulation of body temperature at rest and during exercise. J. Appl. Physiol. 4:245.
72. Kerslake, D. McK. (1955). Factors concerned in the regulation of sweat production in man. J. Physiol. 127:280.
73. Benzinger, T. H. (1959). On physical heat regulation and the sense of temperature in man. Proc. Natl. Acad. Sci. 45:645.
74. Burton, A. C. (1941). The operating characteristics of the human thermoregulatory mechanism. *In* Temperature, Its Measurement and Control in Science and Industry, pp. 521–528. American Institute of Physics. Reinhold Publishing Corporation, New York.
75. Barbour, H. G. (1912). Die Wirkung unmittelbarer Erwärmung und Abkühlung der Wärmezentren auf die Körpertemperaturen. Arch. Exp. Pathol. Pharmakol. 70:1.
76. Bradbury, P. A., Fox, R. H., Goldsmith, R., and Hampton, I. F. G. (1964). The effect of exercise on temperature regulation. J. Physiol. 171:384.
77. Cabanac, M., Cunningham, D. J., and Stolwijk, J. A. J. (1971). Thermoregulatory set point during exercise: a behavioral approach. J. Comp. Physiol. Psychol. 76:94.
78. Jackson, D. C., and Hammel, H. T. (1963). Hypothalamic "set" temperature decreased in exercising dog. Life Sci. 8:554.
79. Mitchell, D., Snellen, J. W., and Atkins, A. R. (1970). Thermoregulation during fever: change of set-point or change of gain. Pfluegers Arch. 321:293.
80. Jéquier, E., Dolivo, M., and Vannotti, A. (1971). Temperature regulation in exercise: the characteristics of proportional control. J. Physiol. (Paris) 63:303.
81. Hayward, J. N., and Baker, M. A. (1969). A comparative study of the role of the cerebral arterial blood in the regulation of brain temperature in five mammals. Brain Res. 16:417.
82. Cabanac, M. (1975). Temperature regulation. Annu. Rev. Physiol. 37:415.
83. Werner, J. (1977). Mathematical treatment of structure and function of the human thermoregulatory system. Biol. Cybern. 25:93.
84. Von Euler, C. (1961). Physiology and pharmacology of temperature regulation. Pharmacol. Rev. 13:361.
85. Von Euler, C., and Söderberg, U. (1958). Co-ordinated changes in temperature thresholds for thermoregulatory reflexes. Acta Physiol. Scand. 42:112.

86. Stoner, H. B., and Elson, P. M. (1971). The effect of injury on monoamine concentrations in the rat hypothalamus. J. Neurochem. 18:1837.
87. Myers, R. D., and Veale, W. L. (1970). Body temperature: possible ionic mechanism in the hypothalamus controlling the set point. Science 170:95.
88. Snellen, J. W. (1969). The discrepancy between thermometry and calorimetry during exercise. Pfluegers Arch. 310:35.
89. Snellen, J. W. (1972). Set point and exercise. *In* J. Bligh and R. E. Moore (eds.), Essays on Temperature Regulation, pp. 139–148. Elsevier-North Holland Publishing Company, New York.
90. Snellen, J. W., Mitchell, D., and Busansky, M. (1972). Calorimetric analysis of the effect of drinking saline solution on whole-body sweating. I. An attempt to measure average body temperature. Pfluegers Arch. 331:124.
91. Houdas, Y., and Guieu, J. D. (1973). Le système thermo-régulateur de l'homme: système régulé ou système asservi? Arch. Sci. Physiol. 27:A311.
92. Grimby, G. (1962). Exercise in man during pyrogen-induced fever. Scand. J. Clin. Lab. Invest. (suppl. 67) 14:1.
93. Maskrey, M., and Bligh, J. (1972). The impairment of the thermoregulatory set-point of a sheep in the apparent absence of any interference with the pathways between temperature sensors and thermoregulatory effectors. Experientia 28:794.
94. Snellen, J. W. (1966). Mean body temperature and the control of thermal sweating. Acta Physiol. Pharmacol. Neerl. 14:99.
95. Senay, L. C., Jr., and Christensen, M. L. (1965). Cardiovascular and sweating responses to water ingestion during dehydration. J. Appl. Physiol. 20:975.
96. Åstrand, I. (1960). Aerobic work capacity in men and women, with special reference to age. Acta Physiol. Scand. (suppl. 169) 49:1.
97. Saltin, B. (1970). Circulatory adjustments and body temperature regulation during exercise. *In* J. D. Hardy, A. P. Gagge, and J. A. J. Stolwijk (eds.), Physiological and Behavioral Temperature Regulation, pp. 316–323. Charles C Thomas, Springfield, Illinois.
98. Calloway, N. O. (1976). Body temperature: thermodynamics of homeothermism. J. Theor. Biol. 57:331.
99. Myers, R. D., and Veale, W. L. (1971). The role of sodium and calcium ions in the hypothalamus in the control of body temperature of the unanaesthetized cat. J. Physiol. 212:411.
100. Myers, R. D., and Yaksh, T. L. (1971). Thermoregulation around a new 'set-point' established in the monkey by altering the ratio of sodium to calcium ions within the hypothalamus. J. Physiol. 218:609.
101. Greenleaf, J. E., Kozłowski, S., Nazar, K., Kaciuba-Uscińko, H., Brzezińska, Z., and Ziemba, A. (1976). Ion-osmotic hyperthermia during exercise in dogs. Am. J. Physiol. 230:74.
102. Myers, R. D., Gisolfi, C. V., and Mora, F. (1977). Role of brain Ca^{++} in central control of body temperature during exercise in the monkey. J. Appl. Physiol.: Respir. Environ. Exercise Physiol. 43:689.
103. Sobocińska, J., and Greenleaf, J. E. (1976). Cerebrospinal fluid [Ca^{2+}] and rectal temperature response during exercise in dogs. Am. J. Physiol. 230:1416.
104. Ladell, W. S. S. (1955). The effects of water and salt intake upon the performance of men working in hot and humid environments. J. Physiol. 127:11.
105. Nielsen, B. (1974). Effects of changes in plasma volume and osmolarity on thermoregulation during exercise. Acta Physiol. Scand. 90:725.
106. Pitts, G. C., Johnson, R. E., and Consolazio, F. C. (1944). Work in heat as affected by intake of water, salt and glucose. Am. J. Physiol. 142:253.
107. Greenleaf, J. E., Brock, P. J., Morse, J. T., Van Beaumont, W., Montgomery,

L. D., Convertino, V. A., and Mangseth, G. R. (1978). Effect of sodium and calcium ingestion on thermoregulation during exercise in man. *In* Y. Houdas and J. D. Guieu (eds.), New Trends in Thermal Physiology, pp. 157–160. Masson, Paris.

108. Nielsen, B. (1974). Effect of changes in plasma Na^{+} and Ca^{++} ion concentration on body temperature during exercise. Acta Physiol. Scand. 91:123.
109. Senay, L. C., Jr. (1970). Movement of water, protein and crystalloids between vascular and extravascular compartments in heat-exposed men during dehydration and following limited relief of dehydration. J. Physiol. 210:617.
110. Senay, L. C., Jr. (1968). Relationship of evaporative rates to serum [Na^{+}], [K^{+}], and osmolarity in acute heat stress. J. Appl. Physiol. 25:149.
111. Greenleaf, J. E., Castle, B. L., and Card, D. H. (1974). Blood electrolytes and temperature regulation during exercise in man. Acta Physiol. Pol. 25:397.
112. Greenleaf, J. E., and Castle, B. L. (1971). Exercise temperature regulation in man during hypohydration and hyperhydration. J. Appl. Physiol. 30:847.
113. McCloskey, D. I., and Mitchell, J. H. (1972). Reflex cardiovascular and respiratory responses originating in exercising muscle. J. Physiol. 224:173.
114. Nielsen, B. (1974). Actions of intravenous Ca^{++} on body temperature in rabbits. Acta Physiol. Scand. 90:445.
115. Crandall, F. M. (1899). Inanition fever. Arch. Pediatr. 16:174.
116. Marriott, W. M. (1923). Anhydremia. Physiol. Rev. 3:275.
117. Freund, H. (1913). Über das Kochsalzfieber. Arch. Exp. Pathol. Pharmakol. 65:225.
118. Barbour, H. G. (1921). The heat-regulating mechanism of the body. Physiol. Rev. 1:295.
119. Dontas, S. (1939). Uber den Mechanismus der Wärmeregulation. Pfluegers Arch. 241:612.
120. Balcar, J. O., Sansum, W. D., and Woodyatt, R. T. (1919). Fever and the water reserve of the body. Arch. Int. Med. 24:116.
121. Keith, N. M. (1924). Experimental dehydration: changes in blood composition and body temperature. Am. J. Physiol. 68:80.
122. Green, C. H., and Rowntree, L. G. (1927). The effect of the administration of excessive amounts of water on body temperature. Am. J. Physiol. 80:230.
123. Rietschel, H., and Beck, E. (1928). Dursthyperthermie und körperliche Arbeit. Dtsch. Med. Wochenschr. 54:1283.
124. Barbour, H. G., and Gilman, A. (1934). Heat regulation and water exchange. XVIII. The subservience of vapor pressure homeostasis to temperature homeostasis. Am. J. Physiol. 107:70.
125. Barbour, H. G., and Gilman, A. (1934). Heat regulation and water exchange. XVII. The relation of serum osmotic pressure to the onset of fever. J. Pharmacol. Exp. Ther. 50:277.
126. Barbour, H. G., and Tolstoi, E. (1924). Heat regulation and water exchange. II. The role of the water content of the blood, and its control by the central nervous system. Am J. Physiol. 67:378.
127. Turlejska-Stelmasiak, E. (1974). The influence of dehydration on heat dissipation mechanisms in the rabbit. J. Physiol. (Paris) 68:5.
128. Koep, L. J., Konigsmark, B. W., and Sperber, E. E. (1970). Cellular changes in the human supraoptic and paraventricular hypothalamic nuclei in dehydration. J. Neuropathol. Exp. Neurol. 29:254.
129. Dontas, S., and Phocas, E. (1939). Recherches expérimentales sur l'influence de l'eau de mer sur les centres thermorégulateurs. Prakt. Akad. Athenon 14:83.
130. Kym, O. (1934). Die Beeinflussung des durch verschiedene fiebererzeugende

Stoffe erregten Temperaturzentrums durch lokale Applikation von Ca, K und Na. Arch. Exp. Pathol. Pharmakol. 176:408.

131. Hasama, B. (1930). Pharmakologische und physiologische Studien über die Schweisszentren. Arch. Exp. Pathol. Pharmakol. 153:291.
132. Marinesco, G., Sager, O., and Kreindler, A. (1929). Experimentelle Untersuchungen zum Problem des Schlafmechanismus. Z. Gesamte. Neurol. Psychiatr. 119:277.
133. Cooper, K. E., Cranston, W. I., and Honour, A. J. (1965). Effects of intraventricular and intrahypothalamic injection of noradrenaline and 5-HT on body temperature in conscious rabbits. J. Physiol. (Lond.) 181:852.
134. Schutz, J. (1916). Zur Kenntnis der Wirkung des Magnesiums auf die Körpertemperatur. Arch. Exp. Pathol. Pharmakol. 79:285.
135. Heagy, F. C., and Burton, A. C. (1948). Effect of intravenous injection of magnesium chloride on the body temperature of the unanesthetized dog, with some observations on magnesium levels and body temperature in man. Am. J. Physiol. 152:407.
136. Myers, R. D., and Buckman, J. E. (1972). Deep hypothermia induced in the golden hamster by altering cerebral calcium levels. Am. J. Physiol. 223:1313.
137. Feldberg, W., Myers, R. D., and Veale, W. L. (1970). Perfusion from cerebral ventricle to cisterna magna in the unanaesthesized cat: effect of calcium on body temperature. J. Physiol. (Lond.) 207:403.
138. Clark, W. G., and Cumby, H. R. (1975). Effects on body temperature of the unanaesthetized cat of sodium chloride solutions of varied pH, injected into a lateral cerebral ventricle. Neuropharmacology 14:313.
139. Myers, R. D., Veale, W. L., and Yaksh, T. L. (1971). Changes in body temperature of the unanaesthetized monkey produced by sodium and calcium ions perfused through the cerebral ventricles. J. Physiol. (Lond.) 217:381.
140. Myers, R. D., and Brophy, P. D. (1972). Temperature changes in the rat produced by altering the sodium-calcium ratio in the cerebral ventricles. Neuropharmacology 11:351.
141. Veale, W. L., Benson, M. J., and Malkinson, T. (1977). Brain calcium in the rabbit: site of action for the alteration of body temperature. Brain Res. Bull. 2:67.
142. Dhumal, V. R., and Gulati, O. D. (1973). Effect on body temperature in dogs of perfusion of cerebral ventricles with artificial CSF deficient in calcium or containing excess of sodium or calcium. Br. J. Pharmacol. 49:699.
143. Hanegan, J. L., and Williams, B. A. (1973). Brain calcium: role in temperature regulation. Science 181:663.
144. Hanegan, J. L., and Williams, B. A. (1975). Ca^{++} induced hypothermia in a hibernator (Citellus Beechyi). Comp. Biochem. Physiol. 50A:247.
145. Myers, R. D., and Tytell, M. (1972). Fever: reciprocal shift in brain sodium to calcium ratio as the set point rises. Science 178:765.
146. Sadowski, B., and Szczepańska-Sadowska, E. (1974). The effect of calcium ions chelation and sodium ions excess in the cerebrospinal fluid on body temperature in conscious dogs. Pfluegers Arch. 352:61.
147. Gisolfi, C. B., Wilson, N. C., Myers, R. D., and Phillips, M. I. (1976). Exercise thermoregulation: hypothalamic perfusion of excess calcium reduces elevated colonic temperature of rats. Brain Res. 101:106.
148. Seoane, J. R., and Baile, C. A. (1973). Ionic changes in cerebrospinal fluid and feeding, drinking, and temperature of sheep. Physiol. Behav. 10:915.
149. Nielsen, B., Schwartz, P., and Alhede, J. (1973). Is fever in man reflected in changes in cerebrospinal fluid concentrations of sodium and calcium ions? Scand. J. Clin. Lab. Invest. 32:309.
150. Gisolfi, C. V., Mora, F., and Myers, R. D. (1977). Diencephalic efflux of

calcium ions in the monkey during exercise, thermal stress and feeding. J. Physiol. (Lond.) 273:617.

151. Myers, R. D., Gisolfi, C. V., and Mora, F. (1977). Calcium levels in the brain underlie temperature control during exercise in the primate. Nature 266:178.
152. Wilson, N. C., Gisolfi, C. V., and Phillips, M. I. (1978). Influence of EGTA on an exercise-induced elevation in the colonic temperature of the rat. Brain Res. Bull. 3:97.
153. Greenleaf, J. E., Kozłowski, S., Nazar, K., Kaciuba-Uscilko, H., and Brzezinska, Z. (1974). Temperature responses of exercising dogs to infusion of electrolytes. Experientia 30:769.
154. Bering, E. A., Jr. (1952). Water exchange of central nervous system and cerebrospinal fluid. J. Neurosurg. 9:275.
155. Gibiński, K., Kumaszka, F., Zmudziński, J., Giec, L., Waclawczyk, J., and Dosiak, J. (1973). Sodium $^{24}Na^+$ and potassium $^{42}K^+$ availability for sweat production after intravenous injection and their handling by sweat glands. Acta Biol. Med. Germ. 30:697.
156. Sato, K. (1977). The physiology, pharmacology, and biochemistry of the eccrine sweat gland. Rev. Physiol. Biochem. Pharmacol. 79:51.
157. Gibiński, K., Zmudziński, J., Kumaszka, F., Giec, L., and Waclawczyk, J. (1974). Calcium transit to thermal sweat. Acta Biol. Med. Germ. 32:199.
158. Ekblom, B., Greenleaf, C. J., Greenleaf, J. E., and Hermansen, L. (1970). Temperature regulation during exercise dehydration in man. Acta Physiol. Scand. 79:475.
159. Gisolfi, C. V., and Copping, J. R. (1974). Thermal effects of prolonged treadmill exercise in the heat. Med. Sci. Sports 6:108.
160. Nielsen, B., Hansen, G., Jorgensen, S. O., and Nielsen, E. (1971). Thermoregulation in exercising man during dehydration and hyperhydration with water and saline. Int. J. Biometeorol. 15:195.
161. Itokawa, Y., Tanaka, C., and Fujiwara, M. (1974). Changes in body temperature and blood pressure in rats with calcium and magnesium deficiencies. J. Appl. Physiol. 37:835.
162. Habel, A. H., and Simpson, H. (1976). Osmolar relation between cerebrospinal fluid and serum in hyperosmolar hypernatraemic dehydration. Arch. Dis. Child. 51:660.
163. Sweet, W. H., Brownell, G. L., Scholl, J. A., Bowsher, D. R., Benda, P., and Stickley, E. E. (1954). The formation, flow and absorption of cerebrospinal fluid; newer concepts based on studies with isotopes. Res. Publ. Assoc. Res. Nerv. Ment. Dis. 34:101.
164. Bering, E. A., Jr. (1974). The cerebrospinal fluid and the extracellular fluid of the brain. Fed. Proc. 33:2061.
165. Bering, E. A., Jr. (1974). Effects of profound hypothermia and circulatory arrest on cerebral oxygen metabolism and cerebrospinal fluid electrolyte composition in dogs. J. Neurosurg. 39:199.
166. DuVal, H. P., and Wilkerson, J. E. The association of arterial and venous plasma sodium-calcium ratios and body temperature during exercise in man. Med. Sci. Sports 10:61 (abstract).
167. Refsum, H. E., Meen, H. D., and Strömme, S. B. (1973). Whole blood, serum and erythrocyte magnesium concentrations after repeated heavy exercise of long duration. Scand. J. Clin. Lab. Invest. 32:123.
168. Strömme, S. B., Gullestad, R., Meen, H. D., Refsum, H. E., and Krog, J. (1976). Serum sodium and calcium and body temperature during prolonged exercise. J. Sports Med. 16:91.
169. DuBois, E. F. (1939). Heat loss from the human body. Bull. N. Y. Acad. Med. 15:143.

170. Altman, P. L., and Dittmer, D. S. (eds.). (1974). Biology Data Book, Ed. 2, Vol. III. Federation of American Societies for Experimental Biology, Bethesda, Maryland.
171. Turbyfill, C. L., Cramer, M. B., Dewes, W. A., and Huguley, J. W., III. (1970). Serum and cerebral spinal fluid chemistry values for the monkey (*Macaca mulatta*). Lab. Anim. Sci. 20:269.
172. Pappenheimer, J. R., Heisey, S. R., Jordon, R. F., Downer, J. deC., and Nicholl, J., Jr. (1962). Perfusion of the cerebral ventricular system in unanaesthetized goats. Am. J. Physiol. 20:763.
173. Greenleaf, J. E. (1978). Thresholds for Na^+ and Ca^{++} effects on thermoregulation. Experientia (suppl. 32):33.

International Review of Physiology
Environmental Physiology III, Volume 20
Edited by D. Robertshaw

4
Temperature Regulation, Fever, and Disease

M. J. KLUGER
University of Michigan Medical School, Ann Arbor, Michigan

TEMPERATURE REGULATION 210
Temperature Regulation as a Reflex 211
Vertebrate Temperature Regulation 211

FEVER IN MAMMALS 214
Nature of Fever 214
Biology of Fever 215
Activators of Fever 216
Endotoxins 216
Gram-positive Bacteria 217
Viruses 217
Hypersensitivity Reactions 217
Tumors 218
Endogenous Pyrogens 218
Site(s?) of Action of Endogenous Pyrogens 219
Prostaglandins and Fever 221
Cyclic AMP and Fever 225
Na^+/Ca^{2+} and Fever 225
Evolution of Fever 227

FEVER IN BIRDS 227

FEVER IN REPTILES 228

FEVER IN AMPHIBIANS 230

FEVER IN FISHES 231

THE FUNCTION OF FEVER 232
Recent Experiments Concerning the Function of Fever 233
Mechanisms Behind the Function of Fever 239

This chapter is concerned primarily with fever and its role in disease. In the time of Hippocrates, fever was viewed as a response of the body that aided in the removal of some harmful substance or "humor" (1). For approximately the next two millenia this view of fever was commonly held by Western civilization (e.g., "Fever is a mighty engine which Nature brings into the world for the conquest of her enemies" (Sydenham, 1666)). But by the late 1800's these ideas about fever began to change. Antipyretic drug therapy became popular and fevers accompanying a variety of infections were attenuated and often reduced completely. In 1960 the subject of whether or not fever was a host defense mechanism was reviewed extensively by Bennett and Nicastri (2). They were unable to uncover any "hard" experimental evidence that would allow generalizations to be made about the function of fever.

Within the past few years many data have been collected that are pertinent to determining the role of fever in disease. However, before these data can be presented, it is necessary to present some background information concerning the biology of temperature regulation and fever.

TEMPERATURE REGULATION

The ability to sense temperature and to respond appropriately in order to regulate tissue temperature is a phylogenetically ancient characteristic. Representative organisms among plants (3–5), protozoans (6, 7), insects (8–12), and others have been shown to be capable of regulating their temperature. The regulation of body temperature has been broken down into two general categories—ectothermy and endothermy. Ectotherms are those organisms that rely on external sources of heat to regulate their body temperature. Among the vertebrates, most fishes, amphibians, and reptiles are ectotherms. When they become too cold, they attempt to move to a warmer microclimate; when too warm, they do the opposite. A fish swimming to deeper water during the midday hours, a frog sitting on a lily pad and then moving into the cooler water, and a snake basking in the sun are all examples of vertebrates thermoregulating by ectothermy. Among the vertebrates, birds and mammals are endotherms. These organisms can generate sufficient amounts of heat to enable them to regulate their body temperature by altering their heat production and heat loss. While endotherms still rely extensively on behavior to regulate their body temperature and are, therefore, similar to ectotherms, the endotherms possess much greater thermoregulatory flexibility than do the ectotherms. Small changes in environmental temperature, which might force an ectotherm to retreat to the shade or cooler water, have little effect on the behavior of endotherms. As a result, the endotherm can inhabit a wider variety of niches and can remain functionally active for greater lengths of time.

Endothermy clearly has had profound effects on the biology of the vertebrates; however, it has had a surprisingly small effect on the thermoregulatory reflex itself, and on the febrile responses of the vertebrates.

Temperature Regulation as a Reflex

The regulation of any parameter, including temperature regulation, can be conceptualized as consisting of the three general components of the reflex arc: a) sensors, b) integrators, and c) effectors. The sensors are capable of sensing different temperatures and converting these stimuli into the appropriate signal. In vertebrates, nerve cells serve as temperature sensors. These sensors convert the thermal stimuli into the appropriate patterns and frequencies of action potentials, which travel along the afferent or sensory nerves. In vertebrates, temperature sensors have been found in the skin (see refs. 13–16), as well as in internal structures, such as the hypothalamus (see refs. 17–19), spinal cord (20–24), abdomen (25–27), and elsewhere. The integration of thermal information in the vertebrates is thought to occur to some extent in the hypothalamus, with some integrative abilities residing in other central nervous areas. Here, all the thermal inputs from the various sensors are somehow weighed and the appropriate efferent information is then passed via nerves and hormones to the effectors. The effectors are those structures that actually lead to the raising or lowering of body temperature. Effectors used in regulating body temperature often have other functions as well. Some examples of effectors used by the thermoregulatory system, along with their effector response (in parentheses), are skeletal muscle (shivering, behavioral responses such as moving into or out of the shade), skin blood vessels (increasing or decreasing skin blood flow), sweat glands (sweating), and respiratory system (panting).

Vertebrate Temperature Regulation

In the space of one chapter it is clearly impossible to go into any detail concerning the regulation of body temperature in the vertebrates. For more detail, the reader is advised to see, for example, Bligh (18), Ingram and Mount (28), or Kluger (29). For the purposes of this chapter, it is sufficient to state that thermal receptors or sensors have been located in the skin, abdomen, veins, hypothalamus, midbrain, and spinal cord of many species of vertebrates. Data obtained by various types of experimental procedures have led to the theory that the regulation of body temperature depends upon neural inputs from thermal sensors located peripherally (skin) as well as from thermal sensors in the core (hypothalamus, spinal cord, etc.).

Within the past dozen years there have been numerous comparative studies concerned with the role of the hypothalamus in thermoregulation in the vertebrates. Thermodes have been placed into the hypothalamus of organisms from fishes through mammals and, with the possible exceptions of the big brown bat (*Eptesicus fuscus*) (30) and the California quail (*Lophortyx californicus*) (31), the hypothalamus has been found to be thermally sensitive in all these groups. In an endotherm like the house sparrow, warming the hypothalamus led to physiological and behavioral responses that reduced core temperature (32). Cooling the hypothalamus led to a rise in core temperature. In an ectotherm such as the blue-tongued lizard, warming the rostral brainstem (an area surrounding the hypothalamus) led to the lizard's selecting a cooler environmental temperature, resulting in a reduction in body temperature. Cooling the rostral brainstem led to the opposite results (33). In all these studies, peripheral inputs affected the responses initiated by hypothalamic heating and cooling.

Based on these types of data, models of thermoregulation were developed that attempted to predict the thermal responses of a vertebrate based on inputs from the skin and the hypothalamus. The relative importance or weighting given to these two areas apparently varies depending upon what specific thermoregulatory effector response one is measuring, as well as upon the species being studied. These models, however, often fail to take into account information obtained over the past few years that has shown that many internal areas other than the hypothalamus are also thermally sensitive.

In Table 1, I have summarized our present understanding of the comparative aspects of thermal sensation in the vertebrates. The locations of temperature sensors throughout the vertebrates are remarkably similar. The skin, the hypothalamus, the spinal cord, and perhaps other areas are temperature sensitive in virtually all the vertebrates. Based on these similarities, it is obvious that the distinction between the endothermic vertebrates (birds and mammals) and the ectothermic vertebrates (most

Table 1. Location of thermally sensitive areas in the vertebrates

	Skin	Hypothalamus	Spinal cord	Abdomen	Other CNS areas
Mammals	+[a]	+	+	+	+
Birds	+	+	+	?	?
Reptiles	+	+	+	?	?
Amphibians	+	+	+	?	?
Fishes	+	+	?	?	?

From Kluger (29).

[a] +, area shown to be thermally sensitive; ?, area not yet investigated.

fishes, amphibians, and reptiles) is not on the sensory side of the thermoregulatory reflex.

Little is known about how the thermoregulatory integrator actually works. Oftentimes the integration of afferent information for any regulatory system is conceptualized as a black box with arrows entering it (afferent signals) and others leaving (efferent signals). Our knowledge of the integration is at about this level of sophistication.

The area that is most often assigned the role of the thermoregulatory integrator is the hypothalamus. This region contains neurons that are sensitive to local temperature changes as well as to changes in temperature in other regions of the body (34, 35).

Data based on neuropharmacological studies have also implicated the hypothalamus as an important component of the thermoregulatory reflex. For example, infusions of drugs such as norepinephrine, 5-hydroxytryptamine, acetylcholine, and prostaglandins into the hypothalamus have led to changes in body temperature in many species of vertebrates (see, for example, refs. 36 and 37). It is not known whether these substances have any physiological role under natural conditions. Furthermore, it cannot be determined at present whether these drugs are affecting the known temperature sensors that are found in the hypothalamus or the "hypothetical" thermoregulatory integrators. Other studies, such as those involving lesioning (or ablating) of hypothalamic tissue, are similarly difficult to interpret.

Despite the difficulties in interpreting the data obtained from different experimental manipulations, it is generally believed that the hypothalamus does play a crucial role in the integration of body temperature. Other areas outside the hypothalamus are probably also important in the integration of thermoregulatory information (e.g., the spinal cord and medulla). Fairly sophisticated models of thermoregulation, generally based on the hypothalamus as the sole integrator, have been described and are reviewed in some detail by Bligh (18). These models are all theoretical and tend to underscore the fact that we know very little about how organisms actually integrate thermal information. In none of these models, however, is there any fundamental distinction between endotherms and ectotherms concerning the integration of body temperature. It is generally assumed that the sole distinction between endothermy and ectothermy resides in the effector component of the thermoregulatory reflex and not in the sensory or integrating limbs of this reflex.

The effectors, the third arm of the thermoregulatory reflex, are all those responses by the organism that can lead to changes in its rate of heat loss, heat gain, or heat production. These might involve physiological or behavioral responses or both. Without going into detail, some of the effector responses available to many vertebrates are: 1) changes in skin blood flow, 2) changes in evaporative water loss, 3) a variety of behavioral responses, and 4) changes in metabolic heat production. One cannot differentiate between endotherms and ectotherms based on effector responses 1 through 3 (29). The major distinction between endotherms and ectotherms involves response 4, the ability by endotherms to generate significant amounts of heat internally. Endothermy is probably one of the major forces that have affected vertebrate evolution; however, based only on the thermoregulatory reflex, it can be seen that the transition of vertebrates from ectothermy to endothermy would not have affected the sensory or integrating limbs of this reflex, and would have affected only one of the four general types of thermoregulatory effector responses.

This information concerning the thermoregulatory reflex becomes important when the biology and evolution of fever are discussed in the next sections of this chapter. This is because pyrogenic, or fever-inducing, substances do not affect the effector side of the thermoregulatory reflex, but rather the sensory and/or the integrating limbs, the phylogenetically conservative components of this reflex. It was for this reason that my colleagues and I speculated that vertebrates from fishes through mammals should be capable of developing a fever in response to infection with pathogenic organisms.

FEVER IN MAMMALS

The measurement of human body temperature to determine the health status of individuals has been widespread for over a hundred years. For thousands of years physicians have known that during diseases an individual's body temperature often became elevated; however, accurate measuring instruments were not available until the middle 1700's (38). It was Wunderlich in 1871 who, perhaps more than anyone else, firmly established the monitoring of body temperature as a diagnostic tool (39). In his book entitled *Medical Thermometry,* he argued that "the use of the thermometer in disease is . . . an objective, physical method of investigation, which gives exact and accurate results, in signs which can be measured and expressed numerically; which is delicate enough to follow every step of the changing processes of the organism, and places at the disposal of the practitioner a phenomenon dependent upon the sum total of the organic changes in the body." This book was filled with numerous observations of the body temperatures of patients during health and disease and was a major factor leading to the modern practice of systematically recording the body temperature of hospitalized patients.

Nature of Fever

About a hundred years ago Liebermeister suggested that fever was not the result of an inability of the infected organism to regulate its body temperature, but rather that the organism was simply regulating its body temperature at a higher level. This view of fever as a regulated higher body temperature was based on Liebermeister's experimental observations that the body temperatures of febrile subjects returned to their previously raised level after warming or cooling of the body. Liebermeister differentiated between passive rises in body temperature, as might occur during exposure to a hot environment or during heavy physical work, and fever. In fever, the person's heat production became elevated, while at the same time, his heat loss decreased (40).

Our current thoughts concerning fever support Liebermeister's beliefs. For example, Snell and Atkins (41) have defined four categories of body temperature based on the useful concept of thermoregulatory set-point. (The thermoregulatory set-point is the temperature, or temperature range, around which the animal attempts to regulate its body temperature.) These are:

1. Normothermia, which occurs when the thermoregulatory set-point and the actual body temperature are virtually the same (happens most of the time).
2. Hypothermia, which occurs when the thermoregulatory set-point may or may not be normal, but the actual body temperature is below this set-point (can take place in response to disease, drugs, or exposure to the cold).
3. Hyperthermia, which occurs when the thermoregulatory set-point may or may not be normal, but the actual body temperature is above this set-point (can occur in response to disease, drugs, or exposure to the heat).
4. Fever, which occurs when the thermoregulatory set-point is raised above normal and body temperature may or may not be raised to the same level.

Thus, we see from these definitions that an individual can be febrile and either normothermic (body temperature = set-point), hypothermic (body temperature < set-point), or more rarely, hyperthermic (body temperature > set-point). More recent work involving human beings (42, 43), dogs (44, 45), lizards (46), and other vertebrates (47–55) supports this concept of fever as an elevated thermoregulatory set point.

Biology of Fever

The rise in the thermoregulatory set-point is the result of a chain of events that is initiated by contact of the intact organism with some activating agent (56). These activators might be bacterial or viral in origin or they might be any substance that induces an antigen-antibody reaction leading to moderate-to-severe inflammation (e.g., hypersensitivity reactions). These activators are then thought to induce the production of pyrogenic proteins known as endogenous or leukocytic pyrogens. These pyrogenic proteins, which are produced by various types of immunologically active phagocytic cells, circulate to the central nervous system, where they induce some as yet unidentified series of events leading to the elevation of the thermoregulatory set-point. In the next sections of this chapter, these various components of the pathways leading to fever are discussed.

Activators of Fever

Endotoxins Of all substances that are known to serve as activators of fever, endotoxins have been the most intensively studied. Endotoxins are a component of the cell wall of Gram-negative bacteria. Injections with live or dead bacteria (containing endotoxins) lead to numerous effects, including leukopenia followed by leukocytosis, protection against the effects of irradiation, mobilization of interferon, enhancement of nonspecific immunological resistance, reduction in serum iron levels, and fever (57).

While endotoxins are made up of several components, it is thought that the lipopolysaccharide portion is the most important for initiating most of the events attributable to endotoxins. More specifically, it is thought that it is the lipid portion, often called lipid A, which is the most active biologically (58). Phagocytosis of the lipid A component of endotoxins triggers the release of endogenous pyrogen and results in the development of a fever.

For over a hundred years it has been known that dead Gram-negative bacteria are just as potent as live bacteria in initiating a fever (see the review in ref. 59). Even autoclaving a solution known to contain Gram-negative bacteria fails to reduce its pyrogenicity. This is because the lipopolysaccharide component of the bacteria's cell wall is resistant to the temperatures commonly encountered during sterilization. As a result of the common misconception that autoclaving destroys endotoxins, it is not uncommon for experimental biologists to inject their animals with solutions that are sterile, yet still capable of inducing all the effects of endotoxins obtained from living bacteria, such as inflammation and fever. Clearly, the injections of various doses of endotoxins into animals can lead to results that are spurious.

Injection into the laboratory rabbit, the most commonly used experimental animal in fever research (29), of endotoxin or dead or live Gram-negative bacteria results in a fever after a latency of 15–30 min (60). The latency to the onset of fever varies from species to species with a latency of about 1 hr in man and as much as 3 or 4 hr in the lizard and frog (Figure 1).

To compare fevers, Beeson (60) devised an arbitrary scale known as the fever index. This was done by taking as a baseline the animal's body temperature at the time of injection and then integrating the area enclosed by this baseline and the elevated or febrile temperature. The fever index thus yielded arbitrary units, which were useful when comparing fevers between different animals, or between the same animal at different times. Often one comes across the term "fever index$_{60}$" ("F.I.$_{60}$") or "F.I.$_{180}$," etc. This simply means that the integrated rise in body temperature was calculated over a 60- or 180-min period after the injection. Terms such as "an F.I.$_{60}$ of 250" or "of 2,175" were not uncommon. Since the F.I. is nothing more than the average fever (for any given time period) multiplied by some constant, I find it strange that the magnitude and duration of fevers were not simply expressed in terms of the average fever in degrees centigrade, rather than in some arbitrary terms. Clearly, comparisons between different laboratories

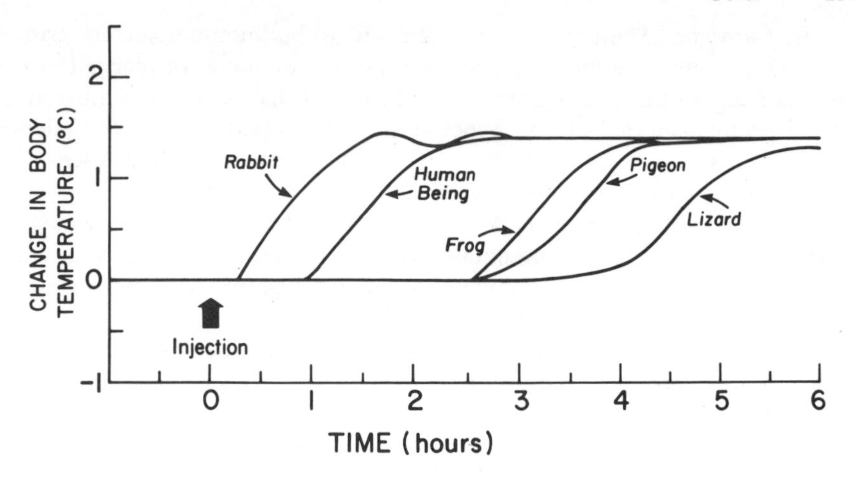

Figure 1. Diagram of the latency in fever response associated with injections of endotoxin into the rabbit, man, frog, pigeon, and lizard. Note that the rabbit develops a fever within 30 min after an injection of endotoxin, whereas this latency can be 4 hr or more in the lizard.

would have been easier with the use of the latter approach. In recent years the use of the arbitrary scale for measuring fever has been less frequent.

Gram-positive Bacteria Gram-positive bacteria are not thought to contain endotoxins. They are, however, strongly pyrogenic (61). A wide variety of Gram-positive organisms have been shown to be pyrogenic in man and in the laboratory rabbit. Atkins and Freedman (61) have shown that, as with Gram-negative bacteria, the Gram-positive ones are also pyrogenic after autoclaving. They found virtually no differences in the fevers induced in rabbits after injections of dead or live Gram-positive bacteria.

One intriguing difference between Gram-negative- and Gram-positive-induced fevers is that the onset or latency to the initiation of the fever is considerably longer with the Gram-positive infections. Hort and Penfold (62) reported that it often took several hours for fevers to develop after inoculation with Gram-positive organisms. Atkins and Freedman (61) found that after an intravenous injection with live or dead Gram-positive bacteria, fevers would develop within 45–60 min; this was considerably longer than the latency to onset of fever after injections with Gram-negative bacteria.

Viruses There have been few experimental studies involving virus-induced fever. I believe this stems from the inherent difficulties of working with viruses. Nevertheless, Wagner et al. (63) have shown that various species of viruses induce fevers in rabbits. These fevers began within 1–2 hr after intravenous injections of viruses and lasted for as long as 24 hr. The pyrogenicity of these viruses (two strains of influenza and one strain of Newcastle disease virus) was destroyed by heating to temperatures as low as 60 °C or 70 °C.

Hypersensitivity Reactions After an initial contact with an antigen, an organism slowly produces antibodies and/or sensitized lymphocytes specific for that antigen. The organism is then said to be immunologically primed (64). Occasionally, upon subsequent exposure to these antigens, the immunological response is heightened to the extent that it becomes harmful to the host organism (65). These hypersensitivity reactions can lead to edema, swelling, tissue necrosis, and many other side effects, including fever.

Hypersensitivity reactions are thought to be responsible for many fevers seen clinically. These types of fevers occur in sensitized patients, who then come in contact with such infectious agents as bacteria, viruses, protozoans, and fungi, or to such noninfectious agents as drugs, toxins present in some ingested food, transplanted tissues, etc. (66).

Tumors Fever is also associated with many types of malignancies. While these fevers are sometimes attributed to secondary bacterial or viral infections, often they occur without any identifiable infectious agent. According to Bodel (67), several hypotheses have been proposed to explain tumor-induced fevers. Some of these are: 1) the production of a toxin from the tumor, 2) tissue necrosis resulting in the release of some pyrogenic substance, and 3) an undiagnosed infection. Therefore, the activator involved in tumor-induced fever might be attributable to some bacterial or viral agent or to a hypersensitivity reaction.

Endogenous Pyrogens

All the activators of fever are thought to induce the formation of protein mediators of fever called endogenous pyrogens. Endogenous pyrogens are known to be produced by many different types of immunologically active phagocytic cells such as neutrophils, eosinophils, monocytes, Kupffer cells, and macrophages (68). (In the past, the name "leucocytic pyrogen" was often used for the pyrogenic protein induced by activators of fever. However, since it is now known that many cells besides leukocytes produce pyrogenic proteins, "endogenous pyrogens" has become the preferred name for these substances.)

After an injection of some activator of fever there is some latency period before the organism becomes febrile. This latency, which can be from 15 min up to several hours, is thought to be related to the time necessary for the various cells listed above to engulf and phagocytize the activators and to then release their endogenous pyrogens.

In order to understand the biological properties of endogenous pyrogens, it is often helpful to compare these protein pyrogens with some of the activators of fever such as endotoxins or Gram-positive bacteria. One major difference between endogenous pyrogens and endotoxins, for example, is that endogenous pyrogens are heat labile. That is, since these substances are proteins, they denature and lose their pyrogenicity at elevated temperatures (as low as 56°C). Endotoxins and Gram-positive bacteria are considerably more heat-stable and can withstand temperatures normally associated with autoclaving (ca. 120°C) without losing their ability to induce fevers.

Another distinction between endogenous pyrogens and the activators of fever is that the latency to the onset of fever is considerably shorter after an injection of endogenous pyrogen. The explanation for this reduced latency with endogenous pyrogens is, of course, that the various activators must first stimulate the production and release of endogenous pyrogen before the fever can develop. Therefore, when endogenous pyrogen is injected, the first step has been removed in the pathway toward the development of fever.

Another important distinction between endogenous pyrogens and the activators of fever is that animals develop little tolerance to repeated injections of endogenous pyrogens. Most species develop some tolerance, or diminished fevers, in response to repeated injections with various activators of fever such as endotoxins and viruses. The question of how tolerance develops appears to be related to how rapidly the pyrogenic materials are cleared, or removed, from the circulation. This subject is discussed in considerably more detail in the reviews by Snell (66), Atkins and Bodel (69), and Kluger (29).

Endogenous pyrogen can be produced in vitro by several methods. One of these involves obtaining white blood cells from the abdominal cavity. In an animal the size of a laboratory rabbit (3 kg) about 400 ml of sterile pyrogen-free saline solution is injected. After about 12 hr this fluid is removed and is generally found to contain a large number of polymorphonuclear leukocytes (70, 71). These exudate leukocytes have already been activated, and incubation of these cells at 37°C for several hours results in the release of endogenous pyrogen into the incubation medium (see, for example, ref. 72). Often an additional stimulant, such as shellfish glycogen, is added to the saline infusate (73), tending to increase the yield of endogenous pyrogen. This peritoneal exudate method of obtaining endogenous pyrogen has been used successfully in rabbits, cats, dogs, goats, and lizards (72, 74).

Another in vitro method for producing endogenous pyrogen uses white blood cells obtained directly from the blood. These leukocytes, unlike those obtained from the exudate method described above, are not activated. As a result, incubation of these cells in saline or some other medium does not yield appreciable amounts of endogenous pyrogen. The addition of some activating agent such as dead or live bacteria results in the increased production and release of endogenous pyrogen (75). With the use of this method, white blood cells obtained from rabbits and human beings have been shown to produce endogenous pyrogens. By using similar methods, Kupffer cells, macrophages, and other immunologically active phagocytic cells have also been shown to produce endogenous pyrogens (68).

Site(s?) of Action of Endogenous Pyrogens

There is considerable evidence that endogenous pyrogens affect the central nervous system. Although this might sound like a trivial statement, until the concept that fever was a raised thermoregulatory set-point was accepted, it could have just as easily been imagined that pyrogenic substances altered the thermoregulatory effectors at the peripheral level. For example, during some types of drug-induced hyperthermia body temperature becomes elevated as a result of the failure to effectively dissipate adequate amounts of heat. The drugs might be acting directly on effectors, such as the sweat glands (to diminish sweating), or the peripheral vasculature (to prevent vasodilation). But experiments with species that thermoregulate by behavioral means, such as lizards (46, 47), frogs (48–50), and fishes (51–55), clearly have shown that during a fever the organism actively raises its body temperature, demonstrating a central nervous involvement in the febrile process.

The first direct evidence that the brain was involved in the development of fever was by King and Wood (76). These investigators thought that, if the brain were involved in fever, then the injection of an endogenous pyrogen into an artery leading toward the brain should produce a fever of shorter latency and greater magnitude than the injection of an endogenous pyrogen into a vein (resulting in the dilution of the endogenous pyrogen). If endogenous pyrogens caused fevers by affecting tissues peripheral to the brain, then there should be either little difference in the fevers produced through the two routes of injection, or alternately, the intravenous route should actually lead to fevers of shorter latency and of greater magnitude. King and Wood found that an injection into the arteries leading to the brain did produce fevers of shorter latency and greater magnitude, confirming that the brain was involved in the development of fever.

Experiments by Adler and Joy (77), Cooper et al. (78), and Jackson (79) have provided direct evidence that the brain is involved in the development of fever. For example, Cooper et al. injected small quantities of endogenous pyrogens into different areas of the brains of rabbits and induced fevers with doses of endogenous pyrogens that were less than one-hundredth of those required to produce fevers by intravenous injections. The areas most sensitive to the endogenous pyrogens were the preoptic area and the anterior hypothalamus (POAH). Rosendorff and Mooney (80) reported that the brainstem of rabbits was also sensitive to endogenous pyrogens. They injected minute amounts of endogenous pyrogens into various regions of the brain and found that, whereas the greatest fevers were produced after microinjections into the POAH, the brainstem was also sensitive to the endogenous pyrogens. Other areas, such as the posterior hypothalamus, were insensitive to these proteins.

Based on these data it has been speculated that endogenous pyrogens cross the blood-brain barrier in the region of the POAH or brainstem or both. In an attempt to confirm this hypothesis, Allen (81) produced endogenous pyrogens in a medium that contained radioactively labeled iodine. An injection of radioactively labeled serum from control animals led to no detectable traces of radioactivity within the brain. However, within 60 min after an injection of serum containing endogenous pyrogens produced in the medium containing the radioactive iodine, an area in the posterior hypothalamus could be seen to be radioactive (using radioautographs). No radioactivity was found in the areas shown to be sensitive to endogenous pyrogens. It was not known, however, whether the radioactivity could be attributed specifically to the endogenous pyrogens.

More refined techniques have recently become available and the answer to the question of where, specifically, endogenous pyrogens act might soon be resolved. The exciting breakthrough to which I am referring is the development of radioimmunoassay for endogenous pyrogens by Dinarello et al. (82). This technique enables one to identify minute amounts of endogenous pyrogens, and should allow Dinarello and his co-workers to determine precisely where these proteins leave the circulation (if at all) and enter the central nervous system. It is possible that under normal physiological conditions endogenous pyrogens exert their effects on the central nervous system via some intermediate messenger, and as such do not actually enter the brain.

To add to the confusion concerning the role of the POAH in the development of fever, there have been a series of papers that has shown that lesioning or removing the POAH has little effect on febrogenesis. For example, Veale and Cooper (83) reported that, after removal of the entire POAH of rabbits, an intravenous injection of endogenous pyrogen still produced fever of similar magnitude to that found in control rabbits. Similar results have been reported by Andersson et al. (84) in goats and by Lipton and Trzcinka (85) in monkeys. Lipton and Trzcinka reviewed much of the literature relating to the febrile responses in POAH-lesioned animals and have concluded that the control of fever cannot be localized solely in the POAH. The results of Rosendorff and Mooney (80), which showed that the brainstem is also sensitive to pyrogens, might provide the explanation for these apparently paradoxical results. It is, nevertheless, unsettling to learn that one can remove such a large chunk of the brain—the entire POAH (a region that has been shown to be thermally sensitive, responsive to pyrogens, and implicated in the integration of thermal information)—and produce only minimum deficits in the febrile response.

Prostaglandins and Fever

Regardless of whether endogenous pyrogens actually enter the central nervous system, there is considerable evidence that these proteins either directly or indirectly trigger an elevation in the thermoregulatory set-point. One proposed group of intermediaries between endogenous pyrogens and this raised set-point is the prostaglandins. Prostaglandins are a group of lipid acids that have been implicated in many physiological processes, including fever.

Milton and Wendlandt (86) were probably the first to propose that pyrogens might induce a fever via the production of specific prostaglandins and that drugs that reduce a fever (antipyretic drugs) might do so by blocking the synthesis and release of prostaglandins. This seemed at the time to be an attractive hypothesis. For example, Milton and Wendlandt showed that injection of minute amounts of prostaglandins E_1 and E_2 into the cerebral ventricular system of cats and rabbits produced a fever within a few minutes. Hales et al. (87) and Feldberg and Saxena (88) reported that intraventricular injections of prostaglandins also induced fevers in sheep and rats. Studies by Feldberg and Saxena (89) and by Stitt (90) showed that microinjections of prostaglandins into the POAH produced fevers in cats and rabbits, but microinjections into other areas of the brain (posterior hypothalamus and midbrain reticular formation) failed to produce a fever. Primates also respond to intracerebral or intrahypothalamic infusions of prostaglandins by developing a fever (85, 91, 92).

Another line of evidence that supports the prostaglandin-fever theory is that prostaglandins are known to be natural constituents of the hypothalamus (93, 94), and that, after injection of bacteria into the ventricular system or intravenously, samples of cerebrospinal fluid contained large amounts of prostaglandins (95, 96).

Perhaps the strongest link between prostaglandins and fever is the fact that antipyretic drugs, such as aspirin and indomethacin, are potent inhibitors of prostaglandin synthesis (97–99).

There have, however, recently been many indications that pathogen-induced fevers are not prostaglandin induced. Perhaps the first inkling that prostaglandins might not be the causative agent in fevers that are responses to pathogenic organisms came from the comparative studies of Baird et al. (100). These investigators reported that the primitive mammal, the echidna, would develop a fever in response to an intravenous injection of bacteria, but became hypothermic in response to intraventricular injections of prostaglandins. Pittman et al. (101) reported similar findings in newborn lambs. Microinjections of prostaglandins into the hypothalamus of lambs failed to produce a fever, whereas injections of bacterial pyrogens into these areas, or given intravenously, did result in a fever. In a subsequent study Pittman et al. (102) showed that injections of prostaglandins directly into the hypothalamus of adult sheep also failed to induce a fever. Injections of prostaglandins into the cerebral ventricles did, however, produce a fever in the adult sheep. These observations, taken by themselves, might not argue strongly against the role of prostaglandins in pathogen-induced fevers. However, as is discussed in a later section, the febrile response is remarkably ubiquitous throughout the vertebrates, and the failure of the prostaglandins to induce fevers in some species, or at some time during their life cycle, is difficult to reconcile with this conservativeness.

Another interesting finding that adds to the controversy over the role of prostaglandins in pathogen-induced fevers is that, after lesioning of the POAH of rabbits, an intravenous injection of endogenous pyrogen still produced a fever; however, injections of prostaglandins into the area of the lesioned POAH no longer produced a rise in body temperature (83). Since injections of prostaglandins into other areas of the central nervous system of rabbits failed to raise body temperature (90), the above finding provides further evidence that prostaglandin-induced fevers and pathogen-induced fevers are not the same. Lipton and Trzcinka (85) reported essentially the opposite results. After POAH lesions in monkeys, intraventricular infusions of prostaglandins still led to fevers. These authors suggest that these different results may be related to species differences. Again, since fever seems to be so conservative, phylogenetically, it seems unlikely that there would be fundamental differences in the final pathway from endogenous pyrogens to the actual raising of the set-point. More likely, prostaglandin fevers simply do not share the etiology of pathogen-induced fevers.

Other data have further clouded the question of prostaglandins and pathogen-induced fevers. It has been shown by several laboratories that pyrogens alter the firing rate of thermosensitive neurons found in the hypothalamus (103–105). These investigators all found that pyrogens decreased the firing rate of warm-sensitive neurons at virtually all physiological temperatures. Cabanac et al. (104) also showed that pyrogens caused cold-sensitive neurons to increase their sensitivity to temperature. The antipyretic drug acetylsalicylate was shown to return the pyrogen-depressed, warm-sensitive neurons back toward their initial sensitivity (103). The results of these studies supported the raised set-point theory of fever, since at any given temperature, the firing rate of the warm-sensitive neurons was depressed and the firing rate of cold-sensitive neurons was elevated by pyrogens. This is what would happen if the organism were hypothermic and, as a result, the animal would initiate those responses that would tend to elevate its body temperature. Stitt and Hardy (106), however, found that microinjections of prostaglandins onto these thermally sensitive neurons had little, if any, effect. In fact, the effects were generally those of mild facilitation of the warm-sensitive neurons, the opposite of what one might have expected based on the results with bacterial pyrogens. Ford (107) did find that microinjections of prostaglandins excited cold-sensitive neurons in the decerebrate cat.

Lastly, a fourth type of experiment has added to the growing evidence that prostaglandins may not be a component of pathogen-induced fevers. Cranston et al. (108) have shown that intravenous infusions of endogenous pyrogens led to a rise in both the rectal temperature and the cerebrospinal fluid prostaglandin levels in rabbits. When an intravenous infusion of a prostaglandin inhibitor (sodium salicylate) was administered, it blocked the elevation in prostaglandin levels in the cerebrospinal fluid, but not the rise in rectal temperature. They then showed that an intraventricular infusion of a drug that was a potent prostaglandin antagonist (SC 19220) blocked prostaglandin-induced fevers, but not fevers in response to endogenous pyrogens.

In 1977, Laburn et al. (109) presented data that might help to explain some of the conflicting evidence presented above. They found that the precursor of prostaglandins, arachidonic acid, actually leads to the production of at least two pyrogenic substances—prostaglandins and either prostaglandin endoperoxide or thromboxanes (Figure 2). An intraventricular injection of the sodium salt of arachidonic acid resulted in a dose-dependent fever in rabbits. As might be expected, this fever was blocked by indomethacin. However, the use of the prostaglandin antagonist SC 19220, which effectively blocked prostaglandin fevers, had little effect on fevers after the injection of arachidonic acid. Laburn et al. concluded that some other breakdown product of arachidonic acid, other than prostaglandins, was also pyrogenic. Additional experiments, similar to those performed above for prostaglandins, must be performed before any of these arachidonic breakdown products can be assigned a role in pathogen-induced fevers.

To summarize, prostaglandins are released in response to many different kinds of insults, including crushing injuries, thermal burns, infections, etc. (111). It is possible that during most pathogen-induced fevers, the blood levels of prostaglandins are too low to have a major role in the development of fever. In these types of fevers, endogenous pyrogens are released, and these proteins could operate through some, as yet unidentified, intermediaries to raise the thermoregulatory set-point. However, during severe trauma, prostaglandins or other metabolites of arachidonic acid, may be released in sufficient quantities to synergistically raise the organism's set-point. In fact, in response to some types of injuries (perhaps internal head wounds), the levels of endogenous pyrogens might be minimal and the bulk of the fever might be attributable to the elevated levels of archidonic acid breakdown products. Clearly more work needs to be done before a specific role for these substances in pathogen-induced fevers can be accepted or rejected.

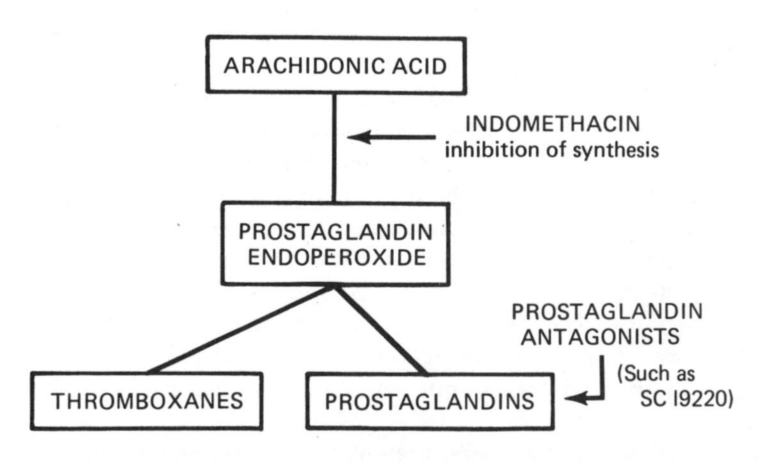

Figure 2. Diagram of the formation of degradation products of arachidonic acid. *Arrows* indicate position of action of indomethacin and of antagonists of prostaglandins. Reproduced from Hamberg et al. (110) and from Laburn et al. (109) with permission of *Journal of Physiology* (London).

Cyclic AMP and Fever

Many hormones and neurotransmitters, including some of those involved in thermoregulation, are thought to exert their effects through the activation or formation of cyclic 3′,5′-adenosine monophosphate (cyclic AMP) (112). There is even some evidence that cyclic AMP is involved in febrogenesis. Injections of pyrogens have been shown to increase the levels of cyclic AMP in the cerebrospinal fluid of cats and rabbits (112, 113). Inhibitors of the nucleotide phosphodiesterase, the enzyme responsible for the destruction of cyclic AMP, lead to a potentiation of fever in rabbits (but not in cats) (114). Similar results have been reported by Woolf et al. (115). When cyclic AMP was injected into the central nervous system of rats and rabbits, a fever generally developed; however, in cats, this generally led to a fall in body temperature (116).

As in the case of arachidonic breakdown products such as prostaglandins, it is not known whether the effect of cyclic AMP on the thermal responses of rats and rabbits has any physiological role in pathogen-induced fevers. The fact that cyclic-AMP does not initiate fevers in cats, but does so in rats and rabbits, decreases the likelihood that this important second messenger for many hormones and neurotransmitters has a specific role in febrogenesis.

Na^+/Ca^{2+} and Fever

Since the early 1900's it has been known that the infusion of various cations such as Ca^{2+} or Na^+ affected body temperature (see review in ref. 117). There was little experimentation in this area concerning the possible ionic control of temperature regulation until 1970 when Feldberg et al. (118) showed that a fever developed when a solution of isotonic sodium chloride was perfused through the cerebral ventricles of unanesthetized cats. If calcium was added to this perfusate (in its normal physiological concentration), the rise in body temperature was blocked. Feldberg et al. suggested that these ions might play a role in adjusting the thermoregulatory set-point upward or downward. These initial observations stimulated a series of experiments in many laboratories to determine whether these ions had a physiological role in thermoregulation and fever.

While space does not allow a lengthy discussion of the role of cations in febrogenesis, I briefly attempt here to summarize some of the arguments for and against this theory. In some, but not all mammals, an alteration in the hypothalamic concentrations of these ions led to predictable changes in body temperature (119–121). Generally, as the ratio of Na^+ to Ca^{2+} increased, body temperature rose; when this ratio decreased, body temperature fell.

While it is not known what specific effects the changes in these cations might have on the neurons responsible for elevating set-point during fever, the work of Hensel and Schafer (122) has raised some interesting possibilities. While investigating peripheral thermal receptors in cats, Hensel and Schafer found that an excess of Ca^{2+} increased the firing rate of peripheral warm-sensitive neurons and decreased the firing rate of peripheral cold-sensitive neurons (Figure 3). Clearly, the removal of Ca^{2+} would then lead to a decrease in the firing rate of the warm-sensitive neurons and an increase in the firing rate of the cold-sensitive neurons. If similar changes occurred in the central nervous areas responsible for febrogenesis, then these changes in the firing rates of neurons could account for the change in set-point during fever.

A potentially serious problem exists in virtually all the experiments that have shown that either body temperature or neuronal activity is affected by changes in Na^{+}/Ca^{2+}. The changes in the Na^{+}/Ca^{2+} necessary to produce these alterations are well above normal physiological levels. Robertshaw and Beier (123) have suggested that it is the ionic levels of calcium that might be critical to the changing of the thermoregulatory set-point and not simply the total calcium levels. In most of the studies in which sodium or calcium ions are infused, what has been measured has been the total sodium and calcium, which includes the bound sodium and calcium as well as the free ions. Whereas almost all the sodium exists in its ionic state, this is not true for calcium (124). Since it is probably the ionic sodium or calcium that affects the temperature-sensitive neurons, the measurements of these ionic concentrations are critical.

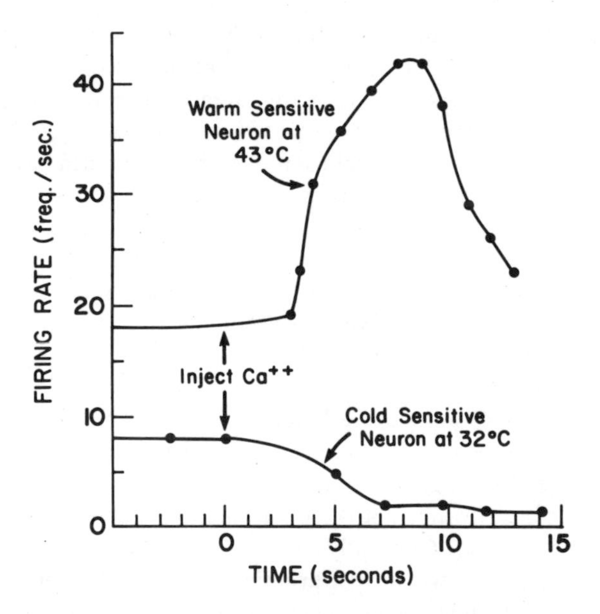

Figure 3. Discharge frequency of representative warm- and cold-sensitive neurons. Within a few seconds after adding Ca^{2+} to the medium surrounding these neurons, the firing rate of the warm-sensitive neurons increased and the firing rate of the cold-sensitive neurons decreased. Reproduced from Hensel and Schafer (122) and from Kluger (29) with permission of Princeton University Press.

Another, often used explanation for why such high levels of cations are required to produce changes in body temperature is related to the hypothesized location of the neurons that are involved in set-point changes. If these neurons are not located near the ventricles of the brain, then changing the cerebrospinal fluid levels of Ca^{2+} (or Na^{+}) severalfold might only slightly raise the concentrations of these ions near these neurons. Clearly more work is needed before the Na^{+}/Ca^{2+} hypothesis of febrogenesis can become generally accepted.

Evolution of Fever

In order to trace the phylogeny of fever, it is first necessary to have some understanding, for comparative purposes, of the nature of fever in mammals. A brief recapitulation will aid in the interpretation of the data obtained from the nonmammalian vertebrates.

Fever in mammals results from the contact of the host's immunologically active phagocytic cells (e.g., leukocytes, Kupffer cells, etc.) with some activating agent. These activators might be bacterial, viral, or fungal in origin, or in special cases might be the result of some type of hypersensitivity reactions. In any event, in response to these activators, the host's own cells release small proteins known as endogenous pyrogens, which circulate to the brain (hypothalamus?, midbrain?) and trigger a rise in the thermoregulatory set-point. How the set-point becomes elevated is still an area of intensive investigation, with substances such as prostaglandins (or other arachidonic acid breakdown products), cyclic AMP, and Na^{+}/Ca^{2+} being some of the putative substances in the link between endogenous pyrogens and the elevation in the set-point for body temperature.

It is presently unclear how antipyretic drugs such as sodium salicylate, indomethacin, and acetaminophen reduce fevers. While this has not been discussed in the preceding pages, I will simply state that the bulk of the evidence supports the hypothesis that these drugs exert their antipyretic actions on some central nervous sites, rather than by working peripherally on the production and/or release of endogenous pyrogens (29).

With this brief summary of fever in mammals, the next sections review what is currently known about the febrile responses in other vertebrate classes.

FEVER IN BIRDS

There have been few papers on the thermal responses of birds after injections with substances that are known to be pyrogenic to mammals. In one of these studies it was reported that injections of endotoxins isolated from *Escherichia coli* bacteria in chickens led to a rise in body temperature of about 0.6 °C within about 3 hr of the injection (125). Pittman et al. (126), however, reported that they were unable to induce fevers in chickens after injections of endotoxins isolated from the bacterium *Salmonella abortus equi.*

In fact, when they injected high doses of the endotoxins the body temperatures of their experimental animals actually fell during the 2nd hour postinjection. The experiments by Pittman et al. were terminated at the end of 3 hr.

D'Alecy and Kluger (127) reported that the bacterium *Pasteurella multocida*, a common avian pathogen, induces long-lasting fevers in pigeons. Since there were few reports concerning the febrile responses of birds, the body temperatures of these animals were continuously monitored for periods of up to 1 week. Injections of low doses of alcohol-killed bacteria (still containing endotoxins) into these birds led to fevers after a latency of about 3 hr. When higher doses of dead bacteria were injected, this led to a fall in body temperature similar to that observed by Pittman et al. This fall in body temperature was then followed by a series of oscillations in body temperature, and then, depending upon the dose, a fever developed within 5–10 hr after the injection. It is not known what caused the initial fall and subsequent series of oscillations in body temperature.

When pigeons were injected with live bacteria, they developed fevers lasting several days. The fever generally began within 3–4 hr after the injection and was maintained until the birds recovered from the infection or died. Oftentimes, just prior to death, their body temperature rose sharply to as high as 45 °C.

There have been, to my knowledge, no carefully controlled studies concerning the thermal effects induced by other activators (e.g., viruses) in birds. It is also not known, at present, whether the immunologically active phagocytic cells of birds produce endogenous pyrogens. However, since both mammals and reptiles have been shown to produce a heat-labile pyrogenic protein in response to activators of fever, it is likely that birds (which also evolved from reptiles) will eventually be shown to produce this substance.

In an attempt to elucidate the intermediaries between the activators of fever and the actual rise in set-point, several laboratories have injected prostaglandins into the hypothalamus of birds. These studies have shown that the injection of prostaglandin E_1 into the hypothalamus of chickens induces a dose-dependent rise in body temperature (126, 128, 129). When low doses of prostaglandins were injected into other areas of the brain, no fever developed. As in mammals, it is not known whether prostaglandin-induced fevers have any relationship to pathogen-induced fevers.

Lastly, the oral administration of antipyretic drugs such as acetylsalicylic acid or sodium salicylate led to the attenuation of bacterially induced fevers in both pigeons and chickens (126, 127).

FEVER IN REPTILES

Most reptiles regulate their body temperatures to varying degrees of precision by behaviorally selecting warmer or cooler microthermal habitats. Often subtle changes in body posture or color can lead to marked changes in the rate of heat transfer between the reptile's body and its environment. Thermoregulation in reptiles has been reviewed many times and the interested reader is referred to the excellent review by Templeton (130).

An extremely important aspect of thermoregulation in most reptiles is that in their natural environment, where the environmental temperatures can often vary by over 30 °C between day and night, the regulation of body temperature is an active process. This means that, in order to maintain a fairly narrow body temperature in the natural world, these animals must either shuttle between the sunlight and shade, elicit subtle or gross changes in their body posture, or employ other behavioral and/or physiological responses to regulate their body temperature.

To study the febrile responses of reptiles, the desert iguana (*Dipsosaurus dorsalis*), a lizard that is known to thermoregulate fairly precisely, was selected (131, 132). In later studies the green iguana (*Iguana iguana*) was also used. In most of these experiments the lizards were in a simulated desert environment of about 2 m^2. During the day the room temperature was controlled at about 30 °C for the desert iguanas and at about 25 °C for the green iguanas. At night, the environmental temperature fell. Suspended above the chamber was a series of heat lamps that were timed to go on and off at different times during the day. As a result, the sand temperature beneath these heat lamps (when they were on) would reach a temperature as high as 50–55 °C. Therefore, during the day, the desert iguanas could select an environmental temperature of from 30–55 °C, and the green iguanas a temperature of from 25–50 °C.

These species of lizards develop fevers in response to injections of various species of bacteria, such as *Aeromonas hydrophila, Pasteurella hemolytica,* and *Citrobacter diversus* (all pathogens of reptiles) (Figure 4). After injections with these bacteria, fevers generally developed within 3–4 hr and often lasted for several days (46, 47, 133). During this time, the lizards would select a higher environmental temperature, and as a result, their body temperatures would become elevated above their afebrile or control levels. If these lizards were placed in a constant temperature chamber, they were not able to raise their body temperatures and, as a result, their body temperatures did not rise after the injections with bacteria. These data indicate that the lizards were behaving as though their thermoregulatory set-point had become elevated (e.g., febrile). As a result, they actively sought out higher environmental temperatures, producing an elevation of their body temperature to match their elevated set-point.

Reptiles can also produce an endogenous pyrogen-like substance in response to activation of leukocytes (74). Using the peritoneal exudate method of producing leukocytes (see under "Endogenous Pyrogens"), Bernheim and Kluger were able to produce a heat-labile pyrogenic substance that induced fevers in lizards that lasted about 5 hr. When this pyrogenic material was denatured (by heating to about 80 °C) it completely lost its pyrogenicity. In this same series of experiments, it was also shown that endogenous pyrogen

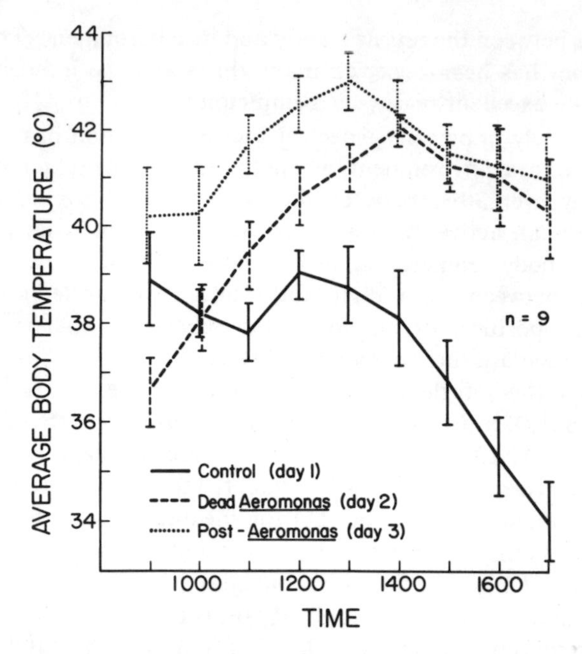

Figure 4. Effects of injection of nine desert iguanas (*D. dorsalis*) with dead *A. hydrophila* on body temperature (mean hourly temperature ± 1 S.E.). On control days lizards had their highest body temperature at midday and their lowest in the late afternoon. After inoculation with *A. hydrophila* (at 0900 hr), lizards developed a fever within 3–4 hr. This led to an elevation in body temperature throughout the 2nd and 3rd days without the normal late afternoon decline in body temperature. Reproduced from Bernheim and Kluger (133) with permission of *American Journal of Physiology*.

isolated from rabbits produced fevers when injected into either rabbits or lizards. These data demonstrated that endogenous pyrogens have cross-class reactivity; that is, they induce a fever when injected into another species from a different class of organisms.

It is currently unknown what the intermediaries are between endogenous pyrogens and the elevation in the thermoregulatory set-point in reptiles. It is known, however, that the drug sodium salicylate is an effective antipyretic in desert iguanas infected with *A. hydrophila.* After the administration of the appropriate dose of sodium salicylate, the lizards selected a cooler environment and, therefore, their body temperatures fell to the control or afebrile levels (133).

FEVER IN AMPHIBIANS

As with reptiles, many species of amphibians regulate their body temperature largely by behavioral means. The regulation of body temperature in amphibians tends to be less precise than in reptiles, and this might be related to problems of salt and water balance in amphibians. For example, many amphibians have a moist skin that is used for, among other things, exchanges of gases. As a result, water is more easily lost from the skin of amphibians than from the tougher, scalier skin of reptiles or from the feathered or furred skin of birds or mammals, respectively. In order for amphibians to regulate their body temperature to within narrow ranges, large amounts of water would often be lost, resulting in the internal concentration of various solutes. As a result, I believe that selection pressures have likely resulted in some compromise between the regulation of body temperature and the regulation of salt and water.

To determine whether amphibians develop fevers in response to infection, tadpoles of a terrestrial species, the green tree frog (*Hyla cinerea*), and two aquatic species, the leopard frog (*Rana pipiens*) and bullfrog (*Rana catesbeiana*), were injected with *A. hydrophila* and their thermal responses were monitored (48, 49). In addition, the frog *Rana esculenta* was injected with various species of killed Mycobacteria (50). All species of frogs developed fevers after the injections of bacteria. As in reptiles, these elevations in body temperature were the result of the febrile organisms' selection of a warmer microthermal habitat.

Myhre et al. (50) obtained preliminary evidence that the blood of injected frogs produces an endogenous pyrogen-like substance. These investigators also have shown that injections of prostaglandins into the diencephalon of frogs led to fevers after a short latency.

FEVER IN FISHES

Many fishes also regulate their body temperatures by behavioral modifications. Reynolds and his colleagues at Pennsylvania State University (51–55) have been investigating the febrile responses of fishes and have found that injections of various species of bacteria into bluegill sunfish, largemouth bass, and goldfish lead to fevers of 2–3 °C. These fevers are attenuated by the drug acetaminophen, a commonly used antipyretic drug for mammals. As occurs in all other vertebrates, the fishes actively raise their body temperatures (in these cases by behavioral means) to the level of the raised, or febrile, thermoregulatory set-point.

To briefly summarize, five of the seven extant classes of vertebrates have been tested for their febrile responses. In all five classes, it has been demonstrated that fevers develop in response to such common activators as Gram-negative bacteria. Table 2 summarizes much of our current understanding of the comparative aspects of fever in the vertebrate subphylum. Based on the many similarities in the febrile responses among the vertebrates, it is likely that the febrile response is a primitive trait, having a long phylogenetic history. Just how far back fever will be traced is presently unknown. Casterlin and Reynolds (134) have recently shown that even crayfish (an invertebrate belonging to the phylum Arthropoda) develop fevers in response to infection with bacteria. It is likely that other species of thermo-

Table 2. Febrile responses of the vertebrates

	Fishes	Amphibians	Reptiles	Birds	Mammals
Live bacteria produce fever	+[a]	?	+	+	+
Dead bacteria produce fever	+	+	+	+	+
Endogenous pyrogen produced	?	+	+	?	+
Prostaglandins produce fever	?	+	?	+	+
Drugs induce antipyresis	+	?	+	+	+

From Kluger (47).
[a] +, positive response; ?, response not yet tested.

regulating invertebrates also develop fevers in response to infection. In fact, the ability of an organism to raise its thermoregulatory set-point in response to some infectious agent might turn out to be a fundamental characteristic of almost all organisms capable of regulating their body temperature.

THE FUNCTION OF FEVER

For thousands of years fever was thought to be beneficial. For example, Hippocrates (460–377 B.C.E.) thought that the fever that accompanied most diseases aided the individual in combating the infection. This belief was founded on the humoral theory of disease. In this theory, the human body was composed of four humors—blood, phlegm, yellow bile, and black bile. According to Yost (1), the humoral theory of disease maintained that illness was caused when one of the four humors was produced in excessive amounts. Once this happened, the body's defenses came into play, as demonstrated by the patient's raised body temperature or fever. This excess humor was then "cooked," separated, and eventually evacuated from the body. So strong was this belief that fever was a beneficial host defense mechanism that the well known physician Rufus of Ephesus (ca. 100 A.C.E.) wrote, "I think that you cannot find another drug which heats in a more penetrating manner than fever; for this reason it is a good remedy for an individual seized with convulsion, and if there were a physician skillful enough to produce a fever, it would be useless to seek any other remedy against disease" (135). Rufus advocated the use of fever therapy to treat many different types of diseases. This practice was to reappear in the late 1800's and early 1900's to treat several diseases, including syphilis and gonorrhea. While space does not permit a detailed discussion of fever therapy, I will simply say that the results obtained from this therapeutic procedure have no bearing on the question of whether fevers that occur during infections have survival value to the host. In fever therapy, the patient is subjected to high environmental temperatures, or injected or infected with fever-inducing organisms (in some cases with malaria). While these treatments are often successful, they do not answer the question of whether the fever that is induced by the initial pathogen results in enhanced host defense mechanisms. For more detail on this subject, see Kluger (29).

The humoral theory of disease greatly simplified the practice of medicine because the physician could help his patient by simply aiding him to cook the humor. This could be done by warming the patient externally, or by administering drugs that raised his body temperature. Another option then available to the physician was to assist the body in evacuating the cooked humor, which could be accomplished by administering purgatives, emetics, sudorifics, or by bleeding (e.g., venesection, cupping). For more information on the history of bloodletting and the humoral theory of disease, see Kluger (136).

The idea that fever is a defense mechanism persisted through the Middle Ages and into the latter part of the 19th century. By the late 1800's the view that fever was adaptive began to change. It is probably not coincidental that the physicians' attitude toward fever changed at just about the same time that antipyretic drugs became readily available (137). I personally believe that the wide usage of such drugs as the salicylates, indomethacin, acetaminophen, etc. is not based on their antipyretic properties, but rather on their analgesic effects.

In any event, based on the present day widespread use of antipyretic drugs to treat all sorts of infections, it might seem that there must be fairly convincing evidence in the scientific literature that indicates that fever is harmful. This is not the case. Until recently there have been no experiments that could help to answer the question of whether fever is beneficial or harmful. The next section reviews some of the more recent literature on this subject.

Recent Experiments Concerning the Function of Fever

The question of whether fever is adaptive has been difficult to answer using endothermic vertebrates because, in order to manipulate the body temperatures of endotherms, one often has to perform rather drastic procedures. For example, a simple experiment would be to infect a population of endotherms with some virulent strain of bacteria or viruses. The population would then be divided into two groups; one group would be allowed to develop the normal fever and one group would be prevented from developing this fever. Differences in survival rate would be recorded and a statement about the role of fever in disease could be generated. The major technical problem in the above experimental design concerns the method of prevention of fever in these endotherms. One could place an animal in a cold environment, inject it with antipyretic drugs, warm its hypothalamus, etc. All of these procedures undoubtedly have side effects that would make interpretation of the survival results difficult to evaluate.

The use of the comparative approach has been a powerful tool to biomedical researchers. It was thought that, if an ectothermic vertebrate were used in the survival studies, the elevation in body temperature could be prevented without the difficulty in interpretation that might occur by using endothermic organisms. Because the febrile response is extremely conservative, in a phylogenetic sense, it was felt that the results obtained from experiments on behavioral thermoregulators, such as reptiles or fishes, could be extrapolated to the higher vertebrates, such as the birds and mammals.

We hypothesized that fever would be beneficial or adaptive for a reason that was not available to earlier investigators. Because the probable evolution of fever can be traced back several hundred million years, it seems likely that this phenomenon has been retained for these many millions of years in present day fishes, amphibians, reptiles, birds, and mammals. Since the development of a fever in either an endotherm or an ectotherm requires considerable expenditure of energy (just based on the Q_{10} effects of temperature on biochemical reactions), it is unlikely that this energy-expensive phenomenon would be so widespread among the vertebrates had it no beneficial role. Clearly, there is no selective pressure for a maladaptation; therefore, if fever were harmful to the infected organism, there is no apparent reason that it would have been retained for so many millions of years.

The first experiments concerned specifically with the role of fever in ectotherms were done with desert iguanas (138). These lizards were infected with live *A. hydrophila* and then placed in temperature-controlled chambers set at 34, 36, 38, 40, or 42 °C. In earlier studies we had shown that these lizards normally preferred about a 38 °C environmental temperature, but when they were infected with *A. hydrophila* they then selected a warmer environmental temperature of 40–42 °C (46). Control lizards were inoculated with saline and were also placed into the environmental chambers set at from 34 °C to 42 °C. The results of this study are shown in Figure 5. The relation between the body temperatures of these lizards and their percentage of survival after the injection with live bacteria was highly significant ($p < 0.005$). These results supported the hypothesis that the fever that occurs during bacterial infections has survival value.

Further support for this hypothesis was obtained by Bernheim and Kluger (139). In this study, the lizards were allowed to select their preferred body temperatures prior to and after injection with live *A. hydrophila*. The infected lizards developed a daytime fever averaging 2.3 °C over the 5-day period after the injection (mean body temperature was 40.6 °C). Their body temperatures returned to control levels by the 6th day. At night, the heat lamps in the simulated desert conditions were turned off and the lizards could not maintain a high body temperature. Even in these more natural conditions, only one of the 12 febrile lizards died. The 92% survival rate (11 out of 12 lizards) of these infected, febrile lizards was similar to that found in the study described above. Based on the data obtained in the simulated desert environment, it is clear that the lizards' body temperatures need not

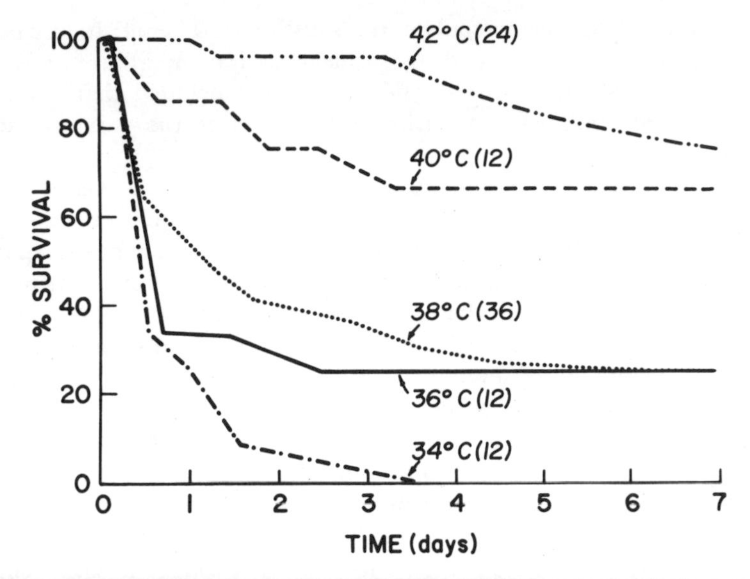

Figure 5. Percentage of survival of desert iguanas infected with *A. hydrophila* and maintained at temperatures of 34–42 °C. The number of lizards in each group is given in parenthesis. Reproduced from Kluger et al. (138) with permission of *Science*. Copyright 1975 by the American Association for the Advancement of Science.

be elevated continuously for them to experience an increased survival rate.

In another part of the study by Bernheim and Kluger (139), the antipyretic drug sodium salicylate was used to attenuate the fevers in these lizards. Lizards were again injected with live *A. hydrophila* and placed in the desert-like chamber. Administration of sodium salicylate led to antipyresis in seven lizards. All seven of these afebrile lizards died. Lizards that were given the same dose of bacteria and sodium salicylate, but were prevented from lowering their body temperature by being placed in a constant temperature chamber set at their febrile temperature during the daytime, had a mortality rate of only 12% (1 out of 8 lizards died). These data indicated that the administration of sodium salicylate to lizards infected with *A. hydrophila* was harmful only when it resulted in a reduction in body temperature. However, before it can be concluded that fever has a general survival function in reptiles, additional work needs to be done with the use of other pathogenic organisms and other species of reptiles.

Covert and Reynolds (140) investigated whether the fevers that developed after infection of goldfish with *A. hydrophila* (see refs. 51 and 53) increased their survival rate. These investigators infected goldfish and then held them at temperatures of 25.5, 28.0, or 30.5 °C. These represented, respectively, hypothermic, normothermic, and febrile temperatures. Goldfish maintained at a febrile temperature of 30.5 °C had a survival rate of 84%; those maintained at 28.0 °C had a survival rate of 64%; those at 25.5 °C had a survival rate of 24%. Another 10 fish were injected with the same dose of live bacteria and were allowed to thermoregulate in a shuttlebox. These fish developed a fever averaging almost 5°C and had a mean body temperature of 32.7°C. None of these fish died. Covert and Reynolds concluded that a fever in response to infection with *A. hydrophila* increases the survival rate of goldfish.

Clearly these few experiments that have used ectotherms as animal models to investigate the function of fever have shown that fever does have survival value to the infected host. Whether these results can be extrapolated to the endotherms is unknown. There have been several experiments that have attempted to ascertain the role of fever in disease by using mammals and birds, and, while these are difficult to interpret, they do provide some insight into the function of fever in endotherms. Some of these experiments and their results are described below.

One type of experiment performed with mammals correlates the magnitude of the fever that develops during an infection with the morbidity or mortality rate. In clinical studies involving human beings there is a tendency for the magnitude of the fever to be associated with the severity of the infection (see review in ref. 2). The difficulty with studies involving human beings is that they are completely uncontrolled. The patients have not been infected with identical doses of pathogens, nor have they been treated with identical doses of drugs. They are not matched according to age, weight, or background. To properly perform these studies on human beings, one must infect similar subjects with identical doses of pathogens, administer no drugs, and then compare the resultant fever with the mortality or morbidity rate. These experiments are obviously unethical and, therefore, animal experimentation has had to suffice.

The correlation between the magnitude of the fever during a bacterial infection and the survival rate has been done in New Zealand white rabbits by Vaughn and Kluger (141–143). Male rabbits that were all about the same age and from the same animal supplier were used in this study. The body temperature of each rabbit was monitored continuously before and after an injection of live *Pasteurella multocida,* a Gram-negative bacterium pathogenic to rabbits. The mortality rate of the rabbits was monitored over a 5-day period and all rabbits that were alive at the end of this time were considered survivors. To determine a rabbit's fever, its average temperature over the 24-hr period beginning 6 hr after injection of bacteria was compared to its average temperature on the control day. The 24-hr period was used because most of the mortality (76%) occurred during this period. The first 6 hr were excluded since the animals were just in the process of becoming severely ill during this time period. The fevers were then correlated with survival rate. The rabbits were grouped into fever ranges of 0.75°C because these were small enough to show clearly the relationship between fever and survival and yet large enough to enable us to perform statistical tests. The results of this study are shown in Figure 6. Most rabbits developed a fever of less than 2.25°C, and within this temperature range there was an increase in

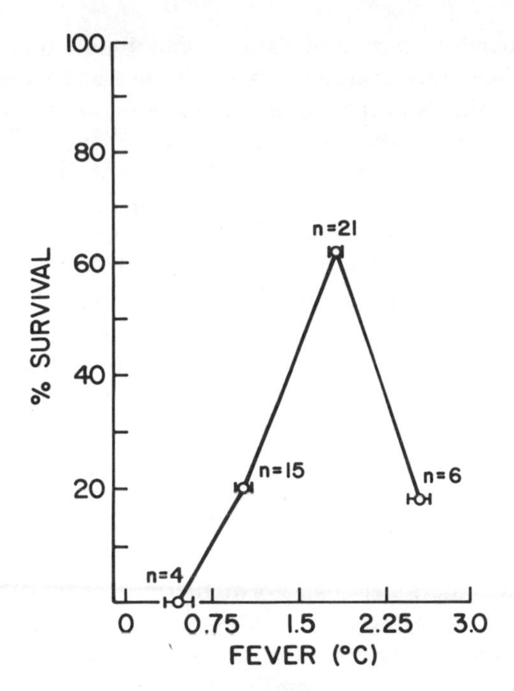

Figure 6. Correlation between the magnitude of fever and the percentage of survival of 46 rabbits infected with *P. multocida*. The difference in survival between rabbits with less than 1.5°C fevers is statistically significant. Reproduced from Vaughn (141).

survival rate as body temperature was elevated. A small number of animals ($n = 6$) developed fevers above 2.25°C and showed a decrease in survival rate. Overall, the difference in survival rate between animals with less than 1.5°C fevers and those with greater than 1.5°C fevers was statistically significant, in favor of those developing the higher fevers ($p < 0.02$, Chi-squared test).

In another correlation study, ferrets (*Mustela* sp.) were infected with different strains of influenza viruses and the resultant fever was correlated with the presence of live viruses in their nasal passages (144). Groups of up to six nonimmune ferrets were inoculated intranasally with a constant dose of virus and at 4-hr intervals the nasal passages were washed and the fluid assayed for the presence of live viruses. Statistically significant negative correlations were found between the ferret's rectal temperatures and the presence of live viruses in the nasal washes ($p < 0.01$). The results of both correlation studies (rabbits infected with *P. multocida* and ferrets infected with influenza virus) are consistent with the theory that fever has an adaptive value.

Another type of experiment that has been used to investigate the function of fever has been to infect an animal and to then induce either hypothermia or hyperthermia. For example, Kass and his associates have injected endotoxins into mice, rats, and rabbits and found that raising the environmental temperature resulted in hyperthermia and an increased susceptibility of these animals to the lethal actions of these endotoxins (145–147). In studies with rabbits, they found that when they induced hypothermia the rabbits became resistant to the lethal effects of the endotoxins. In this often cited study, they conclude that the "induction of fever increases the lethal action of endotoxin and conversely, the prevention of fever exerts a protective effect against such a lethal challenge" (147).

There is an obvious difficulty in attempting to apply the results of the above studies to answering the question of whether fever is beneficial or harmful during disease; the effects of injections of endotoxins are very different from those that occur during infection with bacteria or viruses. For example, one of the effects of large doses of endotoxins is a fall in blood pressure, or endotoxic shock (148). Clearly, the induction of hyperthermia (not fever) by Kass and his colleagues would cause the overheated animal to become peripherally vasodilated, undoubtedly exacerbating the effects of endotoxic shock. Furthermore, these hyperthermic animals would likely be losing considerably more water than normothermic (or hypothermic) ones, as the result of panting, salivation, sweating, etc. This reduction in total body water would further compromise the blood volume and make the return of the animal's blood pressure to preshock levels even more difficult. As a result, the effects of body temperature on endotoxin-induced mortality could have little to do directly with the effects of fever on mortality due to infectious agents.

Several other hyperthermic or hypothermic studies have been described by Bennett and Nicastri in their review article on the subject of the function of fever (2). One of these involved a study on pigeons that were infected with pneumococci (149). When the body temperature of these birds was reduced by ice or by the administration of drugs, their mortality rate increased. Similar results were reported by Muschenheim et al. (150) for pneumococcal infections in rabbits. In this experiment the infected animals were made hypothermic by one of several methods so that their rectal temperatures were between 30 °C and 34 °C. The control rabbits were infected and their body temperatures were maintained at normal-to-moderate febrile levels of 39–41 °C. All of the hypothermic rabbits died, whereas only five of the 31 control animals died.

Another type of hyperthermia-hypothermia experiment has been performed with neonates. For a long time it has been known that many newborn mammals have a fairly labile body temperature during their first few days of life (151). Many newborn mammals have a limited febrile response to infection (152–155). Haahr and Mogensen (152) believe that hyperthermia (or perhaps more precisely, a rise in body temperature) during certain viral infections is beneficial to newborns. To support their hypothesis, they cite several studies that have demonstrated that elevations in body temperature during various viral infections have reduced the mortality rate in newborn mice and dogs. In one of these studies it was found that when 2–3-day-old mice were infected with Coxsackie virus and held at an environmental temperature of 34 °C they had a mean body temperature of almost 36 °C, some 2–3 °C higher than control mice held at normal room temperature (22–24 °C) (156). The mice held at 34 °C had a lower mortality rate than did the control mice. Similar results were found by Carmichael et al. (157) for 2–5-day-old dog pups that were inoculated with canine herpes virus. Based on these results, Haahr and Mogensen (152) suggested that one of the reasons that generalized herpes simplex infections are greatly over-represented in premature babies might be their restricted temperature regulation and poor febrile responses.

While the experiments with endothermic vertebrates are difficult to interpret, most of these tend to support the hypothesis that the fever that accompanies most infections is a host defense mechanism. If this is the case, then the widespread use of antipyretic drugs and other agents, such as sponge baths, to reduce the body temperatures of febrile patients should be re-evaluated. Clearly, when a patient's body temperature is too high (perhaps over 41 °C) there is some danger that the fever itself can cause some harm. However, most fevers seldom reach these critical levels (158). For fevers of less than 40 °C or 41 °C, perhaps it would be best to allow the fever to remain, as this might enhance the patient's defense mechanisms. Just how fever is thought to aid the host is the subject of the next section.

Mechanisms Behind the Function of Fever

An elevation in body temperature can enhance the survival of an infected organism either by a direct effect of temperature on the pathogenic microorganisms or by an indirect effect of temperature on the host's defense mechanisms. Since all organisms have optimal temperature ranges for growth, it is possible that during an infection the accompanying fever might cause these pathogens to move outside their optimal growth temperature. The direct inhibitory effect of temperature on the growth of microorganisms has been demonstrated in numerous experiments.

One example of the effects of temperature on viral infections has been the growth of poliomyelitis virus (159, 160). The yield of these viruses when grown at 37 °C was 250 times greater than when grown at 40 °C. Although the specific mechanism behind the direct effect of temperature on the inhibition of viral growth is not known, it is believed that it might be related to the inhibitory effect of elevated temperature on viral RNA (160). In another study, the growth of a virus (herpes virus, the virus responsible for fever blisters) seemed to be enhanced by an elevated or febrile temperature (161). It is not known, however, whether the growth of these viruses is actually facilitated by the increased temperature or by some other aspect of the febrile episode.

There are many species of bacteria whose growth is known to be inhibited by temperatures often associated with fevers. For example, some strains of pneumococci are actually killed by temperatures as low as 41–41.5 °C (2). Gonococci are another group of bacteria that are particularly sensitive to elevations in temperature. For example, Carpenter et al. (162) found that temperatures of 40–41 °C killed gonococci, results that supported the then popular use of fever as a therapeutic agent in the treatment of gonorrhea.

The spirochetes responsible for neurosyphilis are also killed by elevations in temperature to 41 °C (163), although during the normal course of the infection fevers of this magnitude are seldom, if ever, observed.

In my laboratory we have found that the organism that causes fowl cholera, *P. multocida,* decreases its growth rate fairly abruptly at a temperature of 43 °C (unpublished observations). During most infections with this bacterium the pigeons we have studied develop fevers close to this temperature. Our experiments are done under laboratory conditions with our experimental animals housed in small cages. Under more natural conditions, these birds would be more active and a deep body temperature of close to 45 °C would not be uncommon (164). We suspect that in nature the high temperatures of active birds might serve as a natural form of immunity.

Under natural conditions the direct effect of temperature on the growth of microorganisms might be coupled with some host defense mechanisms. For example, an interesting phenomenon occurs in mammals and reptiles (and perhaps other vertebrates) in response to infection; serum iron levels fall. This appears to be a response of the host to some substance, most likely endogenous pyrogen (165). The iron is stored in various sites, such as the liver, and remains there until the infection is over. Garibaldi (166) and Weinberg (167) have suggested that the growth potential of pathogenic bacteria might be related to the availability of iron or some other nutrient. The reduction of some specific vital nutrient by the host organism might be a defense mechanism, a process Kochan called nutritional immunity (168). Fever fits into this story as follows. Bacteria are usually able to obtain adequate amounts of iron from their growth medium (whether this be a broth solution or a person's plasma) by producing iron chelating substances known as siderophores. The capacity for producing these iron chelators by various species of bacteria has been shown to decrease with increasing temperature in the range of normal-to-febrile body temperatures in mammals (166).

We have tested the nutritional immunity theory in desert iguanas (169). When these lizards were infected with live or dead *A. hydrophila* the levels of serum iron fell by about 30%. These bacteria grow, in vitro, equally well at afebrile (38 °C) and febrile (41 °C) temperatures when there is ample iron available. However, when the iron levels of the growth medium were reduced, the bacteria at 41 °C either failed to grow or grew more slowly than did the bacteria at 38 °C, which grew virtually normally.

Similar results have been found in rabbits infected with *P. multocida* (170). After injection with these pathogenic bacteria, plasma iron levels fell from 246.9 μg/100 ml of plasma to 66 μg/100 ml of plasma. When the bacteria were grown in vitro, it was found that they grew well at afebrile temperatures (39°C or 40°C) in a low iron medium, but not at the febrile temperature of 41 °C. At high levels of iron, corresponding to the levels found in the blood of uninfected rabbits, the bacteria grew well at all three temperatures.

Based on the above data it seems that a reduction in serum iron, coupled with an elevated or febrile body temperature, might constitute an effective host defense mechanism. It is known that other mineral levels vary during infections (e.g., copper and zinc), and it is possible that the changes in these minerals might also participate in the host's defense system. I believe that the entire area of nutritional immunity is an exciting one and should be studied in greater detail.

The other means by which fever might be beneficial is through the effects of temperature on specific aspects of the host's defense system. In fact, several components of the immune response seem to be affected by elevations in temperature.

One of the immune responses that is thought to be affected by temperature is lysosome function (171). Lysosomes are intracellular particles that contain large amounts of hydrolytic enzymes that, when released, are capable of digesting many substances. Lwoff (160) has hypothesized that one of the reasons that fever is beneficial during viral infections is that the combination of an elevation in body temperature and the direct effects of viruses on lysosomes results in the breakdown of these particles. This liberates the lysosomal enzymes and results in the death of the virus and often the cell. Lwoff feels that, whereas the cell often succumbs to these enzymes along with the virus, the overall or net result is beneficial to the organism.

Interferons are another group of proteins that might be affected by temperature. These proteins are produced by many different cell types in response to intracellular contact with various foreign substances, most often viruses. The production of interferon, in turn, results in the nonspecific inhibition of the growth and synthesis of many of these viruses. It has been suggested that an elevated temperature might increase the production of interferon (172). To my knowledge, it is currently unknown whether small elevations in temperature, comparable to that which naturally occurs during infections, increase the production of interferons.

Another component of the immune response that is thought to be affected by temperature is the leukocytes. For example, there is some evidence that the mobility of leukocytes is increased with increasing temperature (173–176). Another aspect of leukocyte function that has been suggested to improve with increases in temperature is its phagocytic activity. Ellingson and Clark (177) reported that human leukocytes, in vitro, ingested more bacteria at febrile temperatures than at afebrile temperatures. These results have not been confirmed in more recent studies (178, 179).

Not only must white blood cells ingest the pathogenic organisms, but they must also digest or kill them. Several investigators have reported that the bactericidal activity of leukocytes is increased with elevations in temperature. For example, Sebag et al. (179) note that elevations in temperature led to increased killing of some species of bacteria, but had little effect on others. In another study it was reported that the percentage of ingested *Staphylococcus* bacteria that were killed increased as the temperature was raised from 26°C to 36°C (180). Above 36°C there was no further increase in the bactericidal activity. As a result of this study, it was suggested that fever does not increase the intracellular killing power of leukocytes, since most fevers in man are in the range of 38–40°C. This might not be technically correct. While it is true that the average rectal temperature of a febrile individual might be 38–40°C during a fever, the average body temperature is considerably less, perhaps only 36–37°C, because the average body temperature incorporates the warmer tissues in the deep body areas (e.g., abdomen, chest cavity, etc.) along with the cooler peripheral areas (e.g., skin, respiratory tract, limbs, etc.). Since many infections actually reside in the cooler areas of the body, an increase in the bactericidal activity of the white blood cells at 36°C might still represent a beneficial effect of fever.

Another area of leukocyte function that might be enhanced by small increases in temperature is lymphocyte transformation. It is well known that lymphocytes (specialized types of leukocytes responsible for cellular and humoral immunity) undergo proliferation and transformation in response to various stimulants. These activated or transformed lymphocytes are then capable of participating in various aspects of the immune response. It has recently been shown that temperatures comparable to moderate fevers in human beings (38.5–39.0°C) result in an enhancement and an acceleration of lymphocyte transformation in response to various types of antigens (181, 182).

Undoubtedly, there are many other aspects of the immune response that are affected by increases in temperature. In some cases it is likely that the immune response is adversely affected by rises in temperature corresponding to normal fevers; in other cases, these immune reactions are aided by normal fevers. Figure 7 summarizes our current understanding of the effects of temperature on the immune system. Not included are those immunological processes that have been found not to be affected by temperature.

When Bennett and Nicastri reviewed the question of the role of fever in disease in 1960, they were, based on the data available to them, unable to determine whether fever was beneficial or harmful. Significant progress has been made in the last two decades, and, while one cannot conclusively answer this ancient question, there is now considerable evidence that fever is often adaptive. Furthermore, many host defense mechanisms have been shown to be sensitive to small, physiological changes in temperature. I predict that within the next two decades many aspects of the immune response, other than those shown in Figure 7, will also be shown to be enhanced by elevations in temperature corresponding to 1, 2, or 3°C fevers. Clearly, as a result of the

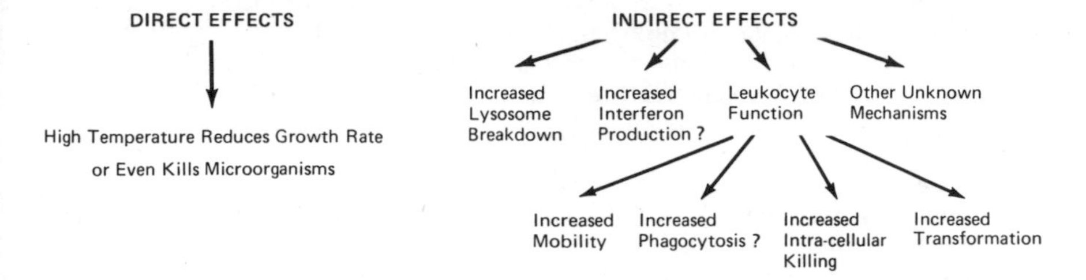

Figure 7. Some of the effects of elevations in temperature on the immune response in vertebrates. Reproduced from Kluger (29) with permission of Princeton University Press.

investigations into the biological function of fever, the use of antipyretic drugs to reduce all fevers should be carefully re-evaluated.

REFERENCES

1. Yost, R. M., Jr. (1950). Sydenham's philosophy of science. Osiris 9:84.
2. Bennett, I. L., and Nicastri, A. (1960). Fever as a mechanism of resistance. Bacteriol. Rev. 24:16.
3. Nagy, K. A., Odell, D. K., and Seymour, R. S. (1972). Temperature regulation by the inflorescence of philodendron. Science 178:1195.
4. Meeuse, B. J. D. (1966). The voodoo lily. Sci. Am. 215:80.
5. Knutson, R. M. (1974). Heat production and temperature regulation in eastern skunk cabbage. Science 186:746.
6. Jennings, H. S. (1906). Behavior of the Lower Organisms. Columbia University Press, New York.
7. Mendelssohn, M. (1902). Recherches sur la thermotaxie des organismes unicellulaires. J. Physiol. Pathol. Gen. 4:393.
8. Fraenkel, G. S., and Gunn, D. L. (1961). The Orientation of Animals: Kineses, Taxes and Compass Reactions. Dover Publishing, Inc., New York.
9. Heath, J. E. (1967). Temperature responses of the periodical "17-year" cicada, *Magicicada cassini* (Homoptera, Cicadidae). Am. Midl. Nat. 77:64.
10. Heath, J. E., Hanegan, J. L., Wilkin, P. J., and Heath, M. S. (1971). Thermoregulation by heat production and behavior in insects. J. Physiol. 63:267.
11. Heinrich, B. (1974). Thermoregulation in endothermic insects. Science 185:747.
12. Heinrich, B. (1977). The physiology of exercise in the bumblebee. Am. Sci. 65:455.
13. Hensel, H., Andres, K. H., and von During, M. (1974). Structure and function of cold receptors. Pfluegers Arch. 352:1.
14. Hensel, H. (1974). Thermoreceptors. Annu. Rev. Physiol. 36:233.
15. Gonzalez, R. R., Kluger, M. J., and Hardy, J. D. (1971). Partitional calorimetry of the New Zealand white rabbit at temperatures 5–35 °C. J. Appl. Physiol. 31:728.
16. Kluger, M. J., Gonzalez, R. R., Mitchell, J. W., and Stolwijk, J. A. J. (1971). The rabbit ear as a temperature sensor. Life Sci. 10:895.

17. Hammel, H. T. (1968). Regulation of internal body temperature. Annu. Rev. Physiol. 30:641.
18. Bligh, J. (1973). Temperature Regulation in Mammals and Other Vertebrates. Elsevier-North Holland Publishing Company, New York.
19. Cabanac, M. (1975). Temperature regulation. Annu. Rev. Physiol. 37:415.
20. Simon, E. (1974). Temperature regulation: the spinal cord as a site of extrahypothalamic thermoregulatory functions. Rev. Physiol. Biochem. Pharmacol. 71:1.
21. Jessen, C., and Mayer, E. T. (1971). Spinal cord and hypothalamus as core sensors of temperature in the conscious dog. I. Equivalence of responses. Pfluegers Arch. 324:189.
22. Duclaux, R., Fantino, M., and Cabanac, M. (1973). Comportement thermoregulateur chez *Rana esculenta.* Pfluegers Arch. 342:347.
23. Gorke, K., Necker, R., and Rautenberg, W. (1975). Neurophysiological investigations of spinal reflexes at different temperatures of the spinal cord in birds and reptiles. Pfluegers Arch. 359:269.
24. Rautenberg, W., Necker, R., and May, B. (1972). Thermoregulatory responses of the pigeon to changes of the brain and the spinal cord temperatures. Pfluegers Arch. 338:31.
25. Rawson, R. O., and Quick, K. P. (1972). Localization of intra-abdominal thermoreceptors in the ewe. J. Physiol. (Lond.) 222:665.
26. Riedel, W., Siaplauras, G., and Simon, E. (1973). Intra-abdominal thermosensitivity in the rabbit as compared with spinal thermosensitivity. Pfluegers Arch. 340:59.
27. Hipskind, S. G., and Hunter, W. S. (1977). Thermoregulatory response to visceral thermal stimulation in unanesthetized cats. Physiologist 20(4):43.
28. Ingram, D. L., and Mount, L. E. (1975). Man and Animals in Hot Environments. Springer-Verlag, New York.
29. Kluger, M. J. Fever: Its Biology, Evolution and Function. Princeton University Press, Princeton, New Jersey. In press.
30. Kluger, M. J., and Heath, J. E. (1971). Effect of preoptic anterior hypothalamic lesions on thermoregulation in the bat. Am. J. Physiol. 221:144.
31. Snapp, B. D., Heller, H. C., and Gospe, S. M., Jr. (1977). Hypothalamic thermosensitivity in California quail (*Lophortyx californicus*). J. Comp. Physiol. 117:345.
32. Mills, S. H., and Heath, J. E. (1972). Responses to thermal stimulation of the preoptic area in the house sparrow, *Passer domesticus.* Am. J. Physiol. 222:914.
33. Myhre, K., and Hammel, H. T. (1969). Behavioral regulation of internal temperature in the lizard *Tiliqua scincoides.* Am. J. Physiol. 217:1490.
34. Hellon, R. F. (1970). The stimulation of hypothalamic neurones by changes in ambient temperature. Pfluegers Arch. 321:56.
35. Boulant, J. A., and Hardy, J. D. (1974). The effect of spinal and skin temperature on the firing rate and thermosensitivity of preoptic neurones. J. Physiol. (Lond.) 240:639.
36. Hellon, R. F. (1975). Monamines, pyrogens and cations: their action on central control of body temperature. Pharmacol. Rev. 26:289.
37. Cox, B., and Lomax, P. (1977). Pharmacologic control of temperature regulation. Annu. Rev. Pharmacol. Toxicol. 17:341.
38. Bolton, H. C. (1900). Evolution of the Thermometer: 1592–1743. Chemical Publishing, Easton.
39. Wunderlich, C. A. (1871). On the Temperature in Diseases: A Manual of Medical Thermometry. The New Sydenham Society, London.

40. Liebermeister, C. (1887). Vorlesungen uber specielle pathologie und therapie. Verlag von F. C. W. Vogel, Leipzig.
41. Snell, E. S., and Atkins, E. (1968). The mechanisms of fever. *In* E. E. Bittar and N. Bittar (eds.), The Biological Basis of Medicine, pp. 397–419. Academic Press, Inc., New York.
42. Cooper, K. E., Cranston, W. I., and Snell, E. S. (1964). Temperature regulation during fever in man. Clin. Sci. 27:345.
43. Cabanac, M., and Massonnet, B. (1974). Temperature regulation during fever: change of set point or change of gain? A tentative answer from a behavioral study in man. J. Physiol. (Lond.) 238:561.
44. Cabanac, M., Duclaux, R., and Gillet, A. (1970). Thermoregulation comportementale chez le chien: effets de la fievre et la thyroxine. Physiol. Behav. 5:697.
45. Sharp, F. R., and Hammel, H. T. (1972). Effects of fever on salivation response in the resting and exercising dog. Am. J. Physiol. 223:77.
46. Vaughn, L. K., Bernheim, H. A., and Kluger, M. J. (1974). Fever in the lizard *Dipsosaurus dorsalis*. Nature 252:473.
47. Kluger, M. J. (1978). The evolution and adaptive value of fever. Am. Sci. 66:38.
48. Kluger, M. J. (1977). Fever in the frog *Hyla cinerea*. J. Thermal Biol. 2:79.
49. Casterlin, M. E., and Reynolds, W. W. (1977). Behavioral fever in anuran amphibian larvae. Life Sci. 20:593.
50. Myhre, K., Cabanac, M., and Myhre, G. (1977). Fever and behavioural temperature regulation in the frog *Rana esculenta*. Acta Physiol. Scand. 101:219.
51. Reynolds, W. W., Casterlin, M. E., and Covert, J. B. (1976). Behavioural fever in teleost fishes. Nature 259:41.
52. Reynolds, W. W. (1977). Fever and antipyresis in the bluegill sunfish, *Lepomis macrochirus*. Comp. Biochem. Physiol. 57C:165.
53. Reynolds, W. W., and Covert, J. B. (1977). Behavioral fever in aquatic ectothermic vertebrates. *In* K. E. Cooper, P. Lomax, and E. Schonbaum (eds.), Drugs, Biogenic Amines and Body Temperature, pp. 108–110. S. Karger, Basel.
54. Reynolds, W. W., Casterlin, M. E., and Covert, J. B. (1978). Febrile responses of bluegill (*Lepomis macrochirus*) to bacterial pyrogens. J. Thermal Biol. 3:129.
55. Reynolds, W. W., Covert, J. B., and Casterlin, M. E. (1978). Febrile responses of goldfish *Carassius auratus* to *Aeromonas hydrophila* and to *Escherichia coli* endotoxin. J. Fish Dis. 1:271.
56. Atkins, E., and Bodel, P. (1972). Fever. N. Engl. J. Med. 286:27.
57. Nowotny, A. (1969). Molecular aspects of endotoxic reactions. Bacteriol. Rev. 33:72.
58. Luderitz, O., Galanos, C., Lehmann, V., Nurminen, M., Rietschel, E. T., Rosenfelder, G., Simon, M., and Westphal, O. (1973). Lipid A: chemical structure and biological activity. J. Infect. Dis. (suppl.) 128:S17.
59. Bennett, I. L., Jr., and Beeson, P. B. (1950). The properties and biologic effects of bacterial pyrogens. Medicine (Balt.) 29:365.
60. Beeson, P. B. (1947). Tolerance to bacterial pyrogens. I. Factors influencing its development. J. Exp. Med. 86:29.
61. Atkins, E., and Freedman, L. R. (1963). Studies in staphylococcal fever. I. Responses to bacterial cells. Yale J. Biol. Med. 63:451.
62. Hort, E. C., and Penfold, W. J. (1912). Microorganisms and their relation to fever. J. Hyg. (Camb.) 12:361.

63. Wagner, R. R., Bennett, I. L., Jr., and LeQuire, V. S. (1949). The production of fever by influenza viruses. I. Factors influencing the febrile response to single injections of virus. J. Exp. Med. 90:321.
64. Roitt, I. (1974). Essential Immunology. Blackwell Scientific Publishing, Oxford.
65. Barrett, J. T. (1974). Textbook of Immunology, An Introduction to Immunochemistry and Immunobiology. C. V. Mosby Company, St. Louis, Missouri.
66. Snell, E. S. (1971). Endotoxin and the pathogenesis of fever. *In* S. Kadis, G. Weinbaum, and S. J. Ajl (eds.), Microbial Toxins, Vol. V, pp. 277–340. Academic Press, Inc., New York.
67. Bodel, P. (1974). Tumors and fever. Part I. Generalized pertubations in host physiology caused by localized tumors. Ann. N. Y. Acad. Sci. 230:6.
68. Dinarello, C. A., and Wolff, S. M. (1978). Pathogenesis of fever in man. N. Engl. J. Med. 298:607.
69. Atkins, E., and Bodel, P. (1974). Fever. *In* B. W. Zweifach, L. Grant, and R. T. McCluskey (eds.), The Inflammatory Process, Ed. 2, Vol. III, pp. 467–514. Academic Press, Inc., New York.
70. Mudd, S., Locke, B., McCutcheon, M., and Strumia, M. (1929). On the mechanism of opsonin and bacteriotropin action. I. Correlation between changes in bacterial surface properties and in phagocytosis by sera of animals under immunization. J. Exp. Med. 49:779.
71. Ponder, E., and MacLeod, J. (1938). Oxygen consumption of white cells from peritoneal exudates. Am. J. Physiol. 123:420.
72. Bornstein, D. L., and Woods, J. W. (1969). Species specificity of leukocyte pyrogens. J. Exp. Med. 130:707.
73. Kozak, M. S., Hahn, H. H., Lennarz, W. J., and Wood, W. B., Jr. (1968). Studies on the pathogenesis of fever. VI. Purification and further characterization of granulocytic pyrogen. J. Exp. Med. 127:341.
74. Bernheim, H. A., and Kluger, M. J. (1977). Endogenous pyrogen-like substance produced by reptiles. J. Physiol. (Lond.) 267:659.
75. Bodel, P., Reynolds, C. R., and Atkins, E. (1973). Lack of effect of salicylate on pyrogen release from human blood leucocytes in vitro. Yale J. Biol. Med. 46:190.
76. King, M. K., and Wood, W. B., Jr. (1958). Studies on the pathogenesis of fever. IV. The site of action of leucocytic and circulating endogenous pyrogen. J. Exp. Med. 107:291.
77. Adler, R. D., and Joy, R. J. T. (1965). Febrile responses to the intracisternal injection of endogenous (leucocytic) pyrogen in the rabbit. Proc. Soc. Exp. Biol. Med. 119:660.
78. Cooper, K. E., Cranston, W. I., and Honour, A. J. (1967). Observations on the site and mode of action of pyrogens in the rabbit brain. J. Physiol. (Lond.) 191:325.
79. Jackson, D. L. (1967). A hypothalamic region responsive to localized injection of pyrogens. J. Neurophysiol. 30:586.
80. Rosendorff, C., and Mooney, J. J. (1971). Central nervous system sites of action of a purified leucocyte pyrogen. Am. J. Physiol. 220:597.
81. Allen, I. V. (1965). The cerebral effects of endogenous serum and granulocytic pyrogen. Br. J. Exp. Pathol. 46:25.
82. Dinarello, C. A., Renfer, L., and Wolff, S. M. (1977). Human leukocytic pyrogen: purification and development of a radioimmunoassay. Proc. Natl. Acad. Sci. 74:4624.
83. Veale, W. L., and Cooper, K. E. (1975). Comparison of sites of action of pros-

taglandin E and leucocyte pyrogen in brain. *In* P. Lomax (ed.), Temperature Regulation and Drug Action, pp. 218–226. S. Karger, Basel.

84. Andersson, B., Gale, C. C., Hokfelt, B., and Larson, B. (1965). Acute and chronic effects of preoptic lesions. Acta Physiol. Scand. 65:45.
85. Lipton, J. M., and Trzcinka, G. P. (1976). Persistence of febrile response to pyrogens after PO/AH lesions in squirrel monkeys. Am. J. Physiol. 231:1638.
86. Milton, A. S., and Wendlandt, S. (1971). Effects on body temperature of prostaglandins of the A,E and F series on injection into the third ventricle of unanesthetized cats and rabbits. J. Physiol. (Lond.) 218:325.
87. Hales, J. R. S., Bennett, J. W., Baird, J. A., and Fawcett, A. A. (1973). Thermoregulatory effects of prostaglandins E_1, E_2, $F_{1\alpha}$ and $F_{2\alpha}$ in the sheep. Pfluegers Arch. 339:125.
88. Feldberg, W., and Saxena, P. N. (1975). Prostaglandins, endotoxins and lipid A on body temperature in rats. J. Physiol. (Lond.) 249:601.
89. Feldberg, W., and Saxena, P. N. (1971). Further studies on prostaglandin E_1 in fever in cats. J. Physiol. (Lond.) 219:739.
90. Stitt, J. T. (1973). Prostaglandin E_1 fever induced in rabbits. J. Physiol. (Lond.) 232:163.
91. Crawshaw, L. I., and Stitt, J. T. (1975). Behavioural and autonomic induction of prostaglandin E_1 fever in squirrel monkeys. J. Physiol. (Lond.) 244:197.
92. Barney, C. C., and Elizondo, R. S. (1978). Effect of ambient temperature on development of prostaglandin E_1 hyperthermia in the rhesus monkey. J. Appl. Physiol. 44(5):751.
93. Holmes, S. W., and Horton, E. W. (1968). The identification of four prostaglandins in the dog brain and their regional distribution in the central nervous system. J. Physiol. (Lond.) 195:731.
94. Ambache, N., Brummer, H. C., Rose, J. G., and Whiting, J. (1966). Thin-layer chromatography of spasmogenic unsaturated hydroxyacids from various tissues. J. Physiol. (Lond.) 185:77P.
95. Feldberg, W., Gupta, K. P., Milton, A. S., and Wendlandt, S. (1973). Effect of pyrogen and antipyretics on prostaglandin activity in cisternal c.s.f. of unanesthetized cats. J. Physiol. (Lond.) 234:279.
96. Philipp-Dormston, W. K., and Seigert, R. (1974). Prostaglandins of the E and F series in rabbit cerebrospinal fluid during fever induced by Newcastle disease virus, *E. coli*-endotoxin, or endogenous pyrogen. Med. Microbiol. Immunol. 159:279.
97. Vane, J. R. (1971). Inhibition of prostaglandin synthesis as a mechanism of action for aspirin-like drugs. Nature (New Biol.) 231:232.
98. Smith, J. B., and Willis, A. L. (1971). Aspirin selectively inhibits prostaglandin production in human platelets. Nature (New Biol.) 231:235.
99. Ferreira, S. H., Moncada, S., and Vane, J. R. (1971). Indomethacin and aspirin abolish prostaglandin release from the spleen. Nature (New Biol.) 231:237.
100. Baird, J. A., Hales, J. R. S., and Lang, W. J. (1974). Thermoregulatory responses to the injection of monamines, acetylcholine and prostaglandins into a lateral cerebral ventricle of the echidna. J. Physiol. (Lond.) 236:539.
101. Pittman, Q. J., Veale, W. L., and Cooper, K. E. (1977). Effect of prostaglandin, pyrogen and noradrenaline, injected into the hypothalamus, on thermoregulation in newborn lambs. Brain Res. 128:473.
102. Pittman, Q. J., Veale, W. L., and Cooper, K. E. (1977). Absence of fever following intrahypothalamic injections of prostaglandins in sheep. Neuropharmacology 16:743.
103. Wit, A., and Wang, S. C. (1968). Temperature-sensitive neurons in

preoptic/anterior hypothalamic region: actions of pyrogens and acetylsalicylate. Am. J. Physiol. 215:1160.

104. Cabanac, M., Stolwijk, J. A. J., and Hardy, J. D. (1968). Effect of temperature and pyrogens on single-unit activity in the rabbit's brain stem. J. Appl. Physiol. 24:645.

105. Eisenman, J. S. (1969). Pyrogen-induced changes in the thermosensitivity of septal and preoptic neurons. Am. J. Physiol. 216:330.

106. Stitt, J. T., and Hardy, J. D. (1975). Microelectrophoresis of PGE_1 onto single units in the rabbit hypothalamus. Am. J. Physiol. 229:240.

107. Ford, D. M. (1974). A selective action of prostaglandin E_1 on hypothalamic neurones in the cat which respond to brain cooling. J. Physiol. (Lond.) 242:142P.

108. Cranston, W. I., Hellon, R. F., and Mitchell. D. (1975). A dissociation between fever and prostaglandin concentration in cerebrospinal fluid. J. Physiol. (Lond.) 253:583.

109. Laburn, H., Mitchell, D., and Rosendorff, C. (1977). Effects of prostaglandin antagonism on sodium arachidonate fever in rabbits. J. Physiol. (Lond.) 267:559.

110. Hamberg, M., Svensson, J., and Samuelson, B. (1975). Thromboxanes: a new group of biologically active compounds derived from prostaglandin endoperoxides. Proc. Natl. Acad. Sci. U.S.A. 72:2994.

111. Douglas, W. W. (1975). Polypeptides-angiotensin, plasma kinins, and other vasoactive agents; prostaglandins. *In* L. S. Goodman and A. Gilman (eds.), The Pharmacological Basis of Therapeutics, pp. 630–652. Macmillan Publishing Company, Inc., New York.

112. Milton, A. S., and Dascombe, M. J. (1977). Cyclic nucleotides in thermoregulation and fever. *In* K. E. Cooper, P. Lomax, and E. Schonbaum (eds.), Drugs, Biogenic Amines and Body Temperature, pp. 129–135. S. Karger, Basel.

113. Philip-Dormston, W. K., and Siegert, R. (1975). Adenosin 3′, 5′ cyclic monophosphate in rabbit cerebrospinal fluid during fever induced by *E. coli* endotoxin. Med. Microbiol. Immunol. 161:11.

114. Dascombe, M. J. (1976). Studies on the possible involvement of adenosine 3′-5′-monophosphate in thermoregulation during fever. Ph.D. thesis, University of Aberdeen, Aberdeen, Scotland.

115. Woolf, C. J., Willies, G. H., Laburn, H., and Rosendorff, C. (1975). Pyrogen and prostaglandin fever in the rabbit. I. Effects of salicylate and the role of cyclic AMP. Neuropharmacology 14:397.

116. Willies, G. H., Woolf, C. J., and Rosendorff, C. (1976). The effect of an inhibitor of adenylate cyclase on the development of pyrogen, prostaglandin and cyclic AMP fevers in the rabbit. Pfluegers Arch. 367:177.

117. Sobocinska, J., and Greenleaf, J. E. (1976). Cerebrospinal fluid [Ca^{2+}] and rectal temperature response during exercise in dogs. Am. J. Physiol. 230:1416.

118. Feldberg, W., Myers, R. D., and Veale, W. L. (1970). Perfusion from cerebral ventricle to cisterna magna in the unanesthetized cat: effect of calcium on body temperature. J. Physiol. (Lond.) 207:403.

119. Myers, R. D., and Veale, W. L. (1971). The role of sodium and calcium ions in the hypothalamus in the control of body temperature of the unanesthetized cat. J. Physiol. (Lond.) 212:411.

120. Seoane, J. R., and Baile, A. C. (1975). Ionic changes in cerebrospinal fluid and feeding, drinking and temperature of sheep. Physiol. Behav. 10:915.

121. Greenleaf, J. E., Kolzowski, S., Nazar, K., Kaciuba-Uscilko, H., Brzeninska, Z., and Ziemba, A. (1976). Ion-osmotic hyperthermia during exercise in dogs. Am. J. Physiol. 230:74.

122. Hensel, H., and Schafer, K. (1974). Effects of calcium on warm and cold receptors. Pfluegers Arch. 352:87.
123. Robertshaw, D., and Beier, C. N. (1977). The relationship between plasma levels of sodium and ionized calcium and thermoregulatory thresholds at different phases of the human menstrual cycle. Physiologist 20(4):80.
124. Davson, H. (1970). Physiology of the Cerebrospinal Fluid. Churchill, London.
125. van Miert, A. S. J. P. A. M., and Frens, J. (1968). The reaction of different animal species to bacterial pyrogens. Zentralbl. Veterinaermed. 15:532.
126. Pittman, Q. J., Veale, W. L., Cockeram, A. W., and Cooper, K. E. (1976). Changes in body temperature produced by prostaglandins and pyrogens in the chicken. Am. J. Physiol. 230:1284.
127. D'Alecy, L. G., and Kluger, M. J. (1975). Avian febrile response. J. Physiol. (Lond.) 253:223.
128. Artunkal, A. A., and Marley, E. (1974). Hyper- and hypothermic effects of prostaglandin E_1 (PGE_1) and their potentiation by indomethacin, in chicks. J. Physiol. (Lond.) 242:141P.
129. Nistico, G., and Marley, E. (1973). Central effects of prostaglandin E_1 in adult fowls. Neuropharmacology 12:1009.
130. Templeton, J. R. (1970). Reptiles. *In* G. C., Whittow (ed.), Comparative Physiology of Thermoregulation, pp. 167-221. Academic Press, Inc., New York.
131. DeWitt, C. B. (1967). Precision of thermoregulation and its relation to environmental temperatures in desert iguanas *Dipsosaurus dorsalis.* Physiol. Zool. 40:49.
132. Kluger, M. J., Tarr, R. S., and Heath, J. E. (1973). Posterior hypothalamic lesions and disturbances in behavioral thermoregulation in the lizard *Dipsosaurus dorsalis.* Physiol. Zool. 46:79.
133. Bernheim, H. A., and Kluger, M. J. (1976). Fever and antipyresis in the lizard *Dipsosaurus dorsalis.* Am. J. Physiol. 231:198.
134. Casterlin, M. E., and Reynolds, W. W. (1977). Behavioral fever in crayfish. Hydrobiologia 56:99.
135. Major, R. H. (1954). A History of Medicine, Vol. I. Charles C Thomas Publisher, Springfield, Illinois.
136. Kluger, M. J. (1978). The history of bloodletting. Nat. Hist. 87:78.
137. Marks, G., and Beatty, W. K. (1971). The Medical Garden. Charles Scribner's Sons, New York.
138. Kluger, M. J., Ringler, D. H., and Anver, M. R. (1975). Fever and survival. Science 188:166.
139. Bernheim, H. A., and Kluger, M. J. (1976). Fever: effect of drug-induced antipyresis on survival. Science 193:237.
140. Covert, J. B., and Reynolds, W. W. (1977). Survival value of fever in fish. Nature 267:43.
141. Vaughn, L. K. (1977). Fever and survival in rabbits infected with *Pasteurella multocida*. Ph.D. dissertation. University of Michigan microfilms, Ann Arbor, Michigan.
142. Vaughn, L. K., and Kluger, M. J. (1977). Fever and survival in bacterially infected rabbits. Fed. Proc. 36(3):511.
143. Kluger, M. J., and Vaughn, L. K. Fever and survival in rabbits infected with *Pasteurella multocida.* J. Physiol. (Lond.) In press.
144. Toms, G. L., Davies, J. A., Woodward, C. G., Sweet, C., and Smith, H. (1977). The relation of pyrexia and nasal inflammatory response to virus levels in nasal washings of ferrets infected with influenza viruses of differing virulence. Br. J. Exp. Pathol. 58:444.

145. Connor, D. G., and Kass, E. H. (1961). Effect of artificial fever in increasing susceptibility to bacterial endotoxin. Nature 190:453.
146. Porter, P. J., and Kass, E. H. (1962). Mediation by the central nervous system of the lethal action of bacterial endotoxin. Clin. Res. 10:185.
147. Atwood, R. P., and Kass, E. H. (1964). Relationship of body temperature to the lethal action of bacterial endotoxin. J. Clin. Invest. 43:151.
148. Greenway, C. V., and Murthy, V. S. (1971). Mesenteric vasoconstriction after endotoxin administration in cats pretreated with aspirin. Br. J. Pharmacol. 43:259.
149. Strouse, S. (1909). Experimental studies on pneumococcus infections. J. Exp. Med. 11:743.
150. Muschenheim, C., Duerschrer, D. R., Hardy, J. D., and Stoll, A. M. (1943). Hypothermia in experimental infections. III. The effects of hypothermia on resistance to experimental pneumococcus infection. J. Infect. Dis. 72:187.
151. Pembrey, M. S. (1895). The effect of variations in external temperature upon the output of carbonic acid and the temperature of young animals. J. Physiol. (Lond.) 18:364.
152. Haahr, S., and Mogensen, S. (1977). Function of fever. Lancet II(8038):613.
153. Satinoff, E., McEwen, G. N., Jr., and Williams, B. A. (1976). Behavioral fever in newborn rabbits. Science 193:1139.
154. Blatteis, C. M. (1977). Comparison of endotoxin and leukocytic pyrogen pyrogenicity in newborn guinea pigs. J. Appl. Physiol. 42:355.
155. Kasting, N. W., Veale, W. L., and Cooper, K. E. (1978). Suppression of fever at term of pregnancy. Nature 271:245.
156. Teisner, B., and Haahr, S. (1974). Poikilothermia and susceptibility of suckling mice to Coxsackie B1 virus. Nature 247:568.
157. Carmichael, L., Barnes, F. D., and Percy, D. H. (1969). Temperature as a factor in resistance of young puppies. J. Infect. Dis. 120:669.
158. DuBois, E. F. (1949). Why are fevers over 106°F rare? Am. J. Med. Sci. 217:361.
159. Lwoff, A. (1959). Factors influencing the evolution of viral diseases at the cellular level and in the organism. Bacteriol. Rev. 23:109.
160. Lwoff, A. (1969). Death and transfiguration of a problem. Bacteriol. Rev. 33:390.
161. Fenner, F., McAuslan, B. R., Mims, C. A., Sambrook, J., and White, D. O. (1974). The Biology of Animal Viruses. Academic Press, Inc., New York.
162. Carpenter, C. M., Boak, R. A., Mucci, L. A., and Warren, S. L. (1933). Studies on the physiologic effects of fever temperatures: the thermal death time of *Neisseria gonorrhoeae* in vitro with special reference to fever temperatures. J. Lab. Clin. Med. 18:981.
163. Bruetsch, W. L. (1949). Why malaria cures general paralysis. Indiana State Med. Assoc. J. 42:211.
164. Hart, J. S., and Roy, O. Z. (1967). Temperature regulation during flight in pigeons. Am. J. Physiol. 213:1311.
165. Merriman, C. R., Pulliam, L. A., and Kampschmidt, R. F. (1977). Comparison of leukocytic pyrogen and leukocytic endogenous mediator. Proc. Soc. Exp. Biol. Med. 154:224.
166. Garibaldi, J. A. (1972). Influence of temperature on the biosynthesis of iron transport compounds by *Salmonella typhimurium*. J. Bacteriol. 110:262.
167. Weinberg, E. D. (1974). Iron and susceptibility to infectious disease. Science 184:952.
168. Kochan, I. (1973). The role of iron in bacterial infections, with special consideration of host-tubercle *Bacillus* interaction. Curr. Top. Microbiol. Immunol. 60:1.

169. Grieger, T. A., and Kluger, M. J. (1978). Fever and survival: the role of serum iron. J. Physiol. (Lond.) 279:187.
170. Kluger, M. J., and Rothenburg, B. A. (1979). Fever and reduced iron: Their interaction as a host defense response to bacterial infection. Science 203:374.
171. Overgaard, J. (1977). Effect of hyperthermia on malignant cells in vivo. Cancer 39:2637.
172. Ho, M. (1970). Factors influencing the interferon response. Arch. Intern. Med. 126:135.
173. Bryant, R. E., DesPrez, R. M., VanWay, M. H., and Rogers, D. E. (1966). Studies on human leukocyte motility. I. Effects of alterations in pH, electrolyte concentration, and phagocytosis on leukocyte migration, adhesiveness, and aggregation. J. Exp. Med. 124:483.
174. Nahas, G. G., Tannieres, M. L., and Lennon, J. F. (1971). Direct measurement of leukocyte motility: effects of pH and temperature. Proc. Soc. Exp. Biol. Med. 138:350.
175. Phelps, P., and Stanislaw, D. (1969). Polymorphonuclear leukocyte mobility in vitro. I. Effect of pH, temperature, ethyl alcohol and caffeine, using a modified Boyden chamber technique. Arthritis Rheum. 12:181.
176. Bernheim, H. A., Bodel, P., Askenase, P., and Atkins, E. (1978). Effects of fever on host defence mechanisms after infection in the lizard *Dipsosaurus dorsalis*. Brit. J. Exp. Pathol. 59:76.
177. Ellingson, H. V., and Clark, P. F. (1942). The influence of artificial fever on mechanisms of resistance. J. Immunol. 43:65.
178. Mandell, G. L. (1975). Effect of temperature on phagocytosis by human polymorphonuclear neutrophils. Infect. Immunol. 12:221.
179. Sebag, J., Reed, W. P., and Williams, R. C., Jr. (1977). Effect of temperature on bacterial killing by serum and by polymorphonuclear leukocytes. Infect. Immunol. 16:947.
180. Craig, C. P., and Suter, E. (1966). Extracellular factors influencing staphylocidal capacity of human polymorphonuclear leukocytes. J. Immunol. 97:287.
181. Roberts, N. J., Jr., and Steigbigel, R. T. (1977). Hyperthermia and human leukocyte functions: effects on response of lymphocytes to mitogen and antigen and bactericidal capacity of monocytes and neutrophils. Infect. Immunol. 18:673.
182. Ashman, R. B., and Nahmias, A. J., (1977). Enhancement of human lymphocyte responses to phytomitogens in vitro by incubation at elevated temperatures. Clin. Exp. Immunol. 29:464.

International Review of Physiology
Environmental Physiology III, Volume 20
Edited by D. Robertshaw

5
Metabolic Status During Diving and Recovery in Marine Mammals

P. W. HOCHACHKA and B. MURPHY
University of British Columbia, Vancouver, British Columbia, Canada

---

PERIPHERAL ORGANS 256
High Potential for Oxidative Metabolism 256
High Potential for Anaerobic Metabolism 257
Oscillation Between Glycogen and Fat Catabolism 257
Adjustments in Muscle Glycolysis 257
Function of Muscle Fructose Biphosphatase 258
Regulatory Nature of Muscle Pyruvate Kinase in Diving Mammals 259
Why Pyruvate Kinase in Dolphin Muscle is Regulatory 259
How Glycogen is Spared During Aerobic Periods 260
The Roles of Alanine and Aspartate 260
Implications 261
Kidney Metabolism 262
Impact of Diving on Kidney Metabolism 263
Impact of Recovery on Kidney Metabolism 263
Liver Metabolism 264
Impact of Diving on Liver Metabolism 264
Impact of Recovery on Liver Metabolism 264

CENTRAL ORGANS 265
Substrates for the Mammalian Heart 265
Fat Oxidation in the Hypoxic Heart 266

Without a John Simon Guggenheim Fellowship (awarded to P. W. Hochachka), much of the metabolic work on Weddell seals, the literature searches, and the data analyses involved in preparing this chapter would not have been possible. The Weddell seal studies were performed at McMurdo Station, Antarctica, and were supported by a United States Antarctic Research Program grant (to W. Zapol, Harvard Medical School, as principal investigator) and by the N.R.C. (Canada) through an operating grant (to PWH).

Heart Metabolic Biochemistry in Diving Animals 268
Brain Metabolic Rate 269
Glucose as Substrate for the Brain 269
Alternate Substrates for the Brain 269
Circulatory and Metabolic Consequences of Brain Hypoxia 270
Metabolic Consequences of Brain Anoxia 272
Brain Metabolism in Diving Animals 273
Lung Metabolism 275

HEART-LUNG-BRAIN METABOLISM IN THE WEDDELL SEAL 276
Enzymes of Aerobic Metabolism 276
Enzymes of Anaerobic Metabolism 276
Blood Glucose and Lactate Profiles 277
Arteriovenous Gradients Across the Lung 277
Lactate as a Substrate for the Lung 278
Arteriovenous Gradients Across the Brain 278
Metabolic Properties of the Weddell Seal Heart 278
Metabolic Model of the Heart, Lung, and Brain in Divers 279

WHAT THE SEAL FETUS DOES WHEN ITS MOTHER DIVES 280
Direct Evidence for Fetal Bradycardia 280
Glucose, Lactate, and Pyruvate Profiles 280

LONG-TERM DIVING CAPACITY 281
Diving Response Cannot Account for Long-term Capacity 281
Potential Intertissue Cycling of Metabolites 282

LIST OF ABBREVIATIONS

G6P, glucose 6-phosphate
F6P, fructose 6-phosphate
FBP, fructose biphosphate
PEP, phosphoenolpyruvate
OXA, oxaloacetate
2-KGA, 2-ketoglutarate
GABA, γ-aminobutyric acid
ATP, adenosine 5′-triphosphate
ADP, adenosine 5′-diphosphate
(5′) AMP, adenosine 5′-monophosphate
NAD^+, nicotinamide adenine dinucleotide, oxidized form

NADH, nicotinamide adenine dinucleotide, reduced form
FAD^+, flavin adenine dinucleotide, oxidized form
$FADH_2$, flavin adenine dinucleotide, reduced form
CoA, coenzyme A
HK, hexokinase
PFK, phosphofructokinase
PK, phosphokinase
LDH, lactate dehydrogenase
FBPase, fructose biphosphatase
SDH, succinate dehydrogenase
ETS, electron transport system
K_m, Michaelis-Menton constant
$CMRO_2$, cerebral metabolic rate for oxygen
Po_2, partial pressure of oxygen
Pao_2, arterial partial pressure of oxygen
Pvo_2, venous partial pressure of oxygen
EEG, electroencephalogram
AV, arteriovenous difference

In terms of the interest they arouse in layman and biologist alike, marine mammals probably are surpassed by no other group of organisms. Many students, therefore, have worked hard to unravel the myriad of biological secrets that reveal the dramatic underwater capabilities of marine mammals. As a result, the broad outlines of functional and structural specializations supporting the diving habit are now well established (1, 2). Although not as much information is available on metabolic and enzymic adjustments, even in this area it is already clear that diving depends upon tissue-specific and readily adjustable blends of aerobic and anaerobic metabolism. Whereas during short-duration dives only oxidative metabolism appears to be utilized (3), in long-duration dives, both may be needed in varying proportions in different tissues (4). In the latter instance, it is notable that anaerobic metabolism may sustain work functions well before the oxygen reserves of the blood are depleted; i.e., a dependence upon glycolysis during diving does not necessarily arise from a simple oxygen lack in the organism. Rather, it stems from the activation of an O_2-conserving mechanism (termed the diving response), which leads to highly specific partitioning of oxygenated blood between central and peripheral tissues.

There are three components to the diving response: apnea, bradycardia, and peripheral vasoconstriction. Bradycardia leads to a drop in cardiac output, whereas peripheral vasoconstriction serves to maintain arterial blood pressure and redistribute blood flow. Although the antecedents to these phenomena were evident in the earlier literature, they were not formalized into a coherent theory until the classic studies of Scholander (5) nearly four decades ago.

Originally Scholander (1, 5) anticipated that the function of bradycardia and peripheral vasoconstriction was to conserve oxygen for the heart, lung, and brain. More recent studies, using microsphere techniques (6, 7), indicate that the situation is perhaps more complex. In terms of absolute perfusion rates, only the brain experiences unchanged rates of flow of blood and hence unchanged delivery of oxygen and carbon substrates (7); all other organs, at least in the Weddell seal, apparently experience a decrease in absolute flow rates, which may be consistent with the suggestion that the metabolic rate of the animal actually drops somewhat during diving (2). In terms of fractional cardiac output, four organs are identified as receiving an increased relative perfusion: the heart, lung, brain, and the adrenal glands (7). Peripheral organs, such as the muscles and kidneys, sustain a sharp reduction both in absolute perfusion and in fractional cardiac output. For this reason, and because a central versus peripheral dichotomy is emphasized in the formal theory (1), it is convenient to discuss metabolic and enzymic adjustments in the two separately.

PERIPHERAL ORGANS

High Potential for Oxidative Metabolism

In considering the metabolism of animals specialized for extended survival without breathing, it is easy to overemphasize the anaerobic potentials while overlooking the single, perhaps the most important, feature of diving mammals as a group: a basal oxidative metabolism that is substantially higher than that of terrestrial mammals of comparable size (8–10). Since the heart, lung, and brain constitute such a small fraction of the overall mass (in a 500-kg Weddell seal, for example, they constitute 5 kg, i.e., one-hundredth of the total mass), it is probable that the bulk of the high respiratory rate must be due to peripheral organs. Available data are consistent with this interpretation and suggest that in the normal, nondiving state the preferred fuel is fat. Values of *RQ* (the ratio of respiratory CO_2 produced to O_2 consumed), where they have been measured, are typically around 0.7, and the free fatty acid content of the blood is high. Both observations indicate a potentially high fat catabolism. Many diving vertebrates, in fact, display functional and anatomical adaptations favoring an active aerobic metabolism. Per unit of body weight, diving vertebrates display a relatively large lung volume and blood volume, an unusually large blood O_2 capacity, and higher muscle myoglobin concentrations. Diving animals such as seals and whales have unusually large blood storage capacities in their venous system. The proportion of red, mitochondria-rich muscle fibers with high lipase activity is high (11), implying a vigorous fat-based metabolism when not diving. It is fat-based a) because many divers use blubber as thermal insulation and thus may have been "preadapted" for fat utilization, b) because fat is the best fuel to use in long migrations that are typically made by many marine mammals and birds, and c) because on a weight or a molar basis fat is nature's most energetically efficient fuel source.

High Potential for Anaerobic Metabolism

At the same time, it is evident that many peripheral organs must be capable of sustaining significant periods of anaerobiosis. This conclusion was already evident in Scholander's original study (5), since it was clear that in the harbor seal prolonged (15–20 min) diving could not be matched by endogenous (myoglobin-bound) O_2 stores in skeletal muscle. Muscle functions under these conditions are, therefore, supported by anaerobic glycolysis and lactate is known to accumulate (1, 2, 5).

More recently, the isozyme forms of lactate dehydrogenase have been examined in a number of diving animals. They have been found to display kinetic properties suited for lactate accumulation and their overall activity is notably high (11, 12).

Oscillation Between Glycogen and Fat Catabolism

Obviously, the above metabolic organization is a consequence of physiological solutions to problems of long-duration diving. What has not been widely appreciated, however, is that these same solutions set the stage for an intriguing metabolic situation emphasizing the transition between diving and recovery metabolic processes: the nondiving state is supported by a vigorous lipid-based oxidative metabolism, whereas, at least in prolonged diving, a vigorous glycogen-based fermentative metabolism must be activated. An important clue to unraveling the metabolic consequences of diving is this requirement for the controlled oscillation between two kinds of metabolic processes during transition into, and out of, diving. Some indication of the way this is achieved can be obtained by comparing the enzymic machinery of diving and nondiving mammals. This would best be done with an organ-by-organ comparison, but to date, of organs sustaining reduced perfusion during diving, only skeletal muscle in the porpoise has been studied in any depth.

Adjustments in Muscle Glycolysis

As previously argued (12), the basic design of muscle glycolysis is so effective that when the muscle of a diving mammal, such as the porpoise, is compared to that of an animal like the common laboratory rat or man, only a small number of modifications are observed. First, glycogen levels are high (11, 12), implying that a high glycogenolytic potential is available if and when it is needed. This is perhaps the single most simple adjustment that may be expected in a tissue geared up to sustain function during anoxia. Coupled to this adjustment are changes in the steady state levels of a number of glycolytic enzymes. From a comparison of muscle enzyme activities, the major differences between the dolphin and the laboratory rat are 1) increased amounts of phosphoglucomutase, aldolase, α-glycerophosphate dehydrogenase, and lactate dehydrogenase, and 2) decreased amounts of pyruvate kinase. A third notable difference is that muscle has high concentrations of fructose biphosphatase, a nonglycolytic enzyme nonetheless involved in glycolytic control.

On closer examination, one set of observations can be readily explained. Thus, because muscle in marine mammals is cut off from blood circulation during diving, it must rely less on blood glucose and more on muscle glycogen than would muscle in a terrestrial mammal; this is reflected in an increased phosphoglucomutase content. Second, an active aldolase contributes to tight regulation of the concentration of fructose 1,6-biphosphate, a key regulatory metabolite in the overall pathway. Third, muscle cells in divers must be prevented from becoming highly reduced; this requirement is reflected in higher titers of lactate dehydrogenase and α-glycerophosphate dehydrogenase. The high fructose biphosphatase and low pyruvate kinase titers, however, are not as easily explained.

Function of Muscle Fructose Biphosphatase

In physiological terms, fructose biphosphatase catalyzes a reaction that is the reverse of that catalyzed by phosphofructokinase.

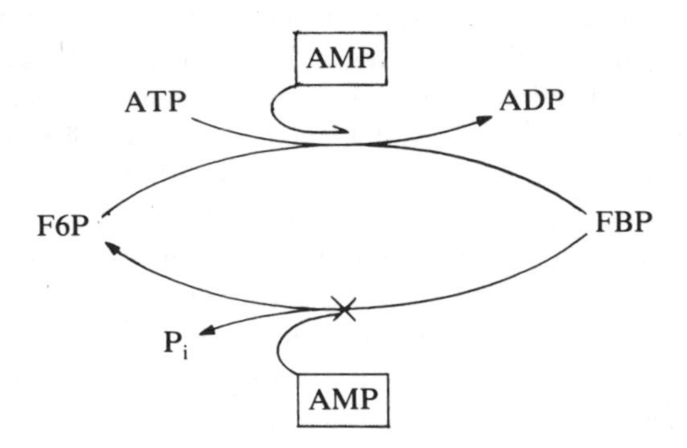

Although phosphofructokinase is regulated by a large number of modulators, fructose biphosphatase seems to be under the regulation of only one effector compound: AMP. Adenosine monophosphate is a potent inhibitor of all fructose biphosphatases thus far examined, except for that in bumble bee flight muscle, where the enzyme is involved in a thermogenic function. When fructose biphosphatase was first discovered in muscle its function was unknown, and its presence posed a perplexing problem, since the simultaneous function of both enzymes in the same cell sets up a futile carbon cycle and a net ATP hydrolysis. That perhaps is precisely the point.

Since both phosphofructokinase and fructose biphosphatase respond (in opposite manner) to at least one signal in common (AMP), the sensitivity of the overall fructose 6-phosphate → fructose 1,6-biphosphate conversion is far greater than if only phosphofructokinase (or only fructose biphosphatase) were AMP sensitive, or if only phosphofructokinase were present in the cell. It is intuitively obvious that the degree of amplification depends in large part on the fructose biphosphatase activity. That may be why there seems to be a rough correlation between high levels of fructose biphosphatase in muscle and high levels of phosphofructokinase. Dolphin muscle fructose biphosphatase levels, for example, are similar to those in the pectoral muscle of "burst" fliers, such as the pheasant and the fowl, and the gastrocnemius of burst runners, such as the cat, but are nearly 5 times higher than the levels in skeletal muscles of long-range fliers like the mallard and other waterfowl. The former muscles are characterized by large percentage swings in glycolytic rate between rest and maximum bursts of activity, a condition that perhaps would most critically call for an important fructose biphosphatase role in the regulation of carbon flow through the F6P → FBP interconversion (12). Ramaiah (13) reviews the current thinking in this area.

Regulatory Nature of Muscle Pyruvate Kinase in Diving Mammals

To explain the comparatively low pyruvate kinase content of porpoise muscle, detailed kinetic studies of the enzyme were needed. Simon et al. (14) had observed that, when diving animals are treated as a single group, the pyruvate kinase content of muscle roughly correlates with the length of diving each species is capable of achieving, and such a correlation between glycolytic capacity and pyruvate kinase titer in fact also holds for normal mammalian tissues (15). However, within diving animals as a group, pyruvate kinase concentrations are lower than in the common laboratory rat. What seems to have occurred at this locus is the elaboration of an enzyme that is sensitive to metabolite regulation. Unlike the enzyme in terrestrial mammals, muscle pyruvate kinase in diving vertebrates is under tight feedback inhibition by ATP, alanine, and probably citrate (12), and under strong feedforward activation by fructose 1,6-biphosphate, which returns the pyruvate kinase maximum potential to the high range expected for highly active glycolysis. Moreover, as with other regulatory pyruvate kinases, fructose 1,6-biphosphate not only directly activates the enzyme (by affecting both enzyme-substrate affinity and the maximum velocity of the enzyme-catalyzed reaction) but also effectively reverses the inhibitory effects of negative modulators. These regulatory characterisitics are, in fact, commonly observed in muscle pyruvate kinases of lower animals (16, 17), but appear to have been lost in most mammals.

Why Pyruvate Kinase in Dolphin Muscle Is Regulatory

The reason diving animals have retained a tight control over pyruvate kinase in part stems from a high reliance on glycolysis during diving. But more importantly, this enzyme fits the control requirements imposed on muscle at the end of the dive—that is, during anaerobic-aerobic transition, when muscle metabolism switches from glycogen to fat as the predominant fuel. Inhibitory control of pyruvate kinase under these conditions would contribute to dampening of glycolysis and thus to a sparing of glycogen during aerobic periods.

How Glycogen Is Spared During Aerobic Periods

In mammals in general, it is now clear that during anaerobic-aerobic transition, when fatty acid oxidation is initiated, a number of profound fluctuations in various Krebs cycle intermediates occur. Of these, the percentage of increase in citrate concentrations is unusually high and this metabolite serves as a feedback inhibitor of phosphofructokinase, effectively blocking glycolysis when that block is appropriate. The same "control loop" is used in various tissues of the vertebrate, but muscle phosphofructokinase is unusually sensitive to citrate (about 10 times more so than, for example, the liver isozyme), and thus in muscle this control interaction is perhaps most effective (18). As far as we know, the same control mechanism at this locus in glycolysis also operates in tissues of diving vertebrates, but whereas this seems to be a sufficient mechanism in muscles of typical terrestrial mammals, it is by no means the only way to turn off glycolysis during aerobic periods in the dolphin because pyruvate kinase here is also an allosteric enzyme. It too is highly sensitive to citrate, and this would supply an additional means for blocking glycolysis during periods of fat catabolism. A third important mechanism for blocking glycolysis at this time involves alanine inhibition of pyruvate kinase, but to fully appreciate the physiological significance of this effect, a brief digression into the nature of Krebs cycle activation during fatty acid oxidation is needed.

The Roles of Alanine and Aspartate

It is important at the outset to realize that the end product of the β oxidation spiral is acetyl coenzyme A, which is channeled into the Krebs cycle at the citrate synthase-catalyzed reaction. In the rat, during the anaerobic-aerobic transition, activation of the β oxidation spiral leads to a momentary piling up of acetyl coenzyme A, as the oxaloacetate reserves for citrate synthesis are inadequate. In a diving animal, such a limitation could be crippling. Thus, at least in the dolphin, muscle the diving habit leads to another important metabolic requirement: a reliable source of oxaloacetate that can be turned on during fatty acid oxidation and Krebs cycle activation. One source of oxaloacetate could be aspartate, via the aspartate aminotransferase step. In marine mammals, such as the porpoise, the activity of this enzyme in muscle is up to 17 times higher than in terrestrial species, and its catalytic properties favor oxaloacetate production in the mitochondria during the anaerobic-aerobic transition (19). Glutamate, produced in the reaction, in turn transaminates with pyruvate to regenerate α-ketoglutarate; this process, catalyzed by alanine aminotransferase, which also occurs in unusually high amounts in muscle of marine mammals, leads to alanine accumulation. The total amount of alanine accumulated under such conditions is equal to the summed increase in concentration of all Krebs cycle intermediates (20). This fact is of fundamental importance because it emphasizes that alanine is perhaps the single best metabolite signal of the degree to which the Krebs cycle is activated. Hence it is not surprising that alanine is such a good inhibitor of pyruvate kinase in dolphin muscle, since the greater the degree of Krebs cycle activation, the greater the degree to which pyruvate kinase is blocked by alanine, and carbohydrate reserves are spared for anaerobic excursions.

Implications

The theme of our analysis is that muscle glycolysis, already a most impressive anaerobic machine, is further improved in its capacity and efficiency in the muscle of the propoise through only a modest number of modifications. Thus, the steady state concentrations of a few glycolytic enzymes are increased, reflecting a higher overall glycolytic potential and an improved capacity to maintain NAD^+:NADH ratios in anoxic stress. To retain control of the higher glycolytic capacity, at least two additional modifications are now known: 1) muscle fructose biphosphatase activity in porpoise muscle is one of the highest thus far reported for any animal species, the enzyme appearing to function in glycolytic control, and 2) muscle pyruvate kinase, although having a lower specific activity, occurs as a regulatory enzyme, sensitive to feedforward activation of fructose biphosphate and feedback inhibition by ATP, alanine, and citrate. The feedforward activation presumably functions during the aerobic-anaerobic transition in the dive, whereas the feedback inhibitions by ATP, alanine, and probably citrate (all acting in effect as end products of aerobic, fatty acid catabolism) appear to function during the anaerobic-aerobic transition at the end of diving. The latter characteristic emphasizes perhaps the most intriguing consequence of the diving habit: a metabolic organization in muscle that swings between an anaerobic, glycogen-based fermentation and an aerobic, fat-based oxidative metabolism. The control requirements imposed on muscle by this metabolic organization seem to call for unusually high levels of aspartate and alanine aminotransferases. The mitochondrial form of aspartate aminotransferase is designed to spark the Krebs cycle by increasing the availability of oxaloacetate at the same time that acetyl coenzyme A is being produced by β oxidation. Alanine aminotransferase regenerates the α-ketoglutarate required for this process and leads to the accumulation of alanine, which plays a key role in turning off glycolysis at this time.

If this theme is valid, at least in overall outline, it leads to some quite specific predictions on transient metabolite concentration changes in muscle during and after diving. Unfortunately, direct sampling of muscle has not been possible to date. On the plus side, however, it is fortunate that muscle perfusion is greatly reduced during diving and thus on reperfusion in recovery it is possible to see "pulses" of at least those metabolites that can diffuse out of the tissue into blood. Dominant among these is lactate, and it is satisfying that a hallmark feature of diving (lactate accumulation in peripheral tissues such as muscle and subsequent washout into the blood during recovery) would be predicted even if it had not already been repeatedly documented (1, 2, 4, 5). A corollary of that observation, that blood pyruvate levels should also rise, was established only recently (21).

Another expectation is that after diving alanine concentrations in muscle should increase, whereas aspartate levels should drop. There is no information on either amino acid. However, such an effect would explain the rise in whole blood alanine levels after diving (21). With respect to alanine and amino nitrogen metabolism in general, it is important to recall that glutamine is the chief form in which excess nitrogen is moved from muscle. In the rat, for example, 3 times as much amino nitrogen leaves muscle in the form of glutamine as it does in the form of alanine (22). It is probable, therefore, that a large part of intracellular alanine, formed in part during transition to aerobic fat oxidation, is quickly deposited in a glutamine sink, a process probably contributing to the rise in blood glutamine levels after diving. In the antarctic Weddell seal blood glutamine levels rise by about 150% from initial levels in the 200–300 μmol/ml range to about 500 μmol/ml of blood at 5–10 min after diving (B. Murphy and P. W. Hochachka, unpublished data).

Other predictions stemming from our view of muscle metabolic organization in diving animals (e.g., changing muscle levels of citrate, acetyl-CoA, long chain acyl-CoA, free CoA, carnitine, and acyl carnitine derivatives) must remain untested for the time being since there are no available data.

It is evident that, in terms of a complete understanding of muscle metabolism through a diving cycle, we are still very much in the dark; but the above discussion should supply a heuristic framework for further work. With respect to the other peripheral organs we know even less; two of these, however, the liver and kidney, are so well understood in other mammals that their probable metabolic situation in diving animals should be briefly discussed.

Kidney Metabolism

In the kidney of terrestrial mammals, the dominant and apparently preferred sources of carbon and energy are glutamine and lactate (23). During acidosis, glutamine contributes approximately twice as much (40%) as does lactate (22%) to total energy requirements of the kidney, whereas alkalosis results in the reverse of this pattern. Glucose and fatty acids are utilized as substrates, but each appears to contribute less than 20% to total respiration.

The above figures apply to the kidney as a whole but not to all of its parts. Oxidative metabolism proceeds without appreciable glycolysis in the cortex, whereas glycolysis with little oxidative activity characterizes the papillary tip, where the Pasteur effect is not demonstrable. The outer medulla, which contains the thick ascending limb, demonstrates both oxidative and anaerobic glycolytic activity and is theoretically capable of generating ATP under either O_2-rich or O_2-depleted conditions. The inner medulla has a large glycolytic capacity, but even its minor oxidative capability can generate as much, or slightly more, ATP than by glycolysis alone (see review in ref. 24).

An additionally important metabolic function of the kidney is to release glucose into the circulation under a variety of metabolic conditions (during starvation, diabetes mellitus, glucocorticoid stimulation, or parathyroid stimulation). Gluconeogenesis appears to be exclusively localized to the proximal tubules since PEP carboxykinase, a requisite enzyme for the process, is known from studies of microdissected tubules to be localized here but in no other regions of the nephron. Fatty acids produce a significant stimulation of gluconeogenesis from lactate as well as an inhibition of glucose oxidation in cortical slices, possibly because of pyruvate dehydrogenase inhibition, coupled with pyruvate carboxylase activation (24).

Under normal conditions, because of a large perfusion but small size, the kidney takes up more substrate than it can possibly completely oxidize to CO_2 and water. As a result, the occurrence of "anaerobic decarboxylations" can lead to CO_2 production without O_2 uptake, to *RQ* values over 1, and to a significant contribution to overall energy production from a rather unusual anaerobic metabolism (25). Yet, even with this potential plus that of glycolysis, the normal mammalian kidney cannot sustain anoxia for prolonged time periods; ATP levels in particular and the adenylate pool in general fall drastically in anoxia (26, 27). It is in this latter characteristic that the kidney of diving animals must most markedly differ from that of terrestrial mammals.

Impact of Diving on Kidney Metabolism In physiological terms, the striking feature of the kidney in diving animals, of course, is that it is treated as a peripheral organ and perfusion is strongly reduced during diving (28–30). In the harbor seal, the kidney can be completely cut off from circulation for at least an hour with no irreversible detrimental effects, and with essentially instantaneous recovery (31). Thus, there is no doubt that the seal kidney possesses a most capable anaerobic metabolism charged entirely by endogenous substrate. However, the details of that metabolism remain unexplored. How is redox balance in the giant mitochondria of the proximal tubules maintained during anoxia? What are the anaerobic sources of intramitochondrial ATP? What happens to hydrogen shuttles and other mitochondrial-cytosolic interactions? The answers to these kinds of questions remain to be ascertained.

Impact of Recovery on Kidney Metabolism During recovery from diving, the metabolic situation is somewhat clearer because at least the following information is available: pH is low, lactate and glutamine levels are high, whereas other parameters (P_{O_2}, P_{CO_2}, glucose levels, fatty acid levels) are probably normal or nearly so, either because they are not expected to change during diving (e.g., fatty acids) or because their return to normal is rapid compared to the lactate and glutamine recovery patterns (e.g., Po_2). Under these conditions, it is probable that glutamine initially contributes the largest fraction to overall energy metabolism, with lactate as the second most important carbon source. As the pH is stabilized, the importance of lactate as a substrate source probably rises (particularly when the washout is unusually large, leading to unusually high lactate levels), whereas that of glutamine gradually declines.

In other mammals, an adaptive increase in ammonia production in acidosis is well known and is thought to be mediated by 2-KGA effects on glutaminase and glutamine uptake; recent evidence suggests that acidosis leads to falling 2-KGA levels in the kidney; which in turn deinhibit the above two key processes in ammonia production (32). If the same events occur in the seal, it would imply that 2-KGA levels should be low at the end of prolonged diving, and that the return to normal levels should follow a time course that would parallel the return to a normal acid-base status.

We would expect most of the above events to be completed before lactate levels are fully returned to normal since the latter process can sometimes take 15–30 min or even longer. During this phase of recovery, when lactate levels are still high, we would expect the kidney to be actively involved in gluconeogenesis, a process probably stimulated by free fatty acids. The other major organ in which this process almost certainly is stimulated at this time is the liver.

Liver Metabolism

Impact of Diving on Liver Metabolism The liver in all mammals, including divers, is the major source of blood glucose for use in other parts of the body. During diving, at least in seals (6, 7), perfusion of the liver is greatly reduced, even more so than in other peripheral organs. Under these conditions, the pool of sugar in the blood is like a tub with the plug pulled and the tap turned off; all the tissues that remain perfused represent the potential drain on the blood glucose pool, whereas the vasoconstriction to the liver represents the closed tap. Not surprisingly, under diving conditions blood glucose levels may gradually decrease (4). At the same time, if diving is prolonged, the liver must be sustained by an even more important anaerobic contribution. However, nothing is known of when or how much anaerobic metabolism is needed in the liver of diving mammals, and this too is an obvious area for future research. In recovery, however, the situation, as in the kidney, seems more understandable.

Impact of Recovery on Liver Metabolism The reason we have a fair appreciation of metabolic events that must occur in the liver after diving is that a lot of blood metabolite information is available that bears on this matter. The following modus operandi appears most likely. On recovery, a high Po_2 and the re-establishment of liver perfusion are probably the first events that herald the return to normal oxidative metabolism. In metabolic terms, this situation should lead to energy-saturated cells quite quickly. At the same time, lactate is being washed out of peripheral organs in general and blood lactate levels are gradually rising. Fatty acid levels, being normally high, would favor the flow of lactate carbon into glucose.

At the same time, catecholamines and norepinephrine, in particular (R. Creasy, personal communication) (33), are known to be elevated in the blood (one reason why the adrenal gland remains perfused during diving). Of several metabolic consequences, a critical one in the liver is the activation of adenyl cyclase and the subsequent activation of the glycogen-mobilizing enzyme cascade. This process would favor the release of glucose from glycogen (11, 34).

These two events by themselves are capable of increasing glucose release from the liver, which in turn should be evident in rising blood glucose levels during recovery. Any gluconeogenesis from other precursors (e.g., glutamine) would simply augment the process. Not surprisingly, blood glucose often rises substantially during recovery from diving (4).

CENTRAL ORGANS

At this juncture we may ask what is known of the metabolic biochemistry of the heart-lung-brain "machine" in marine mammals. Incredibly enough, we know very little! We do know that, before a dive, the sea mammal increases the O_2 content of venous blood for use during diving and that the Po_2 falls to 30 mm or lower during prolonged dives (2, 5, 7), with an approximately 2–4-fold increase in lactic acid concentration in the central circulating blood (4). This increase in lactate could be due to anaerobic fermentation in one or more of the central organs and/or leakage of lactate out of tissues (such as the muscles) that have been isolated from the main circulation. Questions such as what fuel is burned by which tissues and at what times during a dive remain to be answered. In addition, there is the question of intertissue cycling. Are there any metabolites cycled, and, if so, which one(s) and between which tissues?

To put such questions in their proper perspective, it is useful to briefly review what is known of the metabolic biochemistry of these three tissues.

Substrates for the Mammalian Heart

The mammalian heart is considered to be a completely aerobic organ when in resting metabolism, or when only moderately active (35). This fact is further emphasized by the observation that 40% of the total cellular space is occupied by mitochondria (36). When there is an ample supply of oxygen, the heart preferentially oxidizes fatty acids, mostly in the form of plasma free fatty acids, but if these are unavailable, it can burn glucose or lactate (37, 38). In hypoxic or anoxic hearts the preferred fuel is glucose. Such a situation results in a 10- to 20-fold increase in glycolytic flux (35).

In the well-oxygenated fat-burning heart, glycolysis is inhibited at four loci. First, there appears to be an inhibition of glucose transport by free fatty acids. Second, the combined effects of high levels of ATP, G6P, and decreased levels of 5′-AMP and inorganic phosphate inhibit glycogen phosphorylase and hexokinase. Third, increased concentrations of citrate and ATP and decreased concentrations of fructose biphosphate, NH_4^+, and AMP during fatty acid oxidation essentially turn off glycolysis by inhibiting phosphofructokinase. This effectively directs exogenous glucose into glycogen. And, finally, high levels of acetyl-CoA during fatty acid oxidation inhibit the pyruvate dehydrogenase enzyme complex. Fat oxidation also appears to convert this enzyme into its inactive enzymic form (35). During periods of hypoxia and anoxia cardiac glycogenolysis and glycolysis are activated by a reversal of the above effects, whereas fatty acid oxidation is retarded. How fat oxidation is dampened in hypoxia has only recently been gradually clarified.

Fat Oxidation in the Hypoxic Heart

Experimentally, the ischemic, hypoxic, or anoxic perfused heart has been widely utilized in studies of how β oxidation is regulated during aerobic-anaerobic transitions. In such studies, Hull et al. (39) found a 10-fold accumulation of β-hydroxypalmitate in ischemic perfused hearts, the free acid form of this intermediate presumably having been formed from the true intermediate in β oxidation: β-hydroxypalmityl-CoA. In the anoxic perfused heart, Rabinowitz and Hercker (40) found evidence for a variety of shorter chain hydroxy fatty acids, also presumed to be formed from hydroxyacyl-CoA derivatives, which are the true intermediates in β oxidation. In similar studies of the ischemic and hypoxic heart, Whitmer et al. (41) found a pronounced accumulation of long chain acyl-CoA, concomitant with a potent depletion of acetyl-CoA. These studies all indicate that under certain conditions intermediates of the β oxidation spiral appear to accumulate. Of the two oxidation steps, an initial limitation apparently occurs at the NAD^+-dependent (rather than the FAD^+-dependent) oxidation step in the spiral, an interpretation that in fact predicts the accumulation of hydroxyacyl-CoA (as such, as the free acids, or as the carnitine derivatives). This conclusion is further supported by careful studies of isolated liver mitochondria (42). The latter studies point out that when the $NADH:NAD^+$ ratio is very high (to be expected, for example, under severe hypoxia) β oxidation appears to be blocked at the NAD^+-linked oxidative step; as a consequence, hydroxypalmityl carnitine, presumed to be formed from the true acyl-CoA β oxidation intermediate, is accumulated.

There may be good reason why, in these kinds of experiments, the first "block" to β oxidation appears at the NAD^+-linked dehydrogenase because, under hypoxic conditions, succinate is known to accumulate in the vertebrate heart (43) and this accumulation may serve to maintain an adequately oxidizing potential at the FAD^+-dependent dehydrogenase step. The situation here is complex and deserves careful analysis. One important point to bear in mind is that, in mitochondrial preparations exogenously added succinate may in fact lead to a reduction in β oxidation, and the block appears to occur at the FAD^+-linked acyl-CoA dehydrogenase (42). Because both succinate oxidation and acyl-CoA oxidation are FAD^+ linked and both dehydrogenases are thought to compete for $FADH_2$ oxidation by the ETS, one interpretation of the succinate-mediated block assumes that SDH is out-competing acyl-CoA dehydrogenase for ETS oxidizing power, thus leading to an accumulation of electrons and protons in $FADH_2$-linked acyl-CoA dehydrogenase. Bremer and Wotczak (42) appear to favor this interpretation, and it may be the correct one for their in vitro system.

In vivo, however, this competitive situation seems unlikely. It implies that succinate should be used up, not accumulated. But on this point the data are unequivocal in showing that succinate accumulates in the hypoxic or anoxic heart (43, 44). Cascarano and his co-workers (43) have established that at least one route to succinate, probably the predominant one, involves a reversal of a section of the Krebs cycle:

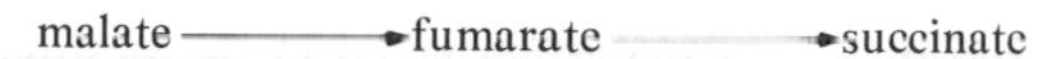

The malate itself may arise from aspartate through a coupled transamination:

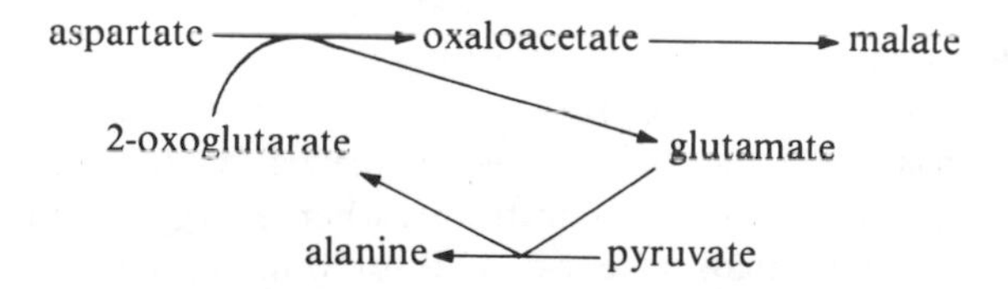

a process that would explain the accumulation of alanine that sometimes occurs during hypoxic or anoxic stress (21, 45). Collicutt and Hochachka (46) have shown that this effective "fermentation" of aspartate to succinate is a very active process in bivalve hearts during anoxia, and there is also some evidence for its contribution to metabolism during the hypoxia associated with diving in marine mammals (19).

Succinate can, of course, also be formed from intermediates in the first span of the Krebs cycle. Indeed, from recent studies we know that acetyl-CoA is quickly depleted in hypoxia, as are citrate and 2-ketoglutarate (47). In the anoxic bivalve heart, the [^{14}C]glutamate that is taken up is converted to succinate (48).

From these different studies, we tentatively conclude that, when O_2 supply is limited, succinate acts as a sink for carbon flowing from both arms of the Krebs cycle. It is easy to see why the first span operates until intermediates (and O_2) are largely depleted. The difficult question is why the second span of the Krebs cycle (malate $\rightarrow$ succinate) should be reversed. We, of course, do not know for certain. One possibility is that "fumarate reductase" and FAD^+-linked acyl-CoA dehydrogenase form a redox couple, in which case fumarate reduction to succinate in effect prevents a block at this level in β oxidation. This redox couple could be linked by the electron transfer of flavoprotein that is associated with acyl-CoA dehydrogenase and the ETS.

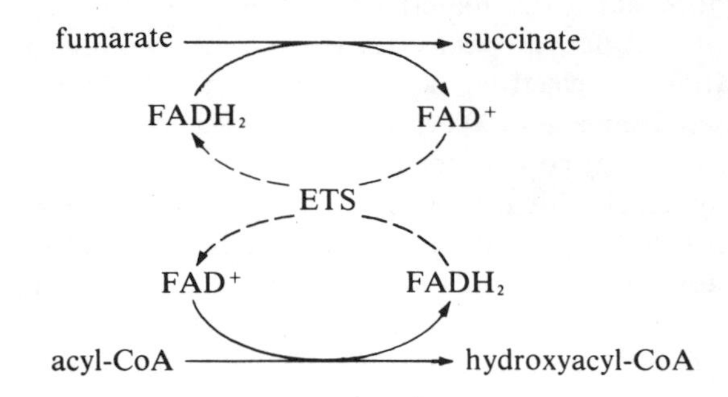

If the above considerations are correct, it is not surprising that the first step to be "turned off" in β oxidation when O_2 is limiting is the NAD^+-linked, and not the FAD^+-linked, dehydrogenase. Moreover, there may be a number of advantages to this arrangement. First, under reducing conditions in the cell, malate dehydrogenase function in the OXA → malate direction would contribute to maintaining redox balance. Second, fumarate reduction to succinate itself contributes to retaining a functional redox balance, as already indicated above. Third, fumarate reduction is thought to be coupled to a phosphorylation, 1 mol of ATP being formed per mole of succinate accumulated, and, under hypoxic conditions when energy production may be at a premium, this small rate of ATP production may be significant. Cascarano and his co-workers (49) believe that it is fundamental to sustaining Ca^{2+} metabolism of the heart during anoxic stress.

Heart Metabolic Biochemistry in Diving Animals

Unlike the well developed field briefly summarized above, there are only scattered reports in the literature on the metabolic biochemistry of the heart in diving animals. Electrophoretic work and pyruvate saturation kinetics of lactate dehydrogenase (LDH) from the hearts of common seals (*Phoca vitulina vitulina* L.), beaver (*Castor fiber*), pond turtle (*Pseudemys scripta elegans*), eider (*Somateria mollissima*), and the hooded seal (*Cystophora cristata*) have been reported (50–53). All studies have demonstrated that the isozyme distribution and/or the pyruvate saturation kinetics in the hearts of diving mammals approach the muscle type of enzyme (LDH-5). The hearts of nondiving mammals are known to contain predominantly heart-type LDH subunits (54). Simon et al. (14) showed that there was increasing pyruvate kinase activity in both the hearts and brains of three different seals with increasing diving time. This, they suggested, implies enhanced glycolytic potential of aerobic tissues and may be important in extending diving time. Kerem et al. (55) found that glycogen concentrations in the cardiac muscle of the Weddell seal are 2–3 times higher than in the hearts of terrestrial mammals, again suggesting an enhanced anaerobic capacity.

Brain Metabolic Rate

Although the mammalian brain is a relatively small organ, its basal oxidative metabolic rate ($CMRO_2$) is high and can account for as much as 20% of the basal metabolic rate of the organism. The human brain utilizes O_2 at an average rate of about 1.5 μmol/g/min at 37°C, with the more active cerebral cortex metabolizing at a substantially higher rate. Cerebral metabolic rate also seems to be influenced by the size of the organism, being high in small animals. The occurrence of so-called vulnerable areas in the brain plus this high oxidative capacity has encouraged the view that the brain is one of the most O_2-dependent organs in the mammalian body (for reviews see refs. 56 and 57).

Glucose as Substrate for the Brain

The unusual oxidative metabolism of the brain is fired predominantly by glucose, upon which the brain is by and large assumed to have an absolute dependence. Glucose reaches the brain cells by diffusing down a concentration gradient, from a plasma level of 6 mM to a CSF and extracellular fluid concentration of 4 mM by means of a stereospecific facilitated mechanism. It then enters cerebral cells, where concentrations are even lower (0.1–3 mM). The K_m for facilitated capillary transfer is approximately 8 mM, for cell uptake 5 mM, and for incorporation into metabolism 0.05 mM. Therefore, the affinity for glucose (measured as $1/K_m$) at each transfer step increases sequentially. A falling concentration and increasing affinity help to maintain a large net flow of glucose from plasma into cerebral metabolic processes.

The rate of cerebral glucose consumption is normally regulated by the rate of conversion of fructose 6-phosphate into fructose 1,6-biphosphate by the phosphofructokinase reaction and not by the delivery rate at cerebral capillaries. The rate of consumption can be accelerated severalfold when demand increases (e.g., in hypoxia). In these cases deinhibition of phosphofructokinase reduces the concentration of both fructose 6-phosphate and its metabolic precursor glucose 6-phosphate. Thus, the net effect is to increase the rate of glucose phosphorylation by the hexokinase reaction (see ref. 58).

Alternate Substrates for the Brain

No substrate can replace glucose in sustaining brain function indefinitely. However, a number of other substrates can be utilized by the brain, and, indeed, some of these are also absolutely essential for certain metabolic processes. Of these, ketone bodies and amino acids are the most important, whereas lactate and pyruvate may take on a significance when blood concentrations are high (59, 60).

The ketone bodies, β-hydroxybutyrate and acetoacetate, can partially support brain function when carbohydrate is in short supply or cannot be utilized. These substrates are generated by the liver and kidney in starvation and diabetic ketoacidosis, and occur in high plasma concentrations in newborn suckling rats. They enter the brain by simple passive diffusion.

At very high blood levels, brain uptake of ketone bodies is limited by metabolic incorporation, which, in turn, depends on intracerebral concentrations of the enzymes involved in acetoacetate-hydroxybutyrate metabolism (61). The concentrations of these enzymes, as well as rates of ketone body consumption, are higher in the brains of neonates than in the brains of adults.

Of the amino acids, glutamate, glutamine, aspartate, *N*-acetylaspartate, and GABA are the predominant amino acids of the mature brain and constitute approximately two-thirds of free α-amino nitrogen. The high concentrations and extensive metabolism of glutamate and its derivatives are key hallmarks of brain metabolism. The product of the α decarboxylation of glutamic acid is GABA, which has been reported to be a major inhibitory transmitter in the vertebrate central nervous system (58).

The large pool of free glutamate in the brain is in equilibrium with the α-ketoglutarate of the Krebs cycle, and aspartate is in equilibrium with oxaloacetate. After injection of labeled glucose, 70% of the isotope present in the soluble fraction of brain is present in amino acids, primarily glutamate, glutamine, aspartate, and GABA (62).

Under normal circumstances, lactate and pyruvate are released from the brain into the circulation at rates that account for the usage of about 10% of the glucose. However, under exceptional conditions, when blood levels are high (for example, after trauma or ischemia), lactate and pyruvate can be utilized by the brain (56). Lactate and pyruvate, as well as other short chain monocarboxylates (such as acetate, proprionate, and butyrate), cross cerebral capillaries by a common facilitated mechanism (63). The lactate transport capacity, however, is low and apparently saturates at 3–4 times the normal plasma concentration of lactate; that presumably is why lactate uptake becomes significant only at high blood lactate levels.

Circulatory and Metabolic Consequences of Brain Hypoxia

According to Siesjo (57) brain hypoxia can be defined as a decrease in O_2 availability of such magnitude that there are measurable changes in metabolism or function of the brain. Since hypoxia may occur in prolonged diving (1, 2, 5, 9) it is important to emphasize that hypoxia differs from anoxia (or complete ischemia) in two ways. First, since tissue P_{O_2} is not reduced to zero, oxidative metabolism continues even if at a reduced rate. Second, since cerebral blood flow (CBF) is maintained, there is a continuous supply of glucose for continued anaerobic metabolism.

It is now well established that hypoxia in mammals leads to a compensatory increase in CBF, whereas hyperoxia leads to a compensatory decrease

(56, 57). Interestingly, variations in Pa_{O_2} from as high as 2,100 mm Hg to as low as 20 mm Hg, with Pv_{O_2} varying between 57 mm and about 15 mm Hg, do not lead to any changes in $CMRO_2$ or the energy states as assessed by adenylate levels. As Siesjo (57) argues, this suggests that the cytochrome oxidase reaction runs at a constant rate despite rather pronounced variations in tissue O_2 levels. Mechanisms underlying this obvious close control of O_2 utilization in the face of wide changes in tissue O_2 availability are not fully clarified (see ref. 57 for a discussion).

Another matter that is unclear is why brain perfusion varies over P_{O_2} ranges that do not cause changes in $CMRO_2$. The earliest detectable metabolic changes in hypoxia are increases in lactate and pyruvate levels and a change in the NADH:NAD^+ ratio, which is expressed by changes in the lactate to pyruvate ratio. These changes occur at a P_{O_2} of about 50 mm Hg, when $CMRO_2$ is still unchanged (57). At face value, the data imply that both glycolytic and aerobic production of ATP are increased with progressive hypoxia, and that the total energy produced in terms of μmol/ATP/g/min must presumably rise. Why this should be so is not at all understood.

A number of metabolic consequences are now well outlined for prolonged hypoxia. Thus, lactate accumulation and glucose utilization both remain elevated. Pyruvate elevation is also sustained and is thought to activate pyruvate carboxylase-catalyzed conversion to oxaloacetate (64) and alanine aminotransferase-catalyzed conversion to alanine (57).

During maintained hypoxia, there also occurs a redistribution of carbon in the Krebs cycle pool as well as a general increase in the size of the pool (57) that is in part initiated by pyruvate and in part by aspartate:

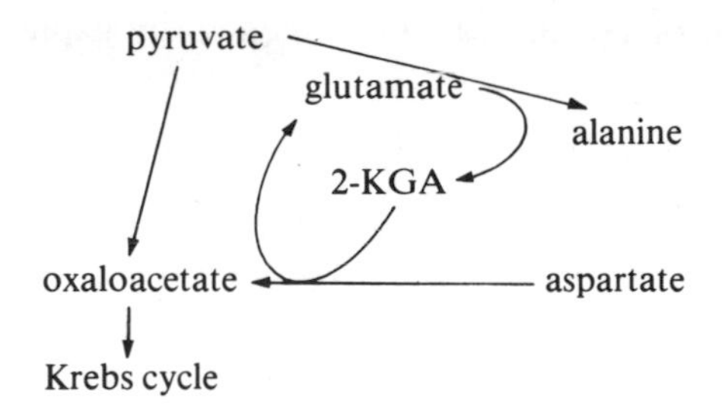

These processes explain rising alanine and dropping aspartate levels coincident with the increase in Krebs cycle pool size observed at this time. Overall, these adjustments are similar to those to be anticipated when the Krebs cycle is activated (20), but they occur in the hypoxic brain, presumably due to a gradual decrease in respiration rates and the increasingly reduced state of mitochondrial metabolism.

Although it has been clearly established that functional changes (e.g., slowing of the EEG) occur in hypoxia, mechanisms underlying the derangement are still speculative. The classical vulnerable regions include small neurons in the neocortex and hippocampus and Purkinje cells in the cerebellum (57). Recent rather provocative studies imply that incomplete ischemia is more damaging to the brain than complete ischemia (65), perhaps because the latter condition prevents O_2-dependent autolytic processes that may damage cell membranes.

Metabolic Consequences of Brain Anoxia

Although it is not useful to summarize the vast literature on anoxia in the mammalian brain, some of the metabolic consequences of brain anoxia should be briefly discussed because they put the above data in sharper focus. If complete ischemia occurs abruptly, aerobic energy production drops to zero almost immediately, in about 4 s in man and in about 1–2 s in the rat cerebral cortex (57). At the same time, phosphofructokinase is activated, probably due to a drop in ATP and creatine phosphate levels and a rise in AMP, ADP, and P_i levels. These changes lead to maximum activation of brain glycolysis, which ultimately leads to complete depletion of glucose and glycogen and the accumulation of lactate, alanine, and succinate. In large animals, the process occurs more slowly than in small ones, presumably due to a lower activity of glycolytic enzymes (see ref. 66).

Interestingly, although lactate accumulates in proportion to the preischemic levels of glucose and glycogen, pyruvate does not build up, probably because the lactate dehydrogenase reaction is shifted toward lactate due to NADH and H^+ accumulation. During complete ischemia, the pool of Krebs cycle intermediates increases substantially and there again occurs a significant redistribution of carbon between the various Krebs cycle components. Key among these are a depletion of 2-ketoglutarate (2-KGA) and oxaloacetate concomitant with an accumulation of succinate.

The depletion of 2-KGA can be satisfactorily explained by reductive amination.

$$\text{2-KGA} + \text{NADH} + H^+ + NH_3 \rightarrow \text{glutamate} + \text{NAD}^+$$

This reaction must be shifted to the right since ischemia is associated with increases in the concentration of NADH, H^+, and NH_3.

The fall in OAA concentration probably is due to a redox-dependent shift in the malate dehydrogenase (MDH) reaction.

$$\text{OXA} + \text{NADH} + H^+ \rightarrow \text{malate} + \text{NAD}^+$$

The depletion of OXA also may be due to carboxylation of PEP and flux of carbon skeletons from the glycolytic chain to the citric acid cycle. Such CO_2 fixation would channel carbon skeletons into the Krebs cycle at the level of OXA and malate at a time when redox conditions favor a reversal of the reactions catalyzed by MDH, fumarase, and succinate dehydrogenase. Succinate accumulates as a result. Succinate thus is viewed as an end product of anaerobic metabolism in the brain, as well as in invertebrate tissues (67).

There are few changes in amino acid concentrations in ischemia except for significant increases in alanine and GABA concentrations. Alanine serves as a pyruvate carbon sink due to a shift in the alanine aminotransferase reaction:

$$\text{pyruvate} + \text{glutamate} \rightarrow \text{2-KGA} + \text{alanine.}$$

Since pyruvate does not accumulate, this shift is probably achieved by the fall in 2-KGA concentration. In the initial phases of ischemia, the increase in the Krebs cycle pool size exceeds the increase in alanine concentration. That is why other anaplerotic reactions must contribute, such as CO_2 fixation at the stage of phosphoenolpyruvate or pyruvate. Although the data do not allow establishing stoichiometries, it is clear that more glycolytically derived carbon flows into the Krebs cycle pool than is indicated by the pool size, since there occurs a simultaneous loss to glutamate via the glutamate dehydrogenase reaction.

Brain Metabolism in Diving Animals

With respect to diving animals, most of the biochemical work reported on cardiac tissue was also performed on the brain with basically the same overall results, all pointing to an increased capacity to sustain periods of hypoxia or anoxia. Physiological studies (55, 68, 69) also have prompted most researchers to infer improved hypoxia tolerance. However, convincing data showing a reliance of the brain on anaerobic metabolism during diving are difficult to come by; in fact, much of the available information implies a relatively normal aerobic brain metabolism. According to Bryan and Jones (70), for example, the $NADH:NAD^+$ ratio in the cerebral hemispheres increases by nearly 40% after 1 min of apneic asphyxia in the fowl, whereas it only increases by 15% in the duck after 2 min of asphyxia. However, at a given level of brain tissue Po_2, both species show the same relative increase in the $NADH:NAD^+$ ratio, implying that both species are equally dependent on an adequate Po_2 for the maintenance of oxidative metabolism in the brain. Moreover, both species show an isoelectric EEG when the $NADH:NAD^+$ ratio increases to within 30–40% of maximum (that occurring at death). Finally, prevention of bradycardia in ducks during apneic asphyxia by atropine causes the $NADH:NAD^+$ ratio to increase at the same rate as in the nonatropinized fowl. The authors, therefore, conclude that O_2-conserving cardiovascular adjustments are responsible for the increased cerebral tolerance to apneic asphyxia in ducks, with no specific biochemical mechanisms being involved. However, the noninvasive fluorometric technique used yields information only upon the redox state of the mitochondria and yields no insight into cytoplasmic (i.e., glycolytic) events. Along with a large increase in cerebral blood flow, there automatically occurs a greatly increased delivery of glucose for cerebral energy metabolism in the duck during diving (or during apneic asphyxia). If aerobic glucose metabolism is unchanged, what happens to the excess supply of this carbon and energy source remains to be clarified. Is it fermented? Is it "dumped" into the free amino acid pool? Or does it have some other fate? Only further work will ascertain the answers.

Other indicators of the metabolic status of the brain of divers are helpful but, thus far at least, are also equivocal. Blix (71), for example, found concentrations of creatine phosphate to be similar in seals and sheep. LDH isozyme patterns in the brain of the Weddell seal (72) and at least two cetaceans (W. Vogel, personal communication) imply that both heart- and muscle-type subunits are involved in generating LDH holoenzymes; however, heart-type subunits definitely predominate and overall LDH isozyme patterns are, therefore, not strikingly different from those observed in terrestrial mammals. In this connection, Shoubridge et al. (73) found a rough correlation between the fraction of muscle-type subunits in brain LDH's and depth (thus duration?) of diving in large whales.

All the above kinds of data, of course, are circumstantial. Rigorous proof of the metabolic status of the brain in diving mammals must come either from direct tissue sampling (thus far unavailable) or from AV gradients across the brain. The latter are in fact available in three groups: the porpoise, the sea lion, and the Weddell seal. Preliminary studies carried out in 1970 by Ridgeway and his collaborators on porpoises and the sea lion show a number of important features (personal communication). First, during breath-hold simulated diving for periods of up to 8 min, blood glucose and lactate levels are unchanged in both the dall and delphinid porpoises. Second, sampling of arterial and central venous sinus ports indicates a substantial AV glucose gradient and thus substantial glucose utilization by the porpoise brain. The gradient drops significantly during diving, with no change in the fraction of glucose released as lactate, implying either that the basal metabolism of the brain drops during diving or more likely that brain perfusion increases during diving, as occurs in the duck (70, 74) and in terrestrial animals during hypoxia (75). Third, a significant fraction of the glucose taken up by the brain is released as lactate into the blood, but the fraction does not seem to change much during diving. Fourth, sampling of venous blood at other ports (e.g., the portal veins) simultaneously with arterial ports indicates substantial glucose utilization by the anterior and posterior quarters of the porpoise; usually these gradients are reduced during diving, as might be anticipated because of reduced perfusion of the peripheral organs.

These are exciting results and, along with similar data for the sea lion (S. H. Ridgeway, personal communication), clearly establish that at least a major fraction of glucose taken up by the brain during these experimental dives must be oxidized fully. Whether or not anaerobic glycolysis is increased cannot, however, be assessed from the data. Since AV gradients for lactate across the brain usually do not change, but blood flow probably rises, it is probable that anaerobic glycolysis is actually somewhat activated. However, only further work can assess this with greater certainty.

Be that as it may, there is a paradox in the porpoise studies that must be emphasized; namely, although the brain, the anterior body, and the posterior body in general are identified (by AV gradients) as utilizing glucose (and producing some lactate) during diving, there occurs no measurable overall decrease in blood glucose concentrations. The two sets of observations can only mean that there is in these species an extremely close regulation of glucose utilization and glucose production, but to date nothing is known of how this is achieved. A similar phenomenon is observed in the Weddell seal, although in this species there always is observed a drop in blood glucose concentration during first phases of diving. The rate of change slows down in later phases and approaches an asymptote toward the end of long dives (4). Again, these data can best be explained by assuming a close coordination of glucose utilization and glucose production. Since the liver is not well perfused during diving, the question remains as to whether or not it can fully account for the required rates of glucose release.

Lung Metabolism

Eight decades ago it was suggested that the lung may play important metabolic roles (76), yet the mapping of metabolic organization in the mammalian lung has only recently been initiated, and surprisingly little is known about it compared to what is known about other tissues, such as the brain or heart. This stems not only from a general lack of interest by biochemists but also because of the many difficulties of working with the lung both in vitro and in vivo. What information is known is still quite sketchy. Tierney and coworkers (77), using perfused rat lungs and tissue slices, demonstrated that under conditions of normal oxygen pressures the lung ferments approximately 30% of the glucose utilized. Glucose has been postulated to play roles in 1) energy production, especially in some of the more anaerobic lung cells; 2) the synthesis of NADPH (via the pentose phosphate shunt) for the rapid reduction of oxidants; and 3) supplying acetyl-CoA and α-glycerol phosphate for lipid synthesis and for the synthesis of surfactants, which control the surface tension of air-liquid interfaces of alveoli.

Whether glucose is absolutely essential for mammalian lung function, however, is still under investigation. Tierney (77) suggests that the isolated rat lung can survive for up to 2 hr without any exogenous glucose, implying utilization of other substrates. Recently, two reports indicate that lactate is a good substrate in the lung, particularly under conditions of high lactate concentrations (78, 79). Cascarano et al. (49) speculate that the lungs of rats, subject to low tensions of oxygen, actually take up the anaerobic end product, succinate, produced by other tissues, and release fumarate and malate for further use in energy metabolism by the hypoxic heart. This process, they suggest, sets up an electron shuttle between the heart and the lungs, but this postulate has not yet been adequately tested, either in diving mammals or in terrestrial ones.

HEART-LUNG-BRAIN METABOLISM IN THE WEDDELL SEAL

In Scholander's original formulation, the diving reflexes were seen to play a primary role in conserving O_2 for the most O_2-dependent tissues. Of these, the heart, lung, and brain were considered the most important. However, in most physiological studies, these have been considered separately, if at all (e.g., in ref. 80), whereas in metabolic studies they have usually been lumped together as a "black box" (1, 12). To date, only one attempt has been made to generate a comprehensive outline of the metabolic properties of the heart, lung, and brain during prolonged diving and recovery. Although the study (72) is ongoing, initial results so strongly influence our view of this area that they need to be briefly described.

In our study of the Weddell seal, a variety of techniques was used to approach the problem at various levels of organization. These included 1) enzyme profiles of all three organs, 2) ultrastructure studies of the heart, 3) tissue slice studies with ^{14}C metabolites, 4) in vivo metabolite infusions with ^{14}C-labeled and -unlabeled metabolites, and finally 5) transient changes of blood metabolite levels prior to, during, and after diving. In addition, in closely collaborative studies, organ perfusion, heart rate pressure, cardiac output, blood gases, and blood pH were simultaneously monitored (7). Thus, physiological parameters could be used to assist in interpreting biochemical data and vice versa.

Enzymes of Aerobic Metabolism

As in the case of skeletal muscle, perhaps the most striking feature of the Weddell seal heart, lung, and brain is not how different, but how similar, the levels of enzyme activities are to those in terrestrial mammals. Four mitochondrial marker enzymes were used to gain an impression of oxidative capacities: acetoacetyl-CoA dehydrogenase and β-hydroxybutyrate dehydrogenase, functioning in β oxidation and ketone body metabolism, respectively; citrate synthase, catalyzing the entry of acetyl-CoA carbon into the Krebs cycle; and glutamate dehydrogenase, catalyzing the entry of glutamate carbon into the Krebs cycle. The activities of these four enzymes, in terms of micromoles of product formed per minute per gram of tissue in the heart, lung, and brain of the Weddell seal, are closely similar to those seen in other mammals (ox and rat tissues being used for comparison). The implication is clear. When O_2 is available, all three organs can sustain oxidative metabolism at levels that may be expected in other species. The same is not true for glycolytic enzymes.

Enzymes of Anaerobic Metabolism

Four enzymes were chosen to gain an impression of anaerobic potentials: HK, PFK, and PK, which are all potential regulating sites in anaerobic glycolysis, and LDH, catalyzing the terminal step in glycolysis. In the Weddell seal brain, HK, PFK, and LDH all occur at 3- to 5-fold higher levels than in the ox brain, whereas PK occurs at only slightly higher activities. In the lung, all glycolytic enzymes assayed, except for LDH, occur in levels similar to those in the ox lung. LDH occurs at slightly higher levels. In the heart, only LDH actively stands out when compared to other species, occurring at levels of about 1,000 μmol of product per minute per gram at 37°C. Except for the lungfish (81), this is the highest level of LDH in any vertebrate heart thus far studied, and it undoubtedly is the highest thus far seen among mammals.

Electrophoretic studies indicate that in all three organs both heart- and muscle-type LDH subunits are synthesized; thus, all five isozymes occur in all three central organs. In the heart and brain, however, the heart-type LDH subunits occur in somewhat higher activities. Nevertheless, all three organs clearly have the potential either for lactate production, catalyzed most effectively by muscle-type LDH's, or for lactate utilization, catalyzed most effectively by heart-type LDH's (for a review of LDH isozyme function and structure, see ref. 82). Although these enzyme profiles yield only an indication of what is or is not emphasized in the heart, lung, and brain of the Weddell seal, it is generally confirmed by in vivo metabolite measurements.

Blood Glucose and Lactate Profiles

A measurable drop in glucose concentration is consistently seen during short- and long-term dives (4) in samples of whole blood taken either from the pulmonary artery (mixed venous) or the aorta. That a notable fraction of the glucose used is being fermented is indicated by two observations: 1) if all the glucose used were being fully oxidized metabolic rates would be far higher than expected and P_{O_2} would drop further than observed; 2) even in dives of relatively short duration, a large fraction (50% or more) of the glucose used may be represented by a concomitant lactate accumulation. In more extreme situations, the fall in blood glucose concentration is approximately equal to the rise in that of lactate (4).

Initially, it was unclear whether lactate in the central circulation was due to glucose fermentation in the lung, heart, or brain, or whether it was due to "leakage" from peripheral organs. One way of settling this issue was to examine arterio-venous (AV) concentrations to assess utilization and/or production of different metabolites.

Arteriovenous Gradients Across the Lung

AV gradients across the lung were determined by sampling in the pulmonary artery and the left ventricle or aorta. These data established that, in the Weddell seal, the lung takes up lactate from the blood during diving and through most of the recovery time; glucose is either not taken up or may be actually released in small amounts (4).

Lactate as a Substrate for the Lung

One fate of lactate taken up by the Weddell seal lung is oxidation. In vivo injection of a bolus of [^{14}C] lactate into the pulmonary artery during both diving and nondiving states leads to an immediate pulse of $^{14}CO_2$ in the left ventricle that can only be generated by lung metabolism. Lung slices isolated in Ringers solution also vigorously oxidize lactate, whereas [6-^{14}C]glucose is only slowly oxidized (72). Similar results, in fact, have also been noted for the rat lung (78, 79). These data clearly rule out the lung as contributing to blood lactate accumulation and glucose depletion during diving. Moreover, they imply that lung metabolism is always essentially aerobic even in prolonged diving.

Arteriovenous Gradients Across the Brain

AV gradients across the brain of seals can be determined by sampling in the left ventricle and the epidural vein close to the occiput. Except in one short (10-min) dive such sampling in the Weddell seal indicates that glucose is taken up by the brain in measurable amounts, with a significant fraction (up to about 20%) of the glucose taken up accounted for by lactate release. Typically, there is no increase in the fraction of glucose that appears as lactate during diving or recovery when compared to the prediving condition. In the rat, the brain releases lactate because of a limitation at the pyruvate dehydrogenase reactions (83) and we assume the same mechanism is operative in the Weddell seal brain.

In extreme situations, however, when the lactate washout from peripheral tissues is high and lactate concentrations rise, the brain turns from net lactate release to net lactate uptake. This novel situation, previously observed only once in rats when lactate to glucose concentration ratios were high (59, 60), can be duplicated by lactate infusion. When blood lactate concentrations are thus artificially raised to over about 7 μmol/ml, the Weddell seal brain turns from net lactate release to net lactate uptake (72).

These data suggest that the brain is definitely involved in the fall noted in blood glucose levels during diving but the lactate released does not seem to be high enough to account for the observed 2- to 4-fold increase seen at this time. This leaves the heart and peripheral leakage as the two remaining possible sources of lactate.

Metabolic Properties of the Weddell Seal Heart

At least four observations are consistent with the idea that the heart may contribute to lactate accumulating centrally during diving (72): 1) heart glycogen stores are the highest thus far found among mammals (55); 2) glycogen is stored as large diameter α rosettes rather than the usual β particles, a storage pattern typically seen only in organs storing unusual levels of glycogen (P. W. Hochachka and W. C. Hulbert, unpublished data); 3) heart LDH activity is the highest known, as already mentioned; 4) heart LDH is potentially bifunctional, the isozymes present favoring either lactate formation or lactate utilization, depending upon metabolic circumstances. All of the above are consistent with a high potential for glycogenolysis whenever O_2 becomes limiting. Whether or not this occurs is not known but it should be emphasized that coronary perfusion appears to drop by a large factor in the Weddell seal (7) as well as in the grey seal (6). Its reduced perfusion may correlate with the greater potential of the heart for anaerobic glycogenolysis, as well as with the reduced work load during diving bradycardia.

Because AV gradients could not be obtained around the heart, we can only surmise the most likely metabolic events during recovery from diving. Since enzymes of aerobic metabolism seem to occur at normal levels, reperfusion and reoxygenation should quickly re-establish oxidative metabolism. During the lactate washout period, lactate undoubtedly serves as at least one carbon and energy source for the heart in the Weddell seal, as it does for the heart under conditions of high blood lactate in other mammalian species as well.

Metabolic Model of the Heart, Lung, and Brain in Divers

From the above data, the outlines of at least some major metabolic processes in the heart, lung, and brain of the Weddell seal can be discerned, allowing us to construct a simple, two-part model of the process during diving and recovery. According to this model, during diving the heart, lung, and brain all display different substrate preferences. The heart appears to preferentially utilize glycogen, releasing CO_2, H_2O, and lactate as end products, the latter becoming ever more important as O_2 supplies to the heart diminish in prolonged diving. Some of the lactate formed here and in the peripheral tissues spills out into the circulation and is utilized by the lung in preference to glucose, a process that dampens somewhat the accumulation of lactate in the actual dive and spares glucose for the brain. The brain utilizes glucose and releases CO_2, H_2O, and some lactate, an organization not too different from that in other mammals; however, the fraction of glucose that can be released as lactate appears to be higher than in terrestrial species.

During diving, the brain and lung are thus seen to be sustained by an aerobic, glucose- and lactate-primed metabolism, whereas the heart (perhaps because of a reduced perfusion) may have to rely more upon anaerobic metabolism, particularly upon glycogenolysis. During this time, many peripheral organs also experience a reduced perfusion and also gradually accumulate lactate.

As recognized over four decades ago, in the recovery process lactate is flushed out of the peripheral tissues into the blood. Probably the largest fraction of this lactate is converted to glucose in the liver (and kidneys), a process that leads to rising glucose levels in the blood. Nevertheless, a large fraction of this lactate is probably fully oxidized, but in this process all three central organs can participate, since the heart, lung, and brain appear metabolically capable of utilizing lactate as a carbon and energy source. Indeed, the heart and brain have unusually high levels of LDH, kinetically suited for uptake (as well as production) of lactate, and this feature can be viewed as an additional enzyme adjustment to the diving habit. Finally, it is interesting to note that this arrangement of lactate utilization by the heart, lung, and brain during recovery spares glucose for use in subsequent diving.

WHAT THE SEAL FETUS DOES WHEN ITS MOTHER DIVES

Direct Evidence for Fetal Bradycardia

As is evident above, the metabolic consequences of diving in the adult are now fairly well appreciated, at least in overall outline. What has not been clarified at all, however, is the metabolic status of the seal fetus during maternal diving. Zapol, Liggins, and their colleagues reasoned that there are two possibilities: either the fetus simply tolerates the consequences of the maternal dive or it too evokes the diving reflex. With respect to bradycardia, direct measurements implicate the latter strategy. Thus, soon after the maternal bradycardia is elicited, a fetal bradycardia also develops, heart rates typically falling from about 120 to 40 beats/min (7). But is the fetal bradycardia associated with a peripheral vasoconstriction?

Unfortunately, direct measurements here are unavailable. However, if the response is utilized by the fetus, then a lactate and pyruvate washout from peripheral tissues similar to that in adult seals should occur. Indeed, Scholander (5) took advantage of the lactate washout in his original study to provide evidence for peripheral vasoconstriction during diving in adult seals.

Glucose, Lactate, and Pyruvate Profiles

Fetal blood glucose levels during diving differ from maternal ones in two important regards. First, fetal blood glucose concentrations are always substantially higher than in maternal blood. Two mechanisms underlying this unusual situation are possible: either the placenta in the Weddell seal can actively transport glucose (from maternal to fetal side) or fetal red blood cells sustain higher steady state concentrations of glucose than do maternal red cells. Of these, the latter is considered the more likely. Be that as it may, what is to be emphasized is the obvious buffering against hypoglycemia offered to the fetus by this arrangement.

A second important difference between fetus and mother is that, in the fetus, blood glucose levels typically rise somewhat as a consequence of diving. Blood glucose levels, of course, fall during diving in the mother. Since the fetus of the species is known to mobilize liver glycogen under stressful conditions, a process leading to increased blood glucose levels (84), it is probable that the same or a similar process is activated in the seal fetus during maternal diving. In any event, there is no doubt that blood glucose regulation of the fetus is relatively independent of that in the mother; otherwise, it would be difficult to understand how fetal concentrations could be rising at the same time that maternal concentrations are falling.

Despite such differences in glucose handling, there is a striking similarity between fetal and maternal profiles of blood lactate and pyruvate. In the fetus, as in the mother, blood lactate increases during the dive, although the fetal response tends to lag somewhat behind the maternal one. Then, after the dive, there occurs the usual washout of lactate and pyruvate from peripheral tissues and hence the recovery spikes in lactate and pyruvate concentrations in the maternal blood. Again, the fetal response lags behind the maternal one. Peak concentrations are usually, but not always, lower in the fetus; however, in the recovery process, the placental gradients for lactate and pyruvate can be reversed. For these reasons, although transplacental diffusion of lactate and pyruvate from the mother to the fetus may complicate the postdiving patterns in fetal blood, it probably cannot account for the observed result. Thus, we tentatively assume that the large concentration changes in fetal blood lactate and pyruvate through diving and recovery are due predominantly to events within the fetus.

Important physiological implications stem from this assumption, since it suggests that the lactate and pyruvate recovery patterns are due to fetal peripheral vasoconstriction during the dive, followed by a washout in recovery. Although direct perfusion measurements are needed to establish this with greater certainty, the metabolite data, taken together with measurements of heart rate, indicate that the two key components of the diving reflex (bradycardia plus peripheral vasoconstriction) are both developed in the near-term seal fetus, and appear to be an integral part of the fetal adaptation response to maternal diving.

LONG-TERM DIVING CAPACITY

Diving Response Cannot Account for Long-term Capacity

Even given the impressive O_2-storage mechanisms, the improved anaerobic capacity of each organ taken individually, and the associated physiological responses, one question still haunts the metabolic biochemist: are these mechanisms adequate to account for the long-term diving capacities that we know from field studies many of these organisms are capable of performing? The question would seem less pressing if there were, for example, a clear-cut relationship between depth of bradycardia and maximum diving time, but such a relationship, even if it may be true in a general sense and within a species (85), is by no means quantitative. The Weddell seal, for example, can dive for 77 min (85), but in simulated dives its heart rate does not drop by as much as does that of a beaver (86) or a gray seal (6). Nor does this acknowledged champion of pinniped divers accumulate blood lactate to any unusual levels. Indeed, the highest blood lactate levels reported in the Weddell seal (3) after over an hour of apneic free diving are substantially lower than in other marine mammals (21). Thus, there still may be something missing that helps to account for long-term capacity. We believe that one such overlooked component of marine mammal metabolism is intertissue metabolic cooperation. At this writing, we can identify at least two ways in which this is achieved: a) by the use of different substrates to minimize competition for potentially limiting carbon sources, and b) by intertissue cycling of metabolites in different oxidation states. The first has already been mentioned above, whereas the second is only poorly explored. However, what data are available should be briefly discussed.

Potential Intertissue Cycling of Metabolites

A necessary characteristic of the diving animal is its ability to sustain a relatively high NAD^+:NADH ratio in the face of prolonged hypoxia. This is well established for several turtle tissues and the duck brain (D. R. Jones, personal communication). But how can this be achieved without large lactate accumulations in the tissue? One interesting possibility involves the transport of lactate, formed in the heart or brain, to the lungs. Because the lungs remain aerobic even in prolonged diving, conditions may be suitable for rapid oxidation of at least some lactate back to pyruvate, which could be returned to the heart and brain for NADH oxidation and reconversion to lactate. Such an intertissue cycling of lactate and pyruvate would readily explain how the redox balance in heart and brain could be maintained in prolonged anoxia without a large lactate accumulation in the tissue or the blood.

It is for this reason that special importance is attached to the lactate and pyruvate profiles through diving and recovery. The importance stems from directions, rather than magnitudes, of change, since, while lactate levels rise continuously during diving, pyruvate levels in fact are falling. That is, lactate production is occurring simultaneously with pyruvate utilization, possibly implicating a pyruvate-lactate-based hydrogen shuttling system between organs that vary in anoxia tolerance (87). The process appears to continue into postdiving recovery as well, which is one reason why the pyruvate washout peaks occur after the lactate washout peaks. Such a process has been reported in extremely hypoxic, perfused heart preparations (88) and, on theoretical grounds, was predicted to be involved in the extended hypoxia tolerance of diving animals (12). Whether or not other metabolites varying in redox state are involved in such intertissue cycling, however, must await further study.

REFERENCES

1. Scholander, P. F. (1962). Physiological consequences of diving in animals and man. Harvey Lect. 57:93.
2. Andersen, H. T. (1966). Physiological adaptation in diving vertebrates. Physiol. Rev. 46:212.
3. Kooyman, G. L., Wahrenbrock, E. A., Sinnett, E. E., Castellini, M. C., and

Davis, R. A. Blood lactic acid levels in Weddell seals after voluntary dives of 10 to 61 minutes. In preparation.

4. Hochachka, P. W., Liggins, G. C., Qvist, J., Schneider, R., Snider, M. Y., Wonders, T. R., and Zapol, W. M. (1978). Pulmonary metabolism during diving: conditioning blood for the brain. Science 198:831.
5. Scholander, P. F. (1940). Experimental investigations in diving mammals and birds. Hvalråd. Skr. 22:1.
6. Blix, A. S., Kjekshus, J. K., Enge, I., and Bergan, A. (1976). Myocardial blood flow in the diving seal. Acta Physiol. Scand. 96:277.
7. Zapol, W. M., Snider, M. Y., Schneider, R., Qvist, J., Liggins, G. C., and Hochachka, P. W. Regional distribution of cardiac output during diving in the Weddell seal. J. Appl. Physiol. (submitted).
8. Ashwell-Erickson, S., and Elsner, R. (1977). Seasonal range of metabolism in the harbor seal. Proceedings of the Second Conference on the Biology of Marine Mammals, p. 41, December, 1977, San Diego.
9. Ridgeway, S. H., Scrance, B. L., and Kanwisher, J. (1969). Respiration and deep diving in the bottlenose porpoise. Science 166:1651.
10. Kanwisher, J., and Sundness, G. (1965). Physiology of a small cetacean. Hvalråd. Skr. 48:45.
11. George, J. C., and Ronald, K. (1975). Metabolic adaptation in pinniped skeletal muscle. Rapports et process-verbeux des Reunions. *In* K. Ronald and A. W. Mansfield (eds.), Biology of the Seal, pp. 432–436. Imprem. Bianco Luna, Copenhagen.
12. Hochachka, P. W., and Storey, K. B. (1975). Metabolic consequences of diving in animals and man. Science 187:613.
13. Ramaiah, A. (1976). Regulation of glycolysis in skeletal muscle. Life Sci. 19:455.
14. Simon, L. M., Robin, E. D., Elsner, R., Van Kessel, Antonius, L. G. J., and Theodore, J. (1974). A biochemical basis for differences in maximal diving time in aquatic mammals. Comp. Biochem. Physiol. 47B:209.
15. Simon, L. M., and Robin, E. D. (1972). Relative anaerobic capacity and pyruvate kinase activity of rabbit tissues. Int. J. Biochem. 3:329.
16. Guderley, H., Fields, J. H. A., Cardenas, J. M., and Hochachka, P. W. (1978). Pyruvate kinase from the liver and kidney of *Arapaima gigas*. Can. J. Zool. 56:852.
17. Mustafa, T., and Hochachka, P. W. (1971). Catalytic and regulatory properties of pyruvate kinases in tissues of a marine bivalve. J. Biol. Chem. 246:3196.
18. Tsai, M. Y., Gonzales, F., and Kemp, R. G. (1975). Physiological significance of phosphofructokinase isozymes. *In* C. L. Markert (ed.), 3rd International Isozyme Conference, pp. 819–835. Academic Press, Inc, New York.
19. Owen, T. G., and Hochachka, P. W. (1974). Purification and properties of dolphin muscle aspartate and alanine transaminases and their possible roles in the energy metabolism of diving mammals. Biochem. J. 143:541.
20. Safer, B., and Williamson, J. R. (1973). Mitochondrial-cytosolic interactions in perfused rat heart: role of coupled transamination in repletion of citric acid cycle intermediates. J. Biol. Chem. 248:2570.
21. Hochachka, P. W., Owen, T. G., Allen, J. F., and Whittow, G. C. (1975). Multiple end products of anaerobiosis in diving vertebrates. Comp. Biochem. Physiol. 50B:17.
22. Ruderman, N. B., and Berger, M. (1974). The formation of glutamine and alanine in skeletal muscle. J. Biol. Chem. 249:5504.
23. Baruch, S. B., Ean, C. K., MacLeod, N., and Pitts, R. F. (1976). Renal CO_2 production from glutamine and lactate as a function of arterial perfusion pressure. Proc. Natl. Acad. Sci. U. S. A. 73:4235.

24. Stoff, J. S., Epstein, F. H., Narins, R., and Relman, A. S. (1976). Recent advances in renal tubular biochemistry. Annu. Rev. Physiol. 38:46.
25. Cohen, J. J. (1968). Renal gaseous and substrate metabolism *in vivo*: relationship to renal function. Proc. Intl. Union Physiol. Sci. 6:233.
26. Bowman, R. H. (1968). Substrate utilization by the isolated perfused rat kidney. Proc. Intl. Union Physiol. Sci. 7:164.
27. Kahng, M. W., Berezesky, I. K., and Trump, B. F. (1978). Metabolic and ultrastructural response of rat kidney cortex to *in vitro* ischemia. Fed. Proc. 37(3):1009.
28. Bradly, S. E., and Bing, R. J. (1942). Renal function in the harbor seal (*Phoca vitulina* L.) during asphyxial ischemia and pyrogenic hyperemia. J. Cell. Comp. Physiol. 19:229.
29. Schmidt-Nielsen, B., Murdaugh, H. U., Jr., O'Dell, R., and Bacsanyi, J. (1959). Urea excretion and diving in the seal (*Phoca vitulina* L.). J. Cell. Comp. Physiol. 53:393.
30. Murdaugh, H. U., Jr., Schmidt-Nielsen, B., Wood, J. W., and Mitchell, W. L. (1961). Cessation of renal function during diving in the trained seal (*Phoca vitulina*). J. Cell. Comp. Physiol. 58:261.
31. Halasz, N. A., Elsner, R., Garvie, R. S., and Grotke, G. T. (1974). Renal recovery from ischemia: a comparative study of harbor seal and dog kidneys. Am. J. Physiol. 227:1331.
32. Boyd, T. A., and Goldstein, L. (1978). Kidney and liver α-ketoglutarate levels in acute acidosis and alkalosis. Fed. Proc. 37(3):407.
33. Huang, H. C., Lung, P. K. L., and Huang, T. F. (1974). Blood volume, lactic acid, and catecholamines in diving responses in ducks. J. Formosan Med. Assoc. 73:203.
34. Hers, H. B. (1976). Control of glycogen metabolism in liver. Annu. Rev. Biochem. 45:167.
35. Neely, J. R., and Morgan, H. E. (1974). Relationship between carbohydrate and lipid metabolism and the energy balance of heart muscle. Annu. Rev. Physiol. 36:413.
36. Fawcett, D. W., and McNutt, N. S. (1969). The ultrastructure of the cat myocardium. I. Ventricular papillary muscle. J. Cell Biol. 42:1.
37. Most, A. S., Brachfeld, N., Gorlin, R., and Wahren, J. (1969). Free fatty acid metabolism of the human heart at rest. J. Clin. Invest. 48:1177.
38. Neely, J. R., Rovetto, M. J., and Oram, J. F. (1972). Myocardial utilization of carbohydrates and lipids. Prog. Cardiovasc. Dis. 15:289.
39. Hull, F. E., Radloff, J. F., and Sweeley, C. C. (1975). Fatty acid oxidation by ischemic myocardium. *In* P. E. Roy and P. Harris (eds.), Recent Advances in Studies on Cardiac Structure, Vol. 8, pp. 153–165. University Park Press, Baltimore.
40. Rabinowitz, J. L., and Hercker, E. S. (1974). Incomplete oxidation of palmitate and leakage of intermediary products during anoxia. Arch. Biochem. Biophys. 161:621.
41. Whitmer, J. T., Wenzer, J. I., Rovetto, M. J., and Neely, J. R. (1978). Control of fatty acid metabolism in ischemic and hypoxic hearts. J. Biol. Chem. 253:4305–4309.
42. Bremer, J., and Wotczak, A. B. (1972). Factors controlling the rate of fatty acid β-oxidation in rat liver mitochondria. Biochim. Biophys. Acta 280:515.
43. Penney, D. G., and Cascarano, J. (1970). Anaerobic rat heart: effects of glucose and carboxylic acid-cycle metabolites on metabolism and physiological performance. Biochem. J. 118:221.
44. Taegtmeyer, H. (1978). Metabolic responses to cardiac hypoxia. I. Increased synthesis of succinate by rabbit papillary muscles. Circ. Res. 43:808.
45. Felig, P., and Wahren, J. (1971). Interrelationship between amino acid and car-

bohydrate metabolism during exercise: the glucose-alanine cycle. *In* B. Pennow and B. Saltin (eds.), Advances in Experimental Medicine and Biology, p. 205. Plenum Press, New York.

46. Collicutt, J. M., and Hochachka, P. W. (1977). The anaerobic oyster heart: coupling of glucose and aspartate fermentation. J. Comp. Physiol. 115:147.
47. Neely, J. R., Whitmer, K. M., and Mochizuki, S. (1976). Effects of mechanical activity and hormones on myocardial glucose and fatty acid utilization. Circ. Res. (suppl. I) 38:22.
48. Collicutt, J. M. (1975). Anaerobic metabolism in the oyster heart. M.Sc. thesis, University of British Columbia.
49. Cascarano, J., Ades, I. Z., and O'Connor, J. D. (1976). Hypoxia: a succinate-fumarate electron shuttle between peripheral cells and lung. J. Exp. Zool. 198:149.
50. Blix, A. S., Berg, T., and Ryhn, H. T. (1970). Lactate dehydrogenase in a diving mammal, the common seal. Int. J. Biochem. 1:292.
51. Messelt, E. B., and Blix, A. S. (1976). The lactate dehydrogenase of the frequently asphyxiated beaver (*Castor fiber*). Comp. Biochem. Physiol. 53B:77.
52. Altman, M., and Robin, E. D. (1969). Survival during prolonged anaerobiosis as a function of an unusual adaptation involving lactate dehydrogenase subunits. Comp. Biochem. Physiol. 30:1179.
53. Blix, A. S., and From, S. H. (1971). Lactate dehydrogenase in diving animals—a comparative study with special reference to the eider (*Somateria mollissima*). Comp. Biochem. Physiol. 40B:579.
54. Appella, F., and Markert, C. L. (1961). Dissociation of lactate dehydrogenase into subunits with guanidine hydrochloride. Biochem. Biophys. Res. Commun. 6:171.
55. Kerem, D., Hammond, D. D., and Elsner, R. (1973). Tissue glycogen levels in the Weddell seal, *Leptonychotes weddelli*: a possible adaptation to asphyxial hypoxia. Comp. Biochem. Physiol. 45A:731.
56. Siesjo, B. J., Carlsson, C., Hagerdal, M., and Nordstrom, C. H. (1976). Brain metabolism in the critically ill. Crit. Care Med. 4:283.
57. Siesjo, B. K., and Nordstrom, C. H. (1977). Brain metabolism in relation to oxygen supply. *In* F. F. Jöbsis (ed.), Oxygen and Physiological Function, p. 459. Professional Information Library, Dallas, Texas.
58. Rappaport, S. I. (1976). Blood-Brain Barrier in Physiology and Medicine, pp. 178–202. Raven Press, New York.
59. Rowe, G. G., Maxwell, G. M., and Castillo, C. A. (1959). A study in man of cerebral blood flow and cerebral glucose, lactate and pyruvate metabolism before and after eating. J. Clin. Invest. 38:2154.
60. Nemoto, E. M., Hoff, J. T., and Severinghouse, J. W. (1974). Lactate uptake and metabolism by brain during hyperlactatemia and hypoglycemia. Stroke 5:48.
61. Sokoloff, L. (1973). Metabolism of ketone bodies by the brain. Annu. Rev. Med. 24:271.
62. Guroff, G. (1972). Transport and metabolism of amino acids. *In* R. W. Albers, G. V. Siegel, R. Katzman, and B. W. Agranoff (ed.), Basic Neurochemistry, pp. 191–201. Little, Brown and Company, Boston.
63. Oldendorf, W. H. (1973). Carrier-mediated blood-brain barrier transport of short chain monocarboxylic organic acids. Am. J. Physiol. 224:1450.
64. Mahan, D. E., Mushawar, I. K., and Koeppe, R. E. (1975). Purification and properties of rat brain pyruvate carboxylase. Biochem. J. 145:25.
65. Nordstrom, C. H., Rehncrona, S., and Siesjo, B. K. (1976). Restitution of cerebral energy state after complete and incomplete ischemia of 30 min duration. Acta Physiol. Scand. 97:270.

66. Michenfelder, J. D., and Theye, R. A. (1970). The effect of anesthesia and hypothermia on canine cerebral ATP and lactate during anoxia produced by decapitation. Anesthesiology 33:430.
67. Hochachka, P. W., and Mustafa, T. (1972). Invertebrate facultative anerobiosis. Science 178:1056.
68. Elsner, R., Shurley, J. T., Hammond, D. D., and Brooks, R. E. (1970). Cerebral tolerance to hypoxemia in asphyxiated Weddell seals. Respir. Physiol. 9:287.
69. Kerem, D., Elsner, R., and Wright, J. (1971). Anaerobic metabolism in the brain of the harbor seal during late stages of a maximum dive. Fed. Proc. 30:484 (abstr.).
70. Bryan, R. M., and Jones, D. R. Cerebral energy metabolism in mallard ducks during apneic asphyxia: the role of oxygen conservation. Can. J. Zool. In press.
71. Blix, A. S. (1971). Creatine in diving animals: a comparative study. Comp. Biochem. Physiol. 40A:805.
72. Hochachka, P. W., and Murphy, B. J. (1978). Metabolic model of the heart-lung-brain "machine" in the Weddell seal. Proceedings of the Second Conference on the Biology of Marine Mammals, p. 44, December, 1977, San Diego.
73. Shoubridge, E. A., Carscadden, J. E., and Leggett, W. C. (1976). LDH isozyme patterns in cetaceans: evidence for a biochemical adaptation to diving. Comp. Biochem. Physiol. 53B:357.
74. Johansen, K. (1964). Regional distribution of circulating blood during submersion asphyxia in the duck. Acta Physiol. Scand. 62:1.
75. Adachi, H., Strauss, H. W., Ochi, H., and Wagner, H. N. (1976). The effect of hypoxia on the regional distribution of cardiac output in the dog. Circ. Res. 39:314.
76. Bohr, C. H., and Henriques, V. (1897). Arch. Sci. Physiol. 9:819.
77. Tierney, D. F. (1974). Intermediary metabolism of the lung. Fed. Proc. 33:2232.
78. Wallace, H. W., Stein, T. P., and Liquori, E. M. (1974). Lactate and lung metabolism. J. Thorac. Cardiovasc. Surg. 68:810.
79. Wolfe, R. R., and Hochachka, P. W. (1978). Lactate oxidation in perfused rat lungs. Fed. Proc. 37(6):2760.
80. Kerem, D., and Elsner, R. (1973). Cerebral tolerance to asphyxial hypoxia in the harbor seal. Respir. Physiol. 19:188.
81. Hochachka, P. W., and Hulbert, W. C. (1978). Glycogen "seas," glycogen bodies, and glycogen granules in heart and skeletal muscle of two air-breathing, burrowing fishes. Can. J. Zool. 56:774.
82. Holbrook, J. J., Liljas, A., Steindel, J. J., and Rossman, M. G. (1975). Lactate dehydrogenase. *In* P. Boyer (ed.), The Enzymes, Vol. II, p. 191. Academic Press, New York.
83. Cremer, J. E., and Teal, H. M. (1974). The activity of pyruvate dehydrogenase in rat brain during postnatal development. FEBS. Lett. 39:17.
84. Dawes, G. S., and Shelley, J. C. (1968). Physiological aspects of carbohydrate metabolism in the fetus and newborn. *In* F. Dickens, P. J. Randle, and W. J. Whelan (eds.), Carbohydrate Metabolism and Its Disorders, Vol. 2, pp. 87–121. Academic Press, Inc., New York.
85. Kooyman, G. L., and Campbell, W. B. (1972). Heart rates in freely diving Weddell seals, *Leptonychotes weddelli.* Comp. Biochem. Physiol. 43:31.
86. Clausen, G., and Ersland, A. (1970/71). Blood O_2 and acid-base changes in the beaver during submersion. Respir. Physiol. 11:104.
87. Hochachka, P. W., Murphy, B. J., Liggins, G. C., Creasy, R., Snider, M., Schneider, R., Qvist, J., and Zapol, W. What the seal foetus does when its mother dives. I. Blood glucose, lactate, and pyruvate profiles. In preparation.

88. Lee, J. C., Halloran, K. H., Taylor, J. F. N., and Downing, S. E. (1973). Coronary flow and myocardial metabolism in newborn lambs: effect of hypoxia and acidemia. Am. J. Physiol. 224:1381.

International Review of Physiology
Environmental Physiology III, Volume 20
Edited by D. Robertshaw

6
Physiological Effects of High Altitude on the Pulmonary Circulation

J. T. REEVES, W. W. WAGNER, JR., I. F. McMURTRY, and R. F. GROVER

Cardiovascular Pulmonary Research Laboratory,
University of Colorado Medical Center, Denver, Colorado

PULMONARY VASCULAR RESPONSE TO HYPOXIA 290

THE MECHANISM OF HYPOXIC PULMONARY VASOCONSTRICTION 293
- **Neural** 297
- **Vasoactive Substances** 297
- **Direct Effect of Hypoxia on Pulmonary Arterial Smooth Muscle** 298

TELEOLOGICAL SPECULATION RELATING TO HYPOXIC PULMONARY PRESSOR RESPONSE 300

When an individual ascends to high altitude, he is exposed to a lower total barometric pressure and a lower partial pressure of oxygen in the ambient air and the alveolar gas (Figure 1). Either acute or chronic exposure to high altitudes (Figure 1) causes an increase in pulmonary arterial pressure, the magnitude of which depends upon such factors as the altitude involved, the duration of the exposure, the age and species of the animal exposed, and the ventilatory and hematopoetic responses of the individual. Pulmonary vasoconstriction with alveolar hypoxia is central to the consideration of high

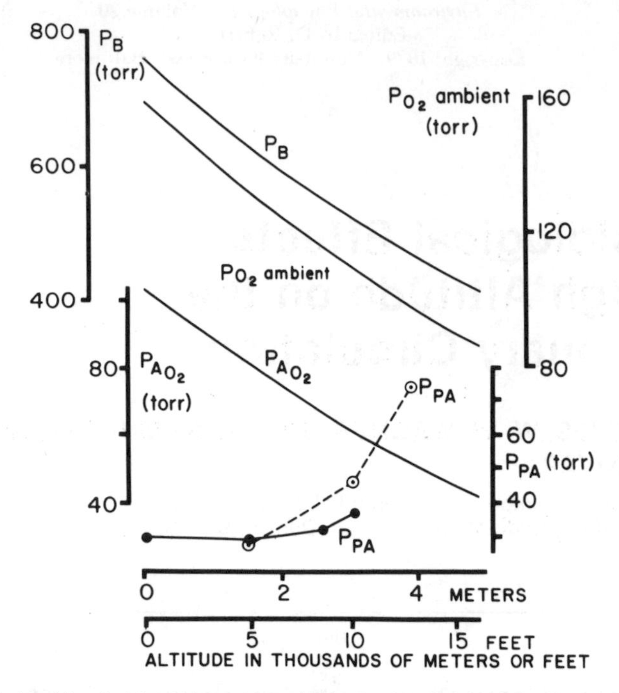

Figure 1. Relationship of altitude (from top to bottom) to barometric pressure (P_B), ambient P_{O_2} (P_{O_2} ambient), estimated alveolar oxygen tension ($P_{A_{O_2}}$), and pulmonary arterial pressure (P_{P_2}). Shown are pulmonary arterial pressures in cattle brought from low to high altitude (○—○) compared to cattle native to various altitudes (●—●).

altitude pulmonary hypertension. This chapter emphasizes our view of the physiological basis of high altitude (hypoxia-induced) pulmonary hypertension. Excellent reviews by other authors present other points of view (1-8). Because we have recently reviewed high altitude pulmonary hypertension and pulmonary edema in normal man (9) we focus here on animal studies.

PULMONARY VASCULAR RESPONSE TO HYPOXIA

The pulmonary vascular response to acute hypoxia has been measured in a large number of animal species (10–18) (Figure 2), although the conditions of the experiments (anesthesia, body positions, size of the species, laboratory of origin) varied considerably from one species to another. All species studied showed a pulmonary pressor response to acute hypoxia. Furthermore, the magnitude of the responses was similar among the various species (except for the small response in the rabbit).

By contrast, when sea level animals were taken to high altitude and exposed to chronic hypoxia for several weeks, certain species developed more pulmonary hypertension than others (14, 19–24) (Figure 3). The sheep, dog, guinea pig, and llama had relatively little pulmonary hypertension in

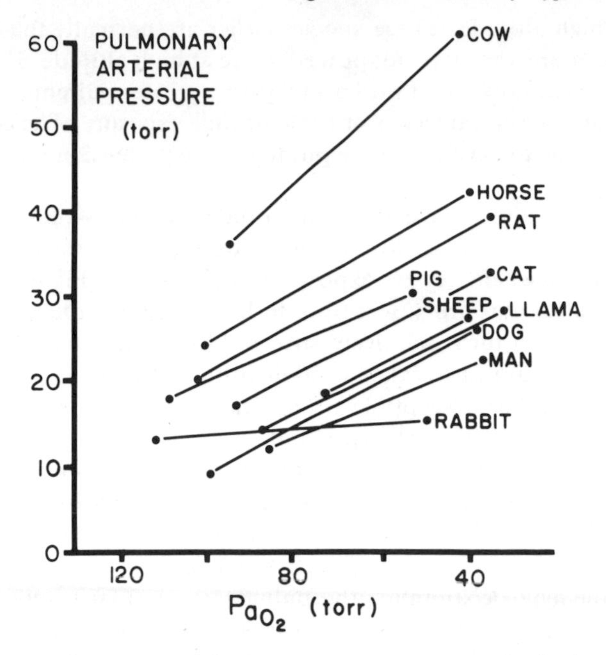

Figure 2. Mean pulmonary arterial pressure responses to acute alveolar hypoxia in various animal species. (For references see text.)

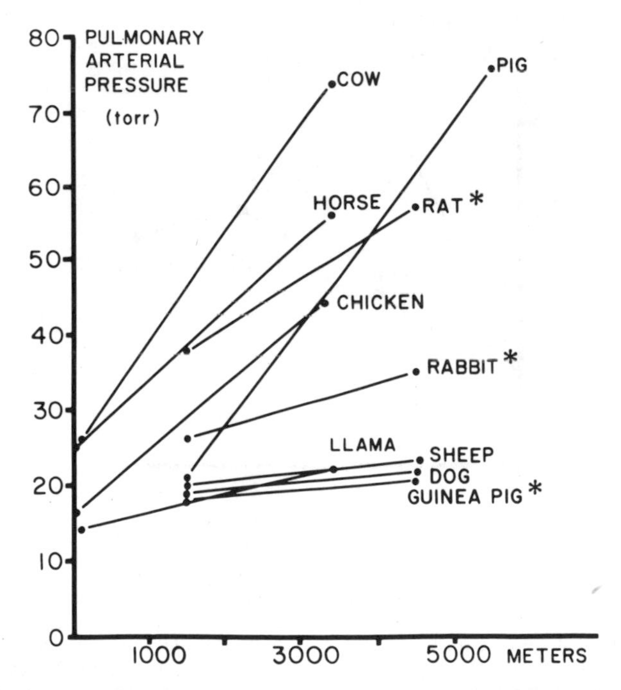

Figure 3. Mean pulmonary arterial pressure responses to chronic residence at high altitude in various animal species. (For references see text.) Asterisk indicates right ventricular systolic pressure.

response to high altitude. These species either are normally found living at high altitude or are known to adapt well to life at high altitude. However, the cow, pig, chicken, horse, and rat had relatively greater pulmonary hypertension in response to several weeks of high altitude exposure. The cow, in particular, was prone to develop severe pulmonary hypertension and right heart failure at high altitude.

However, healthy cattle can live at altitudes of 3,000–4,000 m, and their offspring, even when born at low altitude, are relatively resistant to pulmonary hypertension when they are exposed to high altitude (Figure 4) (25). The offspring of cattle that are susceptible to high-altitude-induced pulmonary hypertension are, in turn, relatively susceptible to pulmonary hypertension and right heart failure when taken to high altitude (Figure 4). Thus, the trait for cattle to be either susceptible or resistant to high altitude pulmonary hypertension may be genetically transmitted.

The maximum pulmonary arterial pressure obtainable during acute hypoxia in animals living at sea level is greatest in the newborn and is lowest in the adult (26, 27). The normoxic pulmonary arterial pressures are also highest in the newborn (28). When the arterial oxygen saturation is used as a measure of the hypoxic stimulus, the pulmonary vascular response becomes approximately linear (29). Calves of various ages had similar pulmonary

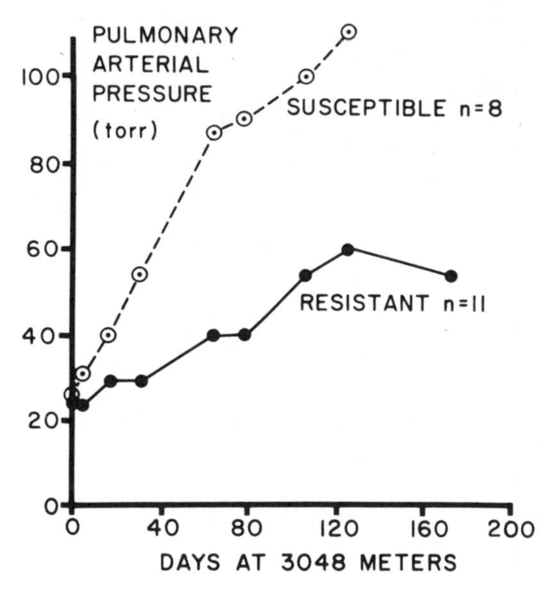

Figure 4. Mean pulmonary arterial pressures in two groups of calves taken from low altitude (1,520 m) to high altitude residence at 3,048 m. Eight susceptible calves were the offspring of cows and bulls that had developed pulmonary hypertension and right heart failure at high altitude. These eight were in turn susceptible to high altitude pulmonary hypertension. The eleven resistant calves were offspring of healthy cattle from 3,040 m altitude, and these eleven developed relatively little pulmonary hypertension at high altitude. The tendency for pulmonary hypertension at high altitude may be genetically transmitted from parent to offspring.

vasopressor responses to acute hypoxia. Only the normoxic pressures differed (Figure 5). When calves were exposed from birth for 5–8 weeks to a simulated altitude of 3,360 m, they developed more severe pulmonary hypertension and an augmented pulmonary vascular response to changes in oxygenation (30) (Figure 5). The calves at high altitude had extensive muscularization of relatively long segments of their pulmonary arterial tree. The results suggested that high altitude exposure after birth remarkably potentiated the hypoxic vasopressor response.

THE MECHANISM OF HYPOXIC PULMONARY VASOCONSTRICTION

When alveolar hypoxia is produced in the intact animal (31) and in the isolated perfused lung (32–35), there is a latent period of a few seconds followed by slow pressor response that stabilizes in 5–10 min (Figure 6). Relief of the hypoxia is promptly followed by a fall in pressure. The slow development and the rapid regression are characteristic of hypoxic pressor responses.

The anatomical locus of the response is considered to be the pulmonary arterioles because they become narrow during acute hypoxia (36) and hypertrophied during chronic hypoxia (37–41), and because there is no evidence of the increased capillary pressure that would be expected should the effective resistance develop in the postcapillary vasculature (42).

The locus of the oxygen sensor is less clear. When the lung is not inflated, as in the fetus (28) or in the atelectatic adult lung (43), the pulmonary

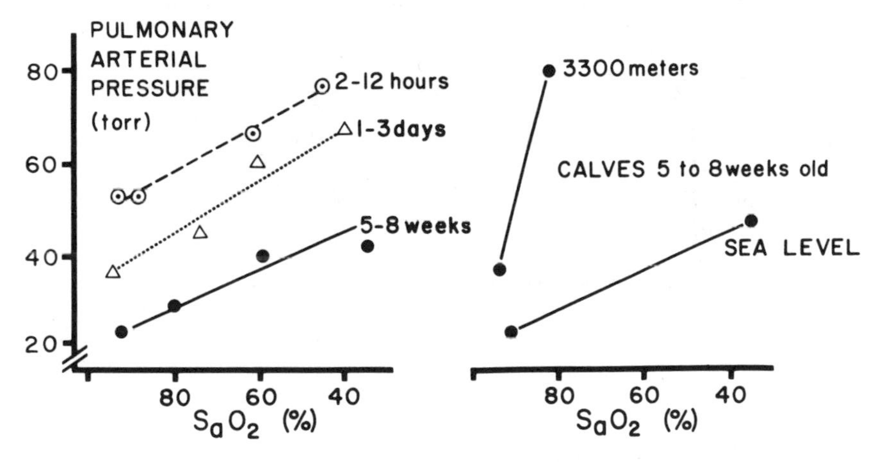

Figure 5. *Left*, mean pulmonary arterial pressure responses to acute hypoxia in calves of different ages. The severity of the hypoxia was judged by the arterial oxygen saturation (Sa_{O_2}). The pulmonary arterial pressures were lower in older calves, but the slopes of the lines representing the pressor responses to hypoxia were not different for the three ages. *Right*, mean pulmonary arterial pressures in response to hypoxia in 5–8-week-old calves residing from birth at sea level (also shown at left) and in calves residing from birth at 3,300 m altitude. High altitude calves have more reactive pulmonary vascular beds to hypoxia than do those at sea level.

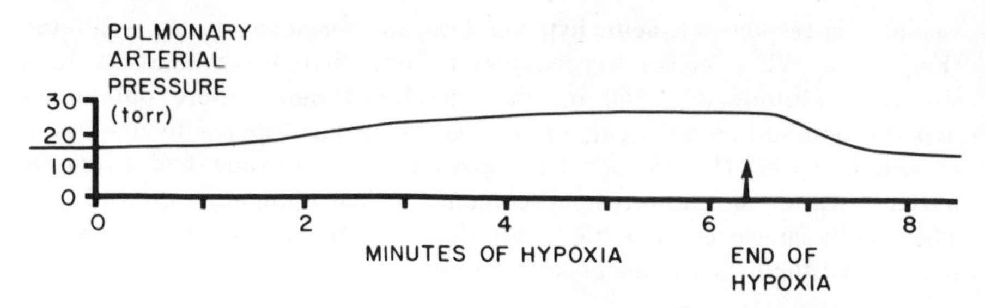

Figure 6. Pulmonary arterial pressor response to ventilation with alveolar hypoxia (3% O_2, 5% CO_2, 92% N_2) in an isolated rat lung perfused with blood at 11 ml/min. After 6 min, when the lung was returned to normoxia (21% O_2, 5% CO_2), the perfusion pressure promptly fell (I. F. McMurtry, unpublished observation).

arterial pressure rises when the P_{O_2} of the perfusing blood falls. When a normoxic inflated lung is ventilated with a hypoxic gas, the pulmonary arterial pressure begins to rise before the inflowing blood becomes hypoxic. When the alveolar hypoxia is relieved, the pressure begins to fall even though the pulmonary arterial blood remains hypoxemic (Figure 7A) (34). Raising the P_{O_2} in the inflowing blood during alveolar hypoxia may attenuate the rise in pressure, but not return the pressure to normoxic levels (44) (Figure 7B). Thus, the sensor can sense the oxygen level in both the pulmonary arterial blood and the alveolar air, but in the inflated lung the main stimulus appears to be the oxygen tension in the alveolar air. Inhaled hydrogen can be detected in the small pulmonary arteries (45), and the outer layer of erythrocytes becomes oxygenated before the pulmonary arterial blood reaches the

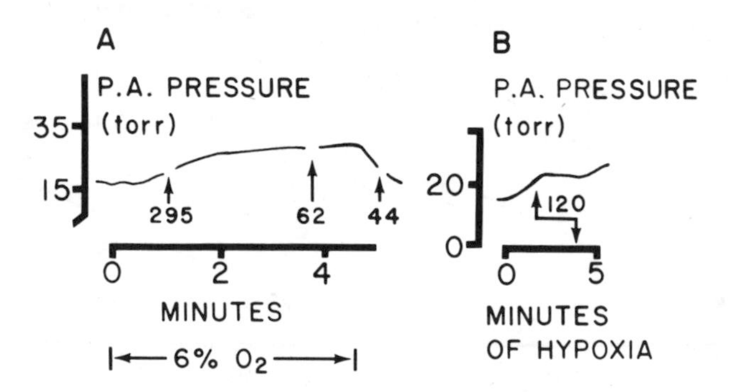

Figure 7. *A*, pulmonary arterial pressor response to hypoxia in an isolated dog lobe in which the blood perfusate recirculated from a large (170 ml) reservoir. Thus, the P_{O_2} in the pulmonary arterial blood remained high (295 torr) after a decrease in the fraction of inspired O_2 to 0.06 had caused the pulmonary arterial pressure to rise. Later, the P_{O_2} in the arterial blood remained low (44 torr) after a return to a high O_2 fraction (0.94) in the inspired air had caused the pulmonary arterial pressure to fall (34). *B*, pulmonary arterial pressor response to alveolar hypoxia (2% O_2) in the isolated perfused rat lung. The pulmonary arterial blood P_{O_2} was low (44 torr). When the arterial inflow P_{O_2} was temporarily raised to 120 torr, the pressor response was slightly blunted (44).

capillary level (36). Furthermore, the small pulmonary arteries are virtually suspended in the alveolar air. Thus, gas exchange begins through the walls of the precapillary lung vessels. We think it likely that the sensor for the hypoxic vasopressor response resides in or very near the walls of the pulmonary arterioles. To date, however, convincing pulmonary pressor responses to hypoxia have been obtained only when lung parenchyma surrounds the vessel. Therefore, proof that both the sensor and the effector reside in the arterial wall itself will require that isolated vessels be shown to constrict with hypoxia.

As pointed out by Torrance (46), the human stimulus-response curve of the lung vessels to hypoxia (9) resembles the stimulus-response curve of ventilation to acute hypoxia (47) and also resembles the stimulus-response curve of the hematopoietic system to chronic hypoxia (48) (Figure 8). Each has a threshold Po_2 of approximately 70–80 torr, below which the response is increasingly strong. The coincidence seems remarkable that each of these functions has an inflection point similar to that of the oxygen-hemoglobin dissociation curve.

In addition, pulmonary vascular resistance during hypoxia is pH sensitive (Figure 9). Acidosis augments and alkalosis inhibits the magnitude of the hypoxic pressor response in intact animals and in isolated perfused lungs (49–53). The intracellular pH may be more important that the blood pH, because raising the CO_2 in the alveolar air without changing the blood pH

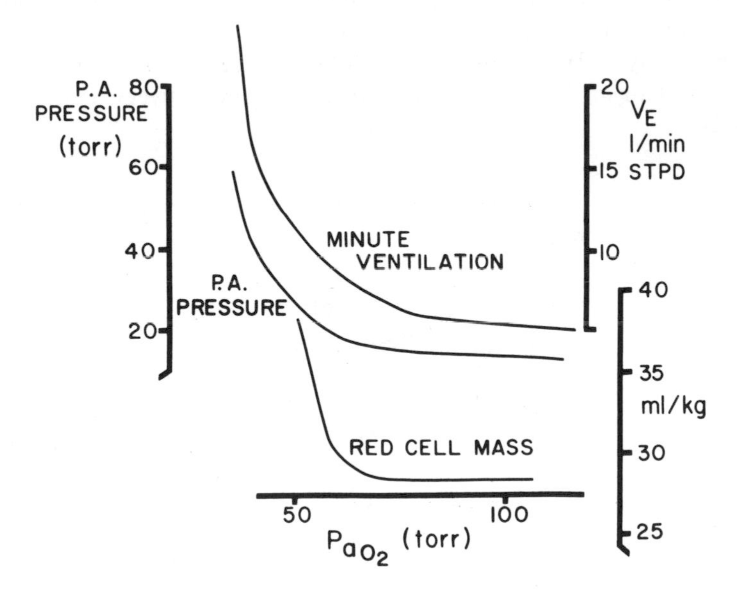

Figure 8. Relationship of arterial oxygen tension (Pa_{O_2}) to (from top to bottom) minute ventilation (V_E) (47), pulmonary arterial (P.A.) pressure (9), and red cell mass in ml/kg (48). The pulmonary ventilation, pulmonary arterial pressure, and red cell mass all begin to increase rapidly when Pa_{O_2} falls below 70–80 torr.

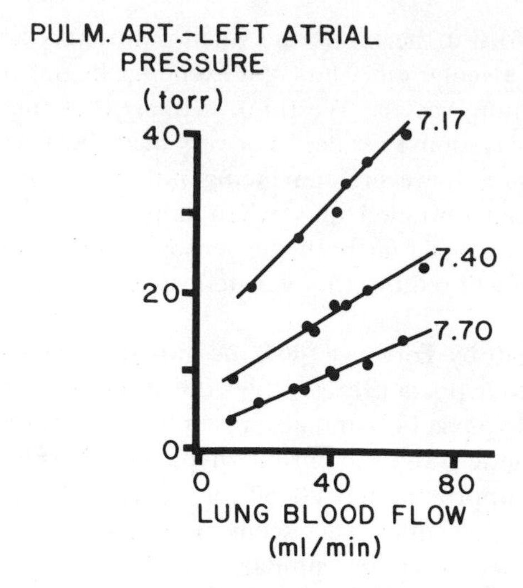

Figure 9. Pressure flow curves from an in vivo autoperfused left lower lobe of a cat during ventilation with 10% O_2 at the different blood pH values of 7.17, 7.40, and 7.70. Acidosis increased and alkalosis decreased the slope and the intercept, suggesting that hypoxia caused greater vasoconstriction during acidosis and less during alkalosis.

can restore a hypoxic response that had been virtually abolished by alkalosis (53) (Figure 10). Intracellular acidosis potentiates the pulmonary vascular response to hypoxia. Acidosis also potentiates the ventilatory response to hypoxia, and it shifts the O_2-Hb dissociation curve to the right, thus decreasing the affinity of Hb for oxygen. Possibly, as suggested by Torrance (46), similar mechanisms are involved in sensing oxygen in diverse systems within the body.

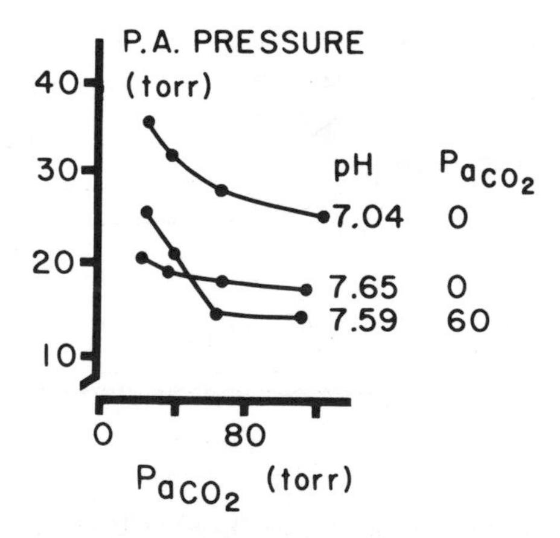

Figure 10. Pulmonary arterial pressure during progressive hypoxia in isolated cat lungs perfused with blood at constant flow. Alkalosis at 0 torr P_{CO_2} virtually abolished the hypoxic pressor response, which could be restored by adding CO_2 to the ventilating gas.

Neural

Despite some evidence to the contrary in the cat (54, 55), the weight of evidence indicates that α adrenergic blocking agents (56–59), β adrenergic blocking agents (57, 59), catecholamine depletion with reserpine (60), sympathetic and parasympathetic nerve section (61), and destruction of sympathetic nerve endings with 6-hyroxydopamine (A. Tucker, unpublished observations) do not block or blunt hypoxic pulmonary vasoconstriction. Furthermore, the finding of the response in the isolated perfused lung is consistent with the concept that hypoxic pulmonary vasoconstriction is not mediated by the sympathetic or parasympathetic nerves.

Vasoactive Substances

The initial enthusiasm for histamine as the mediator of hypoxic pulmonary hypertension (62–64) has not been sustained. The dosages of the antihistamines used were large, and nonspecific effects on the pulmonary vasculature were not ruled out. Histamine is a potent pulmonary vasodilator in the fetal lamb even though hypoxic vasoconstriction is important in maintaining the high tone of the fetal lung arteries (65). Exogenous histamine in the lung stimulates both vasoconstriction (H_1) and vasodilatory (H_2) receptors in the dog lung (66). During hypoxia, however, histamine administration appears to stimulate primarily vasodilatory receptors in the dog (10, 67), cat (68, 69), calf (70), and isolated rat lung (71). The magnitude of hypoxic vasoconstriction appears not to be related to histamine released from the lung by histamine-releasing agents (72, 73). In both dogs and cats (69), H_2-receptor blockade enhanced the pulmonary pressor response to hypoxia. This finding led to the speculation that during hypoxia histamine was indeed released but had a net vasodilating effect that opposed hypoxic vasoconstriction.

Pulmonary histamine stores are largely within mast cells located around small vessels and airways and in the alveolar septa and pleura. High altitude exposure causes perivascular mast cell hyperplasia in rats (74) and calves (75). A relationship between mast cell hyperplasia and pulmonary hypertension has been found for calves, pigs, sheep, and rats (75). However, it is not clear whether the mast cell hyperplasia is the cause or the result of pulmonary hypertension. The fact that mast cell numbers increase around lung vessels when the pressure is increased by administration of crotalaria or by the presence of mitral stenosis suggests that pulmonary hypertension, not hypoxia, is the stimulus for mast cell hyperplasia (76). Thus, the role in pulmonary vascular control of mast cells in general and of histamine in particular remains unknown. Perhaps the mast cell and histamine are involved in processes other than hypoxic pulmonary hypertension, e.g., the inflammatory response, growth, or connective tissue hyperplasia.

Because the lung can synthesize constrictor prostaglandins (77, 78), it was suggested that a constrictor prostaglandin might be synthesized during hypoxia and cause pulmonary hypertension (79). However, subsequent studies have indicated that inhibition of prostaglandin synthesis in dogs (80, 81), calves (80), goats (82), and isolated, perfused rat lungs (83) augment hypoxic pulmonary vasoconstriction. Furthermore, the stimulation of prostaglandin production in the dog by the administration of the precursor, arachidonic acid, causes vasodilation of vessels constricted by hypoxia (84).

Bradykinin is a potent pulmonary vasodilator. The pulmonary production of bradykinin increases after birth when the alveolar oxygenation is improved (85). Weir has proposed the novel hypothesis (86) that bradykinin production during exposure to high oxygen tension may cause pulmonary vessels to lose their normal high tone (as when the fetus takes the first breaths at birth). Substantiation of the hypothesis should account for the facts that isolated, perfused rat lungs that constrict in response to hypoxia also show a pressor response to bradykinin (35), and that hypoxia inhibits the pulmonary destruction of bradykinin (87).

Serotonin is a pulmonary vasoconstrictor in most species and could be considered the mediator of hypoxic pulmonary vasoconstriction. However, depletion of lung serotonin by 93% did not inhibit hypoxic pulmonary vasoconstriction in the rat lung (63). Furthermore, serotonin administration in the pig during an ongoing pulmonary pressor response to hypoxia causes pulmonary vasodilation (J. Mlczoch, unpublished observation). Thus, it is unlikely that serotonin mediates hypoxic pulmonary vasoconstriction.

Angiotensin II has been considered to have a role in hypoxic pulmonary vasoconstriction (88). However, saralasin, a competitive antagonist of angiotensin II, had no effect on the hypoxic pulmonary pressor response in either isolated perfused rat lungs (89), fetal lambs (90), or the intact dog (91).

Substances obtained from tracheal washings of dogs and calves (92) or from the lymph of dogs (93) have been considered possible mediators of the hypoxic response. Such substances have yet to be characterized biochemically, and a cause and effect relationship to hypoxic pulmonary hypertension has not been shown.

Direct Effect of Hypoxia on Pulmonary Arterial Smooth Muscle

The failure to identify a chemical mediator of hypoxic pulmonary vasoconstriction has led to an intensified search for a direct vasoconstrictor effect of oxygen lack on pulmonary vascular smooth muscle. In the systemic arterial bed oxygen lack causes vasodilation, probably by reducing the energy available for the contractile elements in the smooth muscle cells. However, Detar and Bohr (94) showed that aortic strips exposed for 18 hr to nitrogen would contract with epinephrine (adrenaline) but would relax when oxygen was added to the bath. Under these unusual conditions the aorta could be "taught" to behave in a manner similar to that of the pulmonary vasculature. Hypoxic contraction and oxygen-induced relaxation occurred more readily in the isolated pulmonary artery than in the aorta. If hypoxia does directly cause pulmonary arteries to constrict, one wonders what the mechanism might be. After reviewing the evidence, Bohr (1) considered it likely that oxygen lack induced changes in membrane permeability to allow calcium entry and the resultant contraction of the vascular smooth muscle. Haack et al. (95) found increased amounts of calcium associated with the pulmonary vascular smooth muscle of rats and pigs that had been chronically exposed to high altitude. McMurtry et al. (96) and Tucker et al. (97) showed that hypoxic pulmonary vasoconstriction was particularly sensitive to inhibition by substances like verapamil, which inhibit calcium entry into cells. Thus, it is possible that some component of the membrane could undergo a configurational change in response to hypoxia to permit the permeability changes in calcium. One possibility then is that hypoxia directly alters the membrane of the smooth muscle cell to increase the permeability to calcium (without necessarily lowering the energy stores within the cell). The entry of calcium initiates the contraction of the smooth muscle.

An alternate possibility is that hypoxia causes a depletion of high energy phosphate in some critical site within the cell (i.e., the cell membrane) that allows calcium to enter and trigger the contraction. Evidence favoring such a metabolic hypothesis consists of 1) increases in the vascular sensitivity to hypoxia of isolated dog lungs (98) and rat lungs treated with potassium cyanide (99), 2) treatment with dinitrophenol (100), 3) an increase in the lung vascular sensitivity to hypoxia of isolated perfused rat lungs treated with the metabolic inhibitor, 2-deoxyglucose (101) (Figure 11), and 4) an increase in

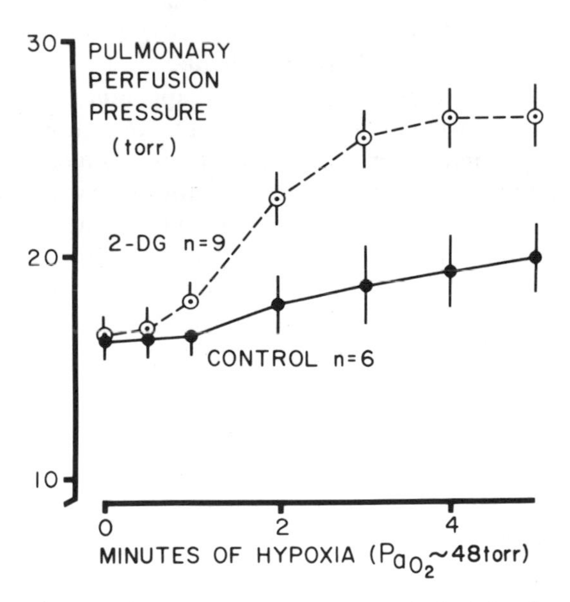

Figure 11. Pulmonary arterial pressor response to hypoxia in isolated perfused rat lungs without (control) and with the addition of the metabolic inhibitor 2 deoxyglucose (2-DG). With the inhibition of glucose metabolism, hypoxia caused the pulmonary arterial pressure to rise more rapidly and to a higher level than when glucose was not inhibited. (I. F. McMurtry, unpublished observation.)

sensitivity to hypoxia in the lung vessels of intact dogs given ethyl alcohol (102). In addition, there is a decrease in vascular sensitivity to hypoxia in lungs that have been cooled (103). Recently, glucose-depleted, isolated, perfused rat lungs have been shown to have brisk pressor responses that are blunted by the restoration of blood glucose levels (104). Time required for energy depletion during hypoxia could account for the latent period between the introduction of hypoxia and the beginning of the pressor response as well as the slow rate of rise of the pressure during hypoxia compared to the rapid fall of the pressure after return to normoxia, as shown in Figure 6. Metabolic adaptation of the membrane compartment to hypoxia with time could account for the reduction of the hypoxic pressor response seen in some species during chronic high altitude exposure.

The concept that hypoxia reduces energy stores to allow calcium to leak into the cells has drawbacks. Vasodilation occurs in isolated cat lungs ventilated with pure carbon monoxide (105). Yet, because of the absence of O_2 in pure CO, one might expect CO to cause pulmonary vasoconstriction. In addition, it is necessary to postulate either a secretory cell with small energy stores or a contractile cell with two energy compartments, a membrane compartment that is easily depleted of energy-rich compounds, and a contractile compartment that is not so easily depleted of energy stores. Sylvester and McGowan (106) have proposed that cytochrome P-450 might be the oxygen sensor although this substance has not yet been identified within the membrane of pulmonary arterial smooth muscle. Finally, the evidence available is indirect for the various hypotheses. Thus at present we see no hypothesis that satisfactorily explains all the evidence available.

Important and unresolved questions are apparent. What factors cause some species (or individuals within a species, i.e., susceptible cattle) to develop progressive pulmonary hypertension as duration of hypoxic exposure increases? What factors in other species (or individuals within a species, i.e., resistant cattle) cause pulmonary arterial pressure during chronic hypoxia to be no greater than during acute hypoxia? The individual and species variations have long puzzled physiologists. Yet, therein may lie clues of the basic mechanism of hypoxic pulmonary vasoconstriction and the ability of the various species to adapt to high altitude existence.

TELEOLOGICAL SPECULATION RELATING TO HYPOXIC PULMONARY PRESSOR RESPONSE

Primitive fishes are able to distribute blood flow to the branchial (gill) arches according to local variations in P_{O_2} (107). One might postulate then that these fish are able to distribute gill perfusion to maximize oxygen transport while minimizing the need to move water across the gills. If so, oxygen tension in water is able to affect gill perfusion and to regulate ventilation perfusion relationships.

Phylogenetically the pulmonary circulation developed in tropical fish some 350 million years ago when the freshwater seas were deficient in oxygen (108). The air-breathing organ supplemented but did not replace organs (skin or gills) for aquatic breathing. Presumably, then, mechanisms were needed to shift perfusion and ventilation away from the air-breathing organ during diving when the air within the organ became hypoxic. Indeed, some reptiles shunt blood to the systemic circulation and away from the lung during diving (109). In these reptiles, pressure in the arterial circulation of the air-breathing organ is nearly equal to the pressure in the systemic arteries. Because of the shunts between the pulmonary and systemic circulations, constriction within the lung arterial bed can reduce the flow to the air-breathing artery to virtually zero.

In other reptiles in which the systemic arterial pressure is much higher than the pulmonary arterial pressure, hypoxia increases pulmonary arterial pressure but not enough to cause a right-to-left shunt (110) (Figure 12). Under these circumstances it seems likely that the increase in pressure causes regional changes in flow within the lung that are advantageous to the organism. The only advantage postulated to date is that of distributing the flow within the lung in favor of better oxygenated areas. Thus, in primitive vertebrates, the fish and reptiles, oxygen levels in the environment exert control over the circulation in the oxygen exchange organs. The control could be for the purposes of improving the local ratio of ventilation to perfusion or of virtually shutting off perfusion to one organ of oxygenation when it is not needed.

In the higher vertebrates such as the mammal, the hypoxic pulmonary vasoconstrictor mechanism can match perfusion to ventilation within the lung in the adult or can direct blood flow away from the entire lung in the fetus. In the fetus the ductus arterial shunt between the pulmonary artery

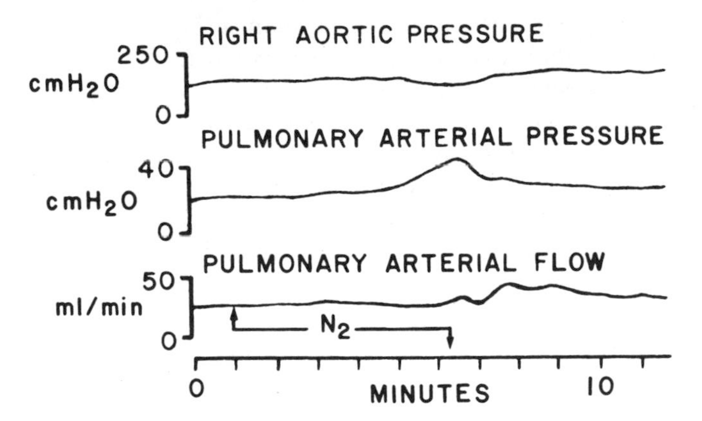

Figure 12. Circulatory effects of nitrogen (N_2) breathing in the lizard, *Varanus niloticus*. From top to bottom: the mean pressure in the right aorta was slightly depressed during hypoxia; the mean pressure in the pulmonary artery was always less than that in the aorta, but during hypoxia it rose slowly, and after hypoxia was terminated, it fell abruptly; the peak systolic pulmonary blood flow did not increase during hypoxia.

and aorta ensures nearly equal pressures, and the placenta is the organ of oxygenation. The hypoxic pulmonary vasoconstrictor mechanism is well developed in the fetus (28) and is a major factor in maintaining the high resistance to flow and hence the low blood flow through the lung. Although it has not been proven that low pulmonary vascular resistance would be detrimental to the fetus, one expects that a large left-to-right shunt, which would accompany a fall in lung resistance, would impair the ability of the fetal heart to maintain a combined left and right ventricular output sufficient to maintain perfusion of the placenta.

One might presume that the hypoxic vasoconstrictor mechanism would be useful to mammals during the first critical hours after birth. Ventilation of the alveoli might be uneven after the initial breaths, and active local hypoxic vasoconstriction would divert pulmonary blood flow to the areas with the highest oxygenation. Later, when lung expansion, regular ventilation, and a low pulmonary arterial pressure are established, the hypoxic mechanism might cause arterial tone to be higher in the more hypoxic areas of the lung, thus preventing overcirculation of the most dependent portions of the lungs. When there is local disease of the lung, as in pneumonia or atelectasis, the hypoxic vasopressor mechanism would be useful in defending the arterial P_{O_2} by shunting blood away from the diseased areas.

W. Wagner has found that those species with the largest pulmonary pressor responses to high altitude, the cow and the pig, also have little or no collateral ventilation and that those species with the smallest responses to high altitude, e.g., dog and sheep, have good collateral ventilation. Perhaps species with poor collateral pulmonary ventilation are poorly equipped to distribute inspired air. Their major defense against regional pulmonary hypoxia appears to be an unusually vigorous pulmonary vasoconstrictor response.

When the hypoxic vasoconstrictor mechanism is absent, as in alcoholic liver cirrhosis in adult man (111), overcirculation of the lung bases, focal edema, and hypoxemia occur even in the absence of lung disease. The absence of the mechanism in liver cirrhotic patients is not lethal. The patients usually pursue a natural history dictated by the severity of their liver disease and not by the presence of mild arterial hypoxemia. In adult mammals and birds we have considered the hypoxic pulmonary vasoconstriction useful, although not necessarily vital, for survival. In the fetus and newborn, however, it may be vital for survival, and thus necessary for the continuation of the species.

High altitude exposure does not augment regional differences in alveolar oxygenation, but rather dictates that all of the alveoli will be hypoxic. Thus, it is not possible for pulmonary blood flow to be shifted from a hypoxic area to a normoxic area with virtually no increase in total perfusing pressure. Rather, there must be an increase in the pulmonary arterial pressure. The increased pulmonary arterial pressure would lead to some redistribution of the pulmonary flow from the most dependent to the most superior portion of the lung in those species, such as erect man, in which the distances from the top to the bottom on the lung are great. Such redistribution of blood flow would, in turn, lead to some increase in diffusing area and hence oxygen-diffusing capacity. It is less clear what advantage might accrue to the mouse (40), chicken, or the rat, where the distance from top to bottom of the lung is so small that relatively even distribution of blood flow might be expected at low altitude. Furthermore, when large animals such as cattle are taken to high altitude and show a progressive increase in pulmonary arterial pressure, there is not a progressive improvement in arterial oxygenation. In addition, cattle with severe pulmonary hypertension at high altitude do not have higher arterial oxygen tensions than do cattle with relatively little pulmonary hypertension (25, 112). Severe pulmonary hypertension not only fails to improve arterial oxygenation, but, instead, places the individual at risk of right heart failure. For example, in the settling of Colorado by the white man in the past 100 years, cattle introduced from low altitude have demonstrated a striking natural selection process. The cattle, which have been bred and raised for many generations at altitudes above 2,500 m, have low pulmonary arterial pressures, and their incidence of right heart failure is strikingly less (0.5%) than that (40%) of cattle brought directly from low altitude (113). Cattle that develop pulmonary hypertension and right heart failure either die or are sold to ranchers at low altitudes. In this way cattle susceptible to hypoxic pulmonary hypertension are eliminated from high altitude herds.

The tendency of natural selection to eliminate pulmonary hypertensive individuals at high altitude is not limited to the bovine species. An examination of the weights of the right ventricles compared to total ventricular weight (Table 1) shows that the range of mean values is 20–24% for the various species at sea level. Sea level species exposed for several weeks to high altitude have evidence of marked right ventricular hypertrophy, with mean ratios varying from 39–76%. These newcomers have evidence of greater right ventricular hypertrophy and, therefore, presumably higher pulmonary arterial pressures than do animals of the same species that have lived for several generations at high altitude. The llama, beaver, marmot, pika, and guinea pig, which have probably been living for many thousands of generations at high altitude, all have only a slight right ventricular hypertrophy despite the rather extreme altitude of residence. Pulmonary arterial pressure in Andean natives (whose ancestors settled at high altitude thousands of years ago) is lower than that in Leadville residents who represent the first, second, and third generations at high altitude (6).

Thus, the evidence suggests that long-term residents of high altitude are best served by an absence of severe pulmonary hypertension. Natural selection of both man and animals leads to less rather than more reactivity of the pulmonary vascular bed to hypoxia. One might suspect then that the purpose of the hypoxic pressor mechanism is not to adapt individuals or species to high altitude life. The mechanism is of doubtful value in diseases with generalized pulmonary hypoxia (i.e., hypoventilation and chronic pulmonary

Table 1. Ratio of right ventricular to total ventricular weights in various species at sea level, in species native to high altitude, and in species brought from low altitude to high altitude

	Sea level native		High altitude measurements[a]			
			Native		Newcomers	
Species	*n*	*RV/T* (Percent)	*n*	*RV/T* (Percent)	*n*	*RV/T* (Percent)
Man	12	21±0.6	10	29±1.5		
Guinea pig	12	20±1.8	10	27±2.3	5	39±2
Rabbit	12	23±1.1	10	31±3	6	45±2
Dog	25	24±1.8	15	29±1.4	5	38±0.4
Sheep	11	22±0.8	20	26±3.9	6	39±1
Pig	10	23±1.9	12	27±1.5	6	57±2
Cow	10	22±1.5	10	26±1	5	76±8
Llama			2	24	1	24
Beaver			30	27±0.7		
Marmot			14	27±0.8		
Pika			40	28±0.6		

From Hultgren et al. (114), Hultgren and Miller (115), Banchero et al. (12), Tucker et al. (23), and W. W. Wagner, unpublished observations.

[a]The high altitude native guinea pigs, rabbits, dogs, sheep, marmots, and pikas lived at 4,300 m. The pigs were native to 3,200 m, cattle to 3,600 m, and man to 3,750 m. The newcomers (except for the llama) were residents of 1,600 m exposed for 3–8 weeks to 4,500 m. The llama had been born at sea level and was exposed for 10 weeks to 3,400 m.

disease) because right heart failure from pulmonary hypertension is a major factor in the mortality and morbidity of such patients. Perhaps the hypoxic pulmonary pressor mechanism serves its most useful purpose in maintaining a high resistance in fetal lung vessels and assisting in defending the arterial oxygenation in the newborn, and in maintaining some tone in the most dependent lung vessels in the adults of large mammalian species. A reactive pulmonary vascular bed appears disadvantageous for high altitude residents.

REFERENCES

1. Bohr, D. F. (1977). The pulmonary hypoxic response: state of the field. Chest (suppl.) 71:244.
2. Fishman, A. P. (1976). Hypoxia on the pulmonary circulation: how and where it acts. Circ. Res. 38:221.
3. Heath, D. (1977). Hypoxia and the pulmonary circulation. J. Clin. Pathol. (suppl.) 11:21.
4. Lloyd, T. C. (1977). Summary of the 19th Aspen Conference, 1977. Chest (suppl. 2) 71:312.
5. Blount, S. G., Jr., and Vogel, J. H. K. (1967). Altitude and the pulmonary circulation. Adv. Intern. Med. 13:11.

6. Hultgren, H. N., and Grover, R. F. (1968). Circulatory adaptation to high altitude. Annu. Rev. Med. 19:119.
7. Severinghaus, J. (1977). Pulmonary vascular function. Am. Rev. Respir. Dis. (suppl.) 115:149.
8. Bergofsky, E. H. (1974). Mechanisms underlying vasomotor regulation of regional pulmonary blood flow in normal and disease states. Am. J. Med. 57:378.
9. Reeves, J. T., and Grover, R. F. (1975). High altitude pulmonary hypertension and pulmonary edema. *In* P. N. Yu and J. F. Goodwin (eds.), Progress in Cardiology, Vol. 4, pp. 99–118. Lea and Febiger, Philadelphia.
10. Tucker, A., Weir, E. K., Reeves, J. T., and Grover, R. F. (1976). Failure of histamine antagonists to prevent hypoxic pulmonary vasoconstriction in dogs. J. Appl. Physiol. 40:496.
11. Martin, L. F., Tucker, A., Monroe, M. L., and Reeves, J. T. (1978). Lung mast cells and hypoxic pulmonary vasoconstriction in cats. Respiration 35:73.
12. Banchero, N., Grover, R. F., and Will, J. A. (1971). High altitude pulmonary hypertension in the llama (*Llama glama*). Am. J. Physiol. 220:422.
13. Owen-Thomas, J. B., and Reeves, J. T. (1969). Hypoxia and pulmonary arterial pressure in the rabbit. J. Physiol. 201:665.
14. Bisgard, G. E., Orr, J. A., and Will, J. A. (1975). Hypoxic pulmonary hypertension in the pony. Am. J. Vet. Res. 36:49.
15. McMurtry, I. F., Frith, C. H., and Will, D. H. (1973). Cardiopulmonary responses of male and female swine to simulated high altitude. J. Appl. Physiol. 35:459.
16. Moore, L. G., McMurtry, I. F., and Reeves, J. T. (1978). Effects of sex hormones on cardiovascular and hematologic responses to chronic hypoxia in rats. Proc. Soc. Exp. Biol. Med. 158:658.
17. Aarseth, P. A., and Karlsen, J. (1977). Blood volume and extra vascular water content in the rat lung during acute alveolar hypoxia. Acta Physiol. Scand. 100:236.
18. Reeves, J. T., Grover, E. B., and Grover, R. F. (1963). Pulmonary circulation and oxygen transport in lambs at high altitude. J. Appl. Physiol. 18:560.
19. Ruiz, A. V., Bisgard, G. E., and Will, J. A. (1973). Hemodynamic responses to hypoxia and hyperoxia in calves at sea level and altitude. Pfluegers Arch. 344:275.
20. Will, J. A., and Bisgard, G. E. (1975). Comparative hemodynamics of domestic animals at high altitude. Prog. Respir. Res. 9:138.
21. Cueva, S., Sillow, H., Valenzuela, A., and Plooz, H. (1974). High altitude induced pulmonary hypertension and right heart failure in broiler chickens. Res. Vet. Sci. 16:370.
22. Burton, R. R., Besch, E. L., and Smith, A. H. (1968). Effect of chronic hypoxia on the pulmonary arterial blood pressure of the chicken. Am. J. Physiol. 214:1438.
23. Tucker, A., McMurtry, I. F., Reeves, J. T., Alexander, A. F., Will, D. H., and Grover, R. F. (1975). Lung vascular smooth muscle as a determinant of pulmonary hypertension at high altitude. Am. J. Physiol. 228:762.
24. Weidman, W. H., Titus, J. L., and Shepherd, J. T. (1965). Effect of chronic hypoxia on the pulmonary circulation of cats. Proc. Soc. Exp. Biol. Med. 118:1158.
25. Will, D. H., Hicks, J. L., Card, C. S., and Alexander, A. F. (1975). Inherited susceptibility of cattle to high altitude induced pulmonary hypertension. J. Appl. Physiol. 38:491.
26. Reeves, J. T., and Leathers, J. E. (1964). Hypoxic pulmonary hypertension of the calf with denervation of the lungs. J. Appl. Physiol. 19:976.

27. Dawes, G. S. (1955–56). Physiological effects of anoxia in the foetal and newborn lamb. Lect. Sci. Basis Med. 5:53.
28. Dawes, G. S. (1968). Foetal and Neonatal Physiology. Year Book Medical Publishers, Inc., Chicago.
29. Harris, P., and Heath, D. (1977). The Human Pulmonary Circulation, p. 452. Churchill-Livingstone, London.
30. Reeves, J. T., and Leathers, J. E. (1967). Postnatal development of the pulmonary and bronchial arterial circulations in the calf and the effects of chronic hypoxia. Anat. Rec. 157:641.
31. von Euler, V. S., and Liljestrand, G. (1946). Observations on the pulmonary arterial blood pressure in the cat. Acta Physiol. Scand. 12:301.
32. Nisell, O. I. (1950). The influence of blood gases on the pulmonary vessels of the cat. Acta Physiol. Scand. (suppl. 73) 21:5.
33. Duke, H. N. (1950–51). Pulmonary vasomotor responses of isolated perfused cat lungs to anoxia and hypercapnia. Q. J. Exp. Physiol. 36:75.
34. Lloyd, T. C. (1964). Effect of alveolar hypoxia upon pulmonary vascular resistance. J. Appl. Physiol. 19:1086.
35. Hauge, A. (1968). Conditions governing the pressor response to ventilation hypoxia in isolated perfused rat lungs. Acta Physiol. Scand. 72:33.
36. Kato, M., and Staub, N. C. (1966). Response of small pulmonary arteries to unilobar hypoxia and hypercapnia. Circ. Res. 19:426.
37. Arias-Stella, J., and Saldana, M. (1963). The terminal portion of the pulmonary arterial tree in people native to high altitude. Circulation 28:915.
38. Jaenke, R. S., and Alexander, A. F. (1973). Fine structural alterations of bovine peripheral pulmonary arteries in hypoxia induced hypertension. Am. J. Pathol. 73:377.
39. Heath, D., Castillo, Y., Arias-Stella, J., and Harris, P. (1969). The small pulmonary arteries of the llama and other domestic animals native to high altitude. Cardiovasc. Res. 3:75.
40. James, W. R. L., and Thomas, A. J. (1968). The effect of hypoxia on the heart and pulmonary arterioles of mice. Cardiovasc. Res. 3:278.
41. Naeye, R. L. (1965). Pulmonary vascular changes with chronic unilateral pulmonary hypoxia. Circ. Res. 17:160.
42. Bland, R. D., Demling, R. H., Selinger, S. L., and Staub, N. C. (1977). Effects of alveolar hypoxia on lung fluid and protein transport in unanesthetized sheep. Circ. Res. 40:269.
43. Howard, P., Barer, G. R., Thompson, B., Warren, P. M., Abbot, C. J., and Mungall, I. P. F. (1975). Factors causing and reversing vasoconstriction in unventilated lung. Respir. Physiol. 24:325.
44. Hauge, A. (1969). Hypoxia and pulmonary vascular resistance: the relative effects of pulmonary arterial and alveolar pO_2. Acta Physiol. Scand. 76:121.
45. Jameson, A. C. (1964). Gaseous diffusion from alveoli into pulmonary arteries. J. Appl. Physiol. 19:448.
46. Torrance, R. W. (1969). *In* A. P. Fishman and H. H. Hecht (eds.), The Idea of a Chemoreceptor in Pulmonary Circulation and Interstitial Space, pp. 223–238. University of Chicago Press, Chicago.
47. Weil, J. V., Byrne-Quinn, E., Sodal, I. E., Frisen, W. O., Underhill, B., Filley, G. F., and Grover, R. F. (1970). Hypoxic ventilatory drive in normal man. J. Clin. Invest. 49:1061.
48. Weil, J. V., Jamieson, G., Brown, D. W., Grover, R. F., Balchum, O. J., and Murray, J. F. (1968). Red cell mass-arterial oxygen relationship in normal man. J. Clin. Invest. 47:1627.
49. Barer, G. R. (1976). The physiology of the pulmonary circulation and methods of study. Pharmacol. Ther. (B). 2:247.

50. Lloyd, T. C. (1966). Influence of blood pH on hypoxic pulmonary vasoconstriction. J. Appl. Physiol. 21:358.
51. Malik, A. B., and Kidd, B. S. L. (1973). Adrenergic blockade and the pulmonary vascular response to hypoxia. Respir. Physiol. 19:96.
52. Rudolph, A. M., and Yuan, S. (1966). Response of the pulmonary vasculature to hypoxia and H^+ ion concentration changes. J. Clin. Invest. 45:399.
53. Viles, P. H., and Shephard, J. T. (1968). Relationship between pH, pO_2 and pCO_2 on the pulmonary vascular bed of the cat. Am. J. Physiol. 215:1170.
54. Barer, G. R. (1966). Reactivity of the vessels of collapsed and ventilated lungs to drugs and hypoxia. Circ. Res. 18:366.
55. Porcelli, R. J., and Bergofsky, E. H. (1973). Adrenergic receptors in pulmonary vasoconstrictor responses to gaseous and humoral agents. J. Appl. Physiol. 34:483.
56. Duke, H. N. (1968). The effect of adrenergic blocking agents on pulmonary vasoconstrictor response to hypoxia in the isolated cats' lungs. J. Physiol. (Lond.) 196:59P.
57. Malik, A. B., and Kidd, B. L. (1973). Independent effects of changes in H^+ and CO_2 concentrations on hypoxic pulmonary vasoconstriction. J. Appl. Physiol. 34:318.
58. Thilenius, O. G., Candiolo, B. M., and Beug, J. L. (1967). Effect of adrenergic blockade on hypoxia-induced pulmonary vasoconstriction in awake dogs. Am. J. Physiol. 213:990.
59. Tucker, A., and Reeves, J. T. (1975). Non-sustained pulmonary vasoconstriction during acute hypoxia in anesthetized dogs. Am. J. Physiol. 228:756.
60. Silove, E. D., and Grover, R. F. (1968). Effects of alpha adrenergic blockade and tissue catecholamine depletion on pulmonary vascular response to hypoxia. J. Clin. Invest. 47:274.
61. Reeves, J. T., and Leathers, J. E. (1964). Circulatory changes following birth of the calf and the effects of hypoxia. Circ. Res. 15:343.
62. Hauge, A. (1968). Role of histamine in hypoxic pulmonary hypertension in the rat. I. Blockade of potentiation of endogenous amines, kinins and ATP. Circ. Res. 22:371.
63. Hauge, A., and Melmon, K. L. (1968). Role of histamine in hypoxic pulmonary hypertension in the rat. II. Depletion of histamine serotonin and catecholamines. Circ. Res. 22:385.
64. Susmano, A., and Carleton, R. A. (1971). Prevention of hypoxic pulmonary hypertension by chlorpheniramine. J. Appl. Physiol. 31:531.
65. Dawes, G. S., and Mott, J. C. (1962). The vascular tone of the foetal lung. J. Physiol. (Lond.) 164:465.
66. Tucker, A., Weir, E. K., Reeves, J. T., and Grover, R. F. (1975). Histamine H_1 and H_2 receptors in the pulmonary and systemic vasculature of the dog. Am. J. Physiol. 229:1008.
67. Hales, C. A., and Kazemi, N. (1975). Role of histamine in the hypoxic vascular response of the lung. Respir. Physiol. 24:8188.
68. Barer, G. R., and McCurrie, J. R. (1969). Pulmonary vasomotor responses in the cat: the effects and interrelationships of drugs, hypoxia and hypercapnia. Q. J. Exp. Physiol. 54:156.
69. Hoffman, E. A., Munroe, M. L., Tucker, A., and Reeves, J. T. (1977). Histamine H_1 and H_2 receptors in the cat and their roles during alveolar hypoxia. Respir. Physiol. 29:255.
70. Silove, E. D., and Simoha, A. J. (1973). Histamine-induced vasodilation in the calf: relationship to hypoxia. J. Appl. Physiol. 38:830.
71. Shaw, J. W. (1971). Pulmonary vasodilator and vasoconstrictor actions of histamine. J. Physiol. 215:34P.

72. Dawson, C. A., Delano, F. A., Hamilton, L. N., and Stekiel, W. J. (1974). Histamine releasers and hypoxic vasoconstriction in isolated cat lungs. J. Appl. Physiol. 37:670.
73. McMurtry, I. F., Dickey, D. W., Souhrada, J. F., Reeves, J. T., and Grover, R. F. (1977). Vascular effects of compound 48/80 in isolated lungs and pulmonary arteries. Artery 3:1.
74. Mungall, I. P. F., and Barer, G. R. (1975). Lung vessels and mast cells in chronically hypoxic rats. Prog. Respir. Res. 9:144.
75. Tucker, A., McMurtry, I. F., Alexander, A. F., Reeves, J. T., and Grover, R. F. (1977). Lung mast cell density and distribution in chronically hypoxic animals. J. Appl. Physiol. 42:174.
76. Kay, J. M., Gillund, T. D., and Heath, D. (1967). Mast cells in the lungs of rats fed on crotalaria spectabilis seeds. Am. J. Pathol. 51:1031.
77. Hyman, A. L., Spannhake, E. W., and Kadowitz, P. J. (1978). Prostaglandins and the lung. Am. Rev. Respir. Dis. 117:111.
78. Weir, E. K., and Grover, R. F. (1978). The role of endogenous prostaglandins in the pulmonary circulation. Anesthesiology 48:201.
79. Said, S. I., and Hara, N. (1975). Prostaglandins and the pulmonary vasoconstrictor response to alveolar hypoxia. Science 189:900.
80. Weir, E. K., McMurtry, I. F., Tucker, A., Reeves, J. T., and Grover, R. F. (1976). Prostaglandin synthetase inhibitors do not decrease hypoxic pulmonary vasoconstriction. J. Appl. Physiol. 41:714.
81. Hales, C. A., Rouse, E. T., and Slate, J. L. (1978). Influence of aspirin and indomethacin on variability of alveolar hypoxic vasoconstriction. J. Appl. Physiol. 45:33.
82. Tyler, T. L., Wallis, R. G., and Leffler, C. W. (1975). The effects of indomethacin on the pulmonary vascular response to hypoxia in the premature and mature newborn goat. Proc. Soc. Exp. Biol. Med. 150:695.
83. Vaage, J., Bjertnaes, L., and Hauge, A. (1975). The pulmonary vasoconstrictor response to hypoxia: effects of inhibitors of prostaglandin biosynthesis. Acta Physiol. Scand. 95:95.
84. Gerber, J. G., Volkel, N., Nies, A. S., Kadowitz, P. J., McMurtry, I., and Reeves, J. T. (1978). Pulmonary vasodilation during hypoxia with infusion of arachidonic acid. Physiologist 21:43.
85. Rudolph, A. M. (1977). Fetal and neonatal pulmonary circulation. Am. Rev. Respir. Dis. 115:11.
86. Weir, E. K. (1978). Does normoxic pulmonary vasodilation rather than hypoxic vasoconstriction account for the pulmonary pressor response to hypoxia. Lancet 426.
87. Stalcup, S. A., Levenberger, P. J., Lipset, J. S., Turino, G. M., and Mellins, R. B. (1978). Decrease in instantaneous pulmonary clearance of bradykinin by acute hypoxia in dogs. Fed. Proc. 37:292.
88. Berkov, S. (1974). Hypoxic pulmonary vasoconstriction in the rat: the necessary role of angiotensin II. Circ. Res. 35:256.
89. McMurtry, I. F., Hiser, W. W., Reeves, J. T., and Grover, R. F. (1975). Disassociation of hypoxia- and angiotensin II-induced pulmonary vasoconstriction by saralasin. Fed. Proc. 34:438.
90. Hyman, A., Heymann, M., Levin, D., and Rudolph, A. (1975). Angiotensin is not the mediator of hypoxia-induced pulmonary vasoconstriction in fetal lambs. Circulation Suppl. 51–52: II–132.
91. Hales, C. A., Rouse, E. T., and Kazemi, H. (1977). Failure of saralasin acetate, a competitive inhibitor of angiotensin II, to diminish alveolar hypoxic vasoconstriction in the dog. Cardiovasc. Res. 11:541.
92. Robin, E. D., Cross, C. E., Millan, J. E., and Murdaugh, H. V. (1967).

Humoral agent from calf lung producing pulmonary arterial vasoconstriction. Science 156:827.
93. Benumof, J. L., Mathers, J. M., and Wahrenbrock, E. A. (1978). The pulmonary interstitial compartment and the mediator of hypoxic pulmonary vasoconstriction. Microvasc. Res. 15:69.
94. Detar, R., and Bohr, D. F. (1972). Contractile responses of isolated vascular smooth muscle during prolonged exposure to anoxia. Am. J. Physiol. 222:1269.
95. Haack, D. W., Abel, J. H., and Jaenke, R. S. (1975). Effects of hypoxia on the distribution of calcium in arterial smooth muscle cells of rats and swine. Cell Tissue Res. 157:125.
96. McMurtry, I. F., Davidson, A. B., Reeves, J. T., and Grover, R. F. (1976). Inhibition of hypoxic pulmonary vasoconstriction by calcium antagonists in isolated rat lungs. Circ. Res. 38:99.
97. Tucker, A., McMurtry, I. F., Grover, R. F., and Reeves, J. T. (1976). Attenuation of hypoxic pulmonary vasoconstriction by verapamil in intact dogs. Proc. Soc. Exp. Biol. Med. 151:611.
98. Lloyd, T. C. (1965). Pulmonary vasoconstriction during histotoxic hypoxia. J. Appl. Physiol. 20:488.
99. McMurtry, I. F., Petrun, M. D., and Reeves, J. T. (1978). Metabolic inhibitors do not reverse the blunted hypoxic pressor response in lungs from high altitude rats. Fed. Proc. 37:293.
100. Bergofsky, E. H., Bass, B. G., Ferretti, R., and Fishman, A. P. (1963). Pulmonary vasoconstriction in response to pre-capillary hypoxemia. J. Clin. Invest. 42:1201.
101. Rounds, S., Hookway, B., and McMurtry, I. F. (1978). Energy metabolism influences hypoxic pulmonary vasoconstriction. Am. Rev. Respir. Dis. 117:387.
102. Doekel, R. C., Weir, E. K., Looga, R., Grover, R. F., Reeves, J. T. (1978). Potentiation of hypoxic pulmonary vasoconstriction by ethyl alcohol in dogs. J. Appl. Physiol. 44:76.
103. Nilsen, K. H., and Hauge, A. (1968). Effects of temperature changes on the pressor response to acute alveolar hypoxia in isolated rat lungs. Acta Physiol. Scand. 73:111.
104. McMurtry, I. F., Hookway, B. W., and Roos, S. D. (1978). Red blood cells but not platelets prolong vascular reactivity of isolated rat lungs. Am. J. Physiol. 234:186.
105. Duke, H. N., and Killick, E. M. (1952). Pulmonary vasomotor responses of isolated perfused lungs to anoxia. J. Physiol. 117:303.
106. Sylvester, J. T., and McGowan, C. (1977). Effect of cytochrome P450 inhibitors on the pulmonary vascular response to hypoxia. Fed. Proc. 36:535.
107. Satchell, G. H., Hanson, D., and Johansen, K. (1970). Differential blood flow through afferent branchial arteries of the skate. Rajarhina. J. Exp. Biol. 52:721.
108. Johansen, K., Lenfant, C., and Hanson, D. (1970). Phylogenetic development of pulmonary circulation. Fed. Proc. 20:1135.
109. Burggren, W. W., Glass, M. L., and Johansen, K. (1977). Pulmonary ventilation: perfusion relationships in terrestrial and acquatic chelonian reptiles. Can. J. Zool. 55:2024.
110. Millard, R. W., and Johansen, K. (1974). Ventricular outflow dynamics in the lizard, *Varanus niloticus*: responses to hypoxia, hypercarbia and diving. J. Exp. Biol. 60:871.
111. Daoud, F. S., Reeves, J. T., and Schaefer, J. W. (1973). Failure of hypoxic pulmonary vasoconstriction in patients with liver cirrhosis. J. Clin. Invest. 51:1076.
112. Will, D. H., Hicks, J. L., Card, C. S., Reeves, J. T., and Alexander, A. F.

(1975). Correlation of acute with chronic hypoxic pulmonary hypertension in cattle. J. Appl. Physiol. 38:494.

113. Will, D. H., Hornell, J. F., Reeves, J. T., and Alexander, A. F. (1975). Influence of altitude and age on pulmonary arterial pressure in cattle. Proc. Soc. Exp. Biol. Med. 150:564.

114. Hultgren, H. N., Marticorena, E., and Miller, H. (1963). Right heart hypertrophy in animals at high altitude. J. Appl. Physiol. 18:913.

115. Hultgren, H. N., and Miller, H. (1967). Human heart weight at high altitude. Circulation 35:207.

International Review of Physiology
Environmental Physiology III, Volume 20
Edited by D. Robertshaw

7
Physiological Effects of Carbon Monoxide

G. R. WRIGHT and R. J. SHEPHARD

Department of Preventive Medicine and Biostatistics,
Faculty of Medicine,
University of Toronto, Toronto, Canada

EXPOSURE TO CARBON MONOXIDE 314
Natural Sources 314
Internal Combustion 314
Tobacco Smoke 316
Resultant COHb Levels 317
Normal Findings 317
Effects of Smoking 318
Occupational Exposures 319

LABORATORY TECHNIQUES 320
Carbon Monoxide Determinations 320
Gas Phase 320
Blood Phase 321
Bloodless Methods 323
Administration of Carbon Monoxide 324
Chamber Exposure 324
Rapid Rebreathing Exposure 324
Epidemiological Studies 324

PHYSIOLOGICAL EFFECTS 325
Cellular Effects 325
Hemoglobin 325

Departmental research relating to carbon monoxide was supported in part by Provincial Research Grant PR-183.

Myoglobin 326
Cytochrome Pigments 326
Uptake and Release of Carbon Monoxide 327
Endogenous Production 327
Uptake 327
Elimination 328
Models 329
Cardiorespiratory Effects 331
Heart Rate and Ventilation at Rest 331
Heart Rate and Ventilation with Exercise 332
Cardiorespiratory Effects of CO During Maximum Exercise 332
Effects of CO on the Central Nervous System 332
General Considerations 332
Animal Experiments 333
Visual System 334
Visual Acuity 334
Intensity Discrimination 334
Temporal Discrimination 334
Dark Adaptation 335
Visual Evoked Response 335
Auditory System 335
Auditory Duration Discrimination 335
Auditory Flutter Fusion 336
Cerebral Efficiency 336
Vigilance 336
Time Estimation 336
Learning and Memory 337
Mental and Cognitive Performance 337
Choice Reaction Time 337
Psychomotor Performance 338
Speed of Movement 338
Precision of Movement 338
Driving of Vehicles 339
Adaptation to CO Exposure 339
Interaction with Other Stressors 340
Severe Exposures 341
Acute Toxicity 341
Chronic Toxicity 342

CONSEQUENCES FOR HEALTH 342
Ischemic Heart Disease 343
Animal Experiments 343
Epidemiological Studies 343
Effects on Diseased Hearts 344

Accidents 345
Litigation 345

AIR QUALITY CRITERIA FOR CARBON MONOXIDE 345

According to Greek legend (1), Prometheus stole fire from Olympus and gave it to man as a symbol of knowledge and progress. From that time man must have been at least partially aware of the adverse effects of combustion. The first writings on the effects of carbon monoxide on health are attributed to the Greek philosopher Aristotle, who lived in the third century B.C. (2). In ancient Rome, the fumes of charcoal were used to commit suicide, and by the time of Cicero (106–43 B.C.) criminals were executed by exposure to charcoal fumes (2, 3).

The combustion of coal for domestic heating increased during the fifteenth century, and cases of domestic carbon monoxide (CO) poisoning showed a parallel rise (4). Deacon (5), in writing of the evils of tobacco smoke, also recognized harmful constituents of coal smoke: "... chimnies (by our Ancestors formerly invented), that thereby, the smoke ... might be dispersed abroad in the aire, for feare of hurting the bodies of men." More recent reports of high CO concentrations in primitive dwellings during the cooking of food (6) illustrate that the poorly ventilated combustion of many fuels is potentially dangerous.

In the sixteenth century, coal was recognized as an important national asset and underground mining flourished. A risk of industrial CO poisoning was one consequence of this development, since the gases encountered (termed "damps" from the middle German *dampf*, or vapor) often included quite high concentrations of CO. During the nineteenth century the hazard of industrial CO poisoning again increased with the introduction of such manufactured gases as illuminating gas or coal gas (7.4% CO by volume), producer gas (33.5% CO by volume), and water gas (43.6% CO by volume) (7). In an attempt to reduce the frequency of accidental CO poisoning, the manufacture of gases containing 30% or more of CO was discontinued in Great Britain (8). During World War II, the conversion of many European vehicles to producer gas revived interest in chronic CO poisoning (9, 10). The use of manufactured gases continued in some parts of New York City until 1956 (11), and in England until the discovery of natural gas in the North Sea in the 1960's.

While carbon monoxide poisoning is still a cause of fatalities (12–15) and remains an interest of clinicians (16–23), the concern in the early part of this century shifted to occupational CO exposures (3). It is now well documented that specific occupational groups such as firemen (24, 25), traffic policemen (26–29), vehicle inspectors (30–32), parking garage attendants (33), and steel mill employees (34, 36) can be exposed to significant doses of carbon monoxide.

There is still concern over domestic air quality (37, 475) and work environments. However, today the scope of investigation has widened to include the ambient air of our cities; this stems from the increased use of the internal combustion engine for urban transportation (38).

EXPOSURE TO CARBON MONOXIDE

Natural Sources

Carbon monoxide is produced in volcanic, marsh, and natural gases and during electrical storms (39). Small quantities of CO are also formed by seed germination, seedling growth (40), and injured, cut, or dried green vegetation, such as crop plants, shrubs, grasses, weeds, and algae (41). Carbon monoxide is both produced and trapped in the float cells of marine hydrozoan jellyfish (42) and in the bladders of a number of marine brown algae (43). The world's oceans may thus contribute CO equaling 5% of that generated by human combustion of fossil fuels (44, 478).

The CO that man releases into the atmosphere is apparently increasing. Extrapolating from the calculations of Bates and Witherspoon (45), Robbins et al. (46) estimated that in 1968 man produced 2×10^{14} g of CO (200 million tons). By 1972, production had risen to 270 million tons per year, more than half of the currently estimated CO content of the troposphere (47).

Natural production has been estimated at 3.5 billion tons of CO in the Northern Hemisphere (47), the most likely source being the atmospheric oxidation of methane (48).

Despite such large natural sources of global CO (472), the content of the rural atmosphere remains low. Readings vary between 0.03 Pa and 0.80 Pa, depending on exposure to continental air masses (46). The global average for the troposphere is estimated at 0.1 PaCO, (0.14 Pa for the Northern Hemisphere and 0.06 Pa for the Southern Hemisphere) (49).

Since background levels of CO in the lower atmosphere remain stable, there must be some atmospheric removal process (39). The reaction $CO + OH \rightarrow CO_2 + H$ is one possible mechanism (50, 471). More significant removal occurs at the earth's surface; anaerobic soil bacteria oxidize CO to CO_2 in the absence of H_2 or reduce CO to methane in the presence of H_2 (51).

Internal Combustion

Carbon monoxide, being one of the products of the internal combustion engine, is discharged into the atmosphere along with water, carbon dioxide, nitrogen, hydrogen, nitric oxide, and hydrocarbons (52). Diesel engine exhaust generally contains less than 0.1% CO (53), whereas carburetted engines emit some 4% by volume of CO (52–56). The exhaust volume depends on engine speed, whereas the CO concentration depends mainly on the air to fuel ratio (4). Diesel engines normally operate with a substantial excess of air (53). Fuel-rich operation increases the concentration of CO in both diesel and carburetted exhaust (56). For example, the driving of business routes that require much stop and start driving yields relatively high contaminant emissions per vehicle km; on the other hand, freeway driving at cruising speeds of 70–100 km hr^{-1} produces quite low contaminant emissions per vehicle km (57, 58).

Engineering improvements have reduced the CO emissions from a typical vehicle by 70% compared to the "uncontrolled" car of 1960. The total estimated automotive emissions for the United States have thus fallen 17% from their 1967 level of 90×10^9 kg per year (59). This reduction has been attained mostly by maintaining an increased air-to-fuel ratio (approximately 14.5:1) in all modes of operation (4). A number of factors may increase emissions from supposedly improved vehicles. First, pollution control devices are tested on new, well tuned engines. It is difficult to sustain acceptable emission limits beyond 15,000–18,000 km of operation (60) and the increased maintenance required by pollution control devices is not generally recognized by the motoring public. About 70% of the vehicles checked in one large North American city had not been adequately maintained (61). Another problem is the increased emissions from a cold engine (52); CO produced during the first few minutes after start up (as in many short urban trips) comprises an important fraction of total emissions (62). Driving patterns are also affected by snow-covered roads, and winter driving conditions cause emission control systems to deteriorate quickly (61). Carbon monoxide emissions increase with altitude, since the decreased atmospheric partial pressure of oxygen results in less complete combustion (63). Motor vehicle use will probably continue to increase, and control devices may thus do little more than maintain the status quo (64) of emission statistics. Finally, it must be stressed that evaluation of air pollution control devices has relied on gross emissions, thereby neglecting the most critical area—population exposure in the cities (65).

The usual approach to urban sampling is the use of a limited network of long beam-path, infrared analyzers. They are generally placed at fixed sampling stations, often away from traffic; the concentrations recorded (3–10 Pa) indicate general background CO levels (66–69) rather than the concentrations met by people in specific areas of a metropolis (48, 70–73). Depending on wind speed and direction, atmospheric stability, traffic density, and the height of nearby buildings, street level CO concentrations are likely to range from 10 Pa to 50 Pa (70–77), with much higher readings (50–200 Pa) in such adverse sites as overhanging entrances, poorly ventilated underpasses, and underground garages (33, 72, 73, 78–81). The motorist who drives with his windows closed encounters less variable CO concentrations (10–20 Pa) than the pedestrian (71, 82). However, when the windows are open, the buffering effect of his cab is lost and higher (20–50 Pa) levels are attained (70, 83).

Tobacco Smoke

Since tobacco smoke is formed by slow combustion, its CO content is not surprising (84, 85). Dudley (84) realized that although CO_2 was first formed as the air was drawn through the "layer of fire," the hot carbon residues reduced the CO_2 to CO. Wahl (86) was the first to determine the CO content of pipe and cigar smoke, finding respective partial pressures of 2.3 and 6.2 kPa. Armstrong (87) suggested that closeness of packing of tobacco and rate of smoking modified the CO pressure (0.9–1.2 kPa for cigarettes, 1.1 kPa for pipes, and 5.8 kPa for cigars); the overall level of his data for cigarettes and pipes is somewhat low relative to more recent studies.

While Baumberger (88) mentioned "burning point" smoke, it was Bogen (89) who introduced the now accepted terms of "mainstream" for the smoke drawn through the cigarette and "sidestream" for the smoke escaping from the burning end. Today the term "sidestream smoke" also includes smoke escaping from the mouthpiece during puff intermission (90) and the small volume that diffuses laterally through the cigarette paper (91). The average cigarette smoker takes a 35-ml puff of mainstream smoke over an interval of 2 s, at a rate of one puff per minute (92, 93).

Early analyses of cigarette smoke needed a sample volume of 100 ml, and thus the smoking of several cigarettes (94). However, gas chromatography (molecular sieve column) has allowed CO determinations in a single puff. Mumpower et al. (93) found a minimum CO pressure in the third puff of an unfiltered king-size cigarette (3.1 kPa), an intermediate value (4.4 kPa) in the eighth puff, and the maximum value in the thirteenth puff (5.0 kPa). The average CO concentration increased with puff volume, corresponding figures for a 60-ml puff being 4.8 kPa, 5.3 kPa, and 7.7 kPa. The CO yields of 15 brands of commercial cigarettes were compared. There were no major differences between filter and nonfilter cigarettes, nor between king-size and regular cigarettes at a standardized butt length. However, the CO delivery per puff tended to be higher for regular size than for king-size cigarettes during the initial puffs (95, 96).

Hoffman and Wynder (97) investigated specific compositional differences between the smoke from cigarettes, cigars, and the newer "little" cigars. All three products were machine smoked to give a 35-ml puff of 2-s duration once every minute, although it is debatable whether cigars are smoked in this way. The CO pressure in the mainstream smoke of one brand of little cigar (5.3 kPa) was comparable to that of cigarette smoke (4.5 kPa). However, the smoke from another brand of "little" cigar and a small cigar yielded the high CO readings of 11.1 kPa and 7.7 kPa, respectively. The CO concentration of cigarette sidestream smoke averaged 4.7 times that for mainstream smoke (91, 98).

The smoke found in enclosed spaces consists of all the sidestream smoke and that part of the mainstream smoke that has not been absorbed in the smoker's mouth or lungs. The extent to which nonsmoking individuals are exposed depends on the proximity of smokers, the amount and type of smoke they produce, the volume and ventilation of the room, and any local obstructions to ventilation (99). The pattern of smoking influences both the composition of exhaled smoke and the proportion of sidestream smoke produced per cigarette. One cigarette smoker in an office-sized room can produce an average CO concentration of 10 Pa (100), although concentrations are higher (20–30 Pa) in close proximity to the smoker (101). CO concentrations of from 30 Pa to 50 Pa were developed when a closed conference room was occupied by 21 cigarette smokers or nine cigar smokers (100, 102). Given a similar occupancy, but moderate ventilation, the concentration of CO reached only 10 Pa (100), and readings were even lower when the room occupants were pipe smokers. Nevertheless, ventilation of public buildings is not always adequate, and CO concentrations of up to 37 Pa have been traced to the accumulation of cigarette smoke (103).

Resultant COHb Levels

Normal Findings The carboxyhemoglobin (COHb) levels of nonsmokers provide the only satisfactory index of urban exposure to carbon monoxide (104). Results are usually expressed as a saturation of the circulating hemoglobin (percent). Gettler and Mattice (85) made the first blood COHb determinations. Their values of 1.8 ± 1.2% for 18 residents of New York City and 1.2 ± 1.0% for 12 country dwellers were "nowhere as low as ... expected" (p. 95). They were unaware of endogenous CO production, stating that "the ideal normal individual should have no carbon monoxide in his blood" (ref. 85, p. 92, see under "Endogenous Production"). A strong association was found between high COHb levels and smoking. Hanson and Hastings (105) confirmed this conclusion, as have numerous other subsequent studies (28, 64, 104, 106–111).

Geographical location influences the COHb level of urban dwellers (85). High automobile densities are associated with high COHb saturations (104, 110, 112, 113). In Chicago and Los Angeles, three-quarters of nonsmoking blood donors had COHb saturations greater than 1.5%, whereas in Milwaukee only 26% exceeded this figure. Values were even lower in small urban areas; in Vermont and New Hampshire only 18% of the people tested had COHb levels above 1.5% saturation (104). In contrast, 17% of urban nonsmokers living near London, Ontario, had COHb levels above 1.0% saturation (114). COHb levels for male nonsmokers were almost twice as high as those for female nonsmokers (27, 110). The probable reason is that the men were working in, or driving through, a polluted area and women were not (115).

The mean COHb level for nonsmoking mothers can vary from 0.4% to 2.6% saturation, with reported mean fetal COHb levels of 0.7–2.5% saturation (107). These "normal" COHb levels for pregnant women are elevated in comparison to other nonsmokers, in part as a result of the approximate doubling of endogenous CO production (0.9 ml hr^{-1}) due to such factors as an increase in the red cell mass, endogenous CO production in the fetus, and increased heme catabolism associated with high plasma progesterone levels (116, 117).

Effects of Smoking Two presumably unrefereed letters to a medical journal (118, 119) recently made the suggestion that the increased COHb levels of cigarette smokers were constitutionally determined. This is an incredible viewpoint, except in the sense that the type and amount of tobacco usage are influenced somewhat by the individual's constitution. However, researchers are agreed that people who smoke tobacco absorb a substantial proportion of the CO produced, even if they say that they do not inhale (120). In 1888, Dudley (84) noted that CO from tobacco smoke that entered the lungs diffused into the bloodstream. Amerson (121) added that smoke from cigars and pipes was so strong that it could not be inhaled. Hartridge (122) used the reversion spectroscope (123) to demonstrate COHb in the blood of habitual smokers, although his methodology was not sensitive enough to yield consistent results. Dixon (124) commented that the blood of a "cigarette inhaler" could contain up to 5% COHb. Baumberger (88) compared the CO content of inhaled smoke to smoke puffed but not inhaled; he calculated that on average 61% of inhaled CO was retained in the lungs, in agreement with the 56% retention obtained by Haldane (125) for air/CO mixtures, more recent figures of 53% established by pulmonary function laboratories (126, 127), and 54% cited for the absorption of CO from cigarette smoke (128). The results of Bokhoven and Niessen (129) (82–87% of inhaled CO absorbed) are inconsistent with these other studies. However, lack of standardization of depth and length of inhalation could account for the discrepancy. Dalhamn et al. (128) have estimated that 3% of mainstream CO is retained in the mouth.

There is good agreement regarding the average level of COHb observed in smokers, reported values including 4.7 ± 1.7% saturation in longshoremen just before commencing work (64), 3.5 ± 1.2% in Swedish police officers in Stockholm (28), 5.3 ± 1.3% in blood donors smoking between one-half and two packs of cigarettes per day (104), 4.7 ± 0.8% in police officers (111), and 4.9 ± 2.0% in male residents of London, Ontario (114). Higher levels (7.6 ± 2.3% saturation) were observed in heavily smoking males reporting to a smoking withdrawal center (130, 131). All of the above studies, plus those of Buchwald (132), Curphey et al. (106), and Ringold et al. (109) have found COHb levels in excess of 7% for those who smoke between one and one-half and two packs of cigarettes per day. Those who smoke one-half to one pack per day average between 3% and 4% saturation, as do those who smoke cigars or a pipe or both. Since the CO burden from smoking cigarettes is not completely eliminated in 24 hr (110), the COHb levels of subjects who do not smoke on the day of testing falls toward, but does not reach, the level observed in nonsmokers.

Different tobaccos yield smoke of differing composition (7, 94, 97). Russell et al. (102) found a greater increase in blood COHb levels after smoking a non-mild compared to an extra mild filter-tipped cigarette. They therefore proposed that the CO yield of cigarettes should be published together with their tar and nicotine yield. Intersubject differences in COHb levels are large (from 2.2% to 21.8%), suggesting the influence of additional variables, possibly including pulmonary function status (133).

Controversy has arisen regarding the generality of the results of Russell et al. (134), since his subjects all smoked at the same controlled rate (135). Ashton and Telford (136) suggested that subjects could unconsciously alter their puffing rate to obtain a given dose of nicotine, and might thus obtain more CO from low than from high nicotine cigarettes. Russell et al. (137) confirmed that regular smokers regulated nicotine intake. A safer cigarette might thus be one with a relatively high nicotine content but a low tar and CO yield. However, Wright et al. (138) observed that for occasional smokers there was a greater increase of blood COHb with high than with low nicotine cigarettes. Therefore, the safer cigarette for social smokers is one low in tar and very low in nicotine. The results of both Russell et al. (134) and Wright et al. (138) suggest that the female response to cigarette smoking is different from that of the male. Cigarette smokers have been advised to change to a pipe or cigars if they are unable to stop smoking. Unfortunately, those who change from cigarettes often continue to inhale (139).

Occupational Exposures Occupational exposure to carbon monoxide happens most frequently when CO forms part of the finished product of an industry; problems include leakage, inadequate installation of equipment, and deficient ventilation (9). The blast furnace smelting of iron ore usually produces about 6 tons of gas (partial pressure of CO, 26 kPa) for every ton of iron produced (34). Thus, even under normal working conditions the COHb levels of nonsmoking blast furnace personnel reach 2.3–14.9% saturation (34–36). Grut (9) lists many other working processes in which high levels of CO can occur: casting, welding, drying or preheating, boiler cleaning, chimney sweeping, and transport of hot coke or slag. Stewart et al. (140) found COHb concentrations $>2\%$ in occupational groups involved with vehicles, metal processing, chemical processing, stone and glass processing, printing, welding, electrical assembly and repair work, and certain types of graphic artwork.

Occupational exposure to CO can clearly result from inhalation of vehicle exhaust fumes. One early study found COHb levels of 5–13% in policemen patroling the most congested mercantile districts of Philadelphia (29). Subsequent studies on nonsmoking police officers (27, 28, 68, 111) and drivers of cabs and delivery vans (27, 141) have not found COHb levels higher than 5% saturation; indeed, the observed values have generally been quite low ($2.3 \pm 0.4\%$ COHb), with only small increases over the working shift ($+0.3\%$ COHb), suggesting that such workers may not be close to traffic as frequently as postulated (111). Vehicle inspectors spend their working hours close to idling cars. Hofreuter et al. (31) compared the mean COHb levels of nonsmoking and smoking vehicle inspectors and found that, after work, the two groups were indistinguishable (saturations of 3.4% and 3.8%, respectively). Cohen et al. (30) observed that at the end of their shift the COHb levels for nonsmoking and smoking United States border inspectors had increased significantly from the prework levels of 1.5% and 4.8% saturation to 3.6% and 6.4% saturation, respectively. Increases of COHb are greater when controlling vehicles indoors, as in parking garages (33, 132), ship cargo holds (142, 143), box cars (144), trailers (145), and confined warehouse areas (144–146). As in garages undertaking vehicle repairs (147–151), inadequate ventilation and a limited cubic capacity are the principal causes of CO accumulation. These principles also apply to ice arenas where resurfacing can expose the operator, the players, and the public to CO concentrations in excess of 50 Pa (152).

LABORATORY TECHNIQUES

Carbon Monoxide Determinations

Gas Phase Considerations influencing choice of methodology include specificity, linearity of response, interference from contaminants, reliability of calibration, ruggedness, portability, and cost (153, 474). Ratings of detector performance should also note sensitivity, response time, and accuracy relative to true gas concentrations (4). Spot measurements are sometimes useful in surveys, but because of the chronic and cumulative nature of CO poisoning, continuously recorded data are of greater value.

Much early work relied on tedious and sometimes rather imprecise chemical procedures for CO determination: the discoloration of palladium-molybdenum indicator tubes (154, 155) and the formation of CO_2 and iodine by passage of the CO over heated iodine pentoxide (156) or a combination of iodine pentoxide, selenium dioxide, and sulfuric acid (157, 158). With this approach, interference could arise from other reducing agents, including acetylene, H_2S, and hydrogen gas. Early electrochemical devices detected the heat produced when a steady flow of gas was passed over hopcalyte (159, 160), used a galvanic cell to detect the iodine liberated from iodine pentoxide (14, 83, 161), or made photometric measurements of the halogen liberated from a palladium halide (162, 163), the molybdenum blue released from a compound of palladium chloride and phosphomolybdic acid (164), the colored complex formed from palladium and nitroso-β-naphthol (165), or the mercury vapor obtained from mercuric oxide (166). A later form of electromechanical detector is the Ecolyser (167); a catalytically active electrode converts CO into CO_2, and this is measured in terms of the associated hydrogen ion production. The apparatus is small and portable (4.1 kg, battery operated), and minimum interference from temperature, humidity, and other air contaminants is claimed (167). Results are closely correlated with those obtained by gas chromatography (168), and the only limitation seems to be a lack of instantaneous response (90% response time, 10 s; decay time, ~20 s (167)).

The most successful devices for continuous monitoring of CO are based on physical principles (169). However, such instruments are usually cumbersome, require a stabilized a.c. voltage, and can be damaged by vibration or frequent movement. They are, therefore, usually operated at fixed sampling sites. Nondispersive infrared analyzers are widely used for environmental monitoring (4, 170). A very long absorption path or high pressure analysis or both are necessary to give the required sensitivity (171, 172); this slows the response speed or increases the sample volume or both. Filters do not always completely eliminate interference from CO_2 and water vapor (173), and readings are also influenced by oxygen pressure (the collision-broadening phenomenon (171). These criticisms are important when examining respiratory gas, but have less relevance when monitoring ambient air.

Gas chromatography can be used to separate carbon monoxide from other air constituents. A single column packed with silica gel (174), silical gel and iodine pentoxide (175), zeolite (176), or a molecular sieve (177) is adequate for this purpose, although better results may be obtained by twin columns—a silical gel to retain CO_2, followed by a molecular sieve to separate the CO (178). Problems of low CO concentrations can be overcome by use of capillary columns and a sensitive flame ionization detector (179–181). Nevertheless, gas chromatographs are more difficult to operate than nondispersive infrared analyzers; stability of carrier gas flow, consistent oven and detector temperature, and consistent sample injection pattern are all mandatory for satisfactory results. Their role, if any, is in the analysis of blood specimens rather than the gas phase (181, 182).

One of the best current methods of gas analysis is mass spectrometry. However, there are several disadvantages in using this technique for carbon monoxide (177): 1) since the apparatus requires a near vacuum, it is inaccurate when detecting gases in low concentration; 2) carbon monoxide tends to be produced in the ion-generating section of the apparatus; and 3) expensive modifications are necessary to distinguish carbon monoxide from nitrogen (which has the same molecular weight).

Blood Phase Direct estimations of carbon monoxide in the blood phase have used a range of techniques similar to those discussed above (178). The classical Van Slyke manometric procedure used an acid ferricyanide procedure to release CO, CO_2, O_2, and N_2; CO was estimated as a residual gas pressure (183) or a pressure change subsequent to absorption in cuprous chloride (184). However, the analysis was lengthy (85), and until the introduction of microgasometric techniques (159, 185–187) substantial samples of venous blood were required.

Subsequent authors extracted CO by the Van Slyke mercury pump, but relied on a chemical (188–191) or a physical (192–194) rather than a volumetric or manometric determination. An accuracy comparable with the classical Van Slyke procedure was claimed, but methods were still time consuming and demanded considerable experience. Investigators have coupled the gas chromatograph to manometric extraction of CO (195, 196). More recently, vortex extraction has been adopted (197), this method allowing 100-μl samples of blood to be analyzed with a coefficient of variation of 1.7%.

Conway (198) developed a chemical procedure whereby CO was liberated from hemoglobin in a microdiffusion cell by using sulfuric acid. The CO was then absorbed by palladium chloride, and released hydrogen ions were titrated with a microburette. The method was accurate, but tedious, and required 2-ml samples of blood.

Biological estimates are based on differences in the spectra of hemoglobin, oxyhemoglobin, and carboxyhemoglobin. Haldane (199) attempted to match the cherry red color of COHb with standard blood specimens to which he had added known amounts of a carmine dye. However, problems arose from variations in the amount of daylight illuminating the tubes (125). The COHb also tended to dissociate in strong sunlight (200), hemoglobin levels varied between subjects (201), and the COHb reaction was affected by room temperature (202). It was generally agreed that the method was inaccurate, except in Haldane's hands, and it may have led him to various erroneous conclusions (9, 124).

A single beam spectroscopic method was introduced by Hufner as early as 1889 (123), and the twin beam reversion spectroscope is attributable to Hartridge (123, 202). The latter works quite well at high COHb levels (18), but lacks sensitivity below 35% carboxyhemoglobin saturation (178). More recently, spectrophotometric methods have been adopted. These have the advantage that small or very small samples (10 μl–0.5 ml) of blood are required, but interference can occur from other pigments such as methemoglobin and myoglobin. Hemolyzed blood is commonly treated with a reducing agent such as sodium thiosulfite; the proportions of Hb and COHb can then be determined from the relative absorption of light at two suitable wavelengths (203, 204). Klendshøj et al. (205) compared various possible wavelengths and concluded that data obtained at 480 nm and 555 nm gave the best correlation with Van Slyke results. Others compared the observed absorption with that obtained after full oxygenation (206, 207), exploited the slow conversion of COHb to methemoglobin (204) or the less ready heat precipitation of COHb relative to HbO_2 (208). Linderholm et al. (209) and the Air Pollution Control Association (4) concluded that none of the available spectrophotometric methods were of sufficient sensitivity for monitoring the normal blood COHb% of urban inhabitants. However, Lily et al. (207) subsequently claimed an S.D. of only 0.13% for COHb levels $<3\%$.

By coupling an absorption spectrophotometer to an analogue computer, it is now possible to make measurements of COHb, HbO_2, and Hb on a single blood sample (210, 211). Measurements of absorbance are made at three wavelengths (548, 568, and 578 nm) and the analogue solves the three corresponding simultaneous equations. The S.D. of COHb% estimated by this method is 0.4% (210), although frequent calibration is required to ensure valid results (212). Another procedure (213) makes use of a precision spectrometer to make measurements at four wavelengths in the near ultraviolet; the procedure is rapid, requires 100-μl blood samples, and has an accuracy of $\pm 0.6\%$ COHb.

Spectrophotometric methods, like the bloodless methods to be discussed below, have the advantage that the value reported is COHb% rather than blood CO content. To convert CO content to COHb%, it is necessary to obtain an accurate hemoglobin reading, and to assume a fixed stoichiometric relationship of CO and hemoglobin; such an assumption is probably warranted, although it has been disputed (214).

Bloodless Methods Bloodless methods of determining COHb have become quite popular, particularly in urban and industrial surveys. The majority are based on a rearrangement of Haldane's equation:

$$\text{COHb\%} = \text{HbO}_2\text{\%}\, M \frac{P_{CO}}{P_{O_2}}$$

Equilibrium is established between alveolar and pulmonary capillary gas by a period of oxygen rebreathing. It is then possible to assume that HbO_2% is ~100, P_{CO} and P_{O_2} can be measured in the rebreathing bag, and a standard value can be assumed for M (160, 215–217). Criticisms of the procedure include 1) incomplete equilibrium between gas and blood phases; 2) nonuniformity of alveolar CO; 3) hyperventilation due to accumulation of CO_2 in the rebreathing system (9); and 4) interindividual differences in values of M (215). The last criticism is the most serious. Although some authors have found almost constant values of M, Dahlström noted figures ranging from 228 to 413; interestingly, in his experiments, M was more constant (218–229) for the formation of additional quantities of COHb%. Carlsten et al. (218) found M to have a standard deviation of ± 45. Despite such potential problems, comparison of the rebreathing method with more direct measurements has shown good agreement. Henderson and Apthorp (216) used a rebreathing method with preliminary nitrogen washout and found all but two such observations fell within 10% of the line of identity relating them to direct blood determinations. Hackney et al. (219) omitted the nitrogen washout stage of the rebreathing procedure, but still found a coefficient of variation of only 4.4% for their indirect estimates (5.0% if the oxygen content of the rebreathing bag was assumed rather than measured). Jones et al. (220) proposed the even simpler approach of 20-s breathholding, after a maximum inspiration from functional residual volume. The COHb% predicted from alveolar P_{CO} at high COHb% readings again fell within 10% of the directly measured value in 90% of samples. Although their data showed disproportionately large increments of alveolar P_{CO} at high COHb% readings, this point was not confirmed by Ringold et al. (109) or McIlvaine et al. (108). The technique becomes inaccurate if there is a high atmospheric concentration of CO and the subject has not yet equilibrated with his environment (221). Several lincar and quadratic equations for the prediction of COHb have been proposed (64, 108, 222), based on equilibrium studies at low ambient partial pressures of CO:

$$\begin{aligned} \text{COHb\%} &= 0.18\,(\text{CO Pa}) \\ &= 0.14\,(\text{CO Pa}) + 1.09 \\ &= 0.16\,(\text{CO Pa}) \\ \text{CO (Pa)} &= 0.14\,(\text{COHb\%})^2 + 3.1\,(\text{COHb\%}) + 4.2 \end{aligned}$$

Administration of Carbon Monoxide

Chamber Exposure There is no fundamental reason why experimental exposures to carbon monoxide should not be carried out with the use of large military-type gas chambers where groups of men can be exposed simultaneously. However, for reasons of expense, exposures are often conducted on single subjects sitting in very small chambers (223–226). Under such circumstances results could be complicated by sensory isolation (in uncontrolled experiments), or a possible interaction between carbon monoxide and sensory isolation (in supposedly controlled experiments). The advantages of chamber exposure are 1) concentrations can mimic those in the urban environment, and 2) the subject is not aware when exposure has commenced.

Rapid Rebreathing Exposure Godin and Shephard (227) developed a rapid rebreathing technique whereby a known amount of pure CO (40–100 ml) was added to a 2-liter bag of oxygen. After 2 min of rebreathing, a relatively constant increment of COHb% was produced, ranging from 2.4% with 40 ml to 5.4% with 100 ml of CO. Control rebreathing bags filled with air and oxygen were prepared in a single or double-blind manner.

The advantages of this approach are 1) safety (it is impossible for the required dose of CO to be exceeded), and 2) a sharply delineated onset of the increase in blood COHb%. The pattern of exposure is somewhat analogous to the smoking of one or more cigarettes. The main disadvantage is that carbon monoxide is usually undergoing redistribution between the various body compartments while its effects are being observed.

Epidemiological Studies Much recent work has been based on the epidemiology of incidental industrial and urban exposures. While it can be argued that the ultimate proof of the toxicity of low doses of carbon monoxide lies in the demonstration of effects upon city dwellers, such research is beset by many pitfalls (228). Apparent correlations between CO exposure and health may be traced to such associations as exposure to pollution and cold weather, exposure to pollution and poor socioeconomic conditions, or exposure to pollution and a high cigarette consumption. Incidental exposure fails to show clear peaks, and it is thus difficult to establish reference points for the detection of a subsequent response. Fortunately, with many of the probable effects of carbon monoxide, difficulties due to a lag period are less than for other pollutants (229). Finally, for more than 40% of the adult population self-administration of carbon monoxide by cigarette smoking far exceeds incidental urban exposure (131).

PHYSIOLOGICAL EFFECTS

The range of carbon monoxide concentrations producing physiological rather than pathological effects depends upon the conditions of exposure (for example, gas concentration and duration and intensity of associated physical activity), the health of the exposed individual (230, 231), and the possible presence of concomitant stress (see under "Interaction with Other Stressors"). The usual experimental range of inhaled CO is from 25 ppm to 3,000 ppm (~25–2,950 Pa), although cigarette smokers inspire small volumes of smoke containing 15,000–40,000 ppm (~1.5–3.9 kPa).

Cellular Effects

Hemoglobin Carbon monoxide binds to protoheme, a ferrous iron complex of protoporphyrin IX in the hemoglobin molecule; the site is the same as for the reaction with oxygen, but the affinity of carbon monoxide for the pigment is much greater (232–234). The two hemoglobin dissociation curves can be virtually superimposed upon each other if allowance is made for respective affinities (235, 236) in accordance with Haldane's law:

$$\frac{[HbCO]}{[HbO_2]} = M \frac{P_{CO}}{P_{O_2}}$$

where HbCO and HbO_2 are the amounts of carbon monoxide and oxygen bound to hemoglobin, P_{CO} and P_{O_2} are the respective partial pressures, and M is the affinity ratio.

Both the reactions of oxygen with hemoglobin (237) and the analogous reactions for carbon monoxide (238–240) can be represented by four consecutive reactions of the following type:

$$Hb_4 + O_2 \underset{K_1}{\overset{K'_1}{\rightleftharpoons}} Hb_4O_2$$

$$Hb_4O_2 + O_2 \underset{K_2}{\overset{K'_2}{\rightleftharpoons}} Hb_4O_4$$

$$Hb_4O_4 + O_2 \underset{K_3}{\overset{K'_3}{\rightleftharpoons}} Hb_4O_6$$

$$Hb_4O_6 + O_2 \underset{K_4}{\overset{K'_4}{\rightleftharpoons}} Hb_4O_8$$

The equilibrium constant M depends on the magnitude of the various reaction constants, particularly those for the fourth reaction

$$\left(M = \frac{l'_4 \cdot K_4}{K'_4 \cdot l_4}\right)$$

where l'_4 and l_4 are the corresponding fourth stage reaction constants for the binding of carbon monoxide. The association velocity constants for CO are about one-fifth to one-sixth of those for oxygen, but the dissociation constants are only 1/1800 of those for the oxygen reaction (176).

When a mixture of oxygen and carbon monoxide is present, the net effect on the oxygen dissociation curve of hemoglobin is a leftward displacement, with a reduction of tissue oxygen pressures for a given oxygen content (8, 241). This is due to an elimination of low concentration binding sites from the hemoglobin dissociation curve, rather than from any more fundamental change in the hemoglobin molecule. The impairment of oxygen transport for a given hemoglobin saturation is much greater than would be anticipated for an equivalent anemia, and there have been suggestions that the resultant tissue hypoxia is sufficient to cause the vascular lesions found in many smokers (242, 476). Unlike the situation at altitude, some authors see no evidence of adaptation to CO-induced hypoxia through an alteration in the oxygen-binding properties of hemoglobin (243); however, others find CO-induced changes of 2,3-diphosphoglycerate (477, 479) (see under "Adaptation to CO Exposure").

Some authors have found a variation of M with pH (162, 236), values rising from about 180 at a plasma pH of 7.15 to a maximum of 230 at pH 7.35, and thereafter falling to 140 at pH 7.70. Others find no change with pH (235, 244), oxygen pressure (235), or 2,3-DPG levels (245), although there is agreement that the affinity ratio is modified by temperature and exposure to light (246).

Myoglobin The affinity of carbon monoxide for intracellular compounds is usually given in terms of the Warburg partition coefficient K, which describes the ratio of carbon monoxide to oxygen required for 50% saturation with carbon monoxide; in essence, it is the reciprocal of M.

The Warburg coefficient for myoglobin is in the range 0.05 (247) to 0.025 (248). The rise of carboxymyoglobin lags behind the rise of carboxyhemoglobin (227). Under normal steady state conditions blood and muscle pigment saturations become equalized (249), but hypoxia can increase the myoglobin to hemoglobin carbon monoxide saturation ratio from 1.0 to about 2.5, both in cardiac (249) and skeletal (250) muscle. The functional consequences are uncertain, although it is known that myoglobin provides a temporary oxygen store for brief bursts of interval work and also facilitates the diffusion of oxygen within the muscle fibers (251).

Cytochrome Pigments Combination of carbon monoxide with the various cytochrome pigments (480) depends not only on the partial pressure of carbon monoxide, but also upon the intracellular partial presure of oxygen. Some work based on carbon monoxide binding to myoglobin and the use of polarographic microelectrodes indicates an oxygen pressure of 0.53-0.80 kPa (249, 252, 253), but at some points within the cell pressures may be lower (254), particularly in the presence of carboxyhemoglobin.

The Warburg coefficient for cytochrome a_3 varies from 2–30 (255, 256). Early workers estimated that the CO pressure needed to produce 50% saturation of cytochrome oxidase (a mixture of cytochrome *a* and cytochrome a_3) was 21–29 Pa (120, 257). The computed tissue P_{CO} with 5% carboxyhemoglobin is about 2.7 Pa, so that in the worst case (K for cytochrome a_3 = 2), the mitochondrial oxygen pressure would need to drop to 1.3 Pa for 50% binding (258). A further rise of carboxyhemoglobin to 10% or severe tissue hypoxia (<1.3 Pa) could make the binding of physiological significance, although oxygen consumption is unaffected by 50% inhibition of this enzyme (177).

The Warburg coefficient for cytochrome P-450 varies from 0.2 to 5.0; the system seems sensitive to carbon monoxide when electron transport is rapid, but with slow electron transport it becomes almost refractory to CO (259). Practical evidence of impaired function includes a decreased hepatic metabolism of 3-OH-benzo-α-pyrene in rats exposed to 60 Pa of CO (260) and a decreased catabolism of hemoglobin-haptoglobin in dogs with a 10–12% carboxyhemoglobin (261). However, it is still unclear how far tissues can compensate for the poisoning of redox enzymes by an increase of perfusion; this seems probable under resting conditions, but might not be possible during vigorous exercise.

Carbon monoxide could theoretically interact with the reduced form of other heme protein enzymes, including catalase and peroxidase. However, the reaction is extremely unlikely to occur in practice, since these compounds are not normally present in a reduced form in the body.

Uptake and Release of Carbon Monoxide

Endogenous Production Under normal circumstances, the urban nonsmoker has a blood carboxyhemoglobin level of ~1% (110, 262). About one-half of this burden is attributable to the endogenous production of CO (263, 264), which has been estimated at 0.42 ml hr^{-1} (265, 266) or $5–10 \times 10^{-3}$ ml kg^{-1} hr^{-1} (117, 267). Some three-quarters of the endogenous CO is derived from the α-methene carbon atom of heme (268) during erythrocyte breakdown (269-272). This process is increased in hemolytic disease (263, 273, 274), during simulated saturation dives (275), and in women with increases of plasma progesterone (117). A part of the remainder arises from catabolism of hepatic heme-containing enzymes (276), whereas one possible industrial source is the breakdown of methylene chloride in the body (277). The turnover of myoglobin is slow, and for this reason it probably contributes little to endogenous CO production (278).

Uptake Almost all of the body CO is stored in combination with heme pigments (279), some 80% being bound to hemoglobin, 15% to myoglobin, and up to 5% to other heme compounds (particularly in the liver); as noted above, hypoxia encourages CO storage in myoglobin at the expense of hemoglobin (250).

The rate of buildup of blood carboxyhemoglobin depends largely on the volume of air breathed relative to body mass (and thus blood volume) (199, 280). Under resting conditions, only about one-half of the respired dose is retained in the body (199, 281), owing to the effects of dead space and backpressure at the alveolar-capillary interface. During light work, the initial retention remains around 50%, but in heavier work (~ 100 watts; heart rate, 135 beats min^{-1}) only 40% of the CO is retained (282). Uptake is slowed by the breathing of CO in oxygen (281) but is uninfluenced by a decrease of barometric pressure, providing that allowance is made for the associated increase of respiratory minute volume (282, 283). The rate of uptake is also reduced by a rising backpressure in the pulmonary capillaries when subjects have absorbed more than a third of the equilibrium quantity (199, 281). Finally, in older subjects it may be necessary to allow for a decline of pulmonary diffusing capacity due to emphysema (284) or other forms of chronic obstructive lung disease (285).

Elimination Carbon monoxide is eliminated almost entirely via the lungs, less than 1% of the gas being oxidized within the body (266, 286, 287). Most authors have assumed a single exponential curve for carbon monoxide elimination. Typical half-times for resting conditions have been 3-4 hr breathing air and about 1 hr breathing oxygen (287-289), with somewhat faster rates at very high carboxyhemoglobin concentrations (199, 290) and slower rates at very low concentrations (291).

Rode et al. (131) pointed out two problems with most reports: 1) half-times were calculated without reference to the normal urban background of ~1% carboxyhemoglobin, and 2) similar rates of CO elimination were assumed for men and women. Their analysis of data for a smoking withdrawal clinic indicated half-times of 3.7 hr for men and 2.5 hr for women.

A further potential variable is the redistribution of carbon monoxide into poorly perfused vascular compartments such as the spleen and bone marrow, plus a reaction with extravascular pigments (20, 227, 287, 292). Roughton and Root (287) commented that 60-70% of the gas leaving the blood in the first hour was expired, whereas Godin and Shephard (227) and Wagner et al. (293) have both commented on a biphasic CO elimination curve. Presumably, much depends on the speed of CO administration, redistribution being likely with the rapid rebreathing exposure of Godin and Shephard (227). A physiological analogue, based on the probable distribution of the cardiac output and the sizes of the various tissue compartments (227), suggested that carboxymyoglobin concentrations rose steadily during 6 min of exposure to a high concentration of CO, whereas there was an "overshoot" of COHb% in the pulmonary capillary blood and the rapidly circulating blood volume. During the first 5 min of the recovery period, equilibrium between the various vascular compartments was completed, but carbon monoxide continued to penetrate into the myoglobin for up to 1 hr.

While most authors continue to advocate exponential clearance models, Wagner et al. (293) found that their data for 15–90 min of postexposure were satisfied by a linear equation; during this period, clearance was determined by the affinity constants for hemoglobin and myoglobin, rather than minute ventilation or cardiac output.

Although the respiratory centers of a poisoned subject are already depressed by hypoxia (294), clearance of carbon monoxide apparently proceeds most rapidly when breathing a mixture of 5% carbon dioxide in oxygen (280, 295, 296). No additional advantage is gained from the use of a 7% CO_2 mixture (297), but the half-time of elimination can be reduced to ~23 min at an oxygen pressure of 3 atm (289, 298).

Models Various models have been developed to predict the uptake and release of carbon monoxide from the body. Forbes et al. (281) proposed the following empirical equation:

$$\text{Rate of CO uptake} = \frac{KP_{CO}(COHb_\infty - COHb_t)}{(COHb_\infty - COHb_0)}$$

with K being a constant proportional to the intensity of physical activity, P_{CO} the partial pressure of CO in inspired air, and COHb the concentration of carboxyhemoglobin initially (0), at any given time (t), and at equilibrium (∞) as indicated by the subscripts. The increase of COHb% was given by the formula $\Delta COHb\% = K'(\% CO)t$, where K' was a constant of 3–11 (increasing with ventilation and heart rate) and t was the duration of exposure in minutes. Lilienthal and Pine (283) proposed the very similar equation $\Delta COHb\% = 0.05\, \dot{V}_E (P_{CO})t$, where $\dot{V}_E$ was the respiratory minute volume in liters per minute.

Pace et al. (299) further developed this equation, adding a term for blood volume:

$$\Delta COHb\% = \frac{100\, \dot{V}_E (\% CO)t}{46.5\ (\text{blood volume})}$$

Their formula was considered valid for partial pressures of 10–200 Pa_{CO}; from the onset of exposure until blood COHb levels reached one-third of equilibrium values, the estimate was 2.3% COHb%, and results were identical with the equation of Forbes et al. (281) if blood volumes of 4.3, 4.1, 4.9, and 5.9 liters were assumed for thc four work loads.

Goldsmith et al. (300) were the first to attempt a dynamic simulation of CO uptake and release. Their equations were

$$\text{Uptake}\,(\Delta COHb\%)_t = (COHb\%_\infty - COHb\%_0)\,(1 - e^{-0.004t})$$
$$\text{Release}\,(\Delta COHb\%)_t = (COHb\%_0 - COHb\%_\infty)\,(1 - e^{-0.0033t})$$

The uptake constant of 0.004 was derived from a graphic analysis of the data of Forbes et al. (281), and the excretion constant of 0.0033 corresponded to a half-time of elimination of 3.5 hr. Their equations give a good description of long-term changes of COHb%, but are much less satisfactory when applied to short-term transients.

Coburn et al. (26) developed a more sophisticated formula that introduced many pertinent physiological variables, including the pulmonary capillary oxygen pressure ($P_{\overline{C,O_2}}$), the affinity ratio (M), the oxygen saturation (HbO_2%), the pulmonary diffusing capacity ($\dot{D}_L$), alveolar ventilation ($\dot{V}_A$), endogenous CO production ($\dot{V}_{CO}$), blood volume (V_b), and total inspired gas pressure (P_I):

$$\frac{A[\text{COHb\%}]t - \dot{V}_{CO}B - P_{I,CO}}{A[\text{COHb\%}]_0 - \dot{V}_{CO}B - P_{I,CO}} = \exp \frac{-tA}{M \cdot Vb \cdot B}$$

where

$$A = \frac{P_{\overline{C,O_2}}}{[M\ HbO_2\%]} \text{ and } B = \frac{1}{\dot{D}_L} + \frac{P_I - P_{H_2O}}{\dot{V}_A}$$

As in the model of Goldsmith et al. (300) the main problem was a rather slow response to changes, the predicted half-time for elimination of CO being 252 min. At equilibrium, the exponential term became zero, and total blood COHb% was given by the sum of endogenous and exogenous terms:

$$[\text{COHb\%}] = \frac{\dot{V}_{CO} M[HbO_2\%]}{P_{\overline{C,O_2}}}\left[\frac{1}{\dot{D}_L} + \frac{P_I - P_{H_2O}}{\dot{V}_A}\right] + \frac{P_{I,CO} M[HbO_2\%]}{P_{\overline{C,O_2}}}$$

Peterson and Stewart (291) used an empirical logarithmic format, as follows: $\log \Delta[\text{COHb\%}] = 0.857 \log C_{I,CO} (\text{Pa}) + 0.629 \log t(\text{min}) - 2.295$. Finding a half-time ($t'$) of 320 min (128–409 min) for CO elimination, they modified their equation to take account of this factor:

$$\Delta[\text{COHb\%}] = \left[\frac{CO^{0.858}(\text{Pa}) \cdot t^{0.63}\,(min)}{197}\right] 10^{-0.00094t'}$$

The standard error of the revised equation (1.18% COHb) was slightly larger than that for the formula of Coburn et al, (262) (0.97% COHb%). An alternative formula (301) was proposed for brief exposures to very high carbon monoxide concentrations: $\log$ (ΔCOHb% per liter of air breathed) = 1.036 $\log (C_{I,CO}\ \text{Pa}) - 4.4793$.

Likely COHb% levels for various carbon monoxide exposures are summarized in Table 1. Hill et al. (302) have recently developed the Coburn equation for exchange of CO between the mother and the fetus. For the mother,

$$\frac{\partial[HbCO]}{\partial t} = \frac{100}{\text{blood volume}_m}$$

$$\times [\dot{V}_{CO_m} + \dot{D}_{L,CO}(P_{A,CO} - P_{\overline{C,CO}}) - \dot{D}_{P,CO}(P_{\overline{m,CO}} - P_{\overline{f,CO}})]$$

and for the fetus

$$\frac{\partial[HbCO]}{\partial t} = \frac{100}{\text{blood volume}_f}[\dot{V}_{CO_f} + \dot{D}_{P,CO}(P_{\overline{m,CO}} - P_{\overline{f,CO}})]$$

Table 1. Anticipated COHb levels of a healthy resting nonsmoker exposed to known CO concentrations for specified period.

	COHb% and duration of exposure						
Concentration of respired CO	10 min	15 min	30 min	1 hr	3 hr	8 hr	equilib-rium
Endogenous CO	0.4	0.4	0.4	0.4	0.4	0.4	0.4
5 Pa				0.5	0.6	0.8	0.9
10 Pa			0.5	0.6	1.1	1.7	1.8
15 Pa			0.6	0.7	1.4	2.4	2.7
20 Pa		0.5	0.7	0.9	1.7	3.0	3.5
30 Pa	0.5	0.6	0.8	1.3	2.4	4.5	5.0
40 Pa	0.6	0.7	1.0	1.5	2.8	5.9	6.1
50 Pa	0.7	0.8	1.3	1.9	3.5	7.0	7.8
75 Pa	0.8	1.2	1.7	2.6	5.0	9.6	11.0
100 Pa	1.1	1.4	2.2	3.0	6.5	12.5	14.0
200 Pa	1.9	2.5	4.0	6.0	12.5	20.0	24.6

Values are estimated from the data of Forbes et al. (325) and Peterson and Stewart (291).

Here, the subscripts m and f refer to maternal and fetal conditions, $\dot{V}_{CO}$ is endogenous production of CO, $\dot{D}_P$ is the diffusing capacity of the placenta, $P_{\overline{m,CO}}$ and $P_{\overline{f,CO}}$ are mean pressures in the maternal and placental capillaries, respectively, and other symbols are as previously defined.

Cardiorespiratory Effects

Heart Rate and Ventilation at Rest Experiments on the cat have shown that substantial levels of carbon monoxide poisoning (COHb% 18–40%) lead to a 2–5-fold increase of impulse traffic from the carotid chemoreceptors (303–305). The response is observed despite an increase of arterial oxygen pressure. It cannot be explained in terms of a drop in arterial oxygen pressure, hypotension, or a direct effect of CO on cytochrome oxidase, and by exclusion it has thus been attributed to the onset of tissue hypoxia. Brain and muscle P_{O_2} decrease by 106 Pa during exposure to 80 Pa of CO (306).

Permutt and Farhi (307) estimated that a 9% increase of COHb% would lower venous P_{O_2} by 0.53–0.80 kPa, more in tissues with a wide arteriovenous oxygen difference. Unfortunately, the precise arteriovenous oxygen difference for the carotid body is unknown; it is probably about 5 ml per liter. On this basis, Comroe (308) calculated that a 50% increase of COHb% would be needed to lower the end-capillary oxygen pressure from 11.7 kPa to 10.6 kPa. The isocapnic response of the carotid body over this range of oxygen pressures is 361 ml min^{-1} per kPa. Thus, in Comroe's calculation, ventilation would increase by 1.1 $\times$ 361 ml min^{-1} (308), and a more modest 5% increase of COHb% should increase ventilation by a detectable 38 ml min^{-1}.

In fact, changes of heart rate and ventilation with exposure to mild doses of carbon monoxide have been small and equivocal (306, 310–312). With severe CO poisoning, some authors have found hyperventilation, both in goats (313) and man (314); this has been attributed to a tissue acidosis rather than to chemoreceptor stimulation and has been advanced as an argument against provision of CO_2 mixtures for patients with CO poisoning (314). Others (315) found no respiratory response to severe poisoning; presumably, in these experiments the respiratory centers were depressed by the hypoxia. A small (16%) increase of cardiac output has been reported in severe CO poisoning (316).

Heart Rate and Ventilation with Exercise Asmussen and Chiodi (317) and Apthorp et al. (318) found that neither respiratory minute volume nor respiratory rate changed during steady state exercise, even at COHb levels in excess of 30% saturation. However, heart rate increased significantly in all subjects. Pitts and Pace (319) reported that, during mild exercise, the heart rate was unchanged at a COHb level of 6% but increased at 13% saturation. Subsequent studies (311, 320–322) have shown that responses depend on both the intensity of exercise and the percentage of COHb. Generally, the higher the percentage of $\dot{V}_{O_2\ max}$ that is required, the lower the COHb level needed to increase respiratory minute volume and heart rate.

Cardiorespiratory Effects of CO During Maximum Exercise During maximum exercise, heart rate, respiratory minute volume, and respiratory rate remain unchanged after exposure to CO (47, 320, 322), but there is a decline of maximum $\dot{V}_{O_2}$ approximately proportional to the increase of blood COHb (323). Raven et al. (324) suggested that the decline was not linear at COHb levels below 6%. Nevertheless, the 3% decrement of $\dot{V}_{O_2}$ observed in their study, although not statistically significant, was as expected for a linear response to the COHb saturation of 3.6%.

Effects of CO on the Central Nervous System

General Considerations Many authors have found no significant effect of CO on the central nervous system (CNS), even at very high blood carboxyhemoglobin levels (473). Forbes et al. (325) reported no alteration of reaction time, binocular vision, and eye-hand coordination despite a COHb level of 30% saturation. Again, Stewart et al. (326) found no change of hand and foot reaction times, hand steadiness, manual dexterity, visual acuity, depth perception, and time estimation with COHb levels of 11–13%, and O'Donnell et al. (223) detected no changes of critical fusion frequency, time estimation, and mental arithmetic ability with COHb levels up to 12.7%.

In contrast, McFarland et al. (327) and Halperin et al. (328) both noted changes of visual intensity discrimination at COHb levels of 3–5%, and Beard and Wertheim (329) reported impairment of auditory duration discrimination with even lower CO exposures. Likewise, Schulte (330) observed a deterioration of choice discrimination (for colors and letters) at a COHb level of less than 5%, with parallel changes in cognitive ability (mental arithmetic, plural noun underlining, and "t" crossing); Horvath et al. (310) saw a worsening of visual vigilance in nonsmokers at an average COHb level of 6.6%, and Ramsey (285) described a small deterioration in reaction times at a COHb level of 5%.

Such disagreement is perhaps not surprising. In a psychomotor experiment, much depends upon incidental details of the working environment, the characteristics of subjects, and the type of performance measure being used. When testing mental function, it is important to utilize spare mental capacity. Firemen (330) are perhaps more likely to show inaccuracies in simple mental arithmetic than are United States Air Force personnel (223), and it seems significant that positive findings have usually involved the general population, whereas negative findings have been based on university students or military personnel where interest and peer pressure may be greater (331). If aging and oxygen want are related (332), performance changes caused by a CO-induced hypoxia may be more readily apparent in those who are older. Changes in sensory performance (such as thresholds and duration discrimination) have also been detected more easily than changes in complex cerebral function or response. Changes in these latter two categories of performance influence the execution of a well learned task (332). McFarland (333–335) further reported differences between sudden exposure (in the order of minutes) to hypoxic hypoxia and prolonged exposure with acclimatization. Finally, both Fodor and Winneke (336) and Weir et al. (337) suggested that the decrement in psychomotor performance may not be linearly related to COHb levels. The results of Beard and Wertheim (329) and Fodor and Winneke (336) suggest that effects are greatest around 90 min after CO exposure.

Animal Experiments The use of nonverbal animals to investigate behavioral changes during CO exposure is questionable, since extrapolations to human responses are then required. For example, does a reduction in the number of bar presses made by white Lewis strain rats exposed to 200 Pa of CO for 2 hr (338) indicate a change in biological function, a modification of a learned response, a reversal of a previously learned discrimination, or a decrease in feeding activity? Despite such uncertainties, studies on animals do allow one to become familiar with the gross effects of CO and to discover changes which may merit further investigation in humans. For example, Beard and Wertheim (329) observed that early in the course of CO exposure, rats exhibited what seemed to be a derangement of time sense. Therefore, a test of auditory duration discrimination was developed for human use (see under "Auditory System"). Again, studies with adult *Macaca mulatta* monkeys permitted Back and Dominiquez (339) and Back (340) to investigate whether long-term exposure to CO influenced CNS responses, an experiment for which there are few volunteers. The task required of the animals (response to visual and auditory stimuli) was not very complicated; indeed, it became almost automatic with practice, and may thus have lacked sensitivity to any behavioral changes induced by the CO exposure. However, the results indicate that animals continued to perform single learned tasks at COHb levels as high as 30% saturation (225).

Visual System Discrimination and adaptation are basic properties of receptor systems. For the visual system, these properties comprise the following functions: 1) discrimination of spatial extent (visual acuity); 2) discrimination of intensity (differential brightness sensitivity); 3) temporal discrimination (critical flicker fusion); and 4) adaptation to low intensities of illumination.

Visual Acuity Stewart et al. (326) found that 8-hr exposure to 100 Pa of CO (11–13% COHb) did not impair the visual acuity of 18 healthy men. However, Beard and Grandstaff (341) found a 5% reduction of visual acuity after 27 min of exposure to 50 Pa of CO (a COHb level of 3% saturation); they used the more precise psychophysical testing technique of limits (342). Greater decrements of acuity (16% and 32%) were found at estimated COHb levels of 6% and 10% saturation. During exposure to hypoxia, McFarland and Halperin (343) found a significant rightward displacement of the visual acuity curve along the illumination (log photon) axis at a simulated altitude of 3,050 m ($\sim$14.1 kPa O_2). However, they noted a progressively smaller decrease of log visual acuity as log intensity increased. It is thus possible that the discrepancy between the two CO studies can be explained by differences in the intensity of illumination.

Intensity Discrimination McFarland et al. (327) and Halperin et al. (328) used the visual discriminometer developed by Crozier and Holway (344) to test differential light sensitivity in a dim and uniformly illuminated field. Ten measurements were made of the least intensity which could be distinguished. Generally, response to a given increment of COHb was about the same as that seen with an equivalent reduction of arterial oxygen saturation produced by breathing low percentages of oxygen; the apparatus was sensitive enough to demonstrate an effect from one cigarette (a 2% COHb level). More recently, Beard and Grandstaff (341) reported a statistically significant increase in the threshold for visual intensity discrimination within 49 min of exposure to 50 Pa of CO (3% COHb). This increase was also observed 17 min after exposure to either 150 or 250 Pa of CO. Weir et al. (337) did not find any effect at 7% and 14% COHb; however, their visual test was brightness matching of three pairs of colored lights.

Temporal Discrimination Temporal discrimination, or alternatively the illusion of steady light, depends on such physical relationships as the intensity of the source relative to surrounding illumination and the area, shape, and location of the test patch. Anatomical and physiological determinants include physique, size of pupillary opening, chronological age, body temperature, and fatigue. Psychological variables include practice, learning, and motivation (345). In view of the number of factors involved, it is not surprising that there is some disagreement concerning the effects of CO on critical flicker fusion (CFF). Lilienthal and Fugitt (346) found that combined exposure to a simulated altitude of 1,524–1,829 m and an increase in COHb of 5–9% saturation produced an impairment of CFF; however, COHb levels of 13.8–17.4% at sea level did not change the CFF over a 20-min test period. Vollmer et al. (347) found an insignificantly greater effect with 12% COHb and altitude (4,724 m) than with altitude alone. Given the persistence of CO in comparison to hypoxia (328), their technique of baseline determination (average of initial and final performance) could have obscured a CO response. Neither Guest et al. (348) nor O'Donnell et al. (223) observed any changes in CFF at COHb levels of up to 12.7% saturation.

Nicotine and CO have opposite effects on CFF. Most studies of smoking (349–353) examined the effect of nicotine on CFF and did not measure blood COHb levels. Of the CO studies reviewed, only Guest et al. (348) and O'Donnell et al. (223) specifically mention controlling smoking prior to testing.

Dark Adaptation Abramson and Heyman (354) measured visual sensitivity before and after 20 min of CO exposure. Threshold values increased in seven of nine subjects at 10% COHb, but individual variability was such that no relationship between the CO content of the blood and diminution of dark adaptation could be described. Intersubject and intrasubject variability also contributed to the negative findings of Weir et al. (337) at 7% and 14% COHb and McFarland (355) at 11% and 17% COHb.

Visual Evoked Response Xintaras et al. (356) reported alterations in the visual evoked response of rats with CO exposure. In man, no changes were observed until COHb levels were greater than 20% (357). COHb levels of 33% did not alter spontaneous EEG activity (358).

Auditory System Frequency and intensity are the physical attributes which together or separately arouse the sensations associated with sound (359). Changes in the intensity threshold, intensity discrimination, frequency discrimination, or sound localization with CO exposure have not been investigated. However, studies have been conducted on auditory duration discrimination and auditory flutter fusion (AFF).

Auditory Duration Discrimination The auditory duration discrimination task designed by Beard and Wertheim (329) required the comparison of paired presentations of pure 1,000-Hz tones. The mean percentage of correct responses decreased after 90-min of exposure to 50 Pa of CO (a predicted COHb level of less than 3% saturation). Beard and Grandstaff (360) were unable to replicate this result with paired visual signals, despite exposure to 250 Pa of CO. In this second experiment they avoided the range ±125 ms where discrimination would be difficult, if not impossible (226).

Both O'Donnell et al. (224) and Theodore et al. (225) suggested that the effect reported by Beard and Wertheim was due to sensory isolation. Stewart et al. (361) found no change of performance after exposures to 50, 100, 200, or 500 Pa of CO for 2.5 or 5.0 hr. However, the task that they used was not an exact replicate of Beard and Wertheim's, since the second tone followed the first by 1.5 s rather than by 0.5 s. In addition the test was performed only once per hour (rather than every 20 min), with other tasks being performed in the intervening period. The former point is important, since systematic fading of an auditory image occurs over time (342, 362, 363), whereas increased sensory stimulation sustains arousal (331, 336) with beneficial effects on vigilance (364). Wright and Shephard (226) repeated Beard and Wertheim's 1967 experiment and found that the most significant variable was time on the task. Their analysis indicated a significant increase of mean errors at blood COHb levels as low as 3.2%.

Beard and Wertheim's 1967 test protocol presented the stimuli to the subjects by using this psychophysical method of constant stimulus differences. The method is a more appropriate measure of auditory duration discrimination than an analysis based on the percentage of correct responses. When Wright and Shephard (226) applied this method of analysis to their data, the previously observed CO-related changes in performance were not evident. Thus, the generality of Beard and Wertheim's 1967 results cannot be supported. Indeed, our results confirm the claim that auditory duration discrimination is not impaired at COHb levels below 8% saturation.

Auditory Flutter Fusion Variables affecting detection of auditory flutter are similar to those for the detection of visual flicker, namely, intensity, rate, and sound-time fraction (365). No reduction of AFF threshold has been observed with CO exposure (348).

Cerebral Efficiency Cerebral efficiency denotes the ease with which sensory information is processed, that is, the responsiveness of the central nervous system to sensory information, the processing of this information, and a decision about the direction in which a response should occur. This rubric, therefore, encompasses vigilance, time estimation, learning and memory, mental and cognitive performance, and choice reaction time.

Vigilance Lewis et al. (366) made no attempt to define the factors responsible for a decline of auditory vigilance observed when subjects breathed air pumped from the side of a moderately busy arterial road. Horvath et al. (310) found that while breathing 111 Pa of CO in air, significantly fewer signals were identified during a 60-min visual vigilance task. The average COHb level for the 10 nonsmoking subjects was 6.6%. Fodor and Winneke (336) found a decrement of auditory vigilance during the first 45 min of CO exposure; their final COHb was 3.1%. However, during the two subsequent 45-min periods significant differences were no longer observed. A similar time and performance trend was reported by Beard and Grandstaff (331); the signal detection parameter d' (discriminability) decreased with CO exposure, whereas B (subject's criterion) increased at COHb levels of 1.8% and 5.2% saturation, but returned almost to normal at 7.5% saturation.

Time Estimation Sollberger (367) distinguished between a time duration sense (which would need only interval timers) and a time orientation sense (which would use true chronometers). On this basis, time estimation would be distinct from auditory duration discrimination.

Beard (12) had seven subjects a) hold down a button for an estimated 30 s and b) touch the same button at 10-s intervals. Estimation of the 30-s period was significantly impaired after 75 min of exposure to 50 Pa of CO (a predicted COHb of less than 3% saturation) but deterioration in estimation of 10-s intervals was insignificant. Beard and Grandstaff (331) confirmed the increase in 30 s time estimation but also found a significant increase in 10 s estimation.

Stewart et al. (326), O'Donnell et al. (368), and Weir et al. (338) all included short (i.e., 10 s or less) time estimation tasks in their experimental protocols. Weir et al. (337) was the only group which found a significant change (a reduction rather than an increase, and then only at a COHb level of 20% saturation). Both O'Donnell et al. (223) and Stewart et al. (361) observed a significant increase in 30 s estimation at 9.4% COHb, but only during sensory isolation. Thus, it is likely that Beard's 1969 results and Beard and Grandstaff's 1972 results are attributable to the fact that they had their subjects perform only one repetitive task throughout their experiment.

Learning and Memory Both Beard and Grandstaff (331) and Wright (115) have found that 7.5% COHb does not change the short-term retention of digits. In contrast Bender et al. (369) found a significant decrease in backward (but not forward) retention of digits, as well as a significant increase in the time to memorize meaningless syllables at a COHb level of 7.2%. The meaningless syllable task involves long-term memory stores, as does digit reversal (370). Apparently short-term memory is not impaired by CO, but long-term memory is affected.

Mental and Cognitive Performance There is still much to be resolved concerning the effects of CO on mental and cognitive performance. Schulte (330) found a significant correlation between performance and exposure to CO for such tests as responding to letters and colors, underlining plural nouns, crossing "t" 's, and arithmetic ability, although the accuracy of the measured blood COHb levels has been questioned (12, 368). Similarly, interpretation of the study of Lewis et al. (366) is uncertain, for although they found significant increases in the time required to perform both arithmetic and comprehension, they did not measure the CO content of the roadside air. O'Donnell et al. (223) did not find a significant deterioration of mental arithmetic at COHb levels as high as 12.7%. Nevertheless, their subjects were few (*n* of 4) and specialized (altitude trained) relative to the firemen tested by Schulte (330). Bender et al. (369) found a significant deterioration in both letter identification and a common properties section of an IQ test at 7.2% COHb. However, no differences were observed in a common properties test at a lower COHb level (4.9% saturation) (336). Finally, Beard and Grandstaff (331) did not observe any change in performance of a visual search and selection task at COHb levels of 7% and 8%.

Choice Reaction Time Choice reaction time, unlike simple reaction time, requires a decision before a response can be initiated. Ramsey (285, 371, 372) used this task as a measure of changes in CNS function in both healthy adults and in those with a health handicap. Reaction times were significantly slower (371) after 90 min of exposure to congested traffic (38 Pa of CO, estimated COHb levels of 3.5–4.0%). In a subsequent study (285), the increase in reaction time at 5% COHb was significant only when data from normal, emphysematous and anemic subjects were grouped together. In healthy individuals (university students), Ramsey (372) found a decrement in choice reaction time at a blood COHb level of 8.5%, but Fodor and Winneke (336) found no change at a COHb of 5.3%.

Psychomotor Performance The factorial breakdown of psychomotor performance generally follows that of Fleishman (373), but the order has been changed to accommodate the two main groupings of speed and precision, as suggested by Fodor and Winneke (336).

Speed of Movement

1. *Wrist-finger speed.* No CO-related performance deterioration was observed at COHb levels of up to 20% (336, 337).
2. *Rate of arm movement.* Fodor and Winneke (336) did not find a CO-related effect at COHb levels of up to 5.3%.
3. *Simple reaction time.* No significant effect of CO was demonstrated at COHb levels of up to 5.3% (336).
4. *Complex reaction time.* See under "Cerebral Efficiency."

Precision of Movement

1. *Finger dexterity.* The tests for this factor involve grasping, release or manipulation of small objects, as in the Crawford collar and pin test (326) or the Purdue pegboard test (336, 367). Fodor and Winneke (336) did not find a significant impairment at COHb levels of up to 5.3%, but Bender et al. (369) noted a significant deterioration after 3.5 hr exposure to 100 Pa of CO (an average COHb of 7.2% saturation). Stewart et al. (326) reported a significant correlation between the decrease in performance and CO exposure (25, 50, and 100 Pa of CO for 8 hr, but considered this relationship spurious.
2. *Controlled manual dexterity.* No CO-related effect was observed at COHb levels of up to 20% saturation (336, 337).
3. *Manual dexterity.* Stewart et al. (326) did not find any effect at COHb levels of up to 13% saturation.
4. *Arm-hand steadiness.* No significant effect was demonstrated at COHb levels of up to 13% (326), although replication of this result would be desirable since Fodor and Winneke (336) found a marginal effect ($P < 0.10$) at a COHb level of 5.3%.
5. *Psychomotor coordination.* Fodor and Winneke (336) found no effect at a COHb level of 5.3%. However, the results of O'Donnell et al. (224) indicated a transient deterioration of performance after 2 hr of exposure to 50 or 125 Pa of CO (3% or 6.6% COHb).

6. *Postural discrimination.* No deterioration was observed at COHb levels of up to 20% (224, 337).

Driving of Vehicles Considering that the automobile is the prime source of ambient carbon monoxide (see under "Internal Combustion"), it is not surprising that performance assessments have gradually been extended to include driving simulators and on-the-road tests. There are many limitations to simulators. Nevertheless, if the apparatus requires continuous attention, it places a high level of demand upon the central nervous system, using up spare mental capacity. The importance of spare capacity should not be underestimated. Forbes et al. (325) commented that, although their subjects could concentrate long enough to perform a psychomotor task reasonably well, they were subjectively aware that their driving ability was impaired. The results of Ashton et al. (374, 375) and Wright et al. (113) suggest that the simulated demands of a busy and ever changing traffic pattern are enough to eliminate the spare mental capacity of the average driver, with a deterioration of careful driving habits after an increase in blood COHb of 3.4% (113).

The on-the-road studies of Weir et al. (337) and McFarland (355) support the view that CO exposure can affect road safety. Spare mental capacity was reduced through visual occlusions, and steering wheel reversals provided an objective measure of driver control. The data of Weir et al. revealed perceptual narrowing with increased COHb levels (7% and 12% saturation); there was also a reduction in the number of glances into mirrors prior to leapfrog passing at 12% COHb. McFarland (355) found that subjects required significantly more roadway viewing at 80 km/hr with a COHb level of 17% saturation than they did with a nominal COHb level.

Adaptation to CO Exposure

Killick (295) reported a considerable degree of acclimatization to CO in humans, with a lessening of symptoms and a reduction in equilibrium COHb% during successive exposures to the same concentrations of CO. Surprisingly, she observed no increase of red cell count or hemoglobin concentration. However, polycythemia has commonly been observed (see, for example, refs. 225, 376, 377). Vernot et al. (378) exposed rhesus monkeys and beagle dogs to 50–500 Pa of CO for 11–26 weeks; the red cell counts, hemoglobin concentrations, and hematocrits started to increase during the 1st week of exposure and stabilized at an elevated level after 1 month of exposure. The polycythemia was normocytic, normochromic, and a linear function of the CO concentration used. Jones et al. (379) observed an increase in hemoglobin content and hemoglobin in rats, guinea pigs, monkeys, and dogs exposed to 96 and 200 Pa of CO 8 hr per weekday for a total of 90 days. However, changes did not occur at 51 Pa of CO. The concept of a dose-response threshold is supported by negative findings in monkeys with COHb levels of 8% (380), but positive findings after 3 weeks at 10–15% COHb

(381). In humans, data from smoking withdrawal clinics (382) suggest a development of polycythemia in response to a habitual 7% COHb, and reports from Mexico City indicate that a 7–8% drop in arterial oxygen saturation induces polycythemia over a period of at least 2–3 weeks (383).

The red blood cell concentration of 2,3-diphosphoglycerate (DPG) decreases with quite brief (24-hr) exposure to CO (384). This effect is associated with an increased affinity of hemoglobin for oxygen and a displacement of the O_2Hb dissociation curve to the left (the Haldane effect). With hypoxic hypoxia, the O_2Hb dissociation curve shifts to the right; this change is associated with elevations of red cell 2,3-DPG concentrations (245). The exact mechanism must yet be clarified; COHb levels of 20% and greater produce a decrease in 2,3-DPG, but lower COHb levels (16–20%) increase 2,3-DPG concentrations (385).

Interaction with Other Stressors

Early researchers examined interactions between CO and altitude. Both Forbes et al. (281) and Lilienthal et al. (386) considered COHb equivalents resulting from a combined exposure to altitude and CO. McFarland et al. (327) developed "physiological altitudes" dependent on actual altitude and percentage of COHb. The reduction in visual intensity discrimination produced by a given increment of percentage of COHb was equal to that elicited by the same decrement in percentage of O_2Hb.

At sea level, CO exposure can place individuals with an already reduced oxygen-carrying capacity from anemia, low cardiac output, or decreased arterial P_{O_2} at special risk (288, 307, 387). Blood donors who smoke often provide blood containing significant levels of COHb (104, 388) which persist during storage (365). Inhalation of methylene chloride (a common household aerosol and paint remover) causes a sustained elevation of blood COHb levels (389). Contamination of factory air by methylene chloride has produced in excess of 9% COHb in nonsmoking workers (277).

Exposure to CO is a serious concern in sealed environments such as submarines. The CO is produced from incomplete combustion of diesel fuel, fires, overheated insulating material (390), endogenous CO production, and, above all, cigarette smoking (391, 392). Improvements in catalytic oxidation equipment have now brought ambient CO levels below 10 Pa (393). In space exploration, Sjöstrand (394) pointed out that endogenous CO could accumulate to toxic concentrations. In addition, the space cabin must be cleared of synthetic and natural organic substances which yield CO when exposed to light and oxygen (251).

When considering hyperbaric exposure to CO, it is convenient to talk in terms of surface equivalents. For example, if air contaminated with 20 Pa of CO is used at a depth of 50 m (6 atm absolute), a diver is exposed to a partial pressure of CO that is equivalent to 120 Pa of CO at the surface. Furthermore, since diving is physically demanding (395), the respiratory minute volume is increased and equilibrium levels of COHb are reached much more quickly than in a resting subject. Air containing 20 Pa is likely to produce a 5% COHb with 20 min "bottom time" at 6 atm absolute. Greater contamination of the air supply, for example, from compressor exhaust, can produce dangerous COHb levels quite quickly. For deep dives, gas mixtures of 80% helium and 20% oxygen are usual, and dangerous levels of CO contamination are then unlikely. The pulmonary diffusing capacity for CO is also decreased by 20–30% during He-O_2 breathing (396). In saturation diving, COHb levels are increased (275), although the hemolysis of erythrocytes is not potentiated at pressures of up to 130 atm (397).

Increased partial pressure of oxygen provides a useful basis for the treatment of carbon monoxide poisoning (289). The amount of oxygen in physical solution in the plasma is a linear function of its partial pressure (0.023 ml/100 ml of blood per kPa). Thus, if a patient is given 100% oxygen to breathe at an absolute pressure of 3 atm, more than 6 vol% of oxygen will dissolve in the plasma (298). In addition, the half-time for elimination of COHb is reduced (see under "Elimination").

Severe Exposures

Acute Toxicity Early research (8, 9, 199, 398–401) was concerned with the clinical effects of very severe carbon monoxide poisoning and suicide attempts. More recent reviewers have concentrated on specific hazards in firefighting (402–405), aviation (406), and the home (407).

Both Haldane (199) and Henderson et al. (290) found that the symptoms associated with prolonged exposure to CO were related to the percentage of saturation of hemoglobin with CO; i.e., problems were arising from oxygen want in the tissues. A brief exposure to very high concentrations of CO (35,600 Pa) produces a frontal headache and slight sagging of the ST segment of the ECG. at a COHb% of ~10% (140). In the more usual forms of accidental poisoning, a blood COHb of 20% is usually associated with headache, throbbing in the temples, some loss of manual dexterity, and dyspnea on exercise. At 30–40% COHb, the headache becomes severe. Nausea, vomiting, dizziness, and irritability are seen, vision may be disturbed, and the skin assumes a characteristic pinkish (cherry red) hue, sometimes with areas of erythema. At 40–50% COHb, dimness of vision, chest pain, a euphoric confusion, incoordination, and fainting on exertion are likely, and concentrations >50% are liable to cause unconsciousness, hyperpyrexia, convulsions, and even death. Some 8% of hospital admissions for CO poisoning have a fatal outcome (408). At the highest concentrations (>50,000 Pa), death from cardiac arrhythmia may occur before there has been a substantial increase of the average COHb%, a "slug" of anoxic blood provoking ventricular fibrillation (409). The organs most vulnerable to CO are the heart and the brain, the latter showing hyperemia, edema, and petechial hemorrhages, with symmetrical softening of the lenticular nucleus (258). If the patient survives the acute episode, there may be after effects. The electrocardiogram sometimes shows changes such as premature ventricular complexes, atrial fibrillation, and ST abnormalities for several weeks, and an echocardiogram may reveal abnormal motion of the left ventricular wall (410); all of these changes suggest temporary or permanent myocardial injury. The brain, also, may suffer permanent damage (411, 412), although the usual symptoms of such injury (personality change and memory disturbance) are hard to document in the absence of previous objective information regarding a patient.

During exercise, toxic symptoms can be encountered with a COHb of only 12–15%. Pregnancy also increases susceptibility to carbon monoxide (413). The carbon monoxide penetrates to the fetus unless the mother is killed very rapidly, thus raising the question of whether chronic CO exposure from cigarette smoking is responsible for the high incidence of abortions, still births, and neonatal death among mothers who smoke heavily (258).

Chronic Toxicity The existence of a distinct syndrome of chronic carbon monoxide intoxication remains uncertain, at least for human subjects (398). Some animals such as rabbits are very vulnerable. In this species, pathological changes have been reported with as little as 4 hr of exposure to 100 Pa, including hyaline degeneration of heart muscle with associated ECG abnormalities (414–416) and degenerative changes in the cerebral cortex, limbic system, and brainstem with related disturbances of posture, reflexes, and gait (417).

Symptoms have been reported in man with a COHb% level of 5% (9). However, the complaints mentioned (headache, insomnia, fatigue, irritability, impairment of memory, confusion, dizziness, mood disturbances, cardiac symptoms, digestive disturbances, diarrhea, hyperhidrosis, thirst, and decrease of libido) are difficult to evaluate, since the subjects, all drivers of producer-gas vehicles, knew they had been exposed to the gas. Lindgren (418) compared exposed groups with controls taken from similar work; they also found more frequent headaches in exposed subjects, but the frequency of illness was comparable in exposed and control groups.

The most consistently noted physiological effect of long-term carbon monoxide exposure is a polycythemia (see under "Adaptation to CO Exposure"). Other reported long-term effects of CO include an increased reticulocyte count (419) and serum cholesterol (420), a leakage of albumin from the capillaries (421, 422), capillary platelet thrombosis (423), and possible changes of electrolyte excretion (424, 425), clotting time (426, 427), and urine production (419).

CONSEQUENCES FOR HEALTH

Certain health aspects of acute and chronic poisoning have already been discussed (see previous section). Sensitive groups include people with severe anemia, chronic pulmonary disease, atherosclerosis of the coronary and peripheral vessels, fetuses, and newborn infants. Topics covered below include interactions with ischemic heart disease, accidents, and litigation.

Ischemic Heart Disease

Animal Experiments Rabbits (416, 428–433), squirrel monkeys (434), and pigeons (435) all develop atherosclerosis in response to chronic carbon monoxide exposure. In many, but not all, of the experiments cited it is necessary to feed a high cholesterol diet in order to obtain positive findings. Cholesterol accumulates in the vessel walls, and there is an associated splitting of subintimal structures, with the development of edema. Parallel changes can be induced by hypoxia (436), and, if 28% oxygen is administered in association with the carbon monoxide, lesions do not appear (431). The main effect of the carbon monoxide is probably an increased permeability of the vessel walls to lipid (422, 437), although lipase inhibition may also be involved (420).

Other species (rhesus monkeys, baboons, beagle dogs, rats, and mice) do not show atheromatous change after as much as 168 days of exposure to 400–500 Pa (225), and the application of the positive findings to human subjects is thus doubtful. However, one report (438) suggests that carbon monoxide exposure leads to a marked increase of cholesterol uptake in human coronary arteries perfused at postmortem.

Epidemiological Studies One group frequently exposed to high concentrations of CO are firefighters. An early report (439) suggested this group had the highest mortality index for cardiovascular and renal disease of any occupational group. A specific study of death certificates in 271 Canadian city firemen confirmed a highly significant excess of cardiovascular-renal mortality (405), and stress testing disclosed a substantial proportion (10%) of ischemic exercise ECG's in Californian firemen (402). However, in all of these studies, it is difficult to assess other adverse aspects of firefighting—a sedentary life, periodic stressful emergencies, and exposure to a wide variety of toxic agents.

In the Shinshu region of Japan, an open charcoal fire is commonly used to heat the rooms in which tatami mats are manufactured. The workers often have blood carboxyhemoglobin saturation of 20–30%, and atherosclerotic heart disease is 5 times as common as in other parts of Japan (440).

Another potentially vulnerable occupational group are tunnel traffic police. Ayres et al. (441) found an average COHb% of 2.9 ± 1.4% in nonsmoking tunnel workers. Some 10% of the group showed an ischemic exercise ECG, but the frequency of definite or possible heart attacks was no greater than would have been predicted from the age and constitutional makeup of the sample.

Californian investigators have sought to relate the normally modest urban CO exposures to cardiovascular deaths. No association was found between atmospheric CO and the number of hospital admissions for myocardial infarction (442); however, the case fatality rate was related to atmospheric CO on the day of admission in areas of high pollution (8 hr average >8 Pa of CO). One factor that could have led to a spurious correlation is that both high CO readings and fatal heart attacks tend to be more common in the winter months. To meet this possible criticism, Hexter and Goldsmith (443) related daily deaths to environment by a sophisticated multiple regression equation, using day numbers for trend and harmonic terms for cyclic variations in such variables as season, temperature, and air pollution. They concluded that if the 24-hr average for CO increased from 7.3 Pa to the very high figure of 20.2 Pa, this would produce 11 excess deaths in the Los Angeles area. In a study from the Johns Hopkins hospital, 24-hr CO levels were usually in the range of 0–4 Pa; this level of pollution was insufficient to reveal any clustering of fatal heart attacks on days when CO readings were highest (444).

We may conclude that, although there is some evidence that severe CO exposure increases the risk of developing overt ischemic heart disease in man (445), the threshold concentrations are higher than those encountered in most cities.

Effects on Diseased Hearts Ayers et al. (446–448) studied coronary gas pressures during carbon monoxide exposure. The coronary arteriovenous oxygen difference was widened, both in patients with normal vasculature and in those with coronary atherosclerosis; however, as in animals (449, 450) the normal subjects were able to produce a partial compensation by increase of coronary perfusion, whereas this did not occur in those with coronary disease (161). Surprisingly, some reduction of arterial oxygen tension was seen; one possible explanation would be an increased effect of venous-arterial shunting and ventilation-perfusion imbalance in the lungs, associated with the falling mixed venous oxygen content (451). The myocardium of the anginal patients developed a typical hypoxic response during CO administration; Aronow et al. (452) found an increase of end-diastolic pressures, with a decrease of left ventricular dp/dt, stroke index and cardiac output in response to 2.05% increase of carboxyhemoglobin.

Aronow et al. (230) examined the effects of CO exposure on the exercise performance of anginal patients. Unfortunately, their experiment was not double blind (453). On one day, subjects were driven along the Los Angeles freeways in a vehicle with open windows, and COHb% increased by the surprisingly large amount of 3.96% (83, 453). On this occasion, the endurance of a standard bicycle ergometer test diminished from 249 s to 174 s. On a second occasion, the drive was repeated while breathing compressed air, and performance was then unchanged.

The experiment was later repeated in double-blind fashion in the laboratory, the subjects breathing in random order compressed air and compressed air containing CO. On the days that CO was breathed, COHb% increased by an average of 1.65%, whereas endurance time decreased from 224 s to 188 s, and the systolic pressure/heart rate product at the onset of angina also diminished (454). A similar experiment on patients with intermittent claudication (231) showed a reduction of exercise time from 174 s to 144 s in response to a 1.69% increase of carboxyhemoglobin.

Other authors have commented on the exaggeration of minor ECG abnormalities when middle-aged men, either healthy or with overt ischemic heart disease, have undertaken exercise after exposure to carbon monoxide (219, 455, 456). In animals at least, the electrical threshold for induction of ventricular fibrillation is less after carbon monoxide inhalation (409).

Accidents

From the psychomotor changes discussed in previous sections, an increase of accidents might be anticipated in skilled tasks, particularly vehicle driving, where CO exposure is often substantial. However, little information is available on this point (26, 82, 457, 458). Attempts to demonstrate a deterioration of vehicle driving skills (113, 114) have shown relatively minor changes. Two possible explanations are 1) standard driving tasks fail to tax the spare capacity of the average driver (459), so that this reserve can be exploited during CO poisoning, and 2) a modest depression of cerebral function may optimize the arousal of the driver (460), thus maintaining or even improving performance. Epidemiological research relating accidents to ambient CO levels is hampered by the tremendous excess of alcohol-related accidents and the difficulty in allowing for effects due to cigarette smoking. One study from Los Angeles established a significant relationship between the frequency of accidents and the level of oxidant pollution, but found no relationship between accidents and CO exposure (458).

Litigation

Occasional Workmen's Compensation Board awards are made to patients who sustain a myocardial infarction in association with a history of severe carbon monoxide exposure. In view of the chronic nature of ischemic heart disease, the industrial exposure is usually treated as an aggravating rather than a causative factor (461). Key points in making an award are good documentation of environmental and blood carbon monoxide levels, a careful smoking history, and a close temporal relationship between exposure and the onset of symptoms.

AIR QUALITY CRITERIA FOR CARBON MONOXIDE

Various models of the inter-relationship between air pollution and human health have been developed to emphasize the consequences of adopting a particular air quality criterion (462–466). A comparative analysis of these models reveals five distinct levels. Level A is within the range of homeostatic mechanisms. The Russians have suggested (467, 468) that pollutant concentrations should fall below the threshold causing a reflex response; they regard adaptation as a deleterious reaction and stringent air quality standards are set accordingly, although the degree of enforcement is unknown. The casual impression of the visitor is that vehicle densities are low, but that there is much industrial pollution from low, poorly designed and poorly sited chimneys. In a free enterprise society, the main issue in establishing air quality criteria is the extent to which the physiological reserve can be sacrificed in order to achieve economic benefits (469). At level B, changes are of uncertain physiological significance. At level C, the body's efforts at compensation are noticeable, making exposures to a pollutant tolerable but unpleasant. Levels D and E represent unacceptable outcomes, causing physiological breakdown and death, respectively.

Levels of health are distributed asymmetrically about a median line (464). Thus, even low level exposures to a particular pollutant can have an impact on the health status of a few sensitive members of a population. Goldsmith (464) distinguished urban and industrial air quality, community standards reflecting an "earliest effect" level for sensitive subjects rather than a "no effect" level for healthy persons. Hatch (470) suggested that industrial air quality criteria should be set below the limit of physiological compensation, thereby allowing a reserve for 1) any unintentionally greater exposures and 2) impairment by age, residual illness, other environmental stresses, and possible cumulative effects. Beard and Wertheim (329) argued that the capability to perform a given task is also a valid criterion for air quality standards.

The preferred approach in establishing air quality criteria for carbon monoxide is to specify an unsatisfactory blood COHb level and relate this level to probable patterns of exposure. Based on the arguments above, two criteria of unsatisfactory blood COHb levels can be recognized. The lowest, or urban air quality standard, is a concentration that hastens the onset of exercise-induced angina. In nonsmoking patients with coronary atherosclerosis, this occurs when the blood COHb increases from its baseline to 3% saturation (see under "Consequences for Health"). On the assumption that individuals with pre-existing cardiovascular insufficiencies are screened from specific occupations, the second criterion is the performance deterioration related to changes in the functioning of the central nervous system. This is realized when the blood COHb level of the nonsmoker increases to 6% saturation (see under "Effects of CO on the Central Nervous System"). The corresponding ambient levels can be deduced from Table 1.

REFERENCES

1. Fischer, R. (1967). The biological fabric of time. Ann. N. Y. Acad. Sci. 138: 440.
2. Lewin, L. (1920). Carbon Monoxide Poisoning: A Manual for Physicians, Engineers and Accident Investigators, pp. 1–370. Springer-Verlag, New York.
3. Cooper, A. G. (1966). Carbon monoxide: a bibliography with abstracts. Public Health Service Publication No. 1503. U. S. Department of Health, Education, and Welfare, Washington, D. C.
4. Air Pollution Control Association. (1970). Air quality criteria for carbon monoxide. U. S. Department of Health, Education and Welfare Publication AP-62, March, 1970.
5. Deacon, J. (1916). Tobacco Tortured. Published in facsimile by Da Capo Press, Amsterdam.

6. Sofoluwe, G. (1968). Smoke pollution in dwellings of infants with broncho-pneumonia. Arch. Environ. Health 16:670.
7. Munro, L. A. (1964). Chemistry in Engineering. Prentice-Hall, Inc., Englewood Cliffs, New Jersey.
8. Haldane, J. S., and Priestley, J. G. (1935). Respiration, Ed. 2. Clarendon Press, Oxford.
9. Grut, A. (1949). Chronic Monoxide Poisoning. Ejnar Munksgaard, Copenhagen.
10. Petri, H. (1953). Die chronische Kohlenoxydvergiftung. Johann Ambrosius Barth, Leipzig.
11. Eisenbud, M., and Ehrlich, L. R. (1972). Carbon monoxide concentration trends in urban atmospheres. Science 176:193.
12. Beard, R. R. (1969). Toxicological appraisal of carbon monoxide. J. Air Pollut. Control Assoc. 19:722.
13. Dale, B. T., personal communication.
14. DuBois, L., Zdrojewski, A., and Monkman, J. L. (1966). The analysis of carbon monoxide in urban air at the ppm level, and the normal carbon monoxide value. J. Air Pollut. Control Assoc. 16:135.
15. Zikria, B. A., Weston, G. R., Chodoff, M., and Ferrer, J. M. (1972). Smoke and carbon monoxide poisoning in fire victims. J. Trauma 12:641.
16. Croton, L. M. (1959). Carbon monoxide poisoning: a current controversy. Arch. Environ. Health 1:149 (abstr.).
17. Douglas, T. A., Lawson, D. D., Ledingham, I., Norman, J. N., Sharp, G. R., and Smith, G. (1961). Carbogen in experimental carbon monoxide poisoning. Br. Med. J. (ii):1673.
18. Douglas, T. A., Ledingham, I., Lawson, D. D., and Norman, J. N. (1962). Carbon monoxide poisoning: a comparison between the efficiencies of oxygen at one atmosphere pressure, of oxygen at two atmospheres pressure and of 5% and 7% carbon dioxide in oxygen. Lancet (i):68.
19. Editorial. (1959). Chronic carbon monoxide poisoning. N. Engl. J. Med. 261:1248.
20. Killick, E. M., and Marchant, J. V. (1959). Resuscitation of dogs from severe acute carbon monoxide poisoning. J. Physiol. 147:274.
21. Norman, J. N., Douglas, T. A., and Smith, G. (1966). Respiratory and metabolic changes during carbon monoxide poisoning. J. Appl. Physiol. 21:848.
22. Otis, A. B. (1970). The physiology of carbon monoxide poisoning and evidence for acclimatization. Ann. N. Y. Acad. Sci. 174:242.
23. Zoru, O., and Druger, P. D. (1969). The problem of carbon monoxide poisoning. Ind. Med. Surg. 29:580.
24. Goldsmith, J. R. (1970). Contribution of motor vehicle exhaust, industry and cigarette smoking to community carbon monoxide exposures. Ann. N. Y. Acad. Sci. 174:122.
25. Gregory, K. L., Malinoski, V. F., and Sharp, C. R. (1969). Cleveland Clinic Fire Survivorship Study, 1929–1965. Arch. Environ. Health 18:508.
26. Chovin, P. (1967). Carbon monoxide: analysis of exhaust gas investigations in Paris. Environ. Res. 1:198.
27. De Bruin, A. (1967). Carboxyhemoglobin levels due to traffic exhaust. Arch. Environ. Health. 15:384.
28. Göthe, C. J., Frisfedt, B., Sundell, L., Kolmodin, B., Ehrner-Samuel, H., and Göthe, K. (1969). Carbon monoxide hazard in city traffic. Arch. Environ. Health 19:310.
29. Wilson, E. D., Gates, I., Owen, H. R., and Dawson, W. T. (1926). Street risk of carbon monoxide poisoning. J.A.M.A. 87:319.

30. Cohen, S. I., Dorion, G., Goldsmith, J. R., and Permutt, S. (1971). Carbon monoxide uptake by inspectors at a United States-Mexico Border Station. Arch. Environ. Health 22:47.
31. Hofreuter, D. H., Catcott, E. J., and Xintaras, C. (1962). Carboxyhemoglobin in men exposed to carbon monoxide. Arch. Environ. Health 4:81.
32. Sievers, R. F., Edwards, T. I., Murray, A. L., and Schrenk, H. H. (1942). Effect of exposure to known concentrations of carbon monoxide: a study of traffic officers stationed at the Holland Tunnel for 13 years. J.A.M.A. 118:585.
33. Ramsey, J. M. (1967). Carboxyhemoglobinemia in parking garage employees. Arch. Environ. Health 15:580.
34. Davies, G. M., Jones, J. G., and Warner, C. G. (1965). A continuously recording atmospheric carbon monoxide monitoring system with fully automatic alarms in a blast furnace area. Br. J. Industr. Med. 22:270.
35. Farmer, C. J., and Crittenden, P. J. (1929). A study of the carbon monoxide content of the blood of steel mill operatives. J. Industr. Hyg. 11:329.
36. Jones, J. G., and Walters, D. H. (1962). A study of carboxyhemoglobin levels in employees at an integrated steelworks. Ann. Occup. Hyg. 5:221.
37. Yocom, J. E., Clink, W. L., and Cote, W. A. (1971). Indoor/outdoor air quality relationships. J. Air Pollut. Control Assoc. 21:251.
38. Bowne, N., Boyer, A. E., Trent, K. E., and Cooper, D. G. (1971). An air quality model for Metropolitan Toronto. Paper presented at the 64th Annual Meeting of the Air Pollution Control Association, June 27–July 1, Atlantic City, New Jersey.
39. Jaffe, L. S. (1968). Ambient carbon monoxide and its fate in the atmosphere. J. Air. Pollut. Control Assoc. 18:534.
40. Siegel, S. M., Renwick, G., and Rosen, L. A. (1962). Formation of carbon monoxide during seed germination and seedling growth. Science 137:683.
41. Wilks, S. S. (1959). Carbon monoxide in green plants. Science 129:964.
42. Barham, E. G., and Wilton, J. W. (1964). Carbon monoxide production by a bathypelagic siphonophore. Science 144:860.
43. Chapman, D. J., and Tocher, R. D. (1966). Occurrence and production of carbon monoxide in some brown algae. Can. J. Botany 44:1438.
44. Swinnerton, J. W., Linnenbom, V. J., and Lamontagne, R. A. (1970). The Ocean: a natural source of carbon monoxide. Science 167:984.
45. Bates, D. R., and Witherspoon, A. E. (1952). The photochemistry of some minor constituents of the earth's atmosphere. R. Astr. Soc. Monthly Notices 112:101.
46. Robbins, R. C., Borg, K. M., and Robinson, E. (1968). Carbon monoxide in the atmosphere. J. Air Pollut. Control Assoc. 18:106.
47. Maugh, T. H. (1972). Carbon monoxide: natural sources dwarf man's output. Science 177:338.
48. McCormick, R. A., and Xintaras, C. (1962). Variation of carbon monoxide concentrations as related to sampling interval, traffic and meteorological factors. J. Appl. Meteorol. 1:237.
49. Robinson, E., and Robbins, R. C. (1970). Atmospheric background concentrations of carbon monoxide. Ann. N. Y. Acad. Sci. 174:89.
50. Weinstock, B. (1969). Carbon monoxide: residence time in the atmosphere. Science 166:224.
51. Inman, R. E., Ingersoll, R. B., and Levy, E. A. (1971). Soil: a natural sink for carbon monoxide. Science 172:1229.
52. Hurn, R. W. (1962). Comprehensive analyses of automotive exhausts. Arch. Environ. Health 5:592.
53. Elliott, M. A., Nebel, G. J., and Rounds, F. G. (1955). The composition of ex-

haust gases from diesel, gasoline and propane-powered motor coaches. J. Air Pollut. Control Assoc. 5:103.

54. Chipman, J. C., and Massey, M. T. (1960). Proportional sampling system for the collection of an integrated auto exhaust gas sample. J. Air Pollut. Control Assoc. 10:60.
55. Larson, G. P., Chipman, J. C., and Kauper, E. K. (1955). Study of the distribution and effects of auto exhaust gases. J. Air Pollut. Control Assoc. 5:84.
56. Twiss, S. B., Teague, D. M., Bozek, J. W., and Sink, M. V. (1955). Application of infra-red spectroscopy to exhaust gas analysis. J. Air Pollut. Control Assoc. 5:75.
57. Rose, A. H., and Smith, R. (1962). A direct measurement technique for automobile exhaust. Arch. Environ. Health 5:609.
58. Rose, A. H., Smith, R., McMichael, W. F., and Kruse, R. E. (1965). Comparison of auto exhaust emissions in two major cities. J. Air Pollut. Control Assoc. 15:362.
59. Bowditch, F. W. (1972). Progress and future of automotive emission control—U. S. and Canada. Paper T.A. 2. Presented at the International Conference on Automobile Pollution, June 26–28, Toronto, Canada.
60. Chass, R. L., Hamming, W. J., Dickinson, J. E., and MacBeth, W. G. (1972). Los Angeles photochemical smog: past, present and future. Paper M.A. 5. Presented at the International Conference on Automobile Pollution, June 26–28, Toronto, Canada.
61. Winthrop, S. O. (1972). Long range goals for the control of automobile pollution. Paper TA 4. Proceedings of the International Conference on Automobile Pollution, June 26–28, 1972, Toronto, Canada.
62. Hass, G. C., and Brubacher, M. L. (1962). A test procedure for motor vehicle exhaust emissions. J. Air Pollut. Control Assoc. 12:505.
63. Larsen, R. I., and Konopinski, V. J. (1962). Summer tunnel air quality. Arch. Environ. Health 5:597.
64. Goldsmith, J. R., and Landaw, S. A. (1968). Carbon monoxide and human health. Science 162:1352.
65. Sweet, A. H., Steigerwald, B. J., and Ludwig, J. H. (1968). The need for a pollution-free vehicle. J. Air Pollut. Control Assoc. 18:111.
66. Colucci, J. M., and Begeman, C. R. (1969). Carbon monoxide in Detroit, New York and Los Angeles air. Environ. Sci. Technol. 3:41.
67. Hamming, W. J., MacPhee, R. D., and Taylor, J. R. (1960). Contaminant concentrations in the atmosphere of Los Angeles County. J. Air Pollut. Control Assoc. 10:7.
68. Moureu, H. (1964). Carbon monoxide as a test for air pollution in Paris due to motor vehicle traffic. Proc. R. Soc. Med. 57:1015.
69. Ott, W., and Eliassen, R. (1973). A survey technique for determining the representativeness of urban air monitoring stations with respect to carbon monoxide. J. Air Pollut. Control Assoc. 23:685.
70. Brice, R. M., and Roesler, J. F. (1966). The exposure to carbon monoxide of occupants of vehicles moving in heavy traffic. J. Air Pollut. Control Assoc. 16:597.
71. Godin, G., Wright, G. R., and Shephard, R. J. (1972). Urban exposure to carbon monoxide. Arch. Environ. Health 25:305.
72. Wilkins, E. T. (1956). Some measurements of carbon monoxide in the air of London. J. R. Soc. Health 76:677.
73. Wright, G. R., Jewczyk, S., Onrot, J., Tomlinson, P., and Shephard, R. J. (1975). Carbon monoxide in the urban atmospheres: hazards to the pedestrian and the street-worker. Arch. Environ. Health 30:123.

74. Bove, J. L., and Siebenberg, S. (1970). Airborne lead and carbon monoxide at 45th Street, New York City. Science 167:986.
75. Georgii, H. W., Busch, E., and Weber, E. (1967). Investigation of the temporal and spatial distribution of the emission concentration of carbon monoxide in Frankfurt/Main. Report of the Institute for Meteorology and Geophysics, University of Frankfurt/Main, May, 1967.
76. Ontario Ministry of the Environment. (1972). Report on carbon monoxide sampling in downtown Toronto—Summer, 1972. Toronto, Ontario, Canada.
77. Ramsey, J. M. (1966). Concentrations of carbon monoxide at traffic intersections in Dayton, Ohio. Arch. Environ. Health 13:44.
78. Bithel, L. (1970). Report of carbon monoxide assessment, Toronto International Airport. Occupational Health Services, Ontario Department of Health, Toronto, July 17.
79. Conlee, C. J., Kenline, P. A., Cummins, R. L., and Konopinski, V. J. (1967). Motor vehicle exhaust at three selected sites. Arch. Environ. Health 14:429.
80. Lawther, P. J., Commins, B. T., and Henderson, M. (1962). Carbon monoxide in town air: an interim report. Ann. Occup. Hyg. 5:241.
81. Rispler, X., and Gilbert, C. (1970). Report of carbon monoxide survey, Toronto International Airport. Occupational Health Division, Department of National Health and Welfare, Ottawa, July 15–20, 1970.
82. Clayton, G. D., Cook, W. A., and Frederick, W. G. (1960). A study of the relationship of street level carbon monoxide concentrations to traffic accidents. Am. Ind. Hyg. Assoc. J. 21:46.
83. Haagen-Smit, A. J. (1966). Carbon monoxide levels in city driving. Arch. Environ. Health 12:548.
84. Dudley, W. L. (1888). The poisonous effects of cigarette smoking. Med. News 53:286.
85. Gettler, A. O., and Mattice, M. R. (1933). The "normal" carbon monoxide content of the blood. J.A.M.A. 100:92.
86. Wahl, F. (1899). Ueber den Gehalt des Tabakrauches an Kohlenoxyd. Pfluegers Arch. 88:262.
87. Armstrong, H. E. (1922). Carbonic oxide in tobacco smoke. Br. Med. J. (i):992.
88. Baumberger, J. P. (1923). The carbon monoxide content of tobacco smoke and its absorption on inhalation. J. Pharmacol. Exp. Ther. 21:23.
89. Bogen, E. (1929). The composition of cigarettes and cigarette smoke. J.A.M.A. 93:1110.
90. Wynder, E. L., and Hoffmann, D. (1967). Tobacco and Tobacco Smoke: Studies in Experimental Carcinogenesis. Academic Press, Inc. New York.
91. Hoegg, U. R. (1972). Cigarette smoke in closed spaces. Environ. Health Perspect. 2:117.
92. Guillerm, R., Badre, R., Hee, J., and Masurel, G. (1972). Composition de la Fumée de Tabac. Analyse des Facteurs de Nuisance. Rev. Tuber. Pneumol. 36:187.
93. Mumpower, R. C., Lewis, J. S., and Touey, G. P. (1962). Determination of carbon monoxide in cigarette smoke by gas chromatography. Tobacco Sci. 6:142.
94. Osborne, J. S., Adamek, S., and Hobbs, M. E. (1956). Some components of gas phase of cigarette smoke. Anal. Chem. 28:211.
95. Newsome, J. R., and Keith, C. H. (1965). Variation of the gas phase composition within a burning cigarette. Tobacco Sci. 9:65.
96. Robinson, J. C., and Forbes, W. F. (1975). The role of carbon monoxide in cigarette smoking. I. Carbon monoxide yield from cigarettes. Arch. Environ. Health 30:425.

97. Hoffman, D., and Wynder, E. L. (1972). Smoke of cigarettes and little cigars: an analytic comparison. Science 178:1197.
98. Bridge, D. P., and Corn, M. (1972). Contribution to the assessment of exposure of non-smokers to air pollution from cigarette and cigar smoke in occupied spaces. Environ. Res. 5:192.
99. Steinfeld, J. L. (1972). The Health Consequences of Smoking: A Report of the Surgeon General. U. S. Department of Health, Education and Welfare. Washington, D. C.
100. Harke, H.-P. (1970). Zum Problem des "Passiv-Rauchens." (The problem of "passive smoking.") Munch. Med. Wochenschr. 112:2328.
101. Dublin, W. B. (1972). Secondary smoking: a problem that deserves attention. Pathologist 26:244.
102. Russell, M. A. H., Cole, P. V., and Brown, E. (1973). Absorption by non-smokers of carbon monoxide from room air polluted by tobacco smoke. Lancet (i):576.
103. Sebben, J., Pimm, P., and Shephard, R. J. (1977). Cigarette smoke in enclosed public facilities. Arch. Environ. Health 32:53.
104. Stewart, R. D., Baretta, E. D., Platte, L. R., Stewart, E. B., Dodd, H. C., Donohoo, K. K., Graff, S. A., Kalbfleisch, J. H., Hake, C. L., Yserloo, B. V., Rimm, A. A., and Newton, P. E. (1973). "Normal" carboxyhemoglobin levels of blood donors in the United States. Report ENVIR-MED-MCW CRC-COHb-73-1. Medical College of Wisconsin. Madison, Wisconsin, U.S.A.
105. Hanson, H. B., and Hastings, A. B. (1933). The effect of smoking on the carbon monoxide content of blood. J.A.M.A. 100:1481.
106. Curphey, T. J., Hood, L. P. L., and Perkins, N. M. (1965). Carboxyhemoglobin in relation to air pollution and smoking. Arch. Environ. Health 10:179.
107. Longo, L. D. (1970). Carbon monoxide in the pregnant mother and fetus and its exchange across the placenta. Ann. N. Y. Acad. Sci. 174:313.
108. McIlvaine, P. M., Nelson, W. C., and Bartlett, D. (1969). Temporal variation of carboxyhemoglobin concentrations. Arch. Environ. Health 19:83.
109. Ringold, A., Goldsmith, J. R., Helwig, H. L., Finn, R., and Schiette, F. (1962). Estimating recent carbon monoxide exposure: a rapid method. Arch. Environ. Health 5:308.
110. Rode, A., and Shephard, R. J. (1971). Smoking withdrawal and changes of cardiorespiratory fitness. Am. Rev. Respir. Dis. 104:933.
111. Wright, G. R., and Shephard, R. J. (1978). A note on blood carboxyhemoglobin levels in police officers. Can. J. Public Health. 69:393.
112. Rode, A., and Shephard, R. J. (1973). Pulmonary function of Canadian Eskimos. Scand. J. Respir. Dis. 54:191.
113. Wright, G. R., Randell, P., and Shephard, R. J. (1973). Carbon monoxide and driving skills. Arch. Environ. Health 27:349.
114. Wright, G. R., and Shephard, R. J. (1978). Brake reaction time—effects of age, sex and carbon monoxide. Arch. Environ. Health. 33:141.
115. Wright, G. R. (1976). Effects of carbon monoxide on human performance. Ph. D. thesis, University of Toronto.
116. Delivoria-Papadopoulos, M. (1973). The pregnant woman and the fetus. *In* A. B. Dubois (ed.), Effects of Carbon Monoxide on Man's Health in His Home Environment. U. S. Department of Health, Education, and Welfare, Washington, D.C.
117. Delivoria-Papadopoulos, M., Coburn, R. F., and Forster, R. E. (1974). Cyclic variation of rate of carbon monoxide production in normal women. J. Appl. Physiol. 36:49.
118. Hickey, R. J., Clelland, R. C., and Boyce, D. E. (1973). Carbon monoxide:

smoking, air pollution, cardiovascular disease and physiological homeostasis. (Letter). Lancet (ii):571.

119. Hickey, R. J., Clelland, R. C., Boyce, D. E., and Harner, E. B. (1974). Carbon monoxide and smoking. (Letter). Lancet (i):409.
120. Wald, N., and Smith, P. G. (1974). Carbon monoxide and smoking. Lancet (i):750.
121. Amerson, S. S. (1907). Poison in cigarettes. J.A.M.A. 49:625.
122. Hartridge, H. (1920). CO in tobacco smoke. J. Physiol. 53:83.
123. Hartridge, H. (1912). A spectroscopic method of estimating carbon monoxide. J. Physiol. 44:1.
124. Dixon, F. R. S. (1921). The drug habit. Br. Med. J. (ii):819.
125. Haldane, J. (1895). A method of detecting and estimating carbonic oxide in air. J. Physiol. 18:463.
126. Comroe, J. H., Forster, R. E., DuBois, A. B., Briscoe, W. A., and Carlsen, E. (1970). The Lung. Year Book Medical Publishers, Inc., Chicago.
127. Shephard, R. J. (1958). Breath-holding measurement of carbon monoxide diffusion capacity: comparison of a field test with steady-state and other methods of measurement. J. Physiol. 141:408.
128. Dalhamn, T., Edfors, M.-L., and Rylander, R. (1968). Retention of cigarette smoke components in human lungs. Arch. Environ. Health 17:746.
129. Bokhoven, C., and Niessen, H. J. (1961). Amounts of oxides of nitrogen and carbon monoxide in cigarette smoke, with and without inhalation. Nature 192:458.
130. Delarue, N. C. (1973). A study in smoking withdrawal. Can. J. Public Health 64:S5.
131. Rode, A., Ross, R., and Shephard, R. J. (1972). Personality and changes in the cardiorespiratory fitness of participants in a smoking withdrawal program. Arch. Environ. Health 24:27.
132. Buchwald, H. (1969). A rapid and sensitive method for estimating carbon monoxide in blood and its application in problem areas. Am. Ind. Hyg. Assoc. J. 30:564.
133. Cohen, S. I., Perkins, N. M., Ury, H. K., and Goldsmith, J. R. (1971). Carbon monoxide uptake in cigarette smoking. Arch. Environ. Health 22:55.
134. Russell, M. A. H., Wilson, C., Cole, P. V., Idle, M. and Feyerabend, C. (1973). Comparison of increases in carboxyhaemoglobin after smoking "extra mild" and "non-mild" cigarettes. Lancet (ii):687.
135. Ashton, H., and Watson, D. W. (1970). Puffing frequency and nicotine intake in cigarette smokers. Br. Med. J. (iii):679.
136. Ashton, H., and Telford, R. (1973). Smoking and carboxyhaemoglobin. (Letter). Lancet (ii):857.
137. Russell, M. A. H., Wilson, C., Patel, U. A., Cole, P. V., and Feyerabend, C. (1973). Comparison of effect on tobacco consumption and carbon monoxide absorption of changing to high and low nicotine cigarettes. Br. Med. J. (iv):512.
138. Wright, G. R., Baum, J., Cogan, F., and Shephard, R. J. (1974). Smoking and carboxyhemoglobin levels. Lancet (i):635.
139. Castleden, C. M., and Cole, P. V. (1973). Inhalation of tobacco smoke by pipe and cigar smokers. Lancet (ii):21.
140. Stewart, R. D., Peterson, J. E., Fisher, T. N., Hosko, M. J., Baretta, E. D., Dodd, H. C., and Herrmann, A. A. (1973). Experimental human exposure to high concentrations of carbon monoxide. Arch. Environ. Health 26:1.
141. Jones, R. D., Commins, B. T., and Cernik, A. A. (1972). Blood lead and carboxyhaemoglobin levels in London taxi drivers. Lancet (ii):302.

142. Breysse, P. A., and Bovee, H. H. (1969). Use of expired air-carbon monoxide for carboxyhemoglobin determinations in evaluating carbon monoxide exposures resulting from the operation of gasoline fork lift trucks in holds of ships. Am. Ind. Hyg. Assoc. J. 30:477.
143. Gibbs, G. W., Allard, D., Gray, R., and Corsillo, A. (1973). Carbon monoxide and nitrogen dioxide exposure of Montreal longshoremen. Can. J. Public Health 64:147.
144. Hall, D. A., Miller, F. A., Riley, E. C., and Scherberger, R. F. (1957). Evaluation of the carbon monoxide hazard from indoor use of propane-fueled fork lift trucks. Am. Ind. Hyg. Assoc. Q. 18:355.
145. Linch, A. L., and Pfaff, H. V. (1971). Carbon monoxide—evaluation of exposure potential by personnel monitor surveys. Am. Ind. Hyg. Assoc. J. 32:745.
146. Bloomfield, B. D. (1957). Lift trucks raise carbon monoxide level. Arch. Ind. Health 15:172.
147. Bloomfield, J. J., and Isbell, H. S. (1928). The problem of automobile exhaust gas in streets and repair shops of large cities. Public Health Rep. 43:750.
148. Buchwald, H. (1969). Exposure of garage and service station operatives to carbon monoxide: a survey based on carboxyhemoglobin levels. Am. Ind. Hyg. Assoc. J. 30:570.
149. Ciampolini, E. (1924). Carbon monoxide hazard in public garages. J. Ind. Hyg. 6:102.
150. Henderson, Y., and Haggard, H. W. (1923). Health hazard from automobile exhaust gas in city streets, garages and repair shops. J.A.M.A. 81:385.
151. Katz, S. H., and Frevert, H. W. (1928). Carbon monoxide in two large garages. J. Ind. Eng. Chem. 20:31.
152. Anderson, D. E. (1971). Problems created for ice arenas by engine exhaust. Am. Ind. Hyg. Assoc. J. 32:790.
153. Radford, E. P., and Rodkey, F. L. (1973). Methods for measurement of carbon monoxide. *In* A. B. DuBois (ed.), Effects of Carbon Monoxide on Man's Health in His Home Environment. Bureau of Community Environmental Management, U. S. Department of Health, Education, and Welfare, Washington, D. C.
154. Shepherd, M. (1947). Rapid determination of small amounts of carbon monoxide. Anal. Chem. 19:77.
155. Shepherd, M., Schuhmann, S., and Kilday, M. V. (1955). Determination of carbon monoxide in air pollution studies. Anal. Chem. 27:380.
156. Teague, M. C. (1920). The determination of carbon monoxide in air contaminated with motor exhaust gas. J. Industr. Eng. Chem. 12:964.
157. Silverman, L., and Gardner, G. R. (1965). Potassium pallado-sulfite method for carbon monoxide detection. Am. Ind. Hyg. Assoc. J. 26:97.
158. Stewart, C. P., and Stolman, A. (1961). Toxicology: Mechanisms and Analytical Methods, Vol. II. Academic Press, New York.
159. Linderholm, H. (1965). A micromethod for the determination of carbon monoxide in blood. Acta Physiol. Scand. 64:372.
160. Sjöstrand, T. (1948). A method for the determination of carboxyhemoglobin concentrations by analysis of alveolar air. Acta Physiol. Scand. 16:201.
161. Hersch, P. (1964). Galvanic analysis. *In* C. A. Reilley (ed.), Advances in Analytical Chemistry and Instrumentation, Vol. 3, pp. 183–249. Interscience Publishers, New York.
162. Allen, T. H., and Root, W. S. (1955). Colorimetric determination of CO in air by an improved palladium chloride method. J. Biol. Chem. 216:309.
163. Le Moan, M. G. (1952). Semi-microdosage de l'oxyde de carbone sanguin. Ann. Pharm. Fr. 10:269.

164. Polis, R. D., Berger, L. B., and Schrenk, H. H. (1944). Colorimetric determination of low concentration of CO by use of a Pd Cl_2-phosphomolybdic acid-acetone reagent, Report of Investigations 3785, pp. 13. U. S. Bureau of Mines, Washington, D.C.
165. Fabre, R., Truhaut, R., and Boudène, C. (1952). Dosage de petites quantités de nicotine dans les atmosphères et les milieux biologiques animaux ou vegetaux. Application à l'étude de l'absorption et de l'élimination de l'alcaloide chez les sujets exposés. Ann. Pharm. Fr. 10:579.
166. McCullough, J. D., Crane, R. A., and Beckman, A. O. (1947). Determination of carbon monoxide in air by use of red mercuric oxide. Anal. Chem. 19:999.
167. Bay, H. W., Blurton, K. F., Lieb, H. L., and Oswin, H. G. (1972). Electrochemical measurement of carbon monoxide. Am. Lab. July. Vol. 4, No. 7, pp. 57–58, 60–61.
168. Ayres, S. M., personal communication.
169. Strange, J. P., Ball, K. E., and Barnes, D. O. (1960). Continuous parts per billion recorder for air contaminants. J. Air Pollut. Control Assoc. 10:423.
170. Pierce, J. O., and Collins, R. J. (1971). Calibration of an infrared analyzer for continuous measurement of carbon monoxide. Am. Ind. Hyg. Assoc. J. 32:457.
171. Hill, D. W., and Powell, T. (1968). Non-Dispersive Infra-Red Gas Analysis in Science, Medicine and Industry. Adam Hilger, Ltd., London.
172. Jacobs, M. B., Braverman, M. M., and Hockheiser, S. (1959). Continuous determination of carbon monoxide and hydrocarbons in air by a modified infra-red analyzer. J. Air Pollut. Control Assoc. 9:110.
173. Coburn, R. F., Danielson, G. K., Blakemore, W. S., and Forster, R. E. (1964). Carbon monoxide in blood: analytical method and sources of error. J. Appl. Physiol. 19:510.
174. Szulcewski, D. H., and Higuchi, T. (1957). Gas chromatographic separation of some permanent gases on silica gel at reduced temperatures. Anal. Chem. 29:1541.
175. Smith, R. N., Swineheart, J., and Lesnini, D. G. (1958). Chromatographic analysis of gas mixtures containing nitrogen, nitrous oxide, nitric oxide, carbon monoxide and carbon dioxide. Anal. Chem. 30:1217.
176. Kyryacos, G., and Boord, C. E. (1957). Separation of hydrogen, oxygen, nitrogen, methane and carbon monoxide by gas adsorption chromatography. Anal. Chem. 29:787.
177. Forster, R. E. (1969). Reactions of carbon monoxide with heme proteins. *In* A. B. Dubois (ed.), Effects of Chronic Exposure to Low Levels of Carbon Monoxide on Human Health, Behavior and Performance, pp. 10–13. National Academy of Sciences and National Academy of Engineering, Washington, D. C.
178. Commins, B. T. (1975). Measurement of carbon monoxide in the blood: review of available methods. Ann. Occup. Hyg. 18:69.
179. Ettre, L. S. (1961). Application of gas chromatographic methods for air pollution studies. J. Air Pollut. Control Assoc. 11:34.
180. Porter, K., and Volman, D. H. (1962). Flame ionization detection of carbon monoxide for gas chromatographic analysis. Anal. Chem. 34:748.
181. Rodkey, F. L. (1970). The measurement of carbon monoxide in biological fluids. *In* F. W. Sunderman and F. W. Sunderman, Jr. (eds.), Laboratory Diagnosis of Diseases Caused by Toxic Agents, pp. 278–285. W. H. Green, Inc., St. Louis, Missouri.
182. Maisels, M. J., Pathak, A., Nelson, N. M., Nathan, D. G., and Smith, C. A. (1971). Endogenous production of carbon monoxide in normal and erythroblastotic newborn infants. J. Clin. Invest. 50:1.

183. van Slyke, D. D., and Neill, J. M. (1924). Determination of bases in blood and other solutions by vacuum extraction and manometric measurement. J. Biol. Chem. 61:523.
184. Sendroy, J., and Liu, S. H. (1930). Gasometric determination of oxygen and carbon monoxide in blood. J. Biol. Chem. 89:133.
185. Allen, T. H., and Root, W. S. (1955). Colorimetric determination of CO in blood by an improved palladium chloride method. J. Biol. Chem. 216:319.
186. Scholander, P. F., and Roughton, F. J. W. (1942). A simple micro-gasometric method of estimating carbon monoxide in blood. J. Ind. Hyg. Toxicol. 24:218.
187. Scholander, P. F., and Roughton, F. J. W. (1943). Micro-gasometric estimation of the blood gases. II. Carbon monoxide. J. Biol. Chem. 148:551.
188. Ammundsen, E., and Grut, A. (1943). Determination of carbon monoxide in the blood. Acta Med. Scand. 115:151.
189. Christman, A. A., and Randall, E. L. (1933). A convenient and accurate method for the determination and detection of carbon monoxide in blood. J. Biol. Chem. 102:595.
190. Rice, E. W. (1952). Improved photometric determination of carbon monoxide by means of palladious chloride. Arch. Ind. Hyg. Occup. Med. 6:487.
191. Wennesland, R. (1940). A new method for the determination of carbon monoxide in blood. Acta Physiol. Scand. 1:49.
192. Gaensler, E. A., Cadigan, J. B., Ellicott, M. F., and Jones, R. H. (1957). A new method for rapid precise determination of carbon monoxide in blood. J. Lab. Clin. Med. 49:945.
193. Lawther, P. J., and Apthorp, G. H. (1955). A method for the determination of carbon monoxide in blood. Br. J. Ind. Med. 12:326.
194. Moureau, H., Chovin, P., Truffert, L., and Lebbe, J. (1957). Nouvelle microméthode pour la determination rapide et précise de l'oxycarbonémie, par absorption sélective dans l'infrarouge. Arch. Mal. Prof. 18:116.
195. Ayres, S. M., Criscitiello, A., and Giannelli, S. (1966). Determination of blood carbon monoxide content by gas chromatography. J. Appl. Physiol. 21:1368.
196. McCredie, R. M., and Jose, A. D. (1967). Analysis of blood carbon monoxide and oxygen by gas chromatography. J. Appl. Physiol. 22:863.
197. Dahms, T. E., and Horvath, S. M. (1974). Rapid, accurate technique for determination of carbon monoxide in blood. Clin. Chem. 20:533.
198. Conway, E. J. (1950). Microdiffusion Analysis and Volumetric Error, Ed. 3, pp. 1–391. Crosby, Lockwood and Son, London.
199. Haldane, J. (1895). The action of carbonic oxide on man. J. Physiol. 18:430.
200. Haldane, J., and Smith, J. L. (1896). The oxygen tension of arterial blood. J. Physiol. 20:497.
201. Haldane, J., and Smith, J. L. (1898). The absorption of oxygen by the lungs. J. Physiol. 22:231.
202. Hartridge, H. (1912). The action of various conditions on carbon monoxide haemoglobin. J. Physiol. 44:22.
203. Harper, P. V. (1952). A new spectrophotometric method for the determination of CO in the blood. J. Lab. Clin. Med. 40:634.
204. Paul, K.-G., and Theorell, H. (1942). A colorimetrical carbon-monoxide-hemoglobin method of determination for clinical use. Acta Physiol. Scand. 4:285.
205. Klendshøj, N. C., Feldstein, M., and Sprague, A. L. (1950). The spectrophotometric determination of carbon monoxide. J. Biol. Chem. 183:297.
206. Commins, B. T., and Lawther, P. J. (1965). A sensitive method for the determination of carboxyhaemoglobin in a finger prick sample of blood. Br. J. Ind. Med. 22:139.

207. Lily, R. E., Cole, P. V., and Hawkins, L. H. (1972). Spectrophotometric measurement of carboxyhaemoglobin: an evaluation of the method of Commins and Lawther. Br. J. Ind. Med. 29:454.
208. Whitehead, T. P., and Worthington, S. (1961). The determination of carboxyhaemoglobin. Clin. Chim. Acta 6:356.
209. Linderholm, H., Sjöstrand, T., and Söderström, B. (1966). A method for determination of low carbon monoxide concentration in blood. Acta Physiol. Scand. 66:1.
210. Mass, A. H. J., Hamelink, M. L., and de Leeuw, R. J. M. (1970). An evaluation of the spectrophotometric determination of HbO_2, HbCO and Hb in blood with the CO-Oximeter IL 182. Clin. Chim. Acta 29:303.
211. Malenfant, A. L., Gambino, S. R., Waraska, A. J., and Roe, E. I. (1968). Spectrophotometric determination of haemoglobin concentration and percent oxyhaemoglobin and carboxyhaemoglobin saturation Clin. Chem. 14:789 (abstr.).
212. Cole, P. (1973). Carbon monoxide, environmental pollution and public health. Luxembourg, December 17-19, 1973. Cited by Commins (1975).
213. Small, K. A., Radford, E. P., Frazier, J. M., Rodkey, F. L., and Collison, H. A. (1971). A rapid method for simultaneous measurement of carboxy- and methemoglobin in blood. J. Appl. Physiol. 31:154.
214. Lauwerijs, R. Carbon monoxide, environmental pollution and public health. Luxembourg, December 17-19, 1973. Cited by Commins, ref. 178.
215. Dahlström, H. (1955). A critical study of Sjöstrand's method for determination of COHb, and the total hemoglobin of the human body. Acta Physiol. Scand. 33:296.
216. Henderson, M., and Apthorp, G. H. (1960). Rapid method for estimation of carbon monoxide in blood. Br. Med. J. (ii):1853.
217. Sjöstrand, T. (1948). A method for the determination of the total hemoglobin content of the body. Acta Physiol. Scand. 16:211.
218. Carlsten, A., Holmgren, A., Linroth, K., Sjöstrand, J., and Ström, G. (1954). Relationship between low values of alveolar carbon monoxide concentration and carboxyhemoglobin percentage in human blood. Acta Physiol. Scand. 31:62.
219. Hackney, J. D., Kaufman, G. A., Lashier, H., and Lynn, K. (1962). Rebreathing estimate of carbon monoxide hemoglobin. Arch. Environ. Health 5:300.
220. Jones, R. H., Ellicott, M. F., Cadigan, J. B., and Gaensler, E. A. (1958). The relationship between alveolar and blood carbon monoxide concentrations during breath holding. J. Lab. Clin. Med. 51:553.
221. Peterson, J. E. (1970). Post-exposure relationship of carbon monoxide in blood and expired air. Arch. Environ. Health 21:172.
222. Goldsmith, J. R. (1965). Discussion: epidemiologic studies of chronic respiratory diseases. Arch. Environ. Health 10:383.
223. O'Donnell, R. D., Chikos, P., and Theodore, J. (1971). Effect of carbon monoxide exposure on human sleep and psychomotor performance. J. Appl. Physiol. 31:513.
224. O'Donnell, R. D., Mikulka, P., Heinig, P., and Theodore, J. (1971). Low level carbon monoxide exposure and human psychomotor performance. Toxicol. Appl. Pharmacol. 18:593.
225. Theodore, J., O'Donnell, R. D., and Back, K. C. (1971). Toxicological evaluation of carbon monoxide in humans and other mammalian species. J. Occup. Med. 13:242.
226. Wright, G. R., and Shephard, R. J. (1978). Carbon monoxide exposure and auditory duration discrimination. Arch. Environ. Health. 33:226.

227. Godin, G., and Shephard, R. J. (1972). On the course of carbon monoxide uptake and release. Respiration 29:317.
228. Shephard, R. J. (1974). Objectives of human air pollution studies. pp. 150–153. *In* R. J. Shephard, D. Pengelly, and F. Silverman (eds.), Proceedings of 1st Canadian Conference on Research in Air Pollution, June, 1974, Gravenhurst, Ontario York-Toronto Lung Association, Toronto, Ontario.
229. Shephard, R. J., Turner, J. E., Carey, G. C. R., and Phair, J. J. (1960). Correlation of pulmonary function and domestic micro-environment. J. Appl. Physiol. 15:70.
230. Aronow, W. S., Harris, C. N., Isbell, M. W., Rokaw, S. N., and Imparato, B. (1972). Effect of freeway travel on angina pectoris. Ann. Intern. Med. 77:669.
231. Aronow, W. S., Stemmer, E. A., and Isbell, M. W. (1974). Effect of carbon monoxide exposure on intermittent claudication. Circulation 49:415.
232. Antonini, E. (1967). Hemoglobin and its reactions with ligands. Science 158:1417.
233. Perutz, M. F. (1970). Stereochemistry of cooperative effects in haemoglobin. Nature 228:726.
234. Roughton, F. J. W. (1964). Transport of oxygen and carbon dioxide. *In* W. O. Fenn and H. Rahn (eds.), Handbook of Physiology, Section 3, Respiration I, pp. 767–825. American Physiological Society, Washington, D. C.
235. Douglas, C. G., Haldane, J. S., and Haldane, J. B. S. (1912). The laws of combination of haemoglobin with carbon monoxide and oxygen. J. Physiol. 44:275.
236. Joels, N., and Pugh, L. G. C. E. (1958). The carbon monoxide dissociation curve of human blood. J. Physiol. 142:63.
237. Adair, G. S. (1925). The hemoglobin system. VI. The oxygen dissociation curve of hemoglobin. J. Biol. Chem. 63:529.
238. Holland, R. A. B. (1969). Rate of O_2 dissociation from O_2Hb and relative combination rate of CO and O_2 in mammals at 37°C. Respir. Physiol. 7:30.
239. Holland, R. A. B. (1969). Rate at which CO replace O_2 from O_2Hb in red cells of different species. Respir. Physiol. 7:43.
240. Kernohan, J. C. (1961). Kinetics of the reactions of two sheep haemoglobins with oxygen and carbon monoxide. J. Physiol. 155:580.
241. Roughton, F. J. W., and Darling, R. C. (1944). The effect of carbon monoxide on the oxyhemoglobin dissociation curve. Am. J. Physiol. 141:17.
242. Åstrup, P., Kjeldsen, K., and Wanstrup, J. (1967). Enhancing influence of carbon monoxide on the development of atheromatosis in cholesterol-fed rabbits. J. Atheroscler. Res. 7:343.
243. Mulhausen, R. O., Åstrup, P., and Mellemgaard, K. (1968). Oxygen affinity and acid-base status of human blood during exposure to hypoxia and carbon monoxide. Scand. J. Clin. Lab. Invest. (suppl. 103) 22:9.
244. Rodkey, F. L., O'Neal, J. D., and Collison, H. A. (1969). Oxygen and carbon monoxide equilibria of human adult hemoglobin at atmospheric and elevated pressure. Blood 33:57.
245. Benesch, R., and Benesch, R. E. (1967). The effect of organic phosphates from the human erythrocyte on the allosteric properties of hemoglobin. Biochem. Biophys. Res. Commun. 26:162.
246. Gibson, O. H., and Ainsworth, S. (1957). Photosensitivity of haem compounds. Nature 180:1416.
247. Millikan, G. A. (1939). Muscle haemoglobin. Physiol. Rev. 19:503.
248. Rossi-Fanelli, A., and Antonini, E. (1958). Studies on the oxygen and carbon monoxide equilibria on human myoglobin. Arch. Biochem. 77:478.
249. Coburn, R. F., Ploegmakers, F., Gondrie, P., and Abboud, R. (1973). Myocardial myoglobin oxygen tension. Am. J. Physiol. 224:870.
250. Coburn, R. F., and Mayers, L. B. (1971). Myoglobin oxygen tension deter-

mined from measurements of carboxyhemoglobin in skeletal muscle. Am. J. Physiol. 220:66.
251. Wilks, S. S. (1963). Toxic photo-oxidation products in closed environments. Aerospace Med. 34:838.
252. Kessler, M., Lang, H., Sinagowitz, E., Rink, R., and Höper, J. (1973). Homeostasis of oxygen supply in liver and kidney. *In* H. I. Bicher and D. F. Bruley (eds.), Oxygen Transport to Tissue: Instrumentation, Methods and Physiology, pp. 351–360. Plenum Press, New York.
253. Whalen, W. J. (1971). Intracellular PO_2 in heart and skeletal muscle. Physiologist 14:69.
254. Shephard, R. J. (1977). Endurance Fitness, Ed. 2. University of Toronto Press, Toronto.
255. Ball, E. G., Strittmatter, C. F., and Cooper, O. (1951). The reaction of cytochrome oxidase with carbon monoxide. J. Biol. Chem. 193:635.
256. Coburn, R. F. (ed). (1977). The biologic effects of carbon monoxide on the human organism. Report of a Committee of the National Research Council, National Academy of Sciences, U. S. National Academy of Sciences and Environmental Protection Agency, Washington, D. C.
257. Gibson, O. H., and Greenwood, C. (1964). The spectra and some properties of cytochrome oxidase components. J. Biol. Chem. 239:586.
258. Longo, L. D. (1977). The biological effects of carbon monoxide on the pregnant woman, fetus and newborn infant. Am. J. Obstet. Gynecol. 129:69.
259. Estabrook, R. W., Franklin, M. R., and Hildebrandt, A. G. (1970). Factors influencing the inhibitory effect of carbon monoxide on cytochrome P-450 catalyzed mixed function oxidation reactions. Ann. N. Y. Acad. Sci. 174:218.
260. Rondia, D. (1970). Abaissement de l'activité de la benzopyrène—hydroxylase hépatique in vivo après l'inhalation d'oxyde de carbone. C. R. Acad. Sci. (Paris) 271:617.
261. Coburn, R. F., and Kane, P. B. (1968). Maximal erythrocyte and hemoglobin catabolism. J. Clin. Invest. 47:1435.
262. Coburn, R. F., Forster, R. E., and Kane, P. B. (1965). Considerations of the physiological variables that determine the blood COHb concentrations in man. J. Clin. Invest. 44:1899.
263. Sjöstrand, T. (1949). Endogenous formation of carbon monoxide in man under normal and pathological conditions. Scand. J. Clin. Lab. Invest. 1:201.
264. Sjöstrand, T. (1951). Endogenous formation of carbon monoxide: the CO concentration in the inspired and expired air of hospital patients. Acta Physiol. Scand. 22:137.
265. Coburn, R. F., Blakemore, W. S., and Forster, R. E. (1963). Endogenous carbon monoxide production in man. J. Clin. Invest. 42:1172.
266. Luomanmaki, K. (1966). Studies on the metabolism of carbon monoxide. Ann. Med. Exp. Biol. Fenn. (suppl. 2) 44:1.
267. Lynch, S. R., and Moede, A. L. (1972). Variation in the rate of endogenous carbon monoxide production in normal human beings. J. Lab. Clin. Med. 79:85.
268. Coburn, R. F., Williams, W. J., White, P., and Kahn, S. B. (1967). The production of carbon monoxide from hemoglobin in vivo. J. Clin. Invest. 46:346.
269. Coburn, R. F., Williams, W. J., and Forster, R. E. (1964). Effect of erythrocyte destruction on carbon monoxide production in man. J. Clin. Invest. 43:1098.
270. Lundwig, G. D., and Blakemore, W. S. (1957). Production of carbon monoxide by hemin oxidation. J. Clin. Invest. 36:912 (abstr.).
271. Sjöstrand, T. (1952). The formation of carbon monoxide by the decomposition of haemoglobin in vivo. Acta Physiol. Scand. 26:338.

272. Sjöstrand, T. (1952). The formation of carbon monoxide by in vitro decomposition of haemoglobin in bile pigments. Acta Physiol. Scand. 26:328.
273. Coburn, R. F., Blakemore, W. S., and Forster, R. E. (1966). Endogenous carbon monoxide production in patients with hemolytic anemia. J. Clin. Invest. 45:460.
274. Engel, R. R., Rodkey, F. L., and Krill, C. E. (1971). Carboxyhemoglobin as an index of hemolysis. Pediatrics 47:723.
275. Rodkey, F. L., Raymond, L. W., Collison, H. A., and O'Neal, J. D. (1974). Changes in blood carboxyhemoglobin during simulated saturation diving to 50 ATA. Undersea Biomed. Res. 1:197.
276. Coburn, R. F. (1973). Endogenous carbon monoxide metabolism. Ann. Rev. Med. 24:241.
277. Ratney, R. S., Wegman, D. H., and Elkins, H. B. (1974). In vivo conversion of methylene chloride to carbon monoxide. Arch. Environ. Health 28:223.
278. Forster, R. E. (1969). Carbon monoxide production by organisms. National Academy of Sciences and National Academy of Engineering, Washington, D. C. *In* A. B. DuBois (ed.), pp. 15–16.
279. Coburn, R. F. (1967). Endogenous carbon monoxide production and body CO stores. Acta Med. Scand. (suppl. 472) 269.
280. Henderson, Y., and Haggard, H. W. (1921). The elimination of carbon monoxide from the blood after a dangerous degree of asphyxiation, and a therapy for accelerating elimination. J. Pharmacol. Exp. Ther. 16:11.
281. Forbes, W. H., Sargent, F., and Roughton, F. J. W. (1945). The rate of carbon monoxide uptake by normal men. Am. J. Physiol. 143:594.
282. Forbes, W. H. (1970). CO uptake via the lungs. Ann. N. Y. Acad. Sci. 174:72.
283. Lilienthal, J. L., and Pine, M. B. (1946). The effect of oxygen pressure on the uptake of carbon monoxide by man at sea level and at altitude. Am. J. Physiol. 145:346.
284. Bates, D. V. (1952). The uptake of carbon monoxide in health and in emphysema. Clin. Sci. 11:21.
285. Ramsey, J. M. (1972). Carbon monoxide, tissue hypoxia and sensory psychomotor response in hypoxemia. Clin. Sci. 42:619.
286. Fenn, W. O., and Cobb, D. M. (1932). The burning of carbon monoxide by heart and skeletal muscle. Am. J. Physiol. 102:393.
287. Roughton, F. J. W., and Root, W. S. (1945). The fate of CO in the body during recovery from mild carbon monoxide poisoning in man. Am. J. Physiol. 145:239.
288. Bartlett, O. (1968). Patho-physiology of exposure to low concentrations of CO. Arch. Environ. Health 16:719.
289. Pace, N., Stajman, E., and Walker, E. L. (1950). Acceleration of carbon monoxide elimination in man by high pressure oxygen. Science 111:652.
290. Henderson, Y., Haggard, H. W., Teague, M. C., Prince, A. L., and Wunderlich, R. M. (1921). Physiological effects of automobile exhaust gases and standards of exposure for brief periods. J. Ind. Hyg. 3:79 and 137.
291. Peterson, J. E., and Stewart, R. D. (1970). Absorption and elimination of carbon monoxide by inactive young men. Arch. Environ. Health 21:165.
292. Tobias, C. A., Lawrence, J. H., Roughton, F. J. W., Root, W. S., and Gregersen, M. I. (1945). The elimination of carbon monoxide from the human body with reference to the possible conversion of CO to CO_2. Am. J. Physiol. 145:253.
293. Wagner, J. A., Horvath, S. M., and Dahms, J. E. (1975). Carbon monoxide elimination. Respir. Physiol. 23:41.
294. Donald, K. W., and Paton, W. O. M. (1955). Gases administered in artificial

respiration with particular reference to the use of carbon monoxide. Br. Med. J. (i):313.

295. Killick, E. M. (1936). The acclimitization of the human subject to atmospheres containing low concentrations of carbon monoxide. J. Physiol. 87:41.
296. Medical Research Council. (1958). Carbon monoxide poisoning: use of CO_2-O_2 mixture. Br. Med. J. (ii):1408.
297. Norman, J. N., and Ledingham, I. McA. (1967). Carbon monoxide poisoning: investigations and treatment. Prog. Brain Res. 24:101.
298. Kindwall, E. P. (1977). Carbon monoxide and cyanide poisoning. *In* J. C. Davis and T. K. Hunt (eds.), Hyperbaric Oxygen Therapy, pp. 177–190. Undersea Medical Society, Bethesda, Maryland.
299. Pace, N., Consolazio, W. V., White, W. A., and Behnke, A. R. (1946). Formulation of the principal factors affecting the rate of uptake of carbon monoxide by man. Am. J. Physiol. 147:352.
300. Goldsmith, J. R., Terzaghi, J., and Hackney, J. D. (1963). Evaluation of fluctuating carbon monoxide exposures. Arch. Environ. Health 7:647.
301. Peterson, J. E., and Stewart, R. D. (1972). Human absorption of carbon monoxide from high concentrations in air. Am. Ind. Hyg. Assoc. J. 31:293.
302. Hill, E. P., Hill, J. R., Power, G. G., and Longo, L. D. (1977). Carbon monoxide exchanges between the human fetus and mother: a mathematical model. Am. J. Physiol. 232:H311.
303. Duke, H. N., Green, J. H., and Neil, E. (1952). Carotid chemoreceptor impulse activity during inhalation of CO mixtures. J. Physiol. 118:520.
304. Mills, E., and Edwards, McI. W. (1968). Stimulation of aortic and carotid chemoreceptors during carbon monoxide inhalation. J. Appl. Physiol. 25:494.
305. Paintal, A. S. (1967). Mechanism of stimulation of aortic chemoreceptors by natural stimuli and chemical substances. J. Physiol. 189:63.
306. Weiss, H. R., and Cohen, J. A. (1974). Effects of low levels of carbon monoxide on rat brain and muscle tissue PO_2. Environ. Physiol. Biochem. 4:31.
307. Permutt, S., and Farhi, L. (1969). Tissue hypoxia and carbon monoxide. *In* A. B. DuBois (ed.), Effects of Chronic Exposure to Low Levels of Carbon Monoxide on Human Health, Behavior and Performance, pp. 18–24. National Academy of Sciences and National Academy of Engineering, Washington D. C.
308. Comroe, J. H. (1965). Physiology of Respiration. Year Book Medical Publishers, Inc., Chicago.
309. Loeschke, H. H., and Gertz, K. H. (1958). Einfluss des O_2—Druckes in der Einatmungsluft auf die Atemtätigkeit des Menschen, geprüft unter Konstanthaltung des alveolaren CO_2-Druckes. Pfluegers Arch. 267:460.
310. Horvath, S. M., Dahms, T. E., and O'Hanlon, J. F. (1971). Carbon monoxide and human vigilance—a deleterious effect of present urban concentrations. Arch. Environ. Health 23:343.
311. Klausen, K., Rasmussen, B., Gjellerod, H., Madsen, H., and Petersen, E. (1968). Circulation, metabolism and ventilation during prolonged exposure to CO and to high altitude. Scand. J. Clin. Lab. Invest. (suppl. 103) 22:26.
312. Shephard, R. J. (1972). The influence of small doses of carbon monoxide upon heart rate. Respiration 29:516.
313. Santiago, T. V., and Edelman, N. H. (1976). Mechanism of the ventilatory response to carbon monoxide. J. Clin. Invest. 57:977.
314. Leathart, G. L. (1962). Hyperventilation in carbon monoxide poisoning. Br. Med. J. (ii):511.
315. Chiodi, H., Dill, D. B., Consolazio, F., and Horvath, S. M. (1941). Respiratory and circulatory responses to acute carbon monoxide poisoning. Am. J. Physiol. 134:683.

316. Asmussen, E., and Vinther-Paulsen, N. (1949). On the circulatory adaptations to arterial hypoxemia (CO poisoning). Acta Physiol. Scand. 19:115.
317. Asmussen, E., and Chiodi, H. (1941). The effect of hypoxemia on ventilation and circulation in man. Am. J. Physiol. 132:426.
318. Apthorp, G. H., Bates, D. V., Marshall, R., and Mendel, D. (1958). Effect of acute carbon monoxide poisoning on work capacity. Br. Med. J. (ii):476.
319. Pitts, G. C., and Pace, N. (1947). The effect of blood carboxyhemoglobin concentration on hypoxia tolerance. Am. J. Physiol. 148:139.
320. Ekblom, B., and Huot, R. (1972). Response to submaximal and maximal exercise at different levels of carboxyhemoglobin. Acta Physiol. Scand. 86:474.
321. Pirnay, F., Dujardin, J., Deroanne, R., and Petit, J. M., (1971). Muscular exercise during intoxication by carbon monoxide. J. Appl. Physiol. 31:573.
322. Vogel, J. A., Gleser, M. A., Wheeler, R. C., and Whitten, B. K. (1972). Carbon monoxide and physical work capacity. Arch. Environ. Health 24:198.
323. Vogel, J. A., and Gleser, M. A. (1972). Effect of carbon monoxide on oxygen transport during exercise. J. Appl. Physiol. 32:234.
324. Raven, P. B., Drinkwater, B. L., Ruhling, R. O., Bolduan, N., Taguchi, S., Gliner, J., and Horvath, S. M. (1974). Effect of carbon monoxide and peroxyacetyl nitrate on man's maximal aerobic capacity. J. Appl. Physiol. 36:288.
325. Forbes, W. H., Dill, D. B., De Silva, H., and Van Deventer, F. M. (1937). The influence of moderate carbon monoxide poisoning upon the ability to drive automobiles. J. Ind. Hyg. Toxicol. 19:598.
326. Stewart, R. D., Peterson, J. E., Baretta, E. D., Bachand, R. T., Hosko, M. J., and Herrmann, A. A. (1970). Experimental human exposure to carbon monoxide. Arch. Environ. Health 21:154.
327. McFarland, R. A., Roughton, F. J. W., Halperin, M. H., and Niven, J. J. (1944). The effects of carbon monoxide and altitude on visual thresholds. J. Aviat. Med. 15:381.
328. Halperin, M. H., McFarland, R. A., Niven, J. I., and Roughton, F. J. W. (1959). The time course of the effects of carbon monoxide on visual thresholds. J. Physiol. 146:583.
329. Beard, R. R., and Wertheim, G. A. (1967). Behavioral impairment associated with small doses of carbon monoxide. Am. J. Public Health 57:2012.
330. Schulte, J. H. (1963). Effects of mild carbon monoxide intoxication. Arch. Environ. Health 7:524.
331. Beard, R. R., and Grandstaff, N. W. (1975). Carbon monoxide and human functions. Fifth Rochester International Conference on Environmental Toxicity: Behavioral Toxicology. Plenum Press, New York.
332. McFarland, R. A. (1963). Experimental evidence of the relationship between ageing and oxygen want: in search of a theory of ageing. Ergonomics 6:339.
333. McFarland, R. A. (1932). The psychological effects of oxygen deprivation (anoxemia) on human behavior. Arch. Psychol. 22:1.
334. McFarland, R. A. (1937). Psycho-physiological studies at high altitude in the Andes. I. The effects of rapid ascents by aeroplane and train. J. Comp. Psychol. 23:191.
335. McFarland, R. A. (1937). Psycho-physiological studies at high altitude in the Andes. II. Sensory and motor responses during acclimatization. J. Comp. Psychol. 23:227.
336. Fodor, G. G., and Winneke, G. (1972). Effect of low CO concentrations on resistance to monotony and on psychomotor capacity. Staub-Reinhalt Luft. 32:46.
337. Weir, F. W., Mehta, M. M., Johnson, D. F., Anglen, D. M., Rockwell, T. H., Attwood, D. A., Herrin, G. D., and Safford, R. R. (1973). An investigation of

the effects of carbon monoxide on humans in the driving task. Ohio State University Research Foundation, January, 1973.

338. Goldberg, H. D., and Chappell, M. N. (1967). Behavioral measure of effect of carbon monoxide on rats. Arch. Environ. Health 14:671.

339. Back, K. C., Dominquez, A. M. (1968). Psychopharmacology of carbon monoxide under ambient and altitude conditions. U. S. Aerospace Medical Research Laboratory Report AMRL-TR-68-175, Paper No. 7.

340. Back, K. C. (1969). Effects of carbon monoxide on the performance of monkeys. U. S. Aerospace Medical Research Laboratory Report AMRL-TR-69-130, Paper No. 3.

340a. Baker, S. P., Fisher, R. S., Masemore, W. C., and Sopher, I. M. (1972). Fatal unintentional carbon monoxide poisoning in motor vehicles. Am. J. Public Health 62:1463.

341. Beard, R. R., and Grandstaff, N. (1970). Carbon monoxide exposure and cerebral function. Ann. N. Y. Acad. Sci. 174:385.

342. Corso, J. F. (1967). The Experimental Psychology of Sensory Behavior. Holt, Rinehart and Winston, Inc., New York.

343. McFarland, R. A., and Halperin, M. H. (1940). The relation between foveal visual acuity and illumination under reduced oxygen tension. J. Gen. Physiol. 23:613.

344. Crozier, W. J., and Holway, A. H. (1939). Theory and measurement of visual mechanisms. I. A visual discriminometer. J. Gen. Physiol. 22:341.

345. Landis, C. (1954). Determinants of the critical flicker-fusion threshold. Physiol. Rev. 34:259.

346. Lilienthal, J. L., and Fugitt, C. H. (1946). The effect of low concentrations of COHb on the "altitude tolerance" of man. Am. J. Physiol. 145:359.

347. Vollmer, E. P., King, B. G., Birren, J. E., and Fischer, M. B. (1946). The effects of CO on three types of performance; at simulated altitudes of 10,000 and 15,000 feet. J. Exp. Psychol. 36:244.

348. Guest, A. D., Duncan, C., and Lawther, P. J. (1970). Carbon monoxide and phenobarbitone: a comparison of effects on auditory flutter fusion threshold and critical flicker fusion threshold. Ergonomics 13:587.

349. Barlow, D. H., and Baer, D. J. (1967). Effect of cigarette smoking on the critical flicker frequency of heavy and light smokers. Percept. Mot. Skills 24:151.

350. Fabricant, N. D., and Rose, I. W. (1951). Effect of cigarette smoking on the flicker fusion threshold of normal persons. Eye, Ear, Nose and Throat Month. 30:541.

351. Garner, L. L., Carl, E. F., and Grossman, E. E. (1954). Effect of cigarette smoking on flicker fusion threshold. Arch. Ophthalmol. 51:642.

352. Larson, P. S., Finnegan, J. K., and Haag, H. B. (1950). Observations on the effect of cigarette smoking on the fusion frequency of flicker. J. Clin. Invest. 29:483.

353. Warwick, K. M., and Eysenck, H. J. (1963). The effects of smoking on the CFF threshold. Life Sci. 2:219.

354. Abramson, E., and Heyman, T. (1944). Dark adaptation and inhalation of carbon monoxide. Acta Physiol. Scand. 7:303.

355. McFarland, R. A. (1973). Low level exposure to carbon monoxide and driving performance. Arch. Environ. Health 28:355.

356. Xintaras, C., Johnson, B. L., Ulrich, C. E., Terrill, R. E., and Sobecki, M. F. (1966). Application of the evoked response technique in air pollution toxicology. Toxicol. Appl. Pharmacol. 8:77.

357. Sprout, W. L., Neeld, W. E., and Woessner, W. W. (1975). Management of

chemical cyanosis by oxygen saturation readings. Arch. Environ. Health 30:302.
358. Hosko, M. J. (1970). The effect of carbon monoxide on the visual evoked response in man. Arch. Environ. Health. 21:174.
359. Wyburn, C. M., Pickford, R. W., and Hirst, R. J. (1964). Human Senses and Perception. University of Toronto Press, Toronto.
360. Beard, R. R., and Grandstaff, N. (1970). Behavioral responses to small doses of carbon monoxide. U. S. Aerospace Medical Research Laboratory Report AMRL-TR-70-102, Paper No. 7.
361. Stewart, R. D., Newton, P. E., Hosko, M. J., and Peterson, J. E. (1973). Effect of carbon monoxide on time perception. Arch. Environ. Health 27:155.
362. Massaro, D. W., and Kahn, B. J. (1973). Effects of central processing on auditory recognition. J. Exp. Psychol. 97:51.
363. Plomp, R. (1964). Rate of decay of auditory sensation. J. Acoust. Soc. Am. 36:277.
364. Wright, G. R. (1971). Adaptive control in vigilance. M.Sc. thesis, Department of Industrial Engineering, University of Toronto.
365. Millar, R. A., and Gregory, I. C. (1972). Reduced oxygen content on equilibrated fresh heparinized and ACD-stored blood from cigarette smokers. Br. J. Anaesth. 44:1015.
366. Lewis, J., Baddeley, A. O., Bonham, K. G., and Lovett, D. (1970). Traffic pollution and mental efficiency. Nature 225:95.
367. Sollberger, A. (1967). Biological measurements in time, with particular reference to synchronization mechanisms. Ann. N. Y. Acad. Sci. 138:561.
368. O'Donnell, R. D., Mikulka, P., Heinigs, P., and Theodore, J. (1970). Effects of short-term low level carbon monoxide exposure on human performance. U. S. Aerospace Medical Research Laboratory Report AMRL-TR-70-37. Brooks AFB.
369. Bender, W., Gothert, M., Malorny, G., and Sebbesse, P. (1971). Wirkungsbild niedriger Kohlenoxid—Konzentrationen beim Menschen. (Effects of low carbon monoxide concentrations in man.) Arch. Toxicol. 27:142.
370. Fitts, P. M., and Posner, M. I. (1967). Human Performance. Brooks/Cole Publishing Company, Belmont, California.
371. Ramsey, J. M. (1970). Oxygen reduction and reaction time in hypoxic and normal drivers. Arch. Environ. Health 20:597.
372. Ramsey, J. M. (1973). Effects of single exposures of carbon monoxide on sensory and psychomotor response. Am. Ind. Hyg. Assoc. J. May:212.
373. Fleishman, E. A. (1954). Dimensional analysis of psychomotor abilities. J. Exp. Psychol. 48:437.
374. Ashton, H., Savage, R. D., Thompson, J. W., and Watson, D. W. (1972). A method for measuring human behavioural and physiological responses at different stress levels on a driving simulator. Br. J. Pharmacol. 45:532.
375. Ashton, H., Savage, R. D., Telford, R., Thompson, J. W., and Watson, D. W. (1972). The effects of cigarette smoking on the response to stress in a driving simulator. Br. J. Pharmacol. 45:546.
376. Brieger, H. (1944). Carbon monoxide polycythemia. J. Ind. Hyg. Toxicol. 26:321.
377. Hahn-Pedersen, A. (1962). Investigations of prolonged heavy carbon monoxide-hypoxia in albino-female rats. Ph.D. thesis, University of Copenhagen.
378. Vernot, E. H., MacKenzie, W. F., MacEwen, J. D., Montelcone, P. N., George, M. E., Chikos, P. M., Back, K. C., Thomas, A. A., and Hawn, C. C. (1970). Hematological effects of long-term continuous animal exposure to car-

bon monoxide. U. S. Aerospace Medical Research Laboratory Report AMRL-TR-70-102, Paper No. 1.

379. Jones, R. A., Strickland, J. A., Stunkard, J. A. and Siegel, J. (1971). Effects on experimental animals of long-term inhalation exposure to carbon monoxide. Toxicol. Appl. Pharmacol. 19:46.
380. Eckardt, R. E., MacFarland, H. N., Alarie, Y. C. E., and Busey, W. M. (1972). The biologic effect from long term exposure of primates to carbon monoxide. Arch. Environ. Health 25:381.
381. De Bias, D. A., Banerjee, C. M., Birkhead, N. C., Harrer, W. V., and Kazal, L. A. (1973). Carbon monoxide inhalation effects following myocardial infarction in monkeys. Arch. Environ. Health 27:161.
382. Rode, A. (1970). Acute and chronic effects of smoking on fitness. M.Sc. thesis, School of Hygiene, University of Toronto.
383. Shephard, R. J. (1970). Human endurance and the heart at altitude. J. Sports Med. Phys. Fitness 10:72.
384. Åstrup, P., Kjeldsen, K., and Wanstrup, J. (1970). Effects of carbon monoxide exposure on the arterial walls. Ann. N. Y. Acad. Sci. 174:294.
385. Dinman, B. D., Eaton, J. W., and Brewer, G. J. (1970). Effects of carbon monoxide on DPG concentrations in the erythrocyte. Ann. N. Y. Acad. Sci. 174:246.
386. Lilienthal, J. L., Riley, R. L., Prommel, D. D., and Franke, R. E. (1946). The relationships between carbon monoxide, oxygen and hemoglobin in the blood of man at altitude. Am. J. Physiol. 145:351.
387. Dinman, B. D. (1968). Pathophysiologic determinants of community air quality standards for carbon monoxide. J. Occup. Med. 10:446.
388. Shields, C. E. (1971). Elevated carbon monoxide levels from smoking in blood donors. Transfusion 11:89.
389. Stewart, R. D., Fisher, T. N., Hosko, M. J., Peterson, J. E., Baretta, E. D., and Dodd, H. C. (1972). Experimental human exposure to methylene chloride. Arch. Environ. Health 25:342.
390. Alvis, H. J., and Tanner, C. W. (1952). Carbon monoxide toxicity in submarine operations. Arch. Ind. Hyg. Occup. Med. 6:404.
391. Lightfoot, N. F. (1972). Chronic carbon monoxide exposure. Proc. R. Soc. Med. 65:16.
392. Schulte, J. H. (1964). Sealed environment in relation to health and disease. Arch. Environ. Health 8:438.
393. Bondi, K. R., Very, K. R., and Schaefer, K. E. (1978). Carboxyhemoglobin levels during a submarine patrol. Aviat. Space Environ. Med. 49:851.
394. Sjöstrand, T. (1965). The formation of carbon monoxide in the living organism—a factor to be considered in space flight. Reprinted from Proceedings of Second International Symposium on Man in Space, 1965, Paris.
395. Bell, D. G., and Wright, G. R. Energy expenditure and work stress of divers performing a variety of underwater tasks. Ergonomics. In press.
396. Chaussin, M., and Guenard, H. (1977). Diffusion limitation in pulmonary gas exchange. Bull. Eur. Physiopathol. Respir. 13:153P.
397. Brewster, E., Collins, S., Funnell, G. R., and Smith, E. B. (1976). Undersea Biomed. Res. 3:151.
398. Dinman, B. D. (1974). The management of acute carbon monoxide intoxication. J. Occup. Med. 16:662.
399. Drinker, C. K. (1938). Carbon Monoxide Asphyxia. Oxford University Press, New York.
400. Killick, E. M. (1940). Carbon monoxide anoxemia. Physiol. Rev. 20:313.
401. Silk, S. J. (1975). The threshold limit value for carbon monoxide. Ann. Occup. Hyg. 18:29.

402. Barnard, R. J., Gardner, G. W., and Diaco, N. V. (1976). "Ischemic" heart disease in fire fighters with normal coronary arteries. J. Occup. Med. 18:818.
403. Burgess, W. A., Sidor, R., Lynch, J. J., Buchanan, P., and Clougherty, E. (1977). Minimum protection factors for respiratory protective devices for fire fighters. Am. Ind. Hyg. Assoc. J. 38:18.
404. Mastromatteo, E. (1959). Mortality in city firemen. I. A review. Arch. Ind. Health 20:1.
405. Mastromatteo, E. (1959). Mortality in city firemen. II. A study of mortality in firemen of a city fire department. Arch. Ind. Health 20:227.
406. Howlett, L., and Shephard, R. J. (1973). Carbon monoxide as a hazard in aviation. J. Occup. Med. 15:874.
407. Wright, G. R., Polk, L., and Shephard, R. J. (1974). Carbon monoxide in dwellings. Report on domestic standards of CO exposure, 1974. U. S. Federal Committee, Bureau of Community Environment Management.
408. Davidson, S. (1970). The Principles and Practice of Medicine, Ed. 9. E. and S. Livingstone, Ltd., Edinburgh.
409. De Bias, D. A., Banerjee, C. M., Birkhead, N. C., Greene, C. H., Scott, D., and Harrer, W. V. (1976). Effects of carbon monoxide inhalation on ventricular fibrillation. Arch. Environ. Health 31:42.
410. Corya, B. C., Black, M. J., and McHenry, P. L. (1976). Echocardiographic findings after acute carbon monoxide poisoning. Br. Heart J. 38:712.
411. Shillito, F. H., Drinker, C. K., and Shaugnessy, T. J. (1936). The problem of nervous and mental sequelae in carbon monoxide poisoning. J.A.M.A. 106:669.
412. Smith, J. S., and Brandon, S. (1973). Morbidity from acute carbon monoxide poisoning at three-year follow-up. Br. Med. J. (i):318.
413. Smith, E., McMillan, E., and Mack, L. (1935). Factors influencing the lethal action of illuminating gas. J. Ind. Hyg. 17:18.
414. Ehrich, W. E., Bellet, S., and Lewey, F. H. (1944). Cardiac changes from CO poisoning. Am. J. Med. Sci. 208:511.
415. Kjeldsen, K., Thomsen, H. K., and Åstrup, P. (1974). The effects of carbon monoxide on myocardium: ultrastructural changes in rabbits after a moderate, chronic exposure. Circ. Res. 34:339.
416. Wanstrup, J., Kjeldsen, K., and Åstrup, P. (1969). Acceleration of spontaneous changes in rabbit aorta by a prolonged, moderate, carbon monoxide exposure. Acta Pathol. Microbiol. Scand. 75:353.
417. Lewey, F. H., and Drabkin, D. L. (1944). Experimental chronic carbon monoxide poisoning of dogs. Am. J. Med. Sci. 208:502.
418. Lindgren, S. A. (1961). A study of the effect of protracted occupational exposure to carbon monoxide. Acta Med. Scand. (suppl.) 356:5.
418a. Lingen, M. L. (1955). Effects of low concentrations of carbon monoxide on flicker fusion. Nord. Hyg. Tidskr. 36:202.
419. Kjeldsen, K., and Damgaard, F. (1968). Influence of prolonged carbon monoxide exposure and high altitude on the composition of blood and urine in man. Scand. J. Clin. Lab. Invest. (suppl. 103) 22:20.
420. Kjeldsen, K., and Damgaard, F. (1968). Influence of prolonged carbon monoxide exposure and altitude hypoxia on serum lipids in man. Scand. J. Clin. Lab. Invest. (suppl. 103) 22:16.
421. Åstrup, P. (1972). Some physiological and pathological effects of moderate carbon monoxide exposure. Br. Med. J. (iv):447.
422. Siggaard-Andersen, J., Petersen, F. B., Hansen, J. I., and Mellemgaard, K. (1968). Plasma volume and vascular permeability during hypoxia and carbon monoxide exposure. Scand. J. Clin. Lab. Invest. (suppl. 103) 22:39.

423. Niden, A. H., and Schultz, H. (1965). The ultra-structural effects of carbon monoxide inhalation on the rat lung. Virchows Archiv. 339:283.
424. Pauli, H. G., Truniger, B., Larsen, J. K., and Mulhausen, R. O. (1968). Renal function during prolonged exposure to hypoxia and carbon monoxide. I. Glomerular filtration and plasma flow. Scand. J. Clin. Lab. Invest. (suppl. 103) 22:55.
425. Pauli, H. G., Truniger, B., Larsen, J. K., and Mulhausen, R. O. (1968). Renal function during prolonged exposure to hypoxia and carbon monoxide. II. Electrolyte handling. Scand. J. Clin. Lab. Invest. (suppl. 103) 22:61.
426. Kalmaz, E. V., Cauter, L. W., Nau, C. A., and Hampton, J. W. (1977). Effect of carbon monoxide exposure on blood fibrinolytic activity in rabbits. Environ. Res. 14:194.
427. Sairo, O., and El-Attar, A. (1968). Effect of carbon monoxide on the whole blood fibrinolytic activity. Ind. Med. Surg. 37:774.
428. Åstrup, P. (1969) Effects of hypoxia and of carbon monoxide exposures on experimental atherosclerosis. Ann. Intern. Med. 71:426.
429. Åstrup, P. (1967). Carbon monoxide and peripheral vascular disease. Scand. J. Clin. Lab. Invest. (suppl. 99) 19:193.
430. Birstingl, M., Hawkins, L., and McEwen, T. (1970). Experimental atherosclerosis during chronic exposure to carbon monoxide. Eur. Surg. Res. 2:92.
431. Kjeldsen, K., Wanstrup, J., and Åstrup, P. (1968). Enhancing influence of arterial hypoxia on the development of atheromatosis in cholesterol-fed rabbits. J. Atheroscler. Res. 8:835.
432. Kjeldsen, K., Åstrup, P., and Wanstrup, J. (1972). Ultra-structural intimal changes in the rabbit aorta after a moderate carbon monoxide exposure. Atherosclerosis 16:67.
433. Thomsen, H. K., and Kjeldsen, K. (1974). Threshold limit for carbon monoxide-induced myocardial damage: an electron microscopic study in rabbits. Arch. Environ. Health 29:73.
434. Webster, W. S., Clarkson, T. B., and Lofland, H. B. (1968). Carbon monoxide-aggravated atherosclerosis in the squirrel monkey. Exp. Mol. Pathol. 13:36.
435. Armitage, A. K., Davies, R. F., and Turner, D. M. (1976). The effects of carbon monoxide on the development of atherosclerosis in the white carneau pigeon. Atherosclerosis 23:333.
436. Boatman, J. B., and Carter, S. D. (1973). Hypoxia and the arterial surface. Arch. Environ. Health 27:360.
437. Parving, H.-H. (1972). The effect of hypoxia and carbon monoxide exposure on plasma volume and capillary permeability to albumin. Scand. J. Clin. Lab. Invest. 30:49.
438. Tillmanns, H., Sarma, I. S. M., Seeler, K., et al. (1975). Lipid metabolism in perfused human coronary arteries and saphenous veins. *In* G. Schettler and A. Weizel (eds.), Atherosclerosis, Vol. 3, pp. 118–125. Springer-Verlag, New York.
439. Dublin, L. I., and Vane, R. J. (1947). Occupational mortality experience of insured wage earners. Month. Labor Rev. 64:1003.
440. Komatsu, F. (1955). Shinshu myocardosis. Digest Sci. Labour 10:315.
441. Ayres, S. M., Evans, R., Licht, D., Greisbach, J., Reimold, F., Ferrand, E. F., and Criscitiello, A. (1973). Health effects of exposure to high concentrations of automotive emissions. Arch. Environ. Health 27:168.
442. Cohen, S. I., Deane, M., and Goldsmith, J. R. (1969). Carbon monoxide and survival from myocardial infarction. Arch. Environ. Health 19:510.

443. Hexter, A. C., and Goldsmith, J. R. (1971). Carbon monoxide: association of community air pollution with mortality. Science 172:265.
444. Kuller, L. H., Radford, E. P., Swift, D., Perper, J. A., and Fisher, R. (1975). Carbon monoxide and heart attacks. Arch. Environ. Health 30:477.
445. Goldsmith, J. R., and Aronow, W. S. (1975). Carbon monoxide and coronary heart disease: a review. Environ. Res. 10:236.
446. Ayres, S. M., Giannelli, S., and Armstrong, R. G. (1965). Carboxyhemoglobin: hemodynamic and respiratory responses to small concentrations. Science 149:193.
447. Ayres, S. M., Penny, J., Criscitiello, A., and Giannelli, S. (1968). Effect of abnormal hemoglobin (carboxyhemoglobin) on human myocardial metabolism. Clin. Res. 16:220 (abstr.).
448. Ayres, S. M., Mueller, H. S., Gregory, J. J., Giannelli, S., and Penny, J. L. (1969). Systemic and myocardial hemodynamic responses to relatively small concentrations of carboxyhemoglobin (COHb). Arch. Environ. Health 18:699.
449. Adams, J. D., Erickson, H. H., and Stone, H. L. (1973). Myocardial metabolism during exposure to carbon monoxide in the conscious dog. J. Appl. Physiol. 34:238.
450. Young, S. H., and Stone, H. L. (1976). Effect of a reduction in arterial oxygen content (carbon monoxide) on coronary flow. Aviat. Space Environ. Med. 47:142.
451. Brody, J. S., and Coburn, R. F. (1969). Carbon monoxide-induced arterial hypoxemia. Science 164:1297.
452. Aronow, W. S., Cassidy, J., Vangrow, J. S., March, H., Kern, J. C., Goldsmith, J. R., Khemka, M., Pagano, J., and Vawter, M. (1974). Effect of cigarette smoking and breathing carbon monoxide on cardiovascular hemodynamics in anginal patients. Circulation 50:340.
453. Theodore, J. (1973). Angina pectoris and carbon monoxide. (Letter). Ann. Intern. Med. 78:455.
454. Aronow, W. S., and Isbell, M. W. (1973). Carbon monoxide: effect on exercise-induced angina pectoris. Ann. Intern. Med. 79:392.
455. Anderson, E. W., Strauch, J., Knelson, J., and Fortuin, N. (1971). Effects of carbon monoxide (CO) on exercise electrocardiogram (ECG) and systolic time intervals (STI). Circulation (suppl. II) 44:135.
456. Anderson, E. W., Andelman, R. J., Strauch, J. M., Fortuin, N. J., and Knelson, J. H. (1973). Effect of low-level carbon monoxide exposure on onset and duration of angina pectoris: a study in ten patients with ischemic heart disease. Ann. Intern. Med. 79:46.
457. Goldsmith, J. R., Beard, R. R., and Dinman, B. D. (1969). Epidemiological appraisal of carbon monoxide effects. *In* A. B. DuBois (ed.), Effects of Chronic Exposure to Low Levels of Carbon Monoxide on Human Health, Behavior and Performance, pp. 40–60. National Academy of Sciences and National Academy of Engineering, Washington, D. C.
458. Ury, H. K., Perkins, N. M., and Goldsmith, J. R. (1972). Motor vehicle accidents and vehicular pollution in Los Angeles. Arch. Environ. Health 25:314.
459. Brown, I. D., and Poulton, E. C. (1961). Measuring the spare "mental capacity" of car drivers by a subsidiary task. Ergonomics 4:35.
460. Corcoran, D. W. J. (1965). Personality and the inverted-U relation. Br. J. Psychol. 52:267.
461. Beck, H. G., and Suter, G. M. (1938). Role of carbon monoxide in the causation of myocardial disease. J.A.M.A. 110:1982.
462. Anderson, D. O. (1967). The effects of air contamination on health: Part III. Can. Med. Assoc. J. 97:802.

463. Colucci, A. V., Hammer, D. I., Williams, M. E., Hinners, T. A., Pinkerton, C., Kent, J. L., and Love, G. J. (1973). Pollutant burdens and biological response. Arch. Environ. Health 27:151.
464. Goldsmith, J. R. (1962). Some implications of ambient air quality standards. Arch. Environ. Health 4:151.
465. Hatch, T. F. (1962). Changing objectives in occupational health. Am. Ind. Hyg. Assoc. J. 23:1.
466. Phair, J. J., and Sterling, T. (1961). Epidemiological methods and community air pollution. Arch. Environ. Health 3:267.
467. Glass, R. I. (1975). A perspective on environmental health in the U.S.S.R. Arch. Environ. Health 30:391.
468. Ryazanov, V. A. (1962). Sensory physiology as a basis for air quality standards. Arch. Environ. Health 5:480.
469. Stern, A. C. (1965). Basis for criteria and standards. J. Air Pollut. Control Assoc. 15:281.
470. Hatch, T. F. (1973). Criteria for hazardous exposure limits. Arch. Environ. Health 27:231.
471. Weinstock, B., and Niki, H. (1972). Carbon monoxide balance in nature. Science 176:290.
472. McConnell, J. C., McElroy, M. B., and Wofsy, S. C. (1971). Natural sources of atmospheric CO. Science 233:187.
473. Dinman, B. D. (1969). Effects of long-term exposure to carbon monoxide. *In* A. B. DuBois (ed.), Effects of Chronic Exposure to Low Levels of Carbon Monoxide on Human Health, Behavior and Performance, pp. 25–38. National Academy of Sciences and National Academy of Engineering, Washington, D. C.
474. Forster, R. E. (1969). Appendix: techniques for the analysis of carbon monoxide. *In* A. B. DuBois (ed.), Effects of Chronic Exposure to Low Levels of Carbon Monoxide on Human Health, Behavior and Performance, pp. 61–66. National Academy of Sciences and National Academy of Engineering, Washington, D. C.
475. DuBois, A. B. (ed.). (1973). Effects of Carbon Monoxide on Man's Health in His Home Environment. Bureau of Community Environmental Management, U. S. Department of Health, Education, and Welfare, Washington, D. C.
476. Åstrup, P., Hellung-Larsen, P., Kjeldsen, K., and Mellemgaard, K. (1966). The effect of tobacco smoking on the dissociation curve of oxyhemoglobin. Scand. J. Clin. Lab. Invest. 18:450.
477. Benesch, R. E., Maeda, N., and Benesch, R. (1972). 2,3-Diphosphoglycerate and the relative affinity of adult and fetal hemoglobin for oxygen and carbon monoxide. Biochem. Biophys. Acta 257:178.
478. Wilson, D. F., Swinnerton, J. W., and Lamontagne, R. A. (1970). Production of carbon monoxide and gaseous hydrocarbons in seawater: relation to dissolved organic carbon. Science 168:1577.
479. Åstrup, P. (1970). Intraerythrocytic 2,3-diphosphoglycerate and carbon monoxide exposure. Ann. N. Y. Acad. Sci. 174:252.
480. Wald, G., and Allen, D. W. (1967). Equilibrium between cytochrome oxidase and carbon monoxide. J. Gen. Physiol. 40:593.

Index

Abdomen, temperature sensors in, 211–212
Accidents, carbon monoxide and, 345
Acetaminophen, 231
Acetoacetate, brain substrate, 270
Acetoacetyl-CoA dehydrogenase, 276
Acetylcholine, 213
Acetyl coenzyme A, 260, 266–267
Acetylsalicylate, 223, 228
Acidosis, cold thermogenesis and, 120
Acyl coenzyme A dehydrogenase, 266–268
Adenyl cyclase, cold thermogenesis and, 78–81
Adenyl cyclase-cyclic AMP system, cold thermogenesis and, 115
Adrenaline, 83, 88, 89, 90
Adult animal, nonhibernating, non-shivering thermogenesis in, 64–66
Aerobic metabolism
  in diving mammals, 256–257
  enzymes of, 276
Aeromonas hydrophila, 229, 230, 231, 234–236, 240
Air quality criteria, for carbon monoxide, 345–346
Air
  kinematic viscosity coefficient, 18
  thermal expansion coefficient, 18
Alanine, role in diving mammals, 260–261
Alanine aminotransferase, 261
Aldolase, 258
Altitude, pulmonary circulation and, 289–310
Amino acid, brain substrate, 269
Anaerobic metabolism
  in diving mammals, 257
  enzymes of, 276–277
Anesthesia, summit metabolism and, 54
Angiotensin II, hypoxic pulmonary hypertension and, 298
Animal coat
  heat transfer through, 13–21
  structure, 10–11
Animal coat thermal conductivity, 2, 5, 14
Animal coat thermal resistance, 2, 5, 6
  for clothing types, 6
Animal coat thickness, 2, 5, 28–30
Animal insulators, 1–42
Animal size, insulation effectiveness and, 28–30
Antipyretic drug, action mechanism, 221, 222, 227
Apnea, 255
Arachidonic acid, fever and, 223, 224
Arteriovenous gradient, across brain, 278
Aspartate, role in diving mammals, 260–261
Aspartate aminotransferase, 260
Aspirin, 222
ATP production
  in brown fat, 78–81
  shivering and, 98, 122
Auditory flutter fusion, 336
Auditory system, effect of carbon monoxide on, 335–336

Bacteria, gram-positive, fever and, 217
Blood, substrate source for cold thermogenesis, 109–112
Blood flow, cold thermogenesis and, 100–105
Blood glucose, diving recovery and, 264–265
Blood glucose profile
  in adult seal, 277
  in fetal seal, 280–281
Blood lactate profile
  in adult seal, 277
  in fetal seal, 280–281

Blood pressure, cold thermogenesis and, 120–121
Blood pyruvate profile, in fetal seal, 280–281
Body temperature, *see also* Fever; Hyperthermia; Hypothermia; Thermoregulation
  changes in, during exercise, 160–162, 185–189, 197–198
  limits of, 159–160
  regulation, 162–177, 211–214
  studies on, 177–198
  summit metabolism and, 51–52
Body weight
  nonshivering thermogenesis and, 67–68
  summit metabolism and, 54–56
Brown fat, 68–81
  body content of, factors affecting, 75
  conversion to white fat, 72–74
  distribution, 71–72
  fever and, 220
  in hibernators, 66
  historical considerations, 68
  identification, 68–71
  innervation, 85
  mitochondria of, 80
  in newborns, 60–64, 71–72
  nonshivering thermogenesis and, 75–78
  sympathetic control of nonshivering thermogenesis in, 85–92
  thermogenetic mechanisms in, 78–81
Bradycardia, 255
  in fetal seal, 280
Bradykinin, hypoxic pulmonary hypertension and, 298
Brain anoxia, consequences, 272–273
Brain hypoxia, consequences, 270–272
Brain metabolic rate, 269
Brain metabolism
  during diving, 273–275
  in human diver, model for, 279–280
  substrates for, 269–272

Calcium chloride, temperature regulation and, 181–191
Calcium ion
  fever and, 225–227
  thermoregulation and, 181–199
Canine herpes virus, fever studies and, 239
Carbon monoxide
  acute toxicity, 341–342
  administration methods, 324–235
  air quality criteria, 345–346
  auditory system effects, 335–336
  cardiorespiratory effects, 331–332
  cellular effects, 325–327
  central nervous system and, 332–339
  cerebral efficiency and, 336–338
  chronic toxicity, 342
  driving of vehicles and, 339
  elimination of, 328–329
  endogenous production, 327
  hemoglobin and, 317–320, 325–326
  myoglobin and, 326
  physiological effects of, 325–342
  psychomotor performance and, 338–339
  uptake, 327–328
  uptake and release models, 329–331
  urban sampling, 315
  visual system effects, 334–335
Carbon monoxide determination
  bloodless methods, 323–324
  in blood phase, 322–323
  in gas phase, 321–322
Carbon monoxide exposure
  adaptation to, 339–340
  epidemiological studies, 324–325
  health consequences, 342–345
  severe, 341–342
  sources, 314–320
  techniques, 324–325
Carboxyhemoglobin, blood levels, 317–320
Carnitine, cold thermogenesis and, 115
Catecholamines, cold thermogenesis and, 60, 61, 88–93, 114
Choice reaction time, carbon monoxide and, 337–338
Cigarette smoke, carbon monoxide content, 316–317
Circulatory system, adjustments during cold thermogenesis, 99–105
Citrate synthase, 276
Citrobacter diversus, 229
Clothing, *see also* Insulation, from clothing

color, 21, 22
heat transfer through, 13–21
structure, 11–13
thermal insulation values, 12, 14
ventilation of, heat transfer and, 19–21
clo unit, definition, 6
Cognitive performance, carbon dioxide and, 337
Cold exposure, prolonged, summit metabolism and, 52
Color, clothing temperature and, 21, 22
Conductance, evaporative, 7, 24
Conduction, heat transfer and, 13–15
Convection, heat transfer and, 17–19
Convective heat flux, 2, 4, 5
Convective heat transfer coefficient, 5
Corticosteroids, cold thermogenesis and, 117
Coxsackie virus, fever studies and, 239
Curare, 58
Cyclic AMP, fever and, 225
Cytochrome oxidase, cold acclimation and, 122
Cytochrome P-450, 300
Cytochrome pigments, carbon monoxide binding, 326–327

Dark adaptation, carbon monoxide and, 335
Dehydration, definition, 178
2-Deoxyglucose, 299
Diet, substrate source in cold thermogenesis, 112
Dinitrophenol, 299
2,3-Diphosphoglycerate, 340
Diver, human, metabolic model for, 279–280
Diving metabolism, 253–287
in central organs, 265–275
in heart, 265–269
in kidney, 262–264
in liver, 264–265
in peripheral organs, 256–265
in skeletal muscle, 256–262
Diving capacity, long-term, 281–282
Diving response, 255–256
long-term capacity and, 281–282
Driving of vehicles, carbon monoxide and, 339, 345
Drugs, thermogenesis and, 113–118

Ectotherm, 46, 210–211
Electrolyte ingestion, in body temperature control studies, 191–197
Endotherm, 46, 210–211, 213–214
Endotoxin
cold thermogenesis and, 121
fever and, 216–217, 227, 237–238
Energy balance, insulation and, 4–7
Energy balance equation, 4
Escherichia coli endotoxin, fever studies and, 227
Euhydration, definition, 178
Evaporation rate, 2, 4, 5
Exercise
body thermal responses during, 160–162, 197–198
carbon monoxide and, 332
in cold, summit metabolism and, 53–54
hyperthermia and, 157–208

Fasting, summit metabolism and, 52
Fat catabolism
cold thermogenesis and, 99–100
in diving mammals, 257
in hypoxic heart, 266–268
Febrile response, in vertebrates, 232
Ferret, fever studies in, 237
Fever
activators of, 216–218
in amphibians, 230–231
biology of, 215
in birds, 227–228
cyclic AMP and, 225
endogenous pyrogens and, 218–221
evolution of, 227
in fishes, 231–232
function of, 232–243
hypothesized mechanism, 180
in mammals, 214–227
mechanisms of, 239–243
Na^+/Ca^{2+} and, 225–227
nature of, 214–215
prostaglandins and, 213, 221–224, 231
in reptiles, 228–230
Fever index, 216

Fiber conduction, 15
Fingertip, insulation and, 29–30
Fish, metabolic heating, 45
Fluid-electrolyte hypothesis, 177–198
  historical perspective, 179–181
  research on, 181–198
  in temperature control studies, 191–197
Fourier's Law of Heat Loss, 8
Free fatty acid
  heart substrate, 265–267
  thermogenesis substrate, 109–110
Frog, fever studies in, 231
Fructose biphosphatase, from muscle, function, 258–259
Futile cycle, cold thermogenesis and, 84

Gas phase, insulation properties, 31–32
Glucagon, cold thermogenesis and, 115–116
Glucose
  brain substrate, 269
  thermogenesis substrate, 110–111
Glucose-fatty acid cycle, 99
Glutamate, brain substrate, 270
Glutamate dehydrogenase, 276
Glutamine, kidney substrate, 262–264
Glycogen catabolism
  in diving mammals, 257
  sparing of, 260
  cold thermogenesis and, 98–100
  in diving mammals, 257–258
Glycerol, thermogenesis substrate, 111–112
α-Glycerophosphate dehydrogenase, 258
Goldfish, in fever studies, 231, 235–236
Grashof number, 18, 25, 31

Hair, radiant conductivity, 15–17
Heart disease, carbon monoxide and, 343–345
Heart metabolism
  during diving, 265–269
  in human diver, model for, 279–280
  in Weddell seal, 278–279
Heart rate, carbon monoxide and, 331–332
Heat loss, evaporative, homeothermy and, 46–47
Heat loss, nonevaporative, homeothermy and, 46–47
Heat transfer
  effect of wind on, 25–28
  energy balance equation, 4
  mechanisms of, 11–21
  through animal coats and clothing, 1–42
Heat transfer resistance, 2, 5
hectoPascal, 7
Helium
  insulation properties, 32
  in summit metabolism determinations, 50
Hemoglobin, carbon monoxide binding, 317–320, 325–326
Herpes virus, growth temperature, 239
Hexamethonium, 58, 60
Hexokinase, 277
Hibernator, nonshivering thermogenesis in, 66–67
High altitude pulmonary hypertension, 289–310
Histamine, hypoxic pulmonary hypertension and, 297
Homeotherm, 46
Homeothermy
  climatic limits to, 56–57
  shivering thermogenesis and, 93–94
Hormones, thermogenesis and, 113–118
Human, total insulation, 13
Human comfort, insulation and, 33–35
Humoral theory of disease, 232–233
Hydration states, definitions, 178
β-Hydroxybutyrate, brain substrate, 270
β-Hydroxybutyrate dehydrogenase, 276
6-Hydroxy-dopamine, 91, 297
5-Hydroxytryptamine, 213
Hypercapnia, cold thermogenesis and, 120
Hyperhydration, definition, 178
Hyperoxia, cold thermogenesis and, 120
Hypersensitivity reaction, fever and, 217–218
Hyperthermia, *see also* Body temperature; Fever; Hypothermia

definition, 215
exercise and, 157–208
zone of, 8
Hypohydration, definition, 178
Hypophyseal peptide hormones, cold thermogenesis and, 114–115
Hypothalamic temperature, shivering and, 106–107
Hypothalamus, thermoregulation and, 211–213, 220–221
Hypothermia
anesthetics and, 54
definition, 215
in elderly and infirm, 34–35, 57
zone of, 8
Hypoxia
brain, 270–272
cold thermogenesis and, 119–120
effect on pulmonary arterial smooth muscle, 298–300
pulmonary vascular response to, 290–293

Iguana, in fever studies, 229–230, 234–235
Indomethacin, 222, 224
Influenza virus, fever studies and, 217, 237
Insects, metabolic heating, 45–46
Insulation
animal size and, 28–30
energy balance and, 4–7
human comfort and, 33–35
Insulation, from clothing
human comfort and, 33–35
in hyperbaric environments, 30–33
in hypobaric environments, 30–33
in water, 30–33
Insulation value
of clothing types, 6
definition, 6
Insulin, cold thermogenesis and, 116–117
Interferon, in immune response, 241
Internal combustion engine, carbon monoxide source, 314–316
Intracerebral infusion, in temperature control research, 181–189
Intravenous infusion, in temperature control research, 189–191
Ischemic heart disease, 343–345
Isoprenaline, cold thermogenesis and, 88, 92
2-Ketoglutarate, 264, 271–273
Ketone body
brain substrate, 269, 270
thermogenesis substrate, 112
Kidney, site of nonshivering thermogenesis, 83
Kidney metabolism, during diving and recovery, 262–264
kiloPascal, 7

Lactate
brain substrate, 269, 270
kidney substrate, 262–264
lung substrate, 278
thermogenesis substrate, 111
Lactate dehydrogenase, 258
Lactate dehydrogenase isozymes, 268, 274, 277, 278–279
Latent heat vaporization of water, 2, 4, 5
Learning, carbon monoxide and, 337
Leukocyte function, fever and, 241–243
Leukocytic pyrogen, *see* Pyrogen, endogenous
Lewis number, 24, 32
Lewis Relation, 5
Lipid A, 216
Litigation, carbon monoxide exposure and, 345
Liver, site of nonshivering thermogenesis, 83
Liver metabolism, during diving and recovery, 264–265
Lung metabolism
during diving, 275
in human diver, model for, 279–280
in Weddell seal, 277–278
Lymphocyte transformation, fever and, 242
Lysosome, in immune response, 241

Man, total insulation, 13
Mass transfer coefficient, 5

Maximum metabolism, definition, 49
Memory, carbon monoxide and, 337
Metabolic diagram, for homeotherms, 7–10
Metabolic heat production, 7–10
Metabolic rate, body weight and, 54–56
Minimum metabolism, zone of, 8
Myoglobin, carbon monoxide binding, 326
NAD^+:NADH ratio, in diving mammals, 261, 266, 271, 272, 273, 282
Na^+/K^+ membrane pump, cold thermogenesis and, 81, 118
Net metabolic heat flux, 2, 4, 5, 8–9
Net radiation, 2, 4, 5
Neural control
  of cold thermogenesis, 99–105
  of hypoxic pulmonary vasoconstriction, 293–300
  of nonshivering thermogenesis, 85–93
  of shivering thermogenesis, 94–95
Newborn animal, nonshivering thermogenesis in, 60–64
Newcastle disease virus, fever studies and, 217
Newton's Law of Cooling, 8, 47
Nonadrenaline, cold thermogenesis and, 58, 62, 63, 65, 66, 87–93
Norepinephrine, 213
Normothermia, definition, 215
Nusselt number, 5, 18, 24
  for cylinders, 30
Nutritional immunity, 240

Overhydration, definition, 178
Oxaloacetate, 260

Pasteurella hemolytica, 229
Pasteurella multocida, fever studies, 228, 236–237, 240
Peak metabolic effort, *see* Summit metabolism
Peroxisome, cold thermogenesis and, 81, 84
Phenoxybenzamine, 90
Phentolamine, 60, 88, 89
Phenylephrine, 92
Phosphofructokinase, 258, 277
Phosphoglucomutase, 258
Phosphokinase, 277
Pneumococci, 238, 240
POAH, *see* Preoptic area–anterior hypothalamus
Poikilotherm, 46
Poliomyelitis virus, growth temperature, 239
Posture, summit metabolism and, 52
Potassium cyanide, 299
Prandtl number, 31
Preoptic area–anterior hypothalamus, fever and, 220–223
Propranolol, 60, 61, 88
Prostaglandins
  fever and, 213, 221–224, 231
  hypoxic pulmonary hypertension and, 297–298
Prostaglandin E_1, fever induction and, 222, 228
Prostaglandin E_2, fever induction and, 222
Prostaglandin endoperoxide, 224
Pulmonary arterial smooth muscle, effect of hypoxia on, 298–300
Pulmonary circulation, high altitude and, 289–310
Pulmonary pressor response, hypoxic, teleological speculation and, 300–304
Pulmonary vasoconstriction
  mechanisms, 293–300
  triggering conditions, 290–293
Pulmonary ventilation, cold thermogenesis and, 120
Pyrogen
  endogenous, 218–221
  exogenous, 216–218
Pyruvate, brain substrate, 269, 270
Pyruvate kinase, from muscle, regulatory role, 259–260

Rabbit, fever studies in, 236–237, 238
$^{86}Rb^+$ technique, 101–103
Rectal temperature, work load and, 164, 165, 166, 176, 185–197
Rehydration, definition, 178

Respiratory gases, cold thermogenesis and, 118–120
Reptiles, metabolic compensation, 46
Reserpine, 297
Restraint, summit metabolism and, 54

Salmonella abortus equi endotoxin, fever studies and, 227
Salt fever, 179
SC 19220, 223, 224
Sensible heat transfer, simple resistance analogue for, 8–9
Sensible heat flux, 5, 6
Serotonin
  cold thermogenesis and, 114
  hypoxic pulmonary hypertension and, 298
Serum iron, infection response and, 240–241
Set-point concept, 172–175
Sherwood number, 5, 24
Shivering, *see also* Thermogenesis, shivering
Short-wave radiation, heat transfer and, 21
Size, temperature tolerance and, 9, 28–30
Skeletal muscle
  metabolism of, in diving mammals, 256–261
  nonshivering thermogenesis and, 82–83, 92–93
Skin, temperature sensors in, 211–212
Sodium chloride, temperature regulation and, 179–185, 189–191
Sodium ion
  fever and, 225–227
  thermoregulation and, 181–199
Sodium salicylate, 223, 228, 235
Specific heat, constant pressure, 2, 6, 32
Spinal cord, temperature sensors in, 211–212
Spinal cord temperature, shivering and, 106–107
STP value, 6
Stefan–Boltzmann constant, 2, 16
Substrate
  for cold thermogenesis, 107–112
  of heart muscle metabolism, 265–266
  for shivering thermogenesis, 98, 99
Succinate, hypoxic heart metabolism and, 267–268
Summit metabolism, 49–56
  anesthesia and, 54
  body temperature and, 51–52
  body weight and, 54–56
  definitions, 49
  exercise in cold and, 53–54
  fasting and, 52
  measurement, 49–51
  posture and, 52
  prolonged cold exposure and, 52
  restraint and, 54
  stimulation of, 121–123
Suxamethonium, 60, 61
Sweat, heat loss and, 9–10, 23–25
Sweat rate, work load and, 165–167
Sweat secretion, regulation, 169–172, 179

Temperature–metabolism curves, 46–49
Temperature sensor, sites of, 211–213
Thermal insulation
  fabric values, 12
  heat transfer and, 15–17
  homeothermy and, 46–47
  wind and, 26–28
Thermal resistance, *see* Animal coat thermal resistance
Thermal sensing organ, 169–172
Thermogenesis, cold, 43–155
  acidosis and, 120
  blood pressure and, 120–121
  circulatory adjustments during, 99–105
  components of, 57–60
  constraints, 112–121
  definition, 46–47
  drugs and, 113–118
  endotoxins and, 121
  hormones and, 113–118
  mechanisms of, in brown fat, 78–81
  neural control, 106–107
  pulmonary ventilation and, 120
  respiratory gases and, 118–120
  stimulation of, 121–123

Thermogenesis — *continued*
substrate limitation and, 112–113
substrates for, 107–112
trauma and, 121
Thermogenesis, nonshivering
in adult nonhibernating animals, 64–66
body weight and, 67–68
brown fat and, 75
in hibernators, 66–67
in newborn animals, 60–64
occurrence of, 60–68
quantitation, 58–60
sites of, 81–84
in skeletal muscle, 92–93
sympathetic control of, 85–93
Thermogenesis, shivering, 93–99
energy source, 99
magnitude of, 94–95
mechanism, 98
neural control, 97–98
quantitation, 58–60
role in homeothermy, 93–94
Thermoregulation, *see also* Body temperature
fluid-electrolyte metabolism and, 177–198
nonbehavioral mechanisms, 46–49
as reflex, 211
zones of, 7–10
Thermoregulatory effectors, 213–214
Thermoregulatory set-point
definition, 215
fever and, 215
Thromboxane, 224
Thyroid hormones, cold thermogenesis and, 117–118
Tissue
metabolite cycling within, 282
thermogenesis substrates in, 107–108
Tissue substrate stores, cold thermogenesis and, 107–108
Tobacco smoke, carbon monoxide source, 316–317, 318–319
Trauma, cold thermogenesis and, 121
Triglyceride, brown fat thermogenesis and, 78–79
Tumor, fever and, 218

Vasoactive substances, 297–298
Vasoconstriction, peripheral, 255–256
Virus, fever and, 217
Visual system, effects of carbon monoxide on, 334–335
Volumetric specific heat of medium, 5, 6

Water
body temperature control and, 176–177
heat loss in, 32–33
ingestion, temperature regulation and, 191–197
physical properties, survival value of, 158–159
Water vapor concentration, 2, 5
Water vapor diffusivity, 5
Water vapor permeability index, 24
Water vapor transfer, 21–25
Weddell seal
anaerobic metabolism potential, 257
fetal metabolism during diving, 280–281
heat-lung-brain metabolism, 276–279
muscle glycogen levels, 257
oxidative metabolism potential, 256–257
White fat
identification, 68–71
production from brown fat, 72–74
Wind, effect on heat transfer, 25–28
Wind speed, 2, 26
Woman, total insulation, 13